Geodetic Monitoring for Land Deformation

Geodetic Monitoring for Land Deformation

Editors

Alex Hay-Man Ng
Linlin Ge
Hsing-Chung Chang
Zheyuan Du

MDPI • Basel • Beijing • Wuhan • Barcelona • Belgrade • Manchester • Tokyo • Cluj • Tianjin

Editors
Alex Hay-Man Ng
Guangdong University of Technology
China

Linlin Ge
UNSW Australia
Australia

Hsing-Chung Chang
Macquarie University
Australia

Zheyuan Du
Geoscience Australia
Australia

Editorial Office
MDPI
St. Alban-Anlage 66
4052 Basel, Switzerland

This is a reprint of articles from the Special Issue published online in the open access journal *Remote Sensing* (ISSN 2072-4292) (available at: https://www.mdpi.com/journal/remotesensing/special_issues/geodetic_land_deformation).

For citation purposes, cite each article independently as indicated on the article page online and as indicated below:

LastName, A.A.; LastName, B.B.; LastName, C.C. Article Title. *Journal Name* **Year**, *Volume Number*, Page Range.

ISBN 978-3-0365-6442-5 (Hbk)
ISBN 978-3-0365-6443-2 (PDF)

Contents

About the Editors

Alex Hay-Man Ng

Alex Hay-Man Ng, Ph.D., is a Professor for the School of Civil and Transportation Engineering at the Guangdong University of Technology in China. He completed his bachelor's and master's degree in Electrical Engineering in the University of New South Wales (UNSW), Australia, and continued his doctoral study in the Department of Surveying and Spatial Information System. In June 2011, he obtained his doctoral degree. His previous academic positions were at UNSW (Sydney, 2010–2016) and Macquarie University (Sydney, 2016). His research focusses on the satellite remote sensing image-analysis method and its application, InSAR algorithm and software development, ground subsidence monitoring and modeling, and image classification. He shares authorship of 100 papers in international journals and conferences.

Linlin Ge

Linlin Ge, Ph.D., is a professor and the head of discipline in the School of Civil and Environmental Engineering, UNSW (since 2017). From 2017 to now, he has successfully secured three grants including: one UNSW Torch program—Smart Spatial Technology, AUD 400,000 (2018–2020), one short-term Cooperative Research Centres Projects (CRC-P)—All-weather, near real time monitoring of bushfire with satellite InSAR, AUD 97,000 (2021–2022), and one long-term CRC-P—Quantifying the Past and Current Major Australian Floods with SAR and other Satellites, AUD 297,000 (2023–2025).

Hsing-Chung Chang

Hsing-Chung (Michael) Chang, Ph.D., is a senior lecturer in spatial information science in the School of Natural Science at the Macquarie University, Sydney, Australia. Chang's research interests include vegetation and land cover, land use monitoring, change detection, and 3D modeling using earth observation data with the aid of geographic information systems (GIS). He has also contributed to many multi-disciplinary projects on biodiversity conservation, wetland monitoring, natural disaster mitigations (including bushfires, floods, and seismic deformation), public transport planning, geographical mapping of multilingualism, etc., using spatial analyses and modeling.

Zheyuan Du

Zheyuan Du, Ph.D., is currently an InSAR scientist in Geoscience Australia. Dr Du received the B.E. degree in remote sensing and information technology from the China University of Geoscience, China in 2012, Master of Science degree in Geographic Information Science from the University of Edinburgh, UK in 2013, and a Ph.D. degree in Civil and Environmental Engineering from the UNSW in 2018. His research focuses on enhancing the performance of InSAR techniques by exploiting external datasets, such as meteorological/hydrological products, as well as ground in-situ data. He has experience as a Lecturer at the University of Sydney and a Postdoctoral Research Fellow at UNSW from 2018 to 2021. He has published over 25 journal articles, building a promising career in earth observation, in particular using InSAR data.

Preface to "Geodetic Monitoring for Land Deformation"

Land deformation is a hazard that can lead to serious problems. For example, it can increase the risk of flooding in coastal areas, damage buildings and infrastructures, destroy local groundwater systems, generate tension cracks on land, and reactivate faults. It has become a global problem. Consequently, it is essential to monitor land deformation, so that land surface change and/or movement can be better understood and managed for securing the safety of people.

The goal of this Special Issue is to collect papers to provide insights about the use of modern geodetic measurement techniques for land deformation identification, monitoring, and impact assessment. This book covers the up-to-date research in the development and application of modern geodetic technology, including spaceborne radar interferometry (InSAR), Ground-Based Synthetic Aperture Radar (GB-SAR), Global Navigation Satellite Systems (GNSS), Unmanned Aerial Vehicles (UAV), Terrestrial Laser Scanning (TLS), Pixel Offset Tracking (POT), etc., in accurate land deformation mapping and damage evaluations. This book can be used as a reference for expertise in surveyors, environmental engineers, disaster managers, and urban planners. The Guest Editors are grateful to the authors for their contributions of high-quality studies and, specifically, to the MDPI team for their effective assistance and cooperation. Without their devotion and support, it would have been impossible to publish this book

Alex Hay-Man Ng, Linlin Ge, Hsing-Chung Chang, and Zheyuan Du
Editors

remote sensing

Editorial

Geodetic Monitoring for Land Deformation

Alex Hay-Man Ng [1,*], Linlin Ge [2], Hsing-Chung Chang [3] and Zheyuan Du [4]

[1] School of Civil and Transportation Engineering, Guangdong University of Technology, Guangzhou 510006, China
[2] School of Civil and Environmental Engineering, University of New South Wales (UNSW), Sydney, NSW 2052, Australia
[3] School of Natural Sciences, Macquarie University, Sydney, NSW 2109, Australia
[4] Geoscience Australia, Canberra, ACT 2061, Australia
* Correspondence: hayman.ng@gdut.edu.cn

1. Introduction

Land deformation is a pervasive hazard that could lead to serious problems, for example, increasing risk of flooding in coastal areas, damaging buildings and infrastructures, destructing groundwater systems, generating tension cracks on land, and reactivating faults, to name only a few. Hazards caused by land deformation have been reported in many places around the world. Consequently, it is critical to monitor land deformation so that the land surface change and/or movement can be better understood and managed for safeguarding lives.

In the past decades, modern geodetic measurement techniques such as radar interferometry (InSAR), global navigation satellite systems (GNSS), light detection and ranging (LiDAR), close-range photogrammetry (CRP), Robotic Total Station (RTS), digital levelling, etc., have played a very important role in measuring land deformation. With the continuous advancements in geodetic measurement techniques, precise land deformation can be detected in high spatial and temporal scales with the diversification of sensors and processing methods. This has led to extensive applications in the spatio-temporal analysis of areas prone to deformations. These applications have become momentous for surveyors, environmental engineers, disaster managers, urban planners, etc. As such, we organized this Special Issue to cover the latest developments and applications in geodetic measurement techniques.

The Special Issue "Geodetic Monitoring for Land Deformation" of *Remote Sensing* consists of 14 individual works by researchers from various countries and a variety of geodetic measurement techniques and applications, including landslide monitoring, glacier mapping, infrastructure deformation detection, mine subsidence monitoring, and land surface mapping. This editorial provides an overview of the studies presented in this Special Issue. For the ease of access to the readers, a summary of the input data and techniques used in each paper and their applications is provided in Table 1.

Citation: Ng, A.H.-M.; Ge, L.; Chang, H.-C.; Du, Z. Geodetic Monitoring for Land Deformation. *Remote Sens.* **2023**, *15*, 283. https://doi.org/10.3390/rs15010283

Received: 17 November 2022
Accepted: 29 December 2022
Published: 3 January 2023

Table 1. Overview of geodetic monitoring methods, data, and geohazard types that are discussed in the fourteen research papers composing the Special Issue "Geodetic Monitoring for Land Deformation" of *Remote Sensing*. Papers are sorted in ascending order according to their publication date.

Source	Title	Geographic Focus	Applications	Techniques	Data
Wu et al. [1]	Comparative Analysis of the Effect of the Loading Series from GFZ and EOST on Long-Term GPS Height Time Series	Global	GPS Height Time Series	Comparative Analysis	GNSS
Wyszkowska et al [2].	Determination of Terrain Profile from TLS Data by Applying M_{split} Estimation	Not applicable	Determination of Terrain Profile	M_{split} Estimation	TLS
He et al. [3]	Analysis and Discussion on the Optimal Noise Model of Global GNSS Long-Term Coordinate Series Considering Hydrological Loading	Global	GNSS Long-term Coordinate Series	Comparative Analysis	GNSS
Cai et al. [4]	An Accurate Geocoding Method for GB-SAR Images Based on Solution Space Search and Its Application in Landslide Monitoring	Sichuan	Landslide Monitoring	GB-SAR Interferometry	GB-SAR UAV
Jia et al. [5]	A Semi-Automatic Method for Extracting Small Ground Fissures from Loess Areas Using Unmanned Aerial Vehicle Images	Qinghai; Gansu	Ground Fissure detection	Image Classification, Image Segmentation, Feature Extraction	UAV
Gong et al. [6]	Retrieve Ice Velocities and Invert Spatial Rigidity of the Larsen C Ice Shelf Based on Sentinel-1 Interferometric Data	Antarctica	Glacier Flow Monitoring	InSAR, Pixel Offset Tracking	Sentinel-1 SAR
Luo et al. [7]	A Loading Correction Model for GPS Measurements Derived from Multiple-Data Combined Monthly Gravity	Global	GPS Time Series	Comparative Analysis	
Kuang et al. [8]	Displacement Characterization and Spatial–Temporal Evolution of the 2020 Aniangzhai Landslide in Danba County Using Time-Series InSAR and Multi-Temporal Optical Dataset	Sichuan	Landslide Monitoring	InSAR; Pixel Offset Tracking	Sentinel-1 SAR; PlanetScope

Table 1. *Cont.*

Source	Title	Geographic Focus	Applications	Techniques	Data
Yan et al. [9]	Construction of "Space–Sky–Ground" Integrated Collaborative Monitoring Framework for Surface Deformation in Mining Area	Shendong	Mine Subsidence mapping	Collaborative Monitoring; InSAR; TLS; UAV; GNSS CORS; Ground Surveying	UAV infrared; TLS; Sentinel-1 SAR; GNSS; Steel ruler
Zhang et al. [10]	The Current Crustal Vertical Deformation Features of the Sichuan–Yunnan Region Constrained by Fusing the Leveling Data with the GNSS Data	Sichuan–Yunnan	Crustal Movement observation	Data Fusion; GNSS; Ground Surveying	GNSS; leveling
Xing et al. [11]	Measuring Land Surface Deformation over Soft Clay Area Based on an FIPR SAR Interferometry Algorithm—A Case Study of Beijing Capital International Airport (China)	Beijing	Urban Subsidence	InSAR; FIPR	TerraSAR-X SAR
Liu et al. [12]	Quantitative Evaluation of Environmental Loading Products and Thermal Expansion Effect for Correcting GNSS Vertical Coordinate Time Series in Taiwan	Taiwan	GPS Height Time Series	Comparative Analysis	GNSS
Jiao et al. [13]	Comprehensive Remote Sensing Technology for Monitoring Landslide Hazards and Disaster Chain in the Xishan Mining Area of Beijing	Beijing	Landslide Monitoring	InSAR; Change Detection	RadarSAT-2 SAR; Quickbird; GeoEye-1; Worldview-2; Pleiades; BJ-2; Aerial Photo
Han et al. [14]	A Deep Learning Application for Deformation Prediction from Ground-Based InSAR	Sichuan	Landslide Monitoring	GB-InSAR Time Series Analysis	GB-SAR

2. Detection of Land Movement Using Remote Sensing Techniques

The use of modern geodetic monitoring technologies to accurately identify the deformation zones and examine their dynamics is one of the important applications in environmental monitoring. Time-series InSAR (TS-InSAR) has been widely used for mapping long-term deformation in recent years because of its capability for deriving accurate time-series measurements at millimeter accuracy with large spatial coverage. The conventional TS-InSAR technique often requires mathematical empirical models which may not be optimized for describing the nonlinear characteristics of the movement, such as temporal settlement evolution for soft clay. To overcome this limitation, Xing et al. [11] proposed a modified TS-InSAR algorithm, namely FIPR (FastICA Poisson curve reciprocal accumulation method) for mapping infrastructure built on soft clay. This proposed algorithm first separates the original InSAR unwrapped phase information based on FastICA, and then models each extracted deformation component based on its physical characters. The extracted soft soil-related physical deformation component is modelled with Poisson function, and the reciprocal accumulation method (RAM) is utilized to derive the model parameters to generate the total time-series deformation. Xing et al. [11] applied the proposed algorithm over the Beijing Capital International Airport in China, which showed an improvement in modelling accuracy by 36.6% and 16.1%, respectively, for the proposed model compared to the EWA-LM (linear model with equal weight accumulation) algorithm and the EWA-PC (Poisson curve with equal weight accumulation) algorithm.

Gong et al. [6] used a set of Sentinel-1 SAR imagery to retrieve the ice flow velocities of the Larsen C Ice Shelf (LCIS) in Antarctica and then inverted the rigidity of the LCIS based on the deformation data obtained. Since the Sentinel-1 SAR imagery from only one viewing geometry was available for the study, the direct retrieval of 2D velocity field was not possible. Gong et al. [6] proposed to use the artificial neural network (ANN) to recover the azimuth displacement obtained from pixel offset tracking (POT) techniques and integrated with the InSAR results to retrieve the 2D velocity field.

Feature extraction from remote sensing data is widely used to monitor geological hazards caused by land movement. However, conventional extraction techniques are not ideal when dealing with the image of the loess regions due to their rich texture information. Jia et al. [5] proposed an advanced MF-FDOG algorithm for rapid and precise acquisition of ground fissures in the loess areas. Jia et al. [5] applied the proposed data processing scheme with the UAV images acquired at the complex mountainous terrain regions in Qinghai and Gansu for ground fissures extraction. Jia et al. [5] demonstrated that small ground fissures in loess areas can be extracted from high-resolution UAV images with better accuracy and robustness compared to the conventional methods.

3. Mapping Ground Deformation with Fusion of Multiple Datasets

An interesting topic included in this Special Issue is the fusion of multiple geodetic techniques to enhance the mapping capability for various land deformation applications. With the rapid advancement in the diversity of geodetic monitoring techniques, modern geodetic monitoring techniques have the capability to obtain a full of multi-temporal subsidence patterns of the land with various spatial resolutions and coverage. Since different geodetic monitoring techniques have their unique strength, ways to fully utilize these techniques for land subsidence applications is an issue needed to be solved.

GNSS is currently of the most popular geodetic monitoring techniques for surveyors and engineers worldwide. One of the major limitations for GNSS in deformation mapping is that even though the of number stations has increased significantly in the past decades, the spatial resolution of the measurement is still one of its weaknesses compared to other modern geodetic technologies. In order to deal with this issue, Zhang et al. [10] fused the levelling measurements with the GNSS measurements to derive the regional vertical motion field of Sichuan–Yunnan region in China with enhancement in spatial resolution. Zhang et al. [10] used the least squares collocation method to obtain the crustal vertical movements in Sichuan–Yunnan region by combining the high-precision levelling data of 1970s–2010s and the GNSS observation of 320 GNSS stations in the China Crustal Movement Observation Network (CMONOC) and the China Continental Tectonic Environment Monitoring Network (CMTEMN) from 1999 to 2017. Zhang et al. [10] demonstrated that fusion can improve the precision of the vertical subsidence rates in large spatial scales with finer spatial resolution over the region.

Yan et al. [9] established a "space–sky–ground" integrated monitoring framework for mine subsidence monitoring applications. Yan et al. [9] successfully applied the framework for mapping the mine subsidence of a Shengdong coal mine and demonstrated that the framework established, together with the Analytic Hierarchy Process Technique for Order Preference by Similarity to Ideal Solution (AHP-TOPSIS) model, can obtain a more comprehensive information of the subsidence evolution of the site compared to the single monitoring technique.

4. Monitoring Landslide and Slope Stability

Landslides are natural phenomena distributed all around the world. In the past, mapping and monitoring landslides was difficult because many of the landslides occur in remote and rural areas that are often inaccessible. With the advancements in remote sensing technology, the use of remote sensing techniques for monitoring the landslide has become a very popular research topic in recent years, especially in China. Many studies have been conducted on the landslide monitoring methods with various remote sensing sensors and platforms.

Kuang et al. [8] jointly used the Sentinel-1A/B satellite SAR data and the PlanetScope satellites optical images to investigate the deformation time series for the Aniangzhai (ANZ) landslide in Sichuan Province, China. Kuang et al. [8] used optical pixel offset tracking (POT) and InSAR techniques to map the pre- and post-failure surface deformation over the ANZ slope. Kuang et al. [8] was able to identify and characterize several areas of pre-failure movements before the re-activation of the ANZ landslide, and suggested that the most significant triggering factor for the landslide was because of the heavy rainfall. In another article, Jiao et al. [13] collected multi-temporal remote sensing data to extract the deformation information of the landslides in Beijing, China from optical images and time-series InSAR. Jiao et al. [13] demonstrated the capability to use the high-resolution optical images and InSAR data to identify and track the surface deformation of the landslides in Anzigou ditch, Beijing. By analyzing the deformation time-series data together with the temperature and geological data, Jiao et al. [13] determined that the landslide evolution process is closely related to the geological conditions, where the medium and large landslides may occur and trigger a "Quarry–Landslide–Mudflow" disaster chain. The information obtained from multi-platform and multi-sensory remote sensing technologies can be very useful for geological disaster prevention and disaster warning forecasting.

Although the use of InSAR for monitoring ground deformation was first developed for spaceborne application, it has later extended to observations using ground-based SAR sensors because of its capability of high precision as well as high resolution in spatial and temporal domains. Ground-based synthetic aperture radar interferometry (GB-InSAR) has become a popular geodetic technique for monitoring slope instability, especially for landslide deformation monitoring applications. To accurately identify and locate the deformation target for early disaster warning, near real-time processing and high-accuracy geocoding are essential. To tackle the real-time processing problem, Han et al. [14] proposed an improved GB-InSAR time-series processing method based on the LSTM (long short-term memory) model. The method used the deformation and atmospheric parameters obtained from previous acquisitions as the dataset of the LSTM model to predict the deformation and atmospheric delay in the current acquisition. Since only the difference between the current acquisition and the previous one needs to be computed, the method greatly reduced the computer resources and processing time to deliver the time-series data in real time, which significantly improved the processing efficiency for deformation prediction and disaster warning forecasting. Han et al. [14] applied the proposed method to monitor the Guangyuan landslide in Sichuan Province, China. The results show that the proposed method processing time has improved by approximately 7.5 times compared to the traditional method with high deformation prediction accuracy and low memory requirement. To deal with the high-accuracy geocoding issue, Cai et al. [4] proposed an accurate GB-SAR image geocoding method based on solution space search. Cai et al. [4] proposed to use the external high-resolution DSM derived from the UAV photogrammetry to establish a GB-SAR coordinate transformation model to improve the geocoding accuracy. Cai et al. [4] applied the proposed method to monitor the Laoguanjingtai landslide in Sichuan Province, China. The results show that the method is useful in research aiming to overcome the problem of GB-SAR images for selecting the control points in a complex scattering environment. As a result, subpixel geocoding accuracy can be achieved.

5. Improvement of Current Geodetic Measurement Techniques

Another major aspect included in this Special Issue is the research on the improvement of the conventional geodetic measurement techniques.

Wyszkowska et al. [2] proposed an approach to apply the M_{split} estimation in determining the terrain profiles from terrestrial laser scanning (TLS) point clouds. In the study, both squared M_{split} estimation and the absolute M_{split} estimation were considered. Wyszkowska et al. [2] demonstrated that even if there are outliers in an observation set, the proposed approach is still efficient and can provide good terrain profiles compared to the conventional least squares estimation. The result also suggested that better results can be obtained with absolute M_{split} estimation than with the squared M_{split} estimation.

There has been growing interest in positioning techniques based on GNSS for mapping global land deformation with a rapid increase in GNSS continuously operating reference stations in recent years. The study of improving the displacement estimation accuracy has hence become a hot topic, especially in using various loading correction models to improve the accuracy of the GNSS time-series estimation. Wu et al. [1] conducted a comparative analysis to investigate the effect of different loading products from GFZ and EOS, on nonlinear signals in GNSS vertical coordinate time series of 633 global GNSS stations. The results show that the environmental loading corrections are able to reduce the nonlinear deformation signal in most stations around the world, with average reduction rates of 10.6% and 15.4% for all stations after GFZ correction and EOST correction, respectively. Positive root mean square (RME) reduction rates are observed in 82.6% and 87.4% of the stations after GFZ correction and EOST correction, respectively. Wu et al. [1] demonstrated that the effects of environmental loading corrections are inconsistent amongst all stations, the reason for which can be the influence of the annual phase difference between GNSS vertical coordinate time series and the environmental loading products. Liu et al. [12] later conducted a similar study but focused on the GNSS network in Taiwan. Liu et al. [12] quantitatively evaluated the driving factors of nonlinear signals in vertical coordinate time series of stations in the GNSS network, including atmospheric loading (ATML), hydrological loading (HYDL), and non-tidal ocean loading (NTOL) effects. Liu et al. [12] showed that there was no significant difference in correction performance of different environmental loading products. Liu et al. [12] suggested that the combination ATML (GFZ_ECMWF IB) + HYDL (IMLS_MERRA2) + NTOL (IMLS_MPIOM06) is the most suitable for GNSS network in Taiwan Province. Aside from using the common GCM-based loading models, Lu et al. [7] investigated the use of the multiple-data-based combined monthly gravity products LDCmgm90 to provide improved surficial loading models. Lu et al. [7] applied the LDCmgm90 loading model at 249 Global GNSS stations, and compared the results with the three GCM-based loading models, i.e., IMLS, EOST, GFZ models. The result shows that the correction obtained from LDCmgm90 is more effective in attenuating seasonal loading signals compared to the IMLS, EOST, GFZ models. In addition, the use of LDCmgm90 shows a significant improvement to most stations for both data-trend-removed and the data-trend-retained cases. He et al. [3] analyzed and discussed the optimal noise model characteristics of global GNSS time series data of 671 IGS reference stations from 2000 to 2021. He et al. [3] analyzed different noise model combinations and computed the optimal noise model for each station before and after hydrologic loading correction. The results show that optimal noise model characteristics of most stations can be classified into three main categories: white noise + flicker noise, generalized Gauss–Markov noise and white noise + power law noise. The maximum velocity influence value can reach 1.8 mm when hydrological loading is considered.

These selected 14 papers offer a significant contribution to geodetic monitoring for land deformation. With the advance of remote sensing techniques, increase in open earth observation data, and rapid development of machine learning and artificial intelligence, we look forward to upcoming novel research on land deformation mapping.

Author Contributions: Conceptualization, A.H.-M.N. and H.-C.C.; formal analysis, A.H.-M.N.; writing—original draft preparation, A.H.-M.N.; writing—review, and editing, L.G., H.-C.C. and Z.D. All authors have read and agreed to the published version of the manuscript.

Funding: This work was funded by the Program for Guangdong Introducing Innovative and Entrepreneurial Teams (2019ZT08L213), National Natural Science Foundation of China (grant number 42274016), and Natural Science Foundation of Guangdong Province (grant number 2021A1515011483).

Acknowledgments: The Guest Editors of this Special Issue would like to acknowledge all authors who have contributed to this Special Issue for sharing their scientific results. Special gratitude will go to the community of distinguished reviewers for the dedication, time, and expertise to provide their feedback on the submitted manuscripts, helping the authors to enhance the scientific quality of their papers. The Remote Sensing editorial team is sincerely acknowledged for the support during all stages related to the success of publishing this Special Issue.

Conflicts of Interest: The authors declare no conflict of interest.

References

1. Wu, S.; Nie, G.; Meng, X.; Liu, J.; He, Y.; Xue, C.; Li, H. Comparative analysis of the effect of the loading series from gfz and eost on long-term GPS height time series. *Remote Sens.* **2020**, *12*, 2822. [CrossRef]
2. Wyszkowska, P.; Duchnowski, R.; Dumalski, A. Determination of terrain profile from tls data by applying msplit estimation. *Remote Sens.* **2021**, *13*, 31. [CrossRef]
3. He, Y.; Nie, G.; Wu, S.; Li, H. Analysis and discussion on the optimal noise model of global gnss long-term coordinate series considering hydrological loading. *Remote Sens.* **2021**, *13*, 431. [CrossRef]
4. Cai, J.; Jia, H.; Liu, G.; Zhang, B.; Liu, Q.; Fu, Y.; Wang, X.; Zhang, R. An accurate geocoding method for gb-sar images based on solution space search and its application in landslide monitoring. *Remote Sens.* **2021**, *13*, 832. [CrossRef]
5. Jia, H.; Wei, B.; Liu, G.; Zhang, R.; Yu, B.; Wu, S. A semi-automatic method for extracting small ground fissures from loess areas using unmanned aerial vehicle images. *Remote Sens.* **2021**, *13*, 1784. [CrossRef]
6. Gong, F.; Zhang, K.; Liu, S. Retrieve ice velocities and invert spatial rigidity of the larsen c ice shelf based on sentinel-1 interferometric data. *Remote Sens.* **2021**, *13*, 2361. [CrossRef]
7. Luo, J.; Chen, W.; Ray, J.; van Dam, T.; Li, J. A loading correction model for gps measurements derived from multiple-data combined monthly gravity. *Remote Sens.* **2021**, *13*, 4408. [CrossRef]
8. Kuang, J.; Ng, A.H.-M.; Ge, L. Displacement characterization and spatial-temporal evolution of the 2020 aniangzhai landslide in danba county using time-series insar and multi-temporal optical dataset. *Remote Sens.* **2022**, *14*, 68. [CrossRef]
9. Yan, Y.; Li, M.; Dai, L.; Guo, J.; Dai, H.; Tang, W. Construction of space-sky-ground integrated collaborative monitoring framework for surface deformation in mining area. *Remote Sens.* **2022**, *14*, 840. [CrossRef]
10. Zhang, Y.; Xu, C.; Zheng, Z.; Liang, H.; Zhu, S. The current crustal vertical deformation features of the sichuan–yunnan region constrained by fusing the leveling data with the gnss data. *Remote Sens.* **2022**, *14*, 1139.
11. Xing, X.; Zhu, L.; Liu, B.; Peng, W.; Zhang, R.; Ma, X. Measuring land surface deformation over soft clay area based on an fipr sar interferometry algorithm—A case study of beijing capital international airport (China). *Remote Sens.* **2022**, *14*, 4253. [CrossRef]
12. Liu, B.; Ma, X.; Xing, X.; Tan, J.; Peng, W.; Zhang, L. Quantitative evaluation of environmental loading products and thermal expansion effect for correcting gnss vertical coordinate time series in taiwan. *Remote Sens.* **2022**, *14*, 4480. [CrossRef]
13. Jiao, R.; Wang, S.; Yang, H.; Guo, X.; Han, J.; Pei, X.; Yan, C. Comprehensive remote sensing technology for monitoring landslide hazards and disaster chain in the xishan mining area of beijing. *Remote Sens.* **2022**, *14*, 4695. [CrossRef]
14. Han, J.; Yang, H.; Liu, Y.; Lu, Z.; Zeng, K.; Jiao, R. A deep learning application for deformation prediction from ground-based insar. *Remote Sens.* **2022**, *14*, 5067. [CrossRef]

remote sensing

MDPI

Article

Comparative Analysis of the Effect of the Loading Series from GFZ and EOST on Long-Term GPS Height Time Series

Shuguang Wu [1,2], Guigen Nie [1,3,*], Xiaolin Meng [2], Jingnan Liu [1], Yuefan He [1], Changhu Xue [4] and Haiyang Li [1]

1 GNSS Research Center, Wuhan University, Wuhan 430079, China; shgwu@whu.edu.cn (S.W.); jnliu@whu.edu.cn (J.L.); heyuefan@whu.edu.cn (Y.H.); haiyangli@whu.edu.cn (H.L.)
2 Nottingham Geospatial Institute, The University of Nottingham, Nottingham NG7 2TU, UK; xiaolin.meng@nottingham.ac.uk
3 Collaborative Innovation Center for Geospatial Information Technology, Wuhan 430072, China
4 Department of Earth and Space Sciences, Southern University of Science and Technology, Shenzhen 518055, China; xuech@sustech.edu.cn
* Correspondence: ggnie@whu.edu.cn

Received: 4 July 2020; Accepted: 27 August 2020; Published: 31 August 2020

Abstract: In order to investigate the effect of different loading models on the nonlinear variations in Global Positioning System (GPS) height time series, the characteristics of annual signals (amplitude and phase) of GPS time series, loading series from Deutsche GeoForschungsZentrum, Germany (GFZ) and School and Observatory of Earth Sciences, France (EOST) at 633 global GPS stations are processed and analyzed. The change characteristics of the root mean square (RMS) reduction rate, annual amplitude and phase of GPS time series after environmental loading corrections (ELCs) are then detected. Results show that ELCs have a positive effect on the reduction in the nonlinear deformation contained in most GPS stations around the world. RMS reduction rates are positive at 82.6% stations after GFZ correction and 87.4% after EOST correction, and the average reduction rates of all stations are 10.6% and 15.4%, respectively. As for the environmental loading series from GFZ and EOST, their average annual amplitudes are 2.7 and 3.1 mm, which explains ~40% annual amplitude of GPS height time series (7.2 mm). Further analysis of some specific stations indicates that the annual phase difference between GPS height time series and the environmental loading series is an important reason that affects the reduction rates of the RMS and annual amplitude. The linear relationship between the annual phase difference and the annual amplitude reduction rate is significant. The linear fitting results show that when there is no annual phase difference between GPS and loading series, the reduction rates of the RMS and annual amplitude will increase to the maximum of 15.6% and 41.6% for GFZ, and 22.0% and 46.6% for EOST.

Keywords: GPS height time series; surface mass loading; RMS reduction rate; annual amplitude and phase; annual phase difference

1. Introduction

In recent two decades, many researchers have undertaken various topics on earth science by means of Global Positioning System (GPS) time series obtained from continuous GPS (CGPS) stations, including analysis of periodic signals [1–5], detection of stochastic models [6–9], spatial filtering methods [10–14], and surface loading models (SLMs) [15–23].

Existing research results indicate that land mass redistribution is one of the principal causes of the nonlinear motion in GPS time series. By establishing SLMs (including atmospheric loading, nontidal oceanic loading, hydrological loading, etc., and together referred as environmental loadings) and computing the displacement series derived from elastic deformation, effects of geophysical signals on certain GPS stations can be acquired and studied effectively. Characteristic changes of long-term GPS time series before and after the correction of environmental loadings are then compared and analyzed to detect the influence of these loadings, which is helpful to further understand the nonlinear changes included in GPS time series.

van Dam et al. [15] analyzed GPS height time series of 19 stations in the northern hemisphere, and concluded that when atmospheric loading was applied, the total variance of the GPS series could be reduced by 24%, and the atmospheric loading series fluctuated more at higher latitudes. van Dam et al. [17] also found that the nonlinear motion in GPS height time series would be better explained when the influence of unmodeled topographic variability on surface pressure estimates was taken into account. Jiang et al. [19] analyzed the causes of the nonlinear variations in the GPS time series of 11 IGS stations in China. They believed that after environmental loading corrections (ELCs, including atmospheric loading, nontidal oceanic loading, snow and soil moisture loading), the annual amplitude of GPS height series could be effectively weakened, while the effect of ELCs on the horizontal components (north and east) could not.

In the last two years, the effect of ELCs is still a heated topic. Yuan et al. [22] analyzed 235 Crustal Movement Observation Network of China (CMONOC) stations for the nonlinear motion at GPS stations. They calculated the displacement series caused by environmental loadings (including atmospheric loading, nontidal oceanic loading and hydrological loading) and found that after ELCs, the average root mean square (RMS) of GPS height time series could be reduced by ~20%. Andrei et al. [24] reported on the consistent and homogeneous data processing and analysis of a 15 year long time series of ABOA, a Finnish Antarctic research station. Then, they compared the loading series from three service providers GFZ, EOST and International Mass Loading Service (IML), and an obvious difference was found in the nontidal oceanic loading (NTOL) between GFZ and EOST. Li et al. [25] computed the land surface displacements derived from five atmospheric products and compared them with the position time series at 596 global GNSS stations. Results showed that the ERA-Interim model performs best in reducing the scatter of GNSS time series, and the ELC effect at inland stations is better than island stations for all five models. Li et al. [26] investigated the performance of different loading products on the RMS reduction in GNSS time series at CMONOC stations. They concluded that the difference in RMS reduction can reach 20% in the vertical component according to different ELCs applied.

The time span of GPS time series influences noise models, while a long-term series helps to reduce the influence of different noise models on uncertainties of the estimated parameters [27,28]. As the observation data of global GPS stations continue to accumulate, time spans of GPS time series obtained after data processing also increase. Therefore, it is necessary to re-evaluate the parameters of global long-term GPS time series to promote further understanding of the earth surface variation with more accuracy. On the other hand, although the nonlinear motion contained in the GPS time series cannot be completely eliminated by geophysical model corrections, the modeled loading corrections at different GPS stations are quite different. Many previous studies focused on the study of SLMs and the analysis of the ELC effect, while further exploration about why the ELCs perform differently among GPS stations is also important.

Apart from the application in establishing the global terrestrial reference frame, GPS time series can also be applied in some regional geodynamics studies, such as earthquakes, landslides, subsidence and volcanoes. GPS stations located in these geodynamics processes contain not only long-term displacements caused by crustal movement and environmental loadings but also some short-term local variations produced by seismic or volcanic activities. Although both lines of research require different temporal resolutions and time spans of GPS data, it provides an effective method in geodynamics research if the GPS time series keeps a consistent high temporal resolution. The main purpose of this paper was to compare loading products from GFZ and EOST, and detect their effect on the characteristic changes of global GPS height time series; then, to further explore the causes of the annual amplitude and phase differences between the observed and modeled displacement series. Based on 20 year GPS height time series of the globally distributed 633 stations and surface loading data provided by GFZ and EOST, environmental loading effects on GPS height time series of these stations were analyzed, mainly including the spatial distribution characteristics and change of RMS reduction rate, annual amplitude and phase change before and after ELCs. The influence of the colored noise combination, white noise and power law noise (WN+PL) was taken into account, which helped to accurately estimate parameters in the deterministic model of GPS time series. The reason for the large differences in the effect of ELCs on some specific stations was mainly studied. Finally, the relationship between the annual phase difference (between GPS height time series and the environmental loading series), RMS and annual amplitude reduction rate were linearly fitted to obtain the functions between them.

2. Data and Methodology

2.1. GPS Height Time Series

GPS data were obtained from Scripps Orbit and Permanent Array Center (SOPAC), one of the data analysis centers of the International GNSS Service (IGS). The data set included raw and processed GPS time series results. The SOPAC products mainly used in this paper are the linearized mean series after removing the offsets (coseismic or noncoseismic) and outliers, and maintaining the seasonal periodic signals in order to analyze the impact of environmental loadings. More detailed descriptions about the GPS postprocessing and coordinate time series can be found at the SOPAC website—http://sopac-csrc.ucsd.edu/index.php/gambit-globk/.

Before data analysis, it is necessary to select GPS stations according to data quality. The first selection criterion is that the effective time span of a GPS station in two decades (1999~2018) needs to be greater than 5 years, and 705 stations meet this demand from all 1147 stations. The second is that there are no dramatic abnormal variations, including variations caused by large postseismic deformations, single-station-related deformations and others from unknown reasons. We inspect these stations one by one and delete 39 unsatisfactory stations with 666 stations remaining. Third, in order to combine the GPS stations from SOPAC with the loading series of ~6500 stations worldwide from the EOST dataset, we acquire 633 common stations, as shown in Figure 1. The datasets we acquired from SOPAC already excluded most of the stations affected by large earthquakes, and this is expressed by the rare stations along the western coast of the United States where earthquakes happen frequently.

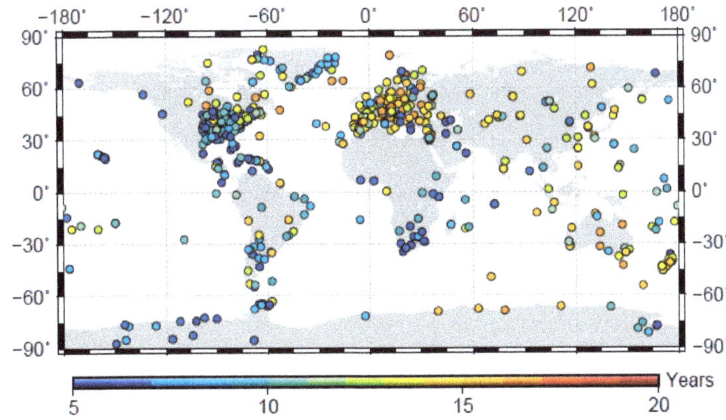

Figure 1. Layout of 633 global GPS stations and their effective time spans in 20 years (1999~2018).

2.2. Environmental Loading Data

Mass redistribution of atmosphere, oceans and terrestrial water storage can cause changes in surface mass loading, resulting in displacements on the Earth's crust [29,30], which can be called environmental loading deformation. Based on different geophysical observations and earth models, SLMs can be established to evaluate the influence on the nonlinear displacement of GPS stations [22].

Different organizations and institutes provide products of environmental loading displacement, including GFZ [31,32] in Germany, EOST (http://loading.u-strasbg.fr/, [33]) Loading Service from University of Strasbourg in France and the Global Geophysical Fluids Center (GGFC, https://geophy.uni.lu/) established by the International Earth Rotation and Reference Systems Service (IERS). Using the environmental loading products provided by GFZ and EOST, this paper estimated displacement series in 633 global GPS stations, and then deducts these displacement series from GPS height time series, so as to analyze the ELC effect derived from products of different organizations.

GFZ provides elastic deformation data of the global land surface derived from atmospheric loading (NTAL), nontidal oceanic loading (NTOL) and hydrological loading (HYDL) on a $0.5 \times 0.5°$ regular global grid, with temporal resolutions of 3 (for NTAL and NTOL) and 24 h (for HYDL). The input geographical data that GFZ adopt to produce products of NTAL, NTOL and HYDL are, respectively, from the European Centre for Medium-Range Weather Forecasts (ECMWF), the Max-Planck Institute Ocean Model (MPIOM, [34]) and the Land Surface Discharge Model (LSDM, [35]). Based on these loading deformation grid data, a bicubic interpolation method was used to obtain displacement series at each station. In order to unify their temporal resolutions, NTAL and NTOL series with 3 h intervals were averaged into daily results. When calculating the sum loading (SUML) series, firstly, we added grid data of the above three loadings, then used the same interpolation method to obtain the SUML series so as to reduce the accuracy loss caused by more interpolations.

EOST provides models of 3D surface displacements, surface gravity and tilt variations at ~6500 global stations, computed using atmospheric, oceanic and hydrological general circulation models. The NTAL loading series employed in this paper were calculated from datasets of ECMWF Reanalysis with a temporal resolution of 6 h (ERA-Interim, [36]). This NTAL model runs from 1979 to present and is updated once per month. The spatial resolution of ERA-Interim is 0.75° in longitude and latitude. This model assumes an inverted barometer ocean response to pressure forcing in model establishment. HYDL loading series were estimated from the MERRA2 model [37] with a temporal resolution of 1 h. The output of the MERRA2 model is a regular $0.625 \times 0.5°$ in longitude by latitude grid data [25]. This hydrological model considers soil moisture and snow depth. The NTOL loading series were estimated from the Estimating the Circulation and Climate of the Ocean Phase II (ECCO2, [38]) ocean bottom pressure with a temporal resolution of 24 h and spatial resolutions of 0.25°. Detailed information about the loading model difference between GFZ and EOST can be found in Li et al. [26].

2.3. Methodology

GPS time series can be expressed as the sum of deterministic and stochastic models [39]. The former consists of polynomial, seasonal signals (annual and semi-annual) and offsets. The stochastic model is commonly known as noises. The functional model of the GPS time series can be expressed as follows:

$$x(t) = \sum_{i=0}^{n_p} p_i(t-t_R)^i + \sum_{j=i}^{n_b} b_j H(t-t_j) + \sum_{k=1}^{n_A} A_k \sin(\omega_k t + \varphi_k) + \sum_{k=1}^{n_L} a_k \log(1 + (t-t_k)/T_k) + \varepsilon \quad (1)$$

where p_i is the coefficient of n_p degree polynomial; t is the epoch of the GPS time series; t_R is the reference epoch; $H(t)$ is the Heaviside step function used to model offsets with amplitudes b_j; n_b is the number of offsets and t_j are the epochs of these offsets; A_k, w_k and φ_k are the amplitudes, angular velocities and initial phases of periodic signals, respectively; n_A is the number of periodic signals, $k = 1$ for annual signal and $k = 2$ for semi-annual signal; n_L is the number of logarithmic functions used to model postseismic deformation; a_k and T_k are two parameters of the logarithmic function; t_k is the epoch of earthquakes that caused logarithmic displacement; ε is the residual series that can be expressed as a combination of varied noise models.

In this paper, we assumed that the polynomial degree was one which is commonly used. Offsets and postseismic deformations of the GPS time series were excluded from the SOPAC products. As for the stochastic models, Hector software [40] was used to estimate the amplitude and phase of the periodic signals in the GPS height time series. This software, like the same type of CATS software [41], is based on maximum likelihood estimation, while the improved algorithms have fastened the computational speed [42]. Therefore, when estimating the periodic signals, we used the noise combination WN+PL as the noise model of the residual series, and it could simultaneously estimate the noninteger spectral index of the residual series.

When assessing the impact of ELCs on GPS height time series, the RMS reduction rate was used as an indicator, expressed as:

$$RMS_{reduction} = \frac{RMS_{GPS} - RMS_{GPS-SUML}}{RMS_{GPS}} \quad (2)$$

In the same way, after deducting the environmental loadings, annual amplitude and phase of GPS height time series also change along with the RMS reduction rate. Annual amplitude reduction rate is expressed as:

$$AMP_{reduction} = \frac{AMP_{GPS} - AMP_{GPS-SUML}}{AMP_{GPS}} \quad (3)$$

3. Results and Analysis

3.1. RMS Reduction Rate

After deducting the SUML displacement series of GFZ and EOST from the GPS height time series, RMS reduction rates are shown in Figure 2. It can be seen from Figure 2a,b that the RMS reduction rates at most stations (523) were positive, accounting for 82.6% of all 633 stations. It indicates that ELCs had a positive effect on correcting the nonlinear deformation of GPS height time series at most stations. As for the ELCs result of GFZ, the number of stations with RMS reduction rates between 0% and 10% was 211, accounting for 33.3% of the total stations. The average RMS reduction rate of all stations was 10.6%. From Figure 2c,d, more GPS stations (553) had decreased RMS values after ELCs of EOST, making up 87.4% of the total stations. The average RMS reduction rate of all stations was 15.4%.

From Figure 2a,c, the impact of ELCs on GPS height time series had obvious spatial distribution characteristics, and were similar to each other. Overall, the ELC effect in the middle and high latitudes was larger than that in low latitudes. In detail, GPS stations in northern North America, eastern Europe, northwestern Asia and Antarctica had a significant ELC effect where the RMS reduction rates were higher. In the southern United States, Central America, southern South America, central and southern Africa, the correction effect of environmental loadings was not obvious, even some stations with RMS increased.

According to the statistics of Figure 2a, the average RMS reduction rates of GPS stations in the low, middle and high latitude regions were 2.4%, 12.7% and 13.9%, respectively, while the counterparts for Figure 2c were 4.2%, 19.8% and 12.7%. In respect to RMS reduction rate, EOST loading series helped more than GFZ products in reducing the nonlinear variation at GPS stations. EOST showed a more concentrated distribution of RMS reduction rates than GFZ (Figure 2b vs. Figure 2d), with less extreme values. It is worth noting that there were gaps in the GPS height time series. In order to keep the same number of observations in the GPS and loading time series, the loading series (consecutive) was processed into nonconsecutive with the same epochs as the GPS series.

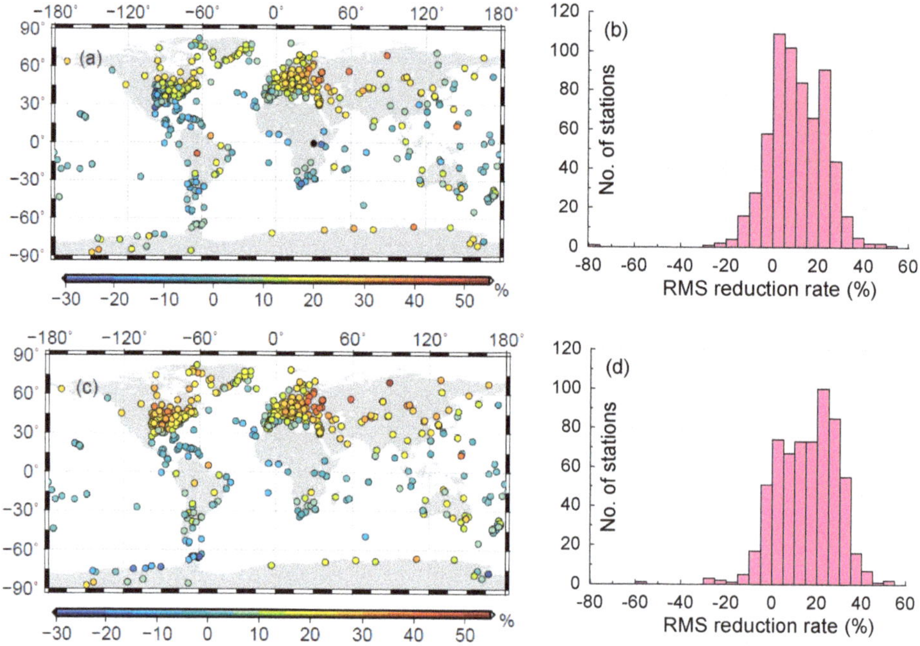

Figure 2. Root mean square (RMS) reduction rates of the 633 global Global Positioning System (GPS) stations after environmental loading corrections (ELCs) and their histograms. Panels (**a**,**b**) show the ELC effect by deducting GFZ loading series, while panels (**c**,**d**) illustrate the effect of EOST loading series. For clear comparison, panel (**d**) keeps the same range of abscissa and ordinate axes as panel (**b**).

As observed from Figure 2a, the two African GPS stations located near equator, MBAR (Uganda, 30.7°E 0.6°S) and RCMN (Kenya, 36.9°E 1.2°S) had the lowest RMS reduction rates of −79.2% and −20.8%, respectively. MBAR is located on the shore of Lake Victoria, the largest African (and third largest worldwide) lake, and RCMN is nearby. However, these two stations did not show the same obvious negative ELC effect in EOST results as in the GFZ results, as shown in Figure 2a,c. The region is controlled by the equatorial low pressure, with annual rainfall and lake water being abundant. The hydro-climatic characteristics are complex and changeable, so different HYDL models in this small-sized area had large variations, as expressed in Figure 3a. In addition, for other GPS stations with negative RMS reduction rates, both GPS and environmental loading series contained obvious annual signals, and the large difference of their annual phases is likely to cause the GPS time series to be more scattered after ELCs. The reason will be further discussed in the next section.

This deviation between loading series from GFZ and EOST can also be represented at stations with positive RMS reduction rates. Observed from Figure 2a, two GPS stations near the equator, POVE (Brazil, 296.1°E 8.7°S) and KOUR (French Guiana, 307.2°E 5.3°N) had RMS reduction rates of 48.4% and 33.1%, respectively, significantly higher than other stations in the low latitude area. From the EOST results in Figure 2c, the counterparts of these two stations were 26.6% and 30.3%. These two stations are located in the Amazon rainforest and also controlled by the equatorial low-pressure zone. There are rich precipitation and lush vegetation that bring huge amounts of transpiration and evaporation. Compared with Lake Victoria, the Amazon rainforest is nearly 80 times larger in acreage, so it is more likely to be considered when establishing hydrological models; therefore, the real hydro-climatic characteristics can be acquired through these models, as shown in Figure 3b. The GPS height time series and loading series of GFZ and EOST at POVE tended to have similar annual amplitude and phases. After ELCs, RMS and the annual amplitude of the GPS-GFZ series greatly decreased. Although there was also an obvious decrease in RMS value for the GPS-EOST series, the residual seasonal terms were still obvious. Therefore, it was necessary to analyze the difference of the annual amplitude and phase with respect to GPS height time series, GFZ and EOST loading series.

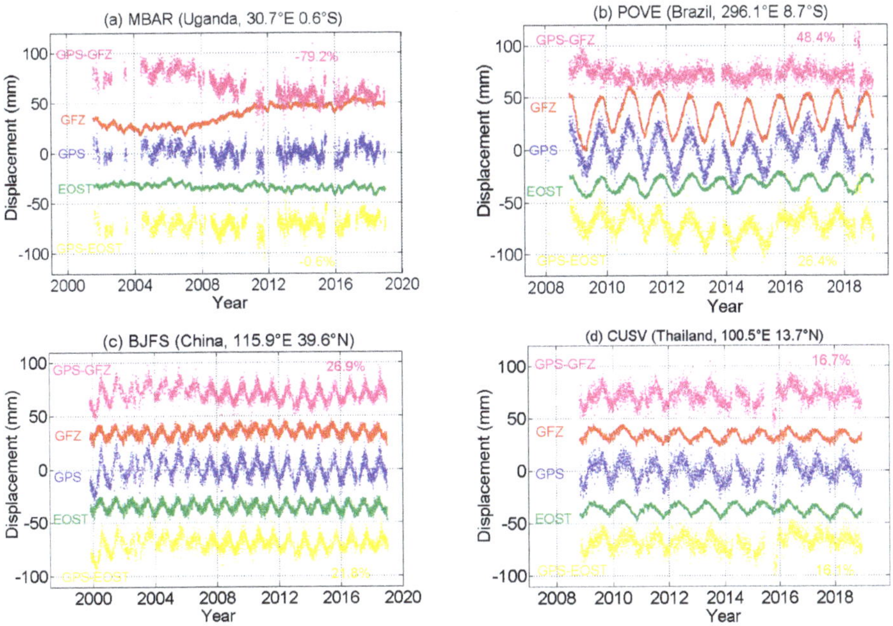

Figure 3. Comparison of ELCs effect in four GPS stations (**a**) MBAR, (**b**) POVE, (**c**) BJFS and (**d**) CUSV. Scatters in blue, red and green represent GPS height series, environmental loading series from GFZ and EOST. Scatters in magenta and yellow are GPS residual series after removing loading series from GFZ and EOST. For clearance, GFZ and EOST series were moved upward and downward 35 mm. GPS-GFZ and GPS-EOST series were moved 70 mm.

3.2. Annual Amplitude and Phase Change

Many studies have demonstrated that there are obvious periodic signals in GPS height time series, and among the signals of various frequencies, the amplitude of annual signal is the largest [43]. In order to study the influence of the loading series from GFZ and EOST on the characteristic changes of annual motion at global GPS stations, based on the WN+PL noise model, the annual amplitude and phase of GPS height time series were calculated before (Figure 4a,b) and after the ELC process (Figure 4g,h for GPS-GFZ; Figure 4i,j for GPS-EOST), and meanwhile the annual amplitude and phase of the loading series (Figure 4c,d for GFZ; Figure 4e,f for EOST).

As can be seen from Figure 4a, the GPS height time series contains strong annual signals with the maximal amplitude of ~20 mm (NOVM, Kazakhstan, 82.9°E 55.0°N). It means that the magnitude of nonlinear motion reaches centimeters at some GPS stations, so it is necessary to take into account the nonlinear variation at the GPS stations and the corresponding geophysical mechanisms when establishing the global terrestrial reference frame. For the spatial layout features of the annual amplitude, GPS stations in the northern hemisphere are generally larger than the southern hemisphere; Eurasia (especially its inland), larger than other continents; and land larger than islands. The average annual amplitude of all 633 stations was 7.2 mm.

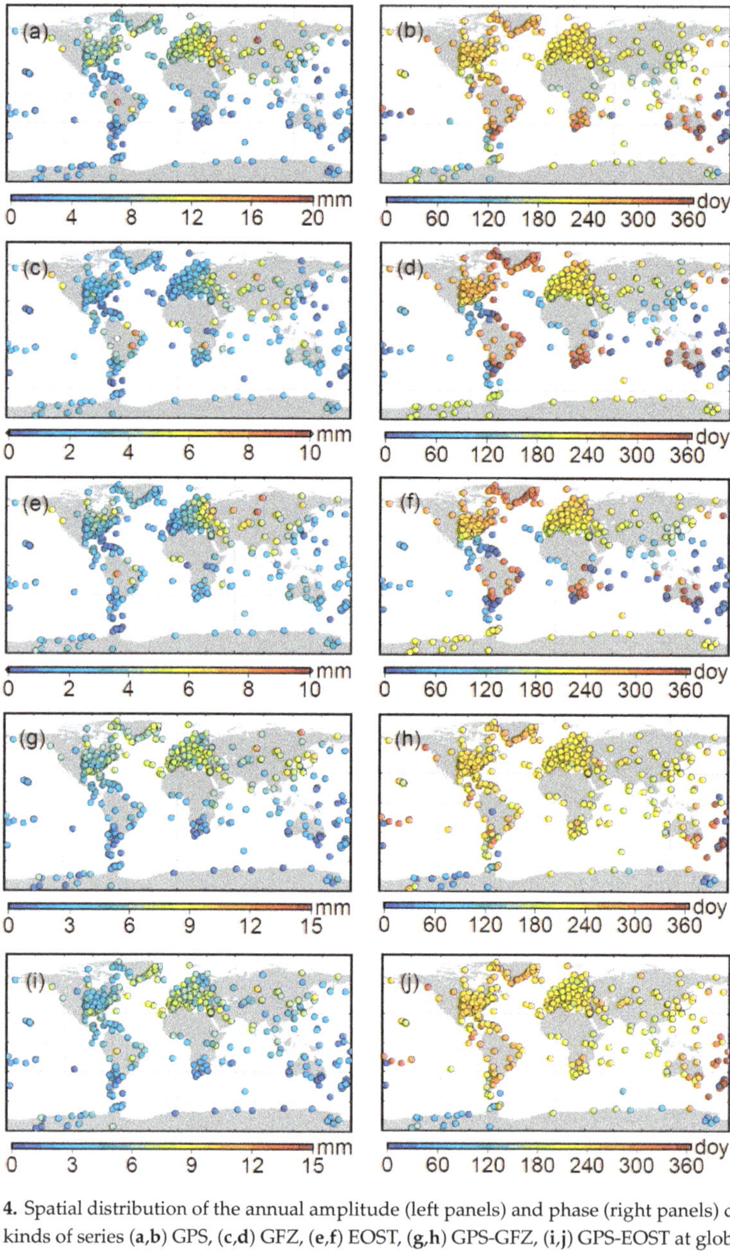

Figure 4. Spatial distribution of the annual amplitude (left panels) and phase (right panels) derived from 5 kinds of series (**a**,**b**) GPS, (**c**,**d**) GFZ, (**e**,**f**) EOST, (**g**,**h**) GPS-GFZ, (**i**,**j**) GPS-EOST at global GPS stations. In order to clearly show the change of distribution features, the scales of the annual amplitude (left panels) are different.

The annual phase of the GPS height time series had more obvious global changes, as shown in Figure 4b. The annual phase of the GPS stations in the northern hemisphere mostly appeared during 180-270 days (expressed as day of year, DOY), that is, July to September. In the southern part of the Earth, annual phases were concentrated in 0-60 and 300-360 days, that is, November to February of next year. For these GPS stations, on the day of their annual phases, they were in the furthest place away from their initial positions and also the Earth's center of figure. It can be explained that the annual phase of GPS height time series was directly related to the latitude of the subsolar point. In other words, when the subsolar point appeared in the northern hemisphere (spring equinox to summer solstice, then to autumn equinox), during this period (approximately 80-266 DOY), 93.9% of the GPS stations in the northern hemisphere responded by reaching their respective annual phases. Similarly, when the subsolar point wandered at the southern hemisphere (autumn equinox to winter solstice, then to spring equinox again), during this period (approximately 266 to 80 of next year), 62.6% of the GPS stations located here also responded. As the GPS stations in the Antarctic and Greenland somehow showed different phases to other regions in their respective hemispheres, the above two ratios would both increase after excluding the GPS stations in these two regions. The relationship between the annual phase of the GPS stations and the latitude of the subsolar point would be more obvious.

The annual amplitude and phase of the environmental loading series are shown in Figure 4c,d (for GFZ) and Figure 4e,f (for EOST). The comparison in Figure 4a,c,e shows that the annual amplitude of the environmental loading series was generally smaller than that of the GPS height time series. The average annual amplitudes of the GFZ and EOST loading series at 633 stations were 2.7 and 3.1 mm, respectively. Compared with the counterpart of the GPS series (7.2 mm), the environmental loading series of GFZ and EOST could explain about 38.4% and 43.3% of the GPS annual amplitude.

There were similar change features of spatial distribution among them, but meanwhile, an obvious difference existed. The annual amplitude of the GPS series in the northern hemisphere was generally larger than that in the southern hemisphere, while this feature seemed unapparent for GFZ and EOST loading series. The annual amplitude of the environmental loading series was larger in the inland of Eurasia and the Brazilian Plateau, such as POVE in Brazil, with the largest annual amplitude of 15.5 mm for the GFZ loading series.

The spatial distribution of the annual phase derived from the environmental loading series had more significant latitude-related and land-sea characteristics. For stations located on ocean islands, the annual amplitudes of GFZ and EOST were mostly distributed from DOY 330 to 120 of next year, that is, December to next April, and they also represented obvious latitude-related differences for stations situated on land. From the equator to 30°N, the annual phases of most stations appeared in DOY 30-120 (excluding island stations in the Caribbean Sea for GFZ), that is, February to April. From the equator to 30°S, annual phases of most stations appeared in 270-360, that is, October to December. From 30 to 60°N, most annual phases were distributed between 210 and 270 (August to September), while from 30 to 60°S, annual phases mostly emerged from 330 to 60 (December to next February). For high-latitude regions, the annual phases of the environmental loading series on the Greenland and Antarctica were concentrated near 210 and 360, standing for July and December, respectively. Compared with the GPS series, the annual phase of the environmental loading series had more obvious latitude-related distribution features.

After the ELC process, spatial distribution of the annual amplitude and phase are displayed in Figure 4g,h (for GPS-GFZ) and Figure 4i,j (for GPS-EOST). To ensure that parameter estimations before and after ELCs are not affected by stochastic models, the same WN+PL noise model was employed. Comparing Figure 4a with Figure 4g,i, the GPS annual amplitudes were decreased and its spatial features were also weakened. The average annual amplitude was 5.3 mm for GPS-GFZ and 4.9 mm for GPS-EOST after ELCs. Compared with the annual amplitude before ELCs (7.2 mm), the amplitude reduction rate was 26.6% for GFZ and 31.6% for EOST.

Particularly, stations NOVM and POVE had the two largest annual amplitudes of 19.5 and 18.1 mm before ELCs. While after ELCs, their annual amplitude decreased to 11.9 and 2.4 mm for GPS-GFZ series, and to 10.3 and 10.3 mm for GPS-EOST series. In GPS-GFZ series, the annual amplitude of TIXI (Russia, 128.9°E 71.6°N) was the largest (12.1 mm). Compared with its counterpart of 12.2 mm before ELCs, its annual amplitude reduction rate was only 0.9%. Even worse, the annual amplitude of 16.9% stations increased after GFZ correction, such as station WUHN (China, 114.4°E 30.5°N) increasing by 3.3 mm. Similarly in GPS-EOST series, the largest annual amplitude was 10.9 mm at station COVX (USA, 284.3°E 36.9°N), and the annual amplitude reduction rate was 9.3%. To facilitate further discussions, the annual amplitude and phase of the above 5 GPS stations are listed in Table 1.

Table 1. Statistics of the annual amplitudes and phases at 5 GPS stations.

Stations	(°E)	(°N)	Country	Annual	GPS	GFZ	EOST	GPS-GFZ	GPS-EOST
NOVM	82.9	55.0	Kazakhstan	Ampli (mm)	19.5	7.5	9.1	11.9	10.3
				Phase (DOY)	211.8	217.0	214.4	208.2	209.3
POVE	296.1	−8.7	Brazil	Ampli	18.1	15.5	8.0	2.4	10.3
				Phase	280.9	277.5	287.4	35.2	274.0
TIXI	128.9	71.6	Russia	Ampli	12.2	0.6	4.9	12.1	7.4
				Phase	221.2	150.9	216.2	224.7	225.5
WUHN	114.4	30.5	China	Ampli	5.9	6.3	4.4	9.2	2.3
				Phase	186.1	92.9	174.8	231.7	213.1
COVX	284.3	36.9	USA	Ampli	12.0	0.6	1.1	11.5	10.9
				Phase	110.8	249.4	228.6	221.7	222.2

Comparing Figure 4b with Figure 4h,j, the GPS annual phase changed significantly after ELCs. The north–south difference of the annual phase was basically eliminated and it appeared at around DOY 230 for most stations. The percentage of stations with an annual phase between DOY 200 to 260 was 80.3% for GPS-GFZ series, and 78.4% for GPS-EOST series. In addition, the annual phase of the GPS stations located in southeastern Australia, New Zealand and nearby islands and the Antarctic were apparently different from other stations, most of which were distributed from DOY 300 to 60 of next year.

Andrei et al. [24] estimated the seasonal signals (including annual and semi-annual components) at station ABOA, a Finnish Antarctic station research station located in Dronning Maud Land, East Antarctic. Figure 5 shows the 32 Antarctic stations of this study and the station ABOA. Although ABOA was not included in this study, the station VESL is relatively near to it, so some comparisons could be made in Table 2. There were differences in the estimations of the annual and semi-annual amplitudes between ABOA and VESL. For station ABOA, the annual amplitude was a little larger than the semi-annual amplitude; while for VESL, the annual signal predominated in its annual motion. For all 32 stations, there were 7 stations with a difference of the annual amplitude and semi-annual amplitude smaller than 0.5 mm, so such differences were probably determined by the local geophysical effects. Overall, the average annual amplitude was 3.05 ± 0.73, nearly twice as much as the semi-annual amplitude. The spectral indexes of ABOA and VESL were nearly equal, which means that their height time series contained the same temporally correlated colored noise. The average index value for the total 32 stations was −1.0, so the GPS height time series at the Antarctic showed obvious flicker noise.

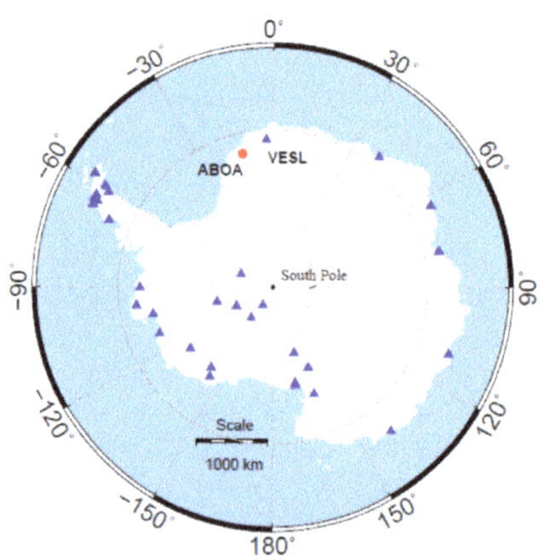

Figure 5. Spatial layout of the Antarctic GPS stations. The 32 blue triangles represent the GPS stations of this study, and the red dot denotes the ABOA station in Andrei et al.

Table 2. Comparisons of the seasonal signals and spectral index between the GPS height time series at stations ABOA and VESL. The column "Mean" displays the average values of the 32 stations. (unit: mm).

	ABOA	VESL	Mean
Annual Ampli	2.43 ± 0.57	3.27 ± 0.59	3.05 ± 0.73
Semi-annual Ampli	2.14 ± 0.42	0.68 ± 0.34	1.45 ± 0.48
Annual Phase	23/24 August	21 July	14 May
Spectral Index	−0.9	−0.9	−1.0

4. Discussions

Comparing the annual amplitude changes of five GPS stations before and after ELCs in Table 1, it can be seen that ELCs had a significant effect on the first two stations (NOVM and POVE). Their annual amplitude reduction rates are 39.2%, 86.5% after GFZ correction, and 47.3% and 47.2% for EOST correction, respectively. Loading series from GFZ and EOST both have a positive role in reducing the GPS annual amplitude. For station TIXI, the effect of GFZ correction is not obvious as its annual amplitude reduction rate is 0.9% only, while the effect of EOST correction is much better with the reduction rate of 39.2%. The difference exists because the EOST annual amplitude and phase is closer to the counterpart of the GPS series than GFZ. For station WUHN, the correction effect of GFZ and EOST loading series seems to have a larger difference, because the annual amplitude reduction rate is −56.1% for GFZ and 61.4% for EOST. Although GFZ annual amplitude (6.3 mm) is closer to the GPS annual amplitude (5.9 mm) than EOST (4.4 mm), its annual phase has a much larger difference with GPS than EOST. As a result, the GFZ loading series plays a negative role in reducing nonlinear motion at this station.

From the above discussions, it is obvious to conclude that ELCs perform effectively at some stations where GPS series and loading series not only have similar annual amplitudes but also similar annual phases. Therefore, after the ELC process, the GPS series contains much less fluctuations, so the nonlinear motion becomes significantly weakened. For station NOVM, although its annual amplitudes of GPS series and loading series are quite different, the annual phase difference is only in a range of a few days, so a good ELC effect can also be obtained. On the other hand, ELCs will result in poor effects at stations where there is a large difference of annual phase between GPS series and loading series, despite a small difference of annual amplitude between them. The peaks of the two series cannot be eliminated in the differencing process, leading to the nonlinear motion in the GPS series increasing, such as the GFZ effect at station WUHN.

In order to visually show the ELC effect at the total number of GPS stations, the annual amplitude reduction rate of GPS series and its histogram were obtained, as shown in Figure 6. The annual amplitude reduction rates at most GPS stations are positive, accounting for 83.1% of total stations for GFZ correction and 82.8% for EOST correction. It indicates that environmental loadings have a positive effect on the annual variations at most GPS stations. The average annual amplitude reduction rate of all stations is 21.1% for GFZ and 23.9% for EOST. Specifically, the annual amplitude reduction rate of GPS stations located in Eurasia, Australia, northern North America and central South America is more obvious, while for stations in Central America, the Caribbean Sea and some marine islands, the annual amplitude reduction rates are comparatively small.

In order to investigate the relationship between the ELC effect and annual phase difference, the absolute phase differences between the GPS series and GFZ, EOST loading series were calculated. The spatial distribution and histograms of the absolute phase difference are shown in Figure 7 (|GPS-GFZ| for Figure 7a,b; |GPS-EOST| for Figure 7c,d). The number of GPS stations with a phase difference of 0-30 days (one month) is 410 for GFZ, and 427 for EOST, accounting for 64.8% and 67.5% of all stations, respectively. Comparing Figures 6 and 7, the annual phase differences and the annual amplitude reduction rates at the GPS stations have a comparatively consistent spatial distribution, similarly showing that the annual phase differences in the GPS stations located in Central America and some ocean islands are obviously different from stations in other regions.

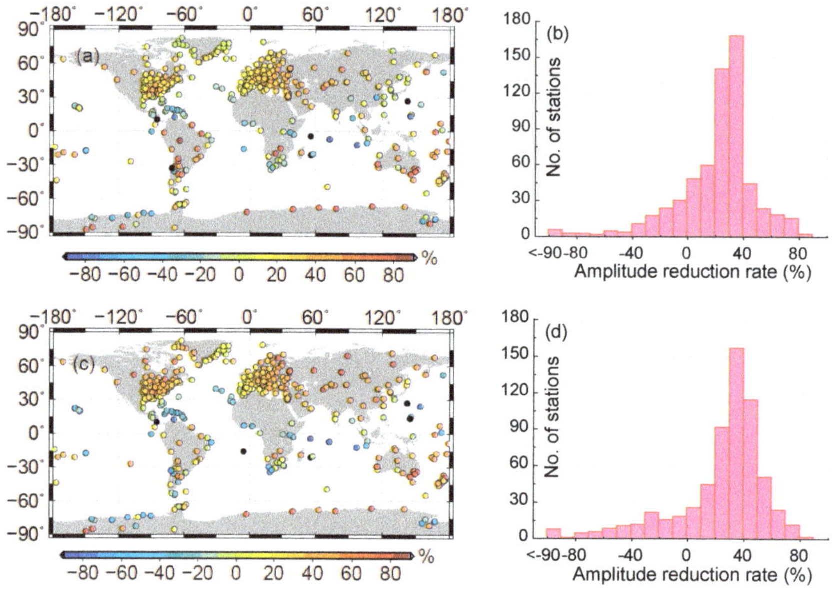

Figure 6. Layout of the annual amplitude reduction rate of GPS height time series after GFZ and EOST loading corrections (panels (**a**,**c**)) and their corresponding histograms (panels (**b**,**d**)).

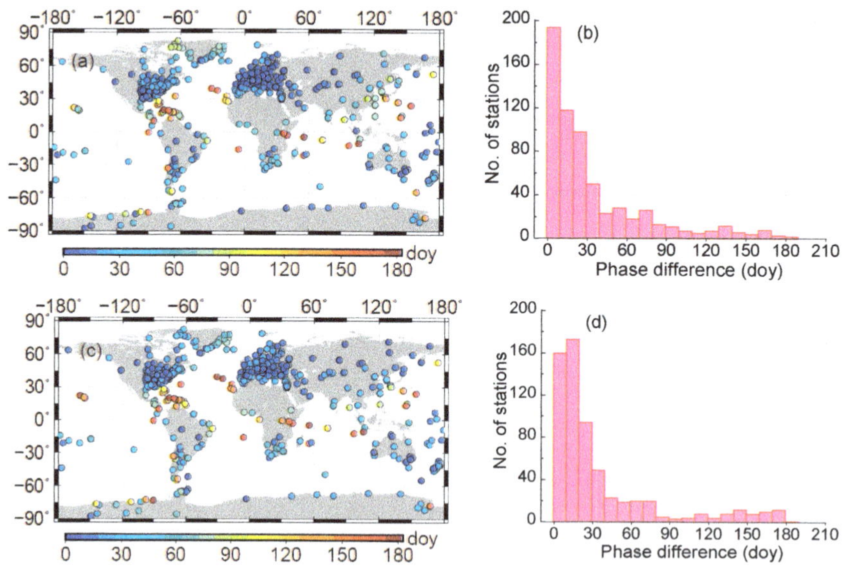

Figure 7. Layout of the absolute differences between the annual phases derived from GPS height time series and environmental loading series of (**a**) GFZ and (**c**) EOST, and their corresponding histograms (**b**,**d**).

Figure 8 illustrates the linear relationship between the annual phase difference (assumed as independent variable X), annual amplitude reduction rate (dependent variable Y1) and RMS reduction rate (dependent variable Y2). It is apparent that there is a strong linear relationship between annual phase difference X, annual amplitude reduction rate Y1 and RMS reduction rate Y2, indicating that the annual phase difference directly affects the reduction rate of the annual amplitude and RMS.

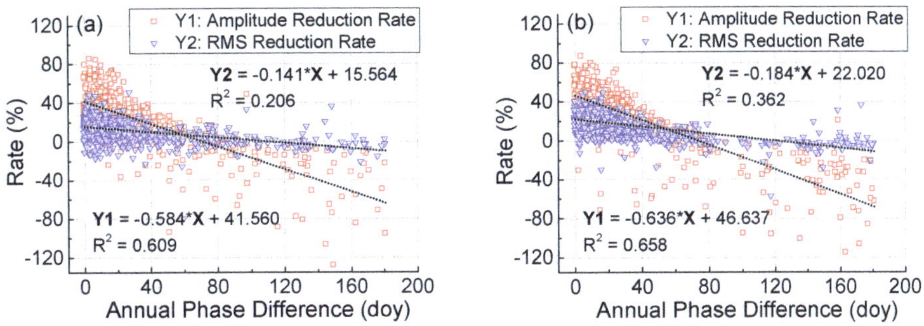

Figure 8. Linear fitting of annual phase differences (X) with reduction rates of annual amplitude (Y1) and RMS (Y2). Panel (**a**) shows GPS vs. GFZ results, and panel (**b**) shows GPS vs. EOST results.

According to the linear fitting results in Figure 8a (GPS vs. GFZ), when the annual phase difference is zero, the reduction rates of annual amplitude and RMS will be the largest, which are 41.6% and 15.6%, respectively. The two ratios are larger in the fitting results of Figure 8b (GPS vs. EOST), which means EOST loading series are possibly more helpful in reducing the nonlinear variations in the GPS time series. In addition, the goodness of fit between X and Y1 is greater than that between X and Y2, indicating that the linear relationship between the annual phase difference and the annual amplitude reduction rate is stronger.

5. Conclusions

This paper analyzes the characteristic changes of 20 year GPS height time series of global 633 stations before and after the environmental loading corrections of GFZ and EOST products, including the reduction rate of RMS, annual amplitude and annual phase difference. Some conclusions were drawn as follows:

Environmental loading corrections have a positive effect on the reduction in the nonlinear motion in GPS height time series of the global GPS stations. RMS reduction rate is positive at 82.6% of stations for GFZ correction and 87.4% for EOST correction, and the average value of all stations is 10.6% and 15.4%, respectively. RMS reduction rate shows an obvious latitude correlation that the ELC effect in the middle and high latitude performs better than in the low latitude.

As for the GPS height time series, the annual amplitudes in the northern hemisphere are generally larger than that in the southern hemisphere; Eurasia (especially its inland), larger than other continents; and land larger than islands. The average annual amplitude of all GPS stations is 7.2 mm. The global layout characteristics of the annual phase are obvious. Most of the annual phases in the northern hemisphere appear from July to September, while in the southern hemisphere, they are concentrated from December to next February. The annual phase of the GPS height time series is affected by the latitude of the subsolar point.

As for the environmental loading series of GFZ and EOST, their average annual amplitude is 2.7 and 3.1 mm, which explains the ~40% annual amplitude of the GPS height time series. The annual amplitude derived from the environmental loading series is generally smaller than that from the GPS height time series, and they have similar characteristics of spatial distribution. Compared with

GPS height time series, the annual phase of environmental loading series shows a more obvious latitude correlation.

After ELCs, the annual amplitudes of most GPS height series are decreased, and their spatial features are also weakened. The average annual amplitude is 5.3 mm for GPS-GFZ and 4.9 mm for GPS-EOST after ELCs. Compared with the annual amplitude before ELCs (7.2 mm), the amplitude reduction rate is ~30%. The annual phases of GPS height time series also change significantly. They are distributed around 230 days in most stations, and are distributed between 200~260 days for ~80% stations.

The effect of ELCs varies greatly among global GPS stations. The largest annual amplitude reduction rate can reach about 87%, while for some stations, the nonlinear motion increases significantly after ELCs. The difference between the annual amplitude and phase of GPS height time series and environmental loading series can be an immediate reason for the unevenness of ELC effects, and the annual phase difference has a greater influence.

According to the fitting results of the RMS reduction rate, annual amplitude reduction rate and annual phase difference, the linear relationship between the annual phase difference and the annual amplitude reduction rate is stronger. Meanwhile, EOST loading series are possibly more helpful in reducing the nonlinear variations in GPS time series.

Author Contributions: Conceptualization, S.W. and G.N.; formal analysis, X.M. and Y.H.; funding acquisition, X.M.; investigation, G.N. and Y.H.; methodology, S.W., G.N., Y.H. and C.X.; resources, X.M. and J.L.; software, S.W.; supervision, C.X.; validation, J.L. and H.L.; visualization, Y.H. and H.L.; writing—original draft, S.W.; writing—review and editing, X.M. and H.L. All authors have read and agreed to the published version of the manuscript.

Funding: This work is supported by the National Natural Science Foundation of China (Grant No. 41074023) and the 2018 Joint R&D and Demonstration Project of the Strategic International Science and Technology Innovation Cooperation Key Project (Project No. SQ2018YFE020091). The Chinese Scholarship Council (CSC) has provided the first author with a scholarship which allows him to visit the University of Nottingham in the UK for one year from September 2019.

Acknowledgments: The dataset we used in this paper includes the GPS height time series of 839 global stations provided by SOPAC (ftp://garner.ucsd.edu/archive/garner/timeseries/measures/ats/), and environmental loading grid data contributed by GFZ (http://rz-vm115.gfz-potsdam.de:8080/repository). We applied the software Hector (http://segal.ubi.pt/hector/) to compute and analyze the GPS time series. In addition, figures in this paper are displayed using GMT, Origin and Matlab. We express our sincere gratitude to these organizations and software providers.

Conflicts of Interest: The authors declare no conflict of interest.

References

1. Blewitt, G.; Lavallee, D.; Clarke, P.; Nurutdinov, K. A new global mode of Earth deformation: Seasonal cycle detected. *Science* **2001**, *294*, 2342–2345. [CrossRef] [PubMed]
2. Wang, M.; Shen, Z.; Dong, D. Effects of non-tectonic crustal deformation on continuous GPS position time series and correction to them. *Chin. J. Geophys.* **2005**, *48*, 1045–1052. [CrossRef]
3. Ray, J.; Altamimi, Z.; Collilieux, X.; van Dam, T. Anomalous harmonics in the spectra of GPS position estimates. *GPS Solut.* **2008**, *12*, 55–64. [CrossRef]
4. Klos, A.; Bos, M.; Bogusz, J. Detecting time-varying seasonal signal in GPS position time series with different noise levels. *GPS Solut.* **2018**, *22*, 21. [CrossRef]
5. Klos, A.; Olivares, G.; Teferle, F.; Hunegnaw, A.; Bogusz, J. On the combined effect of periodic signals and colored noise on velocity uncertainties. *GPS Solut.* **2018**, *22*, 1–13. [CrossRef]
6. Williams, S.; Bock, Y.; Fang, P.; Jamason, P.; Nikolaidis, R.; Prawirodirdjo, L.; Miller, M.; Johnson, D. Error analysis of continuous GPS position time series. *J. Geophys. Res. Solid Earth* **2004**, *109*, B03412. [CrossRef]
7. Beavan, R.J. Noise properties of continuous GPS data from concrete pillar geodetic monuments in New Zealand, and comparison with data from deep drilled braced monuments. *J. Geophys. Res. Solid Earth* **2005**, *110*, B8. [CrossRef]
8. Langbein, J. Noise in GPS displacement measurements from Southern California and Southern Nevada. *J. Geophys. Res. Solid Earth* **2008**, *113*, B05405. [CrossRef]

9. Yuan, L.; Ding, X.; Chen, W.; Guo, Z.; Chen, S.; Hong, B.; Zhou, J. Characteristics of daily position time series from the Hong Kong GPS fiducial network. *Chin. J. Geophys.* **2008**, *51*, 1372–1384. [CrossRef]
10. Dong, D.; Fang, P.; Bock, Y.; Cheng, M.K.; Miyazaki, S. Anatomy of apparent seasonal variations from GPS-derived site position time series. *J. Geophys. Res. Solid Earth* **2002**, *107*, ETG 9-1–ETG 9-16. [CrossRef]
11. Márquez-Azúa, B.; DeMets, C. Crustal velocity field of Mexico from continuous GPS measurements, 1993 to June 2001: Implications for the neotectonics of Mexico. *J. Geophys. Res. Solid Earth* **2003**, *108*, B9. [CrossRef]
12. Tian, Y.; Shen, Z. Extracting the regional common-mode component of GPS station position time series from dense continuous network. *J. Geophys. Res. Solid Earth* **2016**, *121*, 1080–1096. [CrossRef]
13. Ma, C.; Li, F.; Zhang, S.; Lei, J.; E, D.; Hao, W.; Zhang, Q. The coordinate time series analysis of continuous GPS stations in the Antarctic Peninsula with consideration of common mode error. *Chin. J. Geophys.* **2016**, *59*, 2783–2795. [CrossRef]
14. Ming, F.; Yang, Y.; Zeng, A.; Zhao, B. Spatiotemporal filtering for regional GPS network in China using independent component analysis. *J. Geod.* **2017**, *91*, 419–440. [CrossRef]
15. van Dam, T.; Blewitt, G.; Heflin, M. Atmospheric pressure loading effects on Global Positioning System coordinate determinations. *J. Geophys. Res. Solid Earth* **1994**, *99*, 23939–23950. [CrossRef]
16. van Dam, T.; Wahr, J.; Milly, P.; Shmakin, A.; Blewitt, G.; Lavallée, D.; Larson, K. Crustal displacements due to continental water loading. *Geophys. Res. Lett.* **2001**, *28*, 651–654. [CrossRef]
17. van Dam, T.; Altamimi, Z.; Collilieux, X.; Ray, J. Topographically induced height errors in predicted atmospheric loading effects. *J. Geophys. Res. Atmos.* **2010**, *115*, B07415. [CrossRef]
18. Jiang, W.; Li, Z.; van Dam, T.; Ding, W. Comparative analysis of different environmental loading methods and their impacts on the GPS height time series. *J. Geod.* **2013**, *87*, 687–703. [CrossRef]
19. Jiang, W.; Li, Z.; Liu, H.; Zhao, Q. Cause analysis of the non-linear variation of the IGS reference station coordinate time series inside China. *Chin. J. Geophys.* **2013**, *56*, 2228–2237. [CrossRef]
20. Gu, Y.; Yuan, L.; Fan, D.; You, W.; Su, Y. Seasonal crustal vertical deformation induced by environmental mass loading in mainland China derived from GPS, GRACE and surface loading models. *Adv. Space Res.* **2017**, *59*, 88–102. [CrossRef]
21. Chanard, K.; Fleitout, L.; Calais, E.; Rebischung, P.; Avouac, J. Toward a global horizontal and vertical elastic load deforma-tion model derived from GRACE and GNSS station position time series. *J. Geophys. Res. Solid Earth* **2018**, *123*, 3225–3237. [CrossRef]
22. Yuan, P.; Li, Z.; Jiang, W.; Ma, Y.; Chen, W.; Sneeuw, N. Influences of environmental loading corrections on the nonlinear variations and velocity uncertainties for the reprocessed global positioning system height time series of the crustal movement observation network of China. *Remote Sens.* **2018**, *10*, 958. [CrossRef]
23. Klos, A.; Gruszczynska, M.; Bos, M.; Boy, J.; Bogusz, J. Estimates of vertical velocity errors for IGS ITRF2014 stations by applying the improved singular spectrum analysis method and environmental loading models. *Pure Appl. Geophys.* **2018**, *175*, 1823–1840. [CrossRef]
24. Andrei, C.; Lahtinen, S.; Nordman, M.; Näränen, J.; Koivula, H.; Poutanen, M.; Hyyppä, J. GPS time series analysis from aboa the finnish antarctic research station. *Remote Sens.* **2018**, *10*, 1937. [CrossRef]
25. Li, Z.; Chen, W.; van Dam, T.; Rebischung, P.; Altamimi, Z. Comparative analysis of different atmospheric surface pressure models and their impacts on daily ITRF2014 GNSS residual time series. *J. Geod.* **2020**, *94*, 42–61. [CrossRef]
26. Li, C.; Huang, S.; Chen, Q.; Dam, T.; Fok, H.; Zhao, Q.; Wu, W.; Wang, X. Quantitative Evaluation of Environmental Loading Induced Displacement Products for Correcting GNSS Time Series in CMONOC. *Remote Sens.* **2020**, *12*, 594. [CrossRef]
27. Santamaría-Gómez, A.; Bouin, M.; Collilieux, X.; Wöppelmann, G. Correlated errors in GPS position time series: Implications for velocity estimates. *J. Geophys. Res. Solid Earth* **2011**, *116*, B01405. [CrossRef]
28. Jiang, W.; Zhou, X. Effect of the span of Australian GPS coordinate time series in establishing an optimal noise model. *Sci. China Earth Sci.* **2014**, *44*, 2461–2478. [CrossRef]
29. Farrell, W. Deformation of the Earth by surface loads. *Rev. Geophys.* **1972**, *10*, 761–797. [CrossRef]
30. van Dam, T.; Wahr, J. Displacements of the Earth's surface due to atmospheric loading: Effects on gravity and baseline measurements. *J. Geophys. Res. Solid Earth* **1987**, *92*, 1281–1286. [CrossRef]
31. ESMGFZ Product Repository. Available online: http://rz-vm115.gfz-potsdam.de:8080/repository (accessed on 29 August 2020).

32. Dill, R.; Dobslaw, H. Numerical simulations of global scale high-resolution hydrological crustal deformations. *J. Geophys. Res. Solid Earth* **2013**, *118*, 5008–5017. [CrossRef]
33. Mémin, A.; Boy, J.; Santamaría-Gómez, A. Correcting GPS measurements for non-tidal loading. *GPS Solut.* **2020**, *24*, 45. [CrossRef]
34. Marsland, S.; Haak, H.; Jungclaus, J.; Latif, M.; Roske, F. The Max-Planck-Institute global ocean/sea-ice model with orthogonal curvilinear coordinates. *Ocean Model.* **2003**, *5*, 91–127. [CrossRef]
35. Dill, R. Hydrological model LSDM for operational Earth rotation and gravity field variations. *Sci. Tech. Rep.* **2008**. [CrossRef]
36. Berrisford, P.; Dee, D.; Poli, P.; Brugge, R.; Fielding, K.; Fuentes, M.; Kallberg, P.; Kobayashi, S.; Uppala, S.; Simmons, A. *ERA Report Series: The ERA-Interim Archive Version 2.0*; European Centre for Medium Range Weather Forecasts: Shinfield Park, UK, 2011; pp. 1–27.
37. Gelaro, R.; McCarty, W.; Suárez, M.J.; Todling, R.; Molod, A.; Takacs, L.; Randles, C.A.; Darmenov, A.; Bosilovich, M.G.; Reichle, R.; et al. The modern-era retrospective analysis for research and applications, version 2 (MERRA-2). *J. Clim.* **2017**, *30*, 5419–5454. [CrossRef] [PubMed]
38. Menemenlis, D.; Campin, J.; Heimbach, P.; Hill, C.; Lee, T.; Nguyen, A.; Schodlok, M.; Zhang, H. ECCO2: High resolution global ocean and sea ice data synthesis. *Mercat. Ocean Q. Newsl.* **2008**, *31*, 13–21.
39. Bevis, M.; Brown, A. Trajectory models and reference frames for crustal motion geodesy. *J. Geod.* **2014**, *88*, 283–311. [CrossRef]
40. Bos, M.; Fernandes, R. Hector User Manual Version 1.6. 2016. Available online: http://segal.ubi.pt/hector/manual_1.7.2.pdf (accessed on 29 August 2020).
41. Williams, S. CATS: GPS coordinate time series analysis software. *GPS Solut.* **2008**, *12*, 147–153. [CrossRef]
42. Bos, M.; Fernandes, R.; Williams, S.; Bastos, L. Fast error analysis of continuous GNSS observations with missing data. *J. Geod.* **2013**, *87*, 351–360. [CrossRef]
43. Freymueller, J. Seasonal position variations and regional reference frame realization. In *GRF2006 Symposium Proceedings, International Association of Geodesy Symposia*; Bosch, W., Drewes, H., Eds.; Springer: Berlin, Germany, 2009; Volume 134, pp. 191–196. [CrossRef]

remote sensing

MDPI

Article

Determination of Terrain Profile from TLS Data by Applying M_{split} Estimation

Patrycja Wyszkowska *, Robert Duchnowski and Andrzej Dumalski

Department of Geodesy, Institute of Geodesy and Civil Engineering, Faculty of Geoengineering, University of Warmia and Mazury in Olsztyn, 10-719 Olsztyn, Poland; robert.duchnowski@uwm.edu.pl (R.D.); andrzej.dumalski@uwm.edu.pl (A.D.)

* Correspondence: patrycja.wyszkowska@uwm.edu.pl

Abstract: This paper presents an application of an M_{split} estimation in the determination of terrain profiles from terrestrial laser scanning (TLS) data. We consider the squared M_{split} estimation as well as the absolute M_{split} estimation. Both variants have never been used to determine terrain profiles from TLS data (the absolute M_{split} estimation has never been applied in any TLS data processing). The profiles are computed by applying polynomials of a different degree, determining which coefficients are estimated using the method in question. For comparison purposes, the profiles are also determined by applying a conventional least squares estimation. The analyses are based on simulated as well as real TLS data. The actual objects have been chosen to contain terrain details (or obstacles), which provide some measurements which are not referred to as terrain surface; here, they are regarded as outliers. The empirical tests prove that the proposed approach is efficient and can provide good terrain profiles even if there are outliers in an observation set. The best results are obtained when the absolute M_{split} estimation is applied. One can suggest that this method can be used in a vertical displacement analysis in mining damages or ground disasters.

Keywords: M_{split} estimation; TLS measurements; terrain profiles; vertical displacements

Citation: Wyszkowska, P.; Duchnowski, R.; Dumalski, A. Determination of Terrain Profile from TLS Data by Applying M_{split} Estimation. *Remote Sens.* **2021**, *13*, 31. https://doi.org/10.3390/rs13010031

Received: 27 November 2020
Accepted: 21 December 2020
Published: 23 December 2020

1. Introduction

Laser scanning is a popular technique of data acquisition nowadays. Laser scanning data are acquired as point clouds of varying resolution. Depending on the type of platform, there is airborne laser scanning (ALS), terrestrial laser scanning (TLS), satellite laser scanning (SLS) and mobile laser scanning (MLS) [1,2]. All types of laser scanning have a wide range of applications, e.g., terrain surface and vegetation cover measurements and digital terrain model (DTM) generation [3–7], building detection and their condition diagnosis [8–12], displacement detection [13–15], modeling of cultural heritage sites or object structures [2,13,16], registration roads, railways or power lines [17,18], monitoring coastal zones [19], mining damages and ground disasters [20,21].

Here, we consider the classical application of TLS, namely acquiring data of terrain surface. However, grass, roots, fallen leaves, fallen tree branches, shoots (in general vegetation cover), or other terrain details situated on this surface might disturb such TLS measurements. The problem of how to separate such obstacles from the point clouds or how to deal with observations that might be considered as outliers (distinguishing positive and negative outliers) was addressed in several publications, e.g., [22–24]. Generally, TLS data (point cloud) might require a separation of observations related to the terrain surface from observations related to terrain details. Such a separation might be done by segmentation, e.g., region-based methods, Hough transform and Random Sample Consensus (RANSAC) [11,17,25–27]. Obviously, there are also other different methods of point cloud filtering, e.g., surface-based adjustment, morphology-based filtering, Triangulated Irregular Network (TIN)-based refinement or adaptive TIN (ATIN)-based refinement [1,5,7,28]. Another possible approach to such a separation is the application of M_{split} estimation [1].

M_{split} estimation, which is a development of M-estimation, was introduced in [29]. The primary assumption is that an observation set is an unrecognized mixture of realizations of different random variables [29,30]. The idea behind the method is to estimate such variables' location parameters without dividing the observation set into the respective number of subsets (observation aggregations). In many practical cases, such an a priori division would be hard or almost impossible. The basic variant of M_{split} estimation assumes two such variables; thus, the classical functional model is split into two competitive functional models. It is also worth noting that the observations' assignment to the particular variable (and thus to the specific variant of the split functional model) is automatic during the iterative process. In the context of this paper, a point cloud obtained from TLS includes points related to the terrain surface and terrain details. M_{split} estimation allows us to process data without dividing a point cloud into two clouds representing selected surfaces. In general, the method in question will enable us to process a point cloud and approximate two surfaces by estimating surfaces' parameters. In some special cases, one can approximate not the whole surfaces, but rather chosen profiles. Such an approach would be advisable in deformation analysis when one wants to control how the terrain heights are changing between measurement epochs. This might concern, e.g., monitoring landslides or mining damages. In this context, the terrain profile as well as terrain details can be approximated by, e.g., polynomials of a single indeterminate, of which the coefficients can be estimated by different methods, including M_{split} estimation.

Applications of M_{split} estimation concerned so far not only processing of TLS data (presented in [1,15,22,31]) but also deformation analyses [29,30,32–36], direct identification of gross errors [37], linear regression analyses [30,34], robust coordinate transformation [38], S-transformation [39] and marine navigation [40]. Most of these applications applied only the squared M_{split} estimation (SMS), of which the objective functions stem from the assumption that observation errors are normally distributed. However, there is also an alternative variant of M_{split} estimation in which objective functions are based on the L_1 norm condition, which refers to the minimization of L_1 norm of error vector [41,42]. Such a variant is called the absolute M_{split} estimation (AMS) [34,43], and it has not been used in laser scanning data analysis so far. Due to that fact, it seems interesting to apply this method in such a context, especially since some previous papers proved that AMS estimation is generally less sensitive to outliers than SMS estimation [34,35]. Thus, the paper's main objective is to analyze both variants of M_{split} estimation in the context of the determination of terrain and terrain details profiles based on TLS data. The paper presents the squared and absolute M_{split} estimation foundations, their application in profile approximation and the empirical analyses based on simulated and real TLS data.

2. Related Work

As mentioned earlier, one of the variants of M_{split} estimation, namely SMS estimation, was applied in several laser scanning data analyses [1,15,22,31]. We will discuss them in chronological order.

One paper [22] proposed an application of the SMS estimation to fit planes and to detect the edge of adjacent planes. The analyses presented were based on the simulated as well as real ALS data. Good results obtained in both considered cases prove that the method in question can be successfully used in laser scanning data processing.

Another paper presents an application of the SMS estimation for filtering point clouds obtained from ALS [1] and used to determine DTMs. In the first approach (the quality method), the DTM is obtained by applying the terrain surface's estimated parameters. The second way (the quantity method) considers the division of the point cloud into two subsets. The first subset contains observations that are qualified for the creation of DTM, whereas the second subset includes points representing terrain details. Then, DTM is created based on the observations from the first subset. The obtained results prove that SMS estimation can be used for filtering ALS data. The correctness of the results was verified successfully by the ATIN method. This statement is essential for the objective of this study.

One paper [31] showed the SMS estimation application in the case of filtration and aggregation of point clouds that contain a known number of elliptical shapes (e.g., tree trunks, columns, gutters) with preliminary unknown locations and dimensions, respectively. The theoretical considerations related to such a problem are satisfactory.

The SMS estimation's newest application in the context of laser scanning data processing is the detection and dimensioning of the displacement of adjacent planes in a point cloud acquired from TLS [15]. The paper presents tests which concerned the objects representing vertical planes (namely the fragment of the precast concrete wall and the fragment of the concrete slabs of a retaining wall) as well as the objects representing horizontal planes (namely the segment of the asphalt road and the fragment of the pavement). The results show that the SMS estimation can distinguish the planes in a point cloud obtained from TLS. However, in some cases where there is no displacement of the planes relative to other planes, two overlapping planes were determined.

Generally, current research has proved that the SMS estimation could give promising results in laser scanning data processing. Thus, one can suppose that M_{split} estimation might also give satisfactory outcomes in determining the terrain profiles. This paper will discuss both known variants of M_{split} estimation in the context of TLS data analysis.

3. Theoretical Foundations of M_{split} Estimation

In M_{split} estimation, the classical linear functional model is split into two competitive ones [29]:

$$\mathbf{y} = \mathbf{AX} + \mathbf{v} \Rightarrow \begin{cases} \mathbf{y} = \mathbf{AX}_{(1)} + \mathbf{v}_{(1)} \\ \mathbf{y} = \mathbf{AX}_{(2)} + \mathbf{v}_{(2)} \end{cases} \tag{1}$$

where $\mathbf{y}$—observation vector, $\mathbf{A}$—coefficient matrix, $\mathbf{X}$—parameter vector and $\mathbf{v}$—measurement error vector. The split models concern the same observation $\mathbf{y}$, and they differ from each other in the competitive versions $\mathbf{X}_{(1)}$ and $\mathbf{X}_{(2)}$ of the parameter vector $\mathbf{X}$ and also the competitive versions $\mathbf{v}_{(1)}$ and $\mathbf{v}_{(2)}$ of the measurement error vector $\mathbf{v}$. M_{split} estimation can be applied when the observation vector is an unknown mixture of realizations of two random variables that differ from each other at least in location parameters. The assignment of a particular observation to the respective version of the parameter vector is unknown. In other words, there is no supposition about dividing the observations into two aggregations.

The general optimization problem of M_{split} estimation can be formulated as follows:

$$\varphi\left(\mathbf{X}_{(1)}, \mathbf{X}_{(2)}\right) = \sum_{i=1}^{n} \rho\left(v_{i(1)}, v_{i(2)}\right) = \sum_{i=1}^{n} \rho_{(1)}\left(v_{i(1)}\right)\rho_{(2)}\left(v_{i(2)}\right) = \min_{\mathbf{X}_{(1)}, \mathbf{X}_{(2)}} \tag{2}$$

where $\varphi\left(\mathbf{X}_{(1)}, \mathbf{X}_{(2)}\right)$—objective function and $\rho\left(v_{i(1)}, v_{i(2)}\right) = \rho_{(1)}\left(v_{i(1)}\right)\rho_{(2)}\left(v_{i(2)}\right)$—arbitrarily chosen function. Let us introduce two variants of M_{split} estimation: the squared M_{split} estimation (SMS) and the absolute M_{split} estimation (AMS). In the case of the SMS estimation, the objective function can be defined as [29,30]:

$$\varphi\left(\mathbf{X}_{(1)}, \mathbf{X}_{(2)}\right) = \sum_{i=1}^{n} \rho\left(v_{i(1)}, v_{i(2)}\right) = \sum_{i=1}^{n} v_{i(1)}^2 v_{i(2)}^2 \tag{3}$$

Such a formula can be derived from a general assumption that the observation errors are normally distributed. The optimization problem (Equation (2)) of the SMS estimation can be solved by applying the Newton method. The iterative process of the SMS estimation, namely the traditional one [43], is as follows [29]:

$$\begin{aligned} \mathbf{X}^{j}_{(1)} &= \mathbf{X}^{j-1}_{(1)} + d\mathbf{X}^{j}_{(1)} = \mathbf{X}^{j-1}_{(1)} - \left[\mathbf{H}_{(1)}\left(\mathbf{X}^{j-1}_{(1)}, \mathbf{X}^{j-1}_{(2)}\right)\right]^{-1} \mathbf{g}_{(1)}\left(\mathbf{X}^{j-1}_{(1)}, \mathbf{X}^{j-1}_{(2)}\right) \\ \mathbf{X}^{j}_{(2)} &= \mathbf{X}^{j-1}_{(2)} + d\mathbf{X}^{j}_{(2)} = \mathbf{X}^{j-1}_{(2)} - \left[\mathbf{H}_{(2)}\left(\mathbf{X}^{j}_{(1)}, \mathbf{X}^{j-1}_{(2)}\right)\right]^{-1} \mathbf{g}_{(2)}\left(\mathbf{X}^{j}_{(1)}, \mathbf{X}^{j-1}_{(2)}\right) \end{aligned} \tag{4}$$

where Hessians—$\mathbf{H}_{(l)}\left(\mathbf{X}_{(1)}, \mathbf{X}_{(2)}\right)$—and gradients—$\mathbf{g}_{(l)}\left(\mathbf{X}_{(1)}, \mathbf{X}_{(2)}\right)$—are computed in the following way [29]:

$$\begin{cases} \mathbf{H}_{(1)}\left(\mathbf{X}^{j-1}_{(1)}, \mathbf{X}^{j-1}_{(2)}\right) = 2\mathbf{A}^{T}\mathbf{w}_{(1)}\left(\mathbf{v}^{j-1}_{(1)}, \mathbf{v}^{j-1}_{(2)}\right)\mathbf{A} \\ \mathbf{g}_{(1)}\left(\mathbf{X}^{j-1}_{(1)}, \mathbf{X}^{j-1}_{(2)}\right) = -2\mathbf{A}^{T}\mathbf{w}_{(1)}\left(\mathbf{v}^{j-1}_{(1)}, \mathbf{v}^{j-1}_{(2)}\right)\mathbf{v}^{j-1}_{(1)} \\ \mathbf{w}_{(1)}\left(\mathbf{v}^{j-1}_{(1)}, \mathbf{v}^{j-1}_{(2)}\right) = diag\left[w_{(1)}\left(v^{j-1}_{1(1)}, v^{j-1}_{1(2)}\right), \ldots, w_{(1)}\left(v^{j-1}_{n(1)}, v^{j-1}_{n(2)}\right)\right] \end{cases}$$
$$\begin{cases} \mathbf{H}_{(2)}\left(\mathbf{X}^{j}_{(1)}, \mathbf{X}^{j-1}_{(2)}\right) = 2\mathbf{A}^{T}\mathbf{w}_{(2)}\left(\mathbf{v}^{j}_{(1)}, \mathbf{v}^{j-1}_{(2)}\right)\mathbf{A} \\ \mathbf{g}_{(2)}\left(\mathbf{X}^{j}_{(1)}, \mathbf{X}^{j-1}_{(2)}\right) = -2\mathbf{A}^{T}\mathbf{w}_{(2)}\left(\mathbf{v}^{j}_{(1)}, \mathbf{v}^{j-1}_{(2)}\right)\mathbf{v}^{j-1}_{(2)} \\ \mathbf{w}_{(2)}\left(\mathbf{v}^{j}_{(1)}, \mathbf{v}^{j-1}_{(2)}\right) = diag\left[w_{(2)}\left(v^{j}_{1(1)}, v^{j-1}_{1(2)}\right), \ldots, w_{(2)}\left(v^{j}_{n(1)}, v^{j-1}_{n(2)}\right)\right] \end{cases} \tag{5}$$

where $diag(\circ)$—diagonal matrix and $\mathbf{w}_{(l)}\left(\mathbf{v}_{(1)}, \mathbf{v}_{(2)}\right)$—matrices of the weight functions. The objective function of the SMS estimation leads to the following forms of weight functions [29]:

$$w_{(1)}\left(v_{i(1)}, v_{i(2)}\right) = v^2_{i(2)} \quad \text{and} \quad w_{(2)}\left(v_{i(1)}, v_{i(2)}\right) = v^2_{i(1)} \tag{6}$$

Since the first variant's weight function depends only on the second one's errors (and vice versa), one can call this relationship a mutual cross-weighting.

On the other hand, the AMS estimation is based on application L_1 norm condition [41,42]; thus, its objective function is defined in the following way [34,43]:

$$\varphi\left(\mathbf{X}_{(1)}, \mathbf{X}_{(2)}\right) = \sum_{i=1}^{n} \rho\left(v_{i(1)}, v_{i(2)}\right) = \sum_{i=1}^{n} \left|v_{i(1)}\right|\left|v_{i(2)}\right| \tag{7}$$

Such an objective function leads to the following weight functions:

$$w_{(1)}\left(v_{i(1)}, v_{i(2)}\right) = \begin{cases} -\frac{\left|v_{i(2)}\right|}{2v_{i(1)}} & \text{for } v_{i(1)} < 0 \\ \frac{\left|v_{i(2)}\right|}{2v_{i(1)}} & \text{for } v_{i(1)} > 0 \end{cases} \text{ and } w_{(2)}\left(v_{i(1)}, v_{i(2)}\right) = \begin{cases} -\frac{\left|v_{i(1)}\right|}{2v_{i(2)}} & \text{for } v_{i(2)} < 0 \\ \frac{\left|v_{i(1)}\right|}{2v_{i(2)}} & \text{for } v_{i(2)} > 0 \end{cases} \tag{8}$$

They both depend on both variants of the error vectors; hence, there is no mutual cross-weighting in this case. Thus, the iterative process of the AMS estimation is a little bit different than the respective iterative process of the SMS estimation. The iterative process of the AMS estimation can be described as [34,43]:

$$\begin{aligned} \mathbf{X}^{j}_{(1)} &= \mathbf{X}^{j-1}_{(1)} + d\mathbf{X}^{j}_{(1)} = \mathbf{X}^{j-1}_{(1)} - \left[\mathbf{H}_{(1)}\left(\mathbf{X}^{j-1}_{(1)}, \mathbf{X}^{j-1}_{(2)}\right)\right]^{-1} \mathbf{g}_{(1)}\left(\mathbf{X}^{j-1}_{(1)}, \mathbf{X}^{j-1}_{(2)}\right) \\ \mathbf{X}^{j}_{(2)} &= \mathbf{X}^{j-1}_{(2)} + d\mathbf{X}^{j}_{(2)} = \mathbf{X}^{j-1}_{(2)} - \left[\mathbf{H}_{(2)}\left(\mathbf{X}^{j-1}_{(1)}, \mathbf{X}^{j-1}_{(2)}\right)\right]^{-1} \mathbf{g}_{(2)}\left(\mathbf{X}^{j-1}_{(1)}, \mathbf{X}^{j-1}_{(2)}\right) \end{aligned} \tag{9}$$

where:

$$\begin{cases} \mathbf{H}_{(1)}\left(\mathbf{X}_{(1)}^{j-1},\mathbf{X}_{(2)}^{j-1}\right) = 2\mathbf{A}^T\mathbf{w}_{(1)}\left(\mathbf{v}_{(1)}^{j-1},\mathbf{v}_{(2)}^{j-1}\right)\mathbf{A} \\ \mathbf{g}_{(1)}\left(\mathbf{X}_{(1)}^{j-1},\mathbf{X}_{(2)}^{j-1}\right) = -2\mathbf{A}^T\mathbf{w}_{(1)}\left(\mathbf{v}_{(1)}^{j-1},\mathbf{v}_{(2)}^{j-1}\right)\mathbf{v}_{(1)}^{j-1} \\ \mathbf{w}_{(1)}\left(\mathbf{v}_{(1)}^{j-1},\mathbf{v}_{(2)}^{j-1}\right) = diag\left[w_{(1)}\left(v_{1(1)}^{j-1},v_{1(2)}^{j-1}\right),\ldots,w_{(1)}\left(v_{n(1)}^{j-1},v_{n(2)}^{j-1}\right)\right] \\ \mathbf{H}_{(2)}\left(\mathbf{X}_{(1)}^{j-1},\mathbf{X}_{(2)}^{j-1}\right) = 2\mathbf{A}^T\mathbf{w}_{(2)}\left(\mathbf{v}_{(1)}^{j-1},\mathbf{v}_{(2)}^{j-1}\right)\mathbf{A} \\ \mathbf{g}_{(2)}\left(\mathbf{X}_{(1)}^{j-1},\mathbf{X}_{(2)}^{j-1}\right) = -2\mathbf{A}^T\mathbf{w}_{(2)}\left(\mathbf{v}_{(1)}^{j-1},\mathbf{v}_{(2)}^{j-1}\right)\mathbf{v}_{(2)}^{j-1} \\ \mathbf{w}_{(2)}\left(\mathbf{v}_{(1)}^{j-1},\mathbf{v}_{(2)}^{j-1}\right) = diag\left[w_{(2)}\left(v_{1(1)}^{j-1},v_{1(2)}^{j-1}\right),\ldots,w_{(2)}\left(v_{n(1)}^{j-1},v_{n(2)}^{j-1}\right)\right] \end{cases} \tag{10}$$

Such an iterative process might be called parallel [43]. Because the original weight functions of the AMS estimation are not defined for measurement errors equal to zero, which may cause singularity, in practice, we can apply the following modified weight functions:

$$w_{(1)}^*\left(v_{i(1)},v_{i(2)}\right) = \begin{cases} \frac{|v_{i(2)}|}{2b} & \text{for } \left|v_{i(1)}\right| < b \\ \frac{|v_{i(2)}|}{2|v_{i(1)}|} & \text{for } \left|v_{i(1)}\right| \geq b \end{cases} \text{ and } w_{(2)}^*\left(v_{i(1)},v_{i(2)}\right) = \begin{cases} \frac{|v_{i(1)}|}{2b} & \text{for } \left|v_{i(2)}\right| < b \\ \frac{|v_{i(1)}|}{2|v_{i(2)}|} & \text{for } \left|v_{i(2)}\right| \geq b \end{cases} \tag{11}$$

where b—assumed small positive constant.

Both types of iterative processes require a proper choice of starting points [43], and they should both be stopped for such a $j = k$ for which $\mathbf{g}_{(1)}\left(\mathbf{X}_{(1)}^{k-1},\mathbf{X}_{(2)}^{k-1}\right) = 0$, $\mathbf{g}_{(2)}\left(\mathbf{X}_{(1)}^{k-1},\mathbf{X}_{(2)}^{k-1}\right) = 0$ and also $\hat{\mathbf{X}}_{(1)} = \mathbf{X}_{(1)}^{k} = \mathbf{X}_{(1)}^{k-1}$, $\hat{\mathbf{X}}_{(2)} = \mathbf{X}_{(2)}^{k} = \mathbf{X}_{(2)}^{k-1}$ [29,34].

Previous papers [34,35,43] showed that differences in the SMS and AMS estimations' objective functions also result in slightly different general properties. As was shown, both methods' weight functions differ from each other, which leads to the necessity of two different iterative processes. The more significant differences stem from the influence functions [32,39] and conclude that the AMS estimation is less sensitive to the outlying observations of moderate magnitude. This property seems especially interesting in TLS data processing, where observations (points in point clouds) might represent different surfaces and just obstacles.

4. Experiments and Results

Let us apply M_{split} estimation in TLS data processing, namely in the approximation of terrain profiles. Such profiles might describe the terrain surface, but they are also useful in deformation analysis. One can obtain interesting information about vertical movements by comparing terrain profiles determined at different epochs. The essential information is that a point cloud, obtained from TLS measurements, may contain terrain points and points of terrain details or obstacles. Application of M_{split} estimation allows us to get terrain profiles without division of the point cloud in subsets containing points of the mentioned types. In such a context, this approach avoids, for example, errors in the assignment of points to the respective subsets.

To examine the method proposed, we consider simulated as well as real TLS data. In both cases, measurements concern terrain surface and terrain details, and generally, we do not know which points should be assigned to the particular surface.

4.1. Simulated Data

The first test concerns the determination of a terrain profile P based on the simulated TLS data. We assume that a chosen polynomial defines the profile. Let us consider three variants of such polynomials: a second-degree polynomial (a parabola), a third-degree polynomial (a cubic curve) and a fourth-degree polynomial. The simulated polynomial coefficients are presented in Table 1.

Table 1. Simulated coefficients of the polynomials.

Degree of Polynomial	Coefficients of Terms				
	d_i^4	d_i^3	d_i^2	d_i^1	d_i^0
Two	-	-	0.00300	0.04000	1.00000
Three	-	0.00050	−0.00800	−0.02000	1.00000
Four	0.00015	−0.00594	0.07671	−0.33062	0.50000

We assume that each terrain profile is 20 m in length, and we try to approximate it based on 100 points simulated from the laser scanning measurements. The distance d_i $(i = 1, \ldots, 100)$ from the profile beginning to each point is randomly generated from the uniform distribution within the range $(0\,\text{m}, 20\,\text{m})$. The theoretical height H_i of each point is computed using the polynomial chosen that is assumed as a "real" terrain surface. Then, such heights are affected by random errors generated from the normal distribution $N\left(0\,\text{m}, 4 \cdot 10^{-6}\,\text{m}^2\right)$. The analyzes are conducted for several variants that differ in the share of outliers in the simulated observation set. Such types of observations are obtained by adding gross errors to the randomly chosen points. Gross errors follow the uniform distribution within the range of $(0.010\,\text{m}, 0.100\,\text{m})$. The outliers simulate the occurrence of the terrain details. The percentage of outliers in the observation sets in the subsequent cases is equal to 0%, 10%, 20%, 30%, 40% or 50%. All simulated observation sets are presented in Figures 1–3.

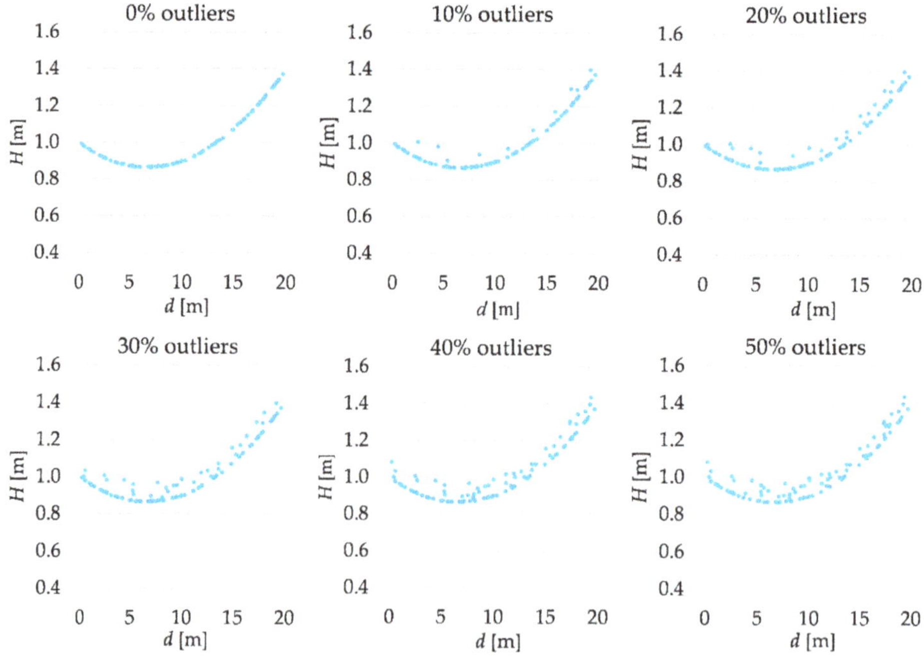

Figure 1. Simulated observation set (the second-degree polynomial).

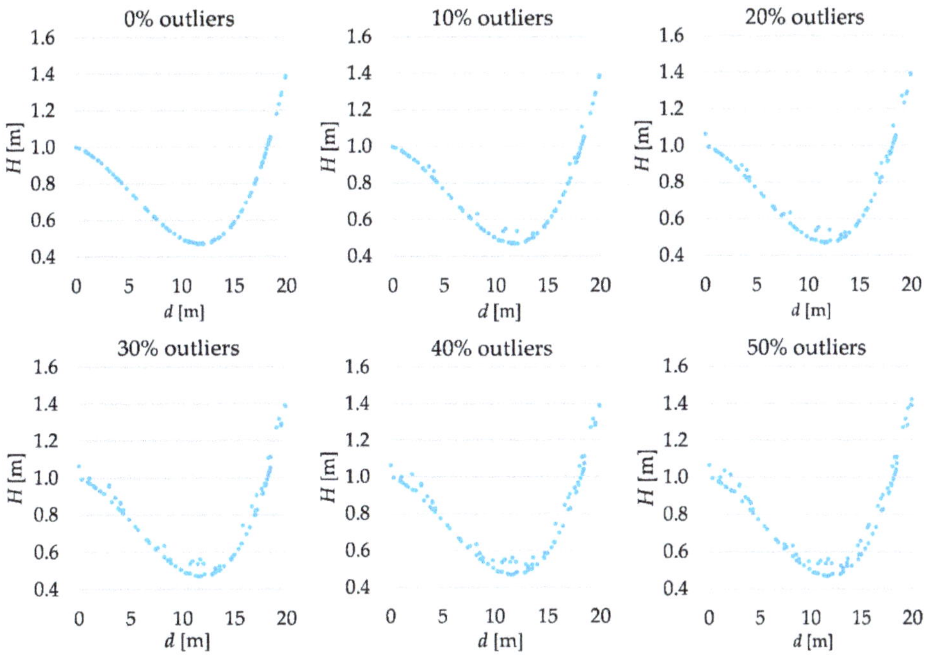

Figure 2. Simulated observation set (third-degree polynomial).

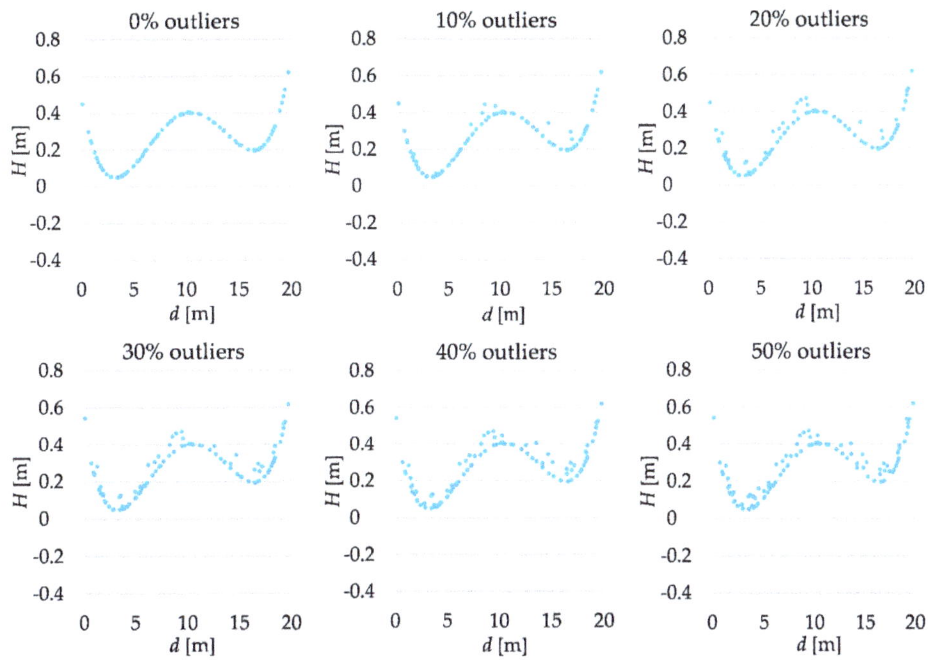

Figure 3. Simulated observation set (fourth-degree polynomial).

In the functional model of Equation (1), the parameter vector **X** (in two competitive variants $\mathbf{X}_{(1)}$, $\mathbf{X}_{(2)}$) includes the respective polynomial coefficients, and the coefficient matrix **A** includes indeterminates d_i, raised to the respective powers. Thus, the terrain profiles are determined based on the polynomial coefficients estimated by applying M_{split} estimation (SMS or AMS estimation) and for the comparison purpose by the least squares method (LS).

The main objective of that test is to examine how the increasing percentage of outliers influences the estimation results. The estimated polynomial coefficients are the basis for the estimated terrain profiles; hence, we can compute the terrain profiles by applying the LS, SMS and AMS estimates, respectively. The estimated profiles are presented in Figure 4 in comparison to simulated profile *P*. It is worth noting that in the SMS as well as the AMS estimation, there is also the second solution (based on the second variant of the estimated parameter vector). Here, in contrast to the typical applications of M_{split} estimation, we are not interested in such a solution. Its task is just "absorbing" the outliers and reducing their influence on the estimated terrain profile; thus, the second estimated profiles are not shown in Figure 4.

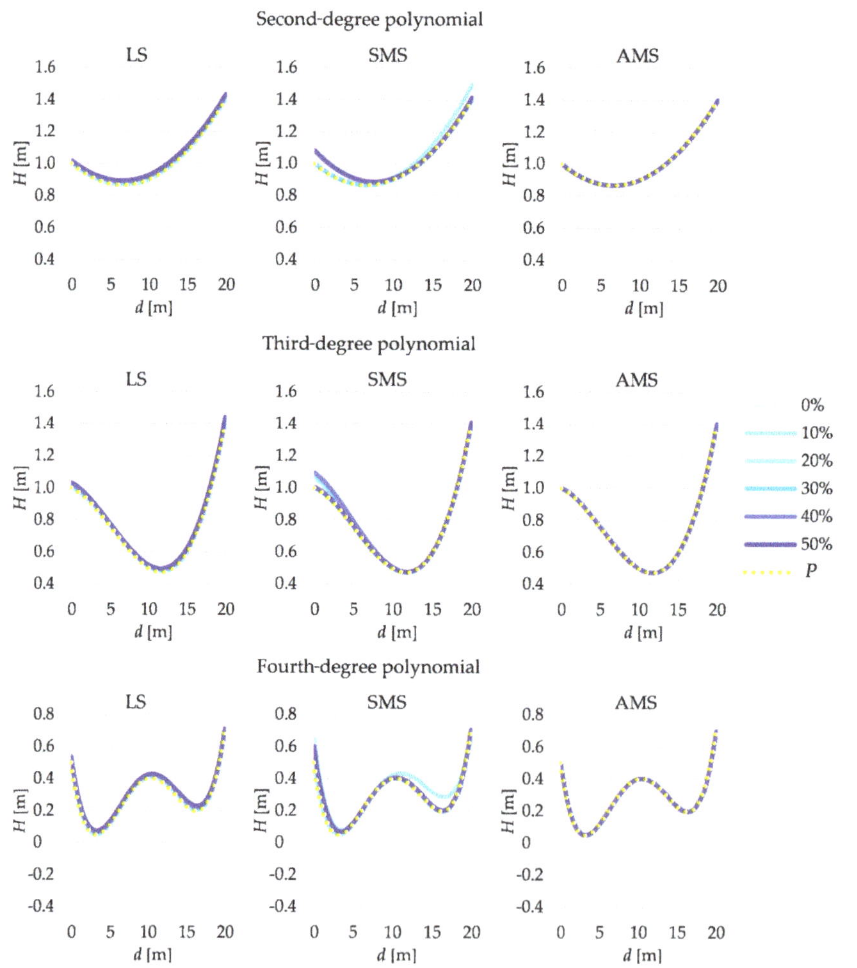

Figure 4. Estimated terrain profiles for LS, SMS and AMS estimates for different percentages of outliers.

Generally, the superior results are obtained for the AMS estimation for all three polynomials simulated. For such a method, even for 50% outliers in the observation set, the estimated terrain profile is comparable to the simulated one. Such good outcomes are not always obtained for SMS estimation. In the case of the second-degree polynomial, the SMS estimation gives good results for the whole terrain profile for the 0%, 20% and 30% outliers in the observation set; in the case of the third-degree polynomial, it gives good results for the 0%, 10% and 50% outliers; in the case of a fourth-degree polynomial, it only gives good results when there is a lack of outliers. It is also worth noting that in the other variants, the estimated profiles are similar to the simulated ones only for the particular values of the distance d. However, these estimated terrain profiles cannot be considered as satisfactory ones. In the LS estimation case, only for 0% and 10% outliers in the observation set, the obtained profiles are similar to the simulated ones. The shape of the estimated profiles is acceptable in the other variants, but the profiles seem to be lifted about the respective simulated terrain profiles. Thus, the results prove that M_{split} estimation, especially the AMS estimation, may give better results than the LS estimation when the observation set is affected by gross errors.

To assess how the estimated profiles fit the simulated ones, we compute root-mean-square deviations $RMSD(\hat{H})$ by applying the following formula:

$$RMSD(\hat{H}) = \sqrt{\frac{\sum_{i=1}^{n}(\hat{H}_j - H_j)}{n}} \quad (12)$$

where: $\hat{H}_j$—estimated height, H_j—simulated height and n = 41—number of points for which heights are determined for $d_j = j \cdot 0.5\,\text{m}$ $(j = 0, \ldots, 40)$. $RMSDs$ which are obtained for all estimation methods and all observation sets are shown in Figure 5. The results confirm the conclusions from the analysis of Figure 4.

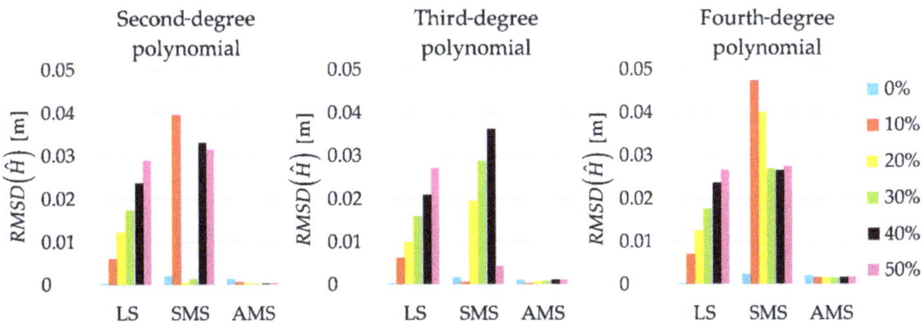

Figure 5. $RMSD(\hat{H})$ for LS, SMS and AMS estimates for different percentages of outliers.

The estimated terrain profiles are best fitted when they are computed by applying the AMS estimates of the polynomial coefficients. The only exception is when the observation sets are free of outliers. Then, unsurprisingly, slightly better fits are obtained by using the LS method. Here, all observations are related to one profile, and hence, the observations do not create two real subsets related to their version of the parameter vector. Thus, the assignment of each observation to either version of the parameters during the M_{split} estimation process seems somewhat random, and it depends on the particular observation error. In such cases, both versions of the SMS and AMS estimates obtain values close to the theoretical values; however, the accuracy of the estimates is lower than the accuracy of the LS estimates [44]. For the growing percentage of outliers, the fit of the estimated profiles based on the LS estimates becomes worse and worse, which is in line with that method's general properties. The most surprising results are those of the SMS estimation.

In some variants, the fit is very good, but in others, the fit is even worse than for the profiles obtained for the LS method. The worst fit is usually obtained in the variants with a higher percentage of outliers, which confirms that the SMS estimation is sensitive to outliers.

Finally, to better describe the fit in question, Figure 6 presents the maximum, mean and median height differences $|\hat{H}_j - H_j|$ between the estimated terrain profiles and the simulated profile.

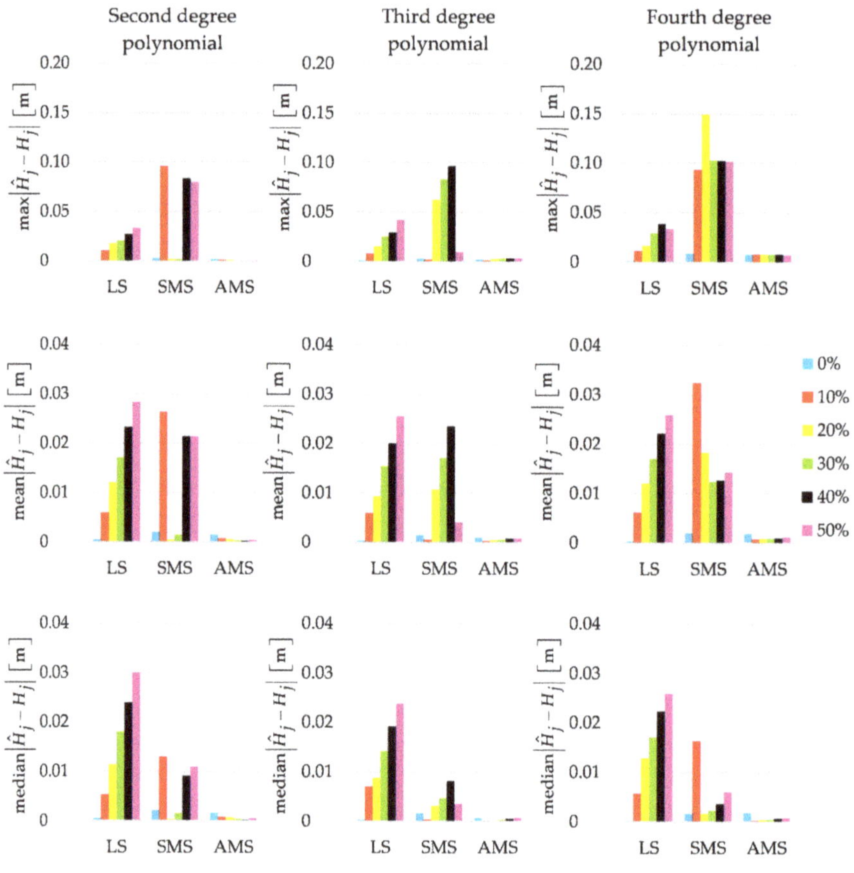

Figure 6. Maximum, mean and median values of $|\hat{H}_j - H_j|$ for LS, SMS and AMS estimates for different percentages of outliers.

The values, especially mean$|\hat{H}_j - H_j|$, confirm that the best fits are for the profiles obtained by applying the AMS estimates of the polynomial coefficients.

Overall, M_{split} estimation proves that it may give satisfactory results in the context of processing TLS data. The estimated terrain profiles by applying the AMS estimation are similar to the simulated one, even for 50% outliers in the observation sets for all three polynomials examined.

4.2. Real TLS Data

Further tests concern real TLS data. We considered two testing objects. The terrain at the first area is covered by small or medium height shrubs, whereas the second area's terrain is covered by fallen leaves, fallen tree branches and low height grass. Both areas were scanned by applying the Leica ScanStation C10 scanner with a resolution of 5 mm at 30 m. We considered two terrain profiles (each 15 m long) at the first object and two at the second one (each 20 m long). Each profile was measured by the electronic level Leica Sprinter to verify the results obtained from TLS data processing. The profile heights were measured every half meter. We regarded the heights obtained from such a leveling as reference heights. The scanning results, namely point clouds, were imported into Leica Cyclone to create the point set related to each profile. The sets obtained contained about 1300 points for the profiles on the first objects or 950 as for the second object profiles (such a difference in results from the scanner location). The final step was to process such observation sets in Mathcad 15.0 by applying the M_{split} estimation to determine terrain profile *P*.

Let us first consider two profiles at the first object. The cutouts of the original point clouds related to the profiles at lines I and II are presented in Figure 7. There were some shrubs on these measurement lines.

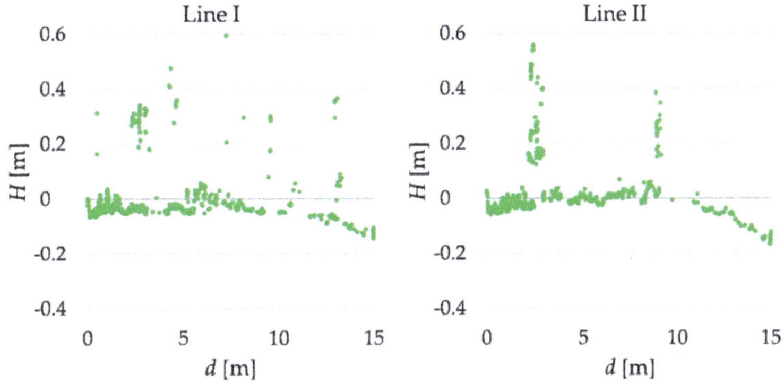

Figure 7. TLS data related to lines I and II.

The profiles are supposed to be approximated by polynomials; hence, we assume two different polynomials, namely a third-degree polynomial and a fourth-degree polynomial. Because of the complexity of the real terrain profiles, we propose to process measurements in arbitrarily chosen intervals of the distances *d*. The first distance intervals are of the length 1.5 m; thus, we consider the intervals [0 m; 1.5 m], [1 m; 2.5 m], etc. In the second variant, the length is 5.5 m. We estimated the polynomial coefficients within each interval, which were the basis for computing the profile heights. Such heights were calculated at the points for which heights were earlier determined by geometric leveling. Hence, we consider four different variants: A—third-degree polynomial, interval length 1.5 m; B—fourth-degree polynomial, interval length 1.5 m; C—third-degree polynomial, interval length 5.5 m; D—fourth-degree polynomial, interval length 5.5 m. The terrain profiles are obtained in these four variants (by applying the SMS, AMS or LS estimation, respectively). Moreover, we determine the terrain profiles based on raw TLS data obtained by using the linear interpolation at every half meter (at the same distances d_j as for the leveling measurements). The terrain profiles at line I are presented in Figure 8 (the second version of the profiles for the SMS and AMS estimations are not shown here).

Generally, the best results are obtained by applying the AMS estimation. These terrain profiles seem to the most similar to the profile obtained from the leveling. SMS estimation gives worse outcomes, but they are still better than the results of the LS estimation (or the profiles based on raw TLS data). The superior results are these of Variant B or A. However, for *d* equal 9.5 m, even the AMS estimation fails. The reason is that almost all measurements from TLS, which are close to that point, concern terrain details (a shrub), not terrain surface. It is worth noting that there are also other intervals where TLS data comes mostly from the terrain details (see the green line and *d* equal 3, 4.5 or 13 m). In these cases, there were also a few points from the terrain, making it possible to estimate the profile acceptably, but only by applying the SMS or especially AMS estimation (the LS estimation provides much worse results). Thus, if the number of points related to terrain is much too small, then the methods examined, even the variants of M_{split} estimation, might not give us satisfactory results (regardless of the variant considered here).

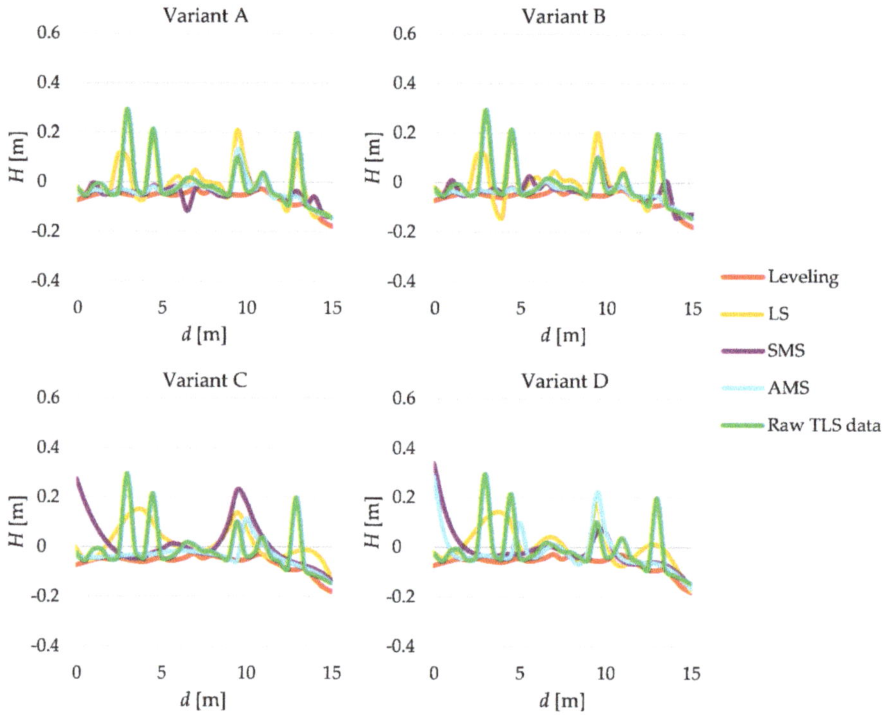

Figure 8. Estimated terrain profiles of line I.

The terrain profiles of line II are shown in Figure 9. Again, the best-fitted profiles are obtained by applying the AMS estimation. Let us focus on Variants A and B. When d is equal to 2.5 m, all estimated profiles represent terrain details rather than the terrain itself. It also means that the number of points related to the terrain were too small. As for the SMS estimation and especially for AMS estimation, such a problem disappears in Variants C and D. Thus, the widening of the interval considered allows us to obtain satisfactory results, especially for the AMS estimation. One can say that in such a wider interval, a few points related to the terrain "appear," which allows the AMS estimation to "catch" terrain and determine the terrain profile in a better way.

The simple graphical analyses of the estimated profiles of lines I and II show that the profiles obtained by applying the AMS estimation are the most fitted in the profile determined by the leveling measurements. However, it is not generally apparent in which variant the AMS estimation gives the best results; in fact, it depends on the shape of the terrain and terrain details and the number of TLS points related to them.

The analyses of the terrain profiles determined at lines I and II are supplemented with the estimated accuracy of the fit, namely, $RMSD(\hat{H})$, which is computed by applying Equation (12) ($\hat{H}_j$—estimated height, H_j—height from leveling measurements, $j = 0,\ldots,30$; $n = 31$). The respective accuracies, calculated for all variants considered, are presented in Figure 10. In the LS estimation case, the differences between the $RMSD(\hat{H})$ values that are obtained for each variant are small and insignificant. As for both variants of M_{split} estimation, such differences are more evident. For line I, the superior accuracy, approximately 0.04 m, is obtained for the AMS estimation in Variants A, B and C; superior accuracy for the SMS estimation is obtained in Variants A and B. Similar values of $RMSD(\hat{H})$ are also obtained for line II, namely, for the SMS estimation in Variants B and C and for the AMS estimation in Variants A and B. The fit's highest accuracy is obtained for the AMS estimation in Variants C and D (approximately 0.02 m, line II). One should realize that *RMSD* is a measure of accuracy that is not robust against outliers. Here, its values are sensitive to the high discrepancies between the estimated and reference profile, even if such discrepancies concern only a single point (or a few points). For that reason, to better describe the fit of the estimated profiles to the reference one, Table 2 shows the maximum, mean and median of the absolute differences $|\hat{H}_j - H_j|$.

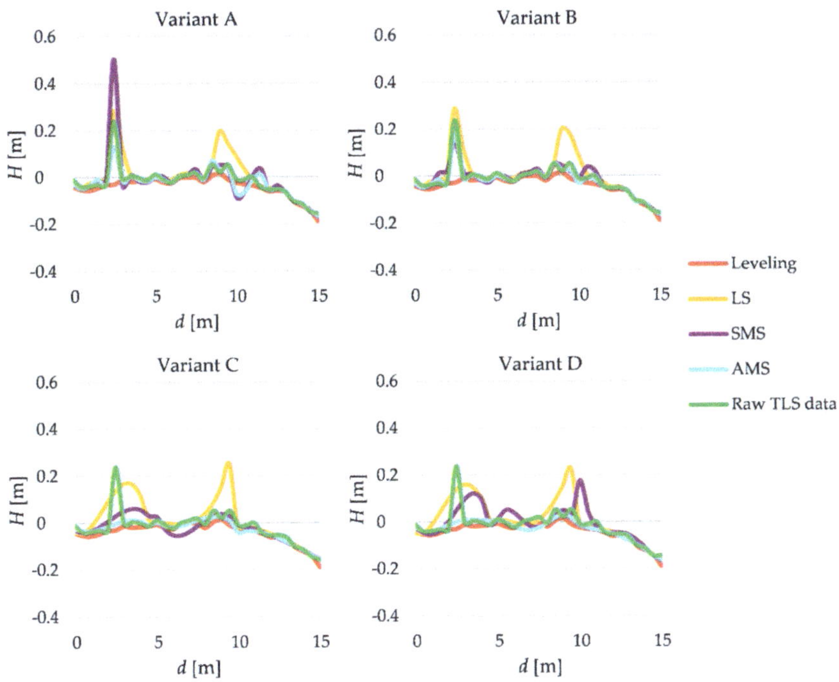

Figure 9. Estimated terrain profiles of line II.

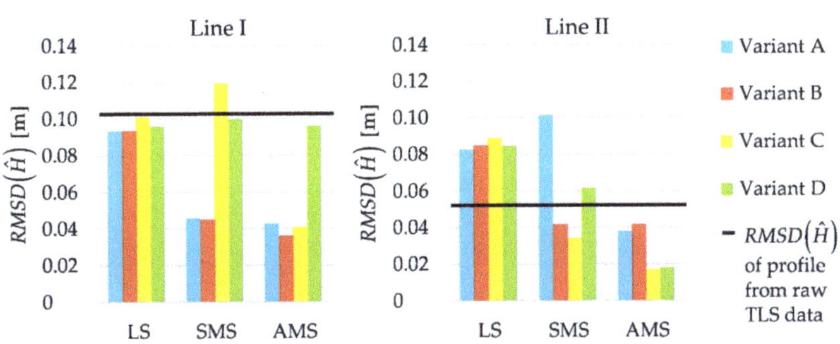

Figure 10. Lines I and II; accuracy of the estimated profiles compared to accuracy of the profile obtained from raw TLS data.

Table 2. Maximum, mean and median of $\left|\hat{H}_j - H_j\right|$ for lines I and II.

Line	Variant	max $\vert\hat{H}_j - H_j\vert$ [m]				mean $\vert\hat{H}_j - H_j\vert$ [m]				median $\vert\hat{H}_j - H_j\vert$ [m]			
		LS	SMS	AMS	Raw TLS Data	LS	SMS	AMS	Raw TLS Data	LS	SMS	AMS	Raw TLS Data
I	A	0.2634	0.1698	0.1875	0.3477	0.0609	0.0303	0.0251	0.0541	0.0367	0.0173	0.0157	0.0180
	B	0.2548	0.1580	0.1569		0.0644	0.0301	0.0229		0.0417	0.0181	0.0172	
	C	0.2083	0.3465	0.1620		0.0796	0.0795	0.0274		0.0707	0.0439	0.0224	
	D	0.2293	0.4133	0.3592		0.0730	0.0558	0.0517		0.0551	0.0309	0.0244	
II	A	0.3124	0.5328	0.1580	0.2670	0.0459	0.0382	0.0218	0.0243	0.0184	0.0211	0.0089	0.0145
	B	0.3161	0.1722	0.2088		0.0460	0.0254	0.0189		0.0180	0.0153	0.0089	
	C	0.2625	0.0792	0.0411		0.0547	0.0270	0.0140		0.0164	0.0238	0.0124	
	D	0.2378	0.1999	0.0346		0.0532	0.0400	0.0156		0.0174	0.0206	0.0159	

Let us now consider terrain profiles at the second object. Let us recall that the object area was covered by fallen leaves and small fallen tree branches. We consider two profiles measured 20 m in length. The cutouts of the original point clouds related to the profiles at lines III and IV are shown in Figure 11.

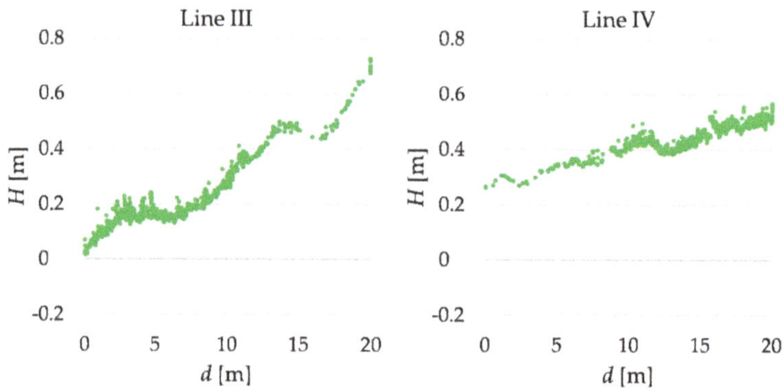

Figure 11. TLS data related to lines III and IV.

The profiles are later estimated in the same way as in the previous test. We also consider the same variants of estimation of the polynomial coefficients, namely A, B, C and D. As on the last test, the profile which was determined by geometric leveling was regarded as a reference.

Figure 12 presents the estimated terrain profiles at the first line, which was considered in that test, i.e., line III.

The terrain profiles obtained from raw TLS data and by applying the LS, SMS and AMS methods are quite similar. They seem to be shifted up a few centimeters above the reference terrain profile. The reason is that the fallen leaves cover the terrain surface almost uniformly. One can also observe the same effect for the estimated profiles at line IV (see, Figure 13). To assess such a shift, we compute the mean of the differences $\left(\hat{H}_j - H_j\right)$, which are presented in Table 3 (in this test $j = 0, \ldots, 40$).

For both lines, the shift values are close to each other; thus, one can conclude that, in fact, it is the thickness of the fallen leave layer. It is also worth noting that TLS data contained some points related to the terrain details (some small "hills," see the green lines in Figures 12 and 13). Usually, such data irregularities are smoothed in the estimated profiles, especially those obtained by applying the AMS estimation in Variants C and D.

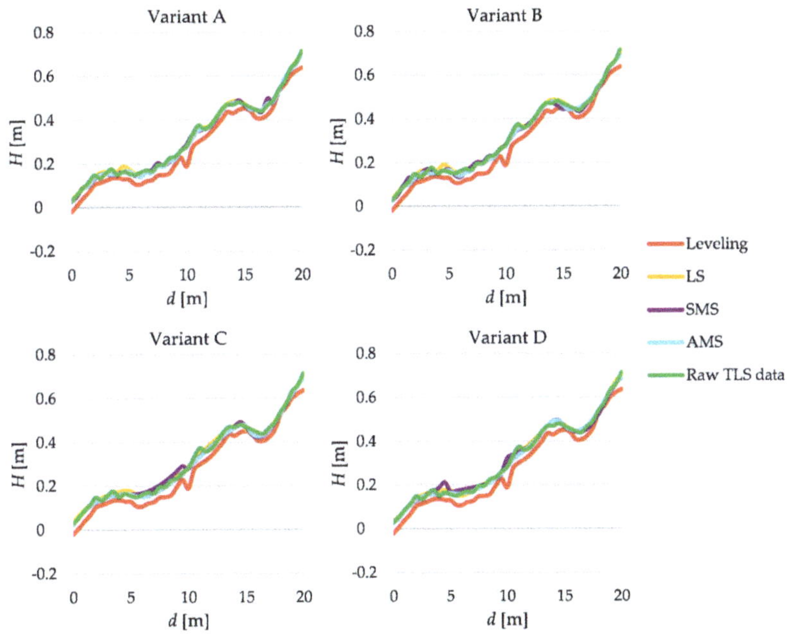

Figure 12. Estimated terrain profiles of line III.

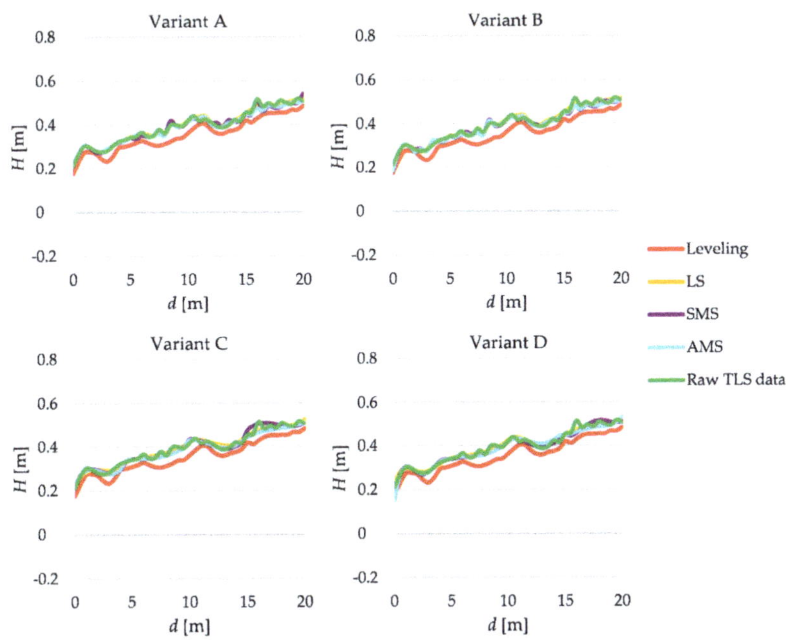

Figure 13. Estimated terrain profiles of line IV.

Table 3. Mean of $\left(\hat{H}_j - H_j\right)$ for lines III and IV.

Variant	mean($\hat{H}_j - H_j$) [m]							
	Line III				Line IV			
	LS	SMS	AMS	Raw TLS Data	LS	SMS	AMS	Raw TLS Data
A	0.0424	0.0374	0.0358	0.0409	0.0411	0.0375	0.0361	0.0403
B	0.0426	0.0376	0.0361		0.0412	0.0359	0.0354	
C	0.0416	0.0389	0.0331		0.0407	0.0387	0.0336	
D	0.0427	0.0431	0.0385		0.0419	0.0354	0.0363	

To compare the estimated profiles with each other, Figure 14 presents *RMSD*s of the profile fits (they are computed by applying Equation (12) and taking $n = 41$).

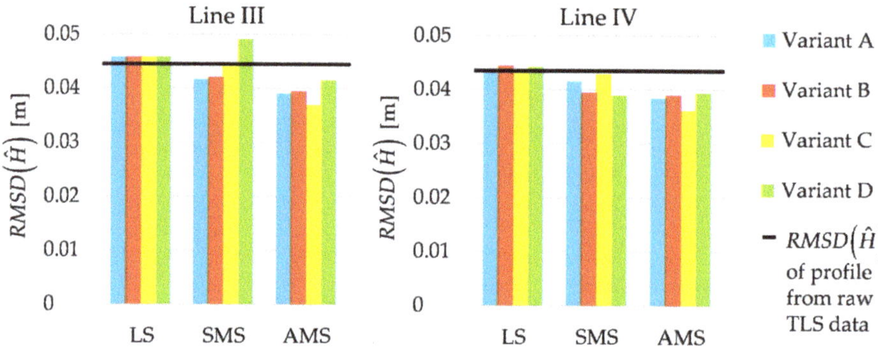

Figure 14. Lines III and IV; accuracy of the estimated profiles compared to accuracy of the profile obtained from raw TLS data.

The differences in *RMSD*s between the estimation methods are much smaller than in the previous tests. This is because there were only a few small terrain details; hence, there was a small number of outlying observations. It is worth noting that the values of *RMSD*s are like those obtained in the previous test (in which the number and the height of the terrain details were more significant) for the profiles estimated by applying M_{split} estimation. Undoubtedly, all *RMSD*s are strongly influenced by the systematic error (the shift of the estimated profiles).

Let us now consider the application of the proposed approach to deformation analysis. Generally, such an application would concern TLS measurements and the estimation of terrain profiles at different epochs. These profiles can then be compared with each other to determine the vertical displacements of the terrain surface. Such an application of M_{split} estimation would be novel. Currently, such a method was used in the deformation analysis in a different context, namely when the observation set was a disordered mixture of the same object's observations from two different measurement epochs [29,32,35,45].

Here, we do not have measurements at two different epochs; however, we do have two different terrain profiles of the same length. We can apply the approach at hand to estimate the vertical differences between the profiles in question. This experiment allows us to assess whether the method proposed in the paper can be used in such a case. Note that the estimation of the vertical displacements (differences between the same profile at different measurement epochs) would be performed very similarly.

Let us analyze the estimated height differences $\Delta\hat{H}_j = \hat{H}_j^{IV} - \hat{H}_j^{III}$ ($\hat{H}_j^{III}$ and $\hat{H}_j^{IV}$—heights at the estimated profile of line III or IV, respectively, at the distance $d_j = j \cdot 0.5\,\mathrm{m}$, $j = 0, \ldots, 40$). Figure 15 presents *RMSD*s computed like in the previous test but related to the estimated height differences. Here, as a reference, we assume the height differences obtained from the geometric leveling.

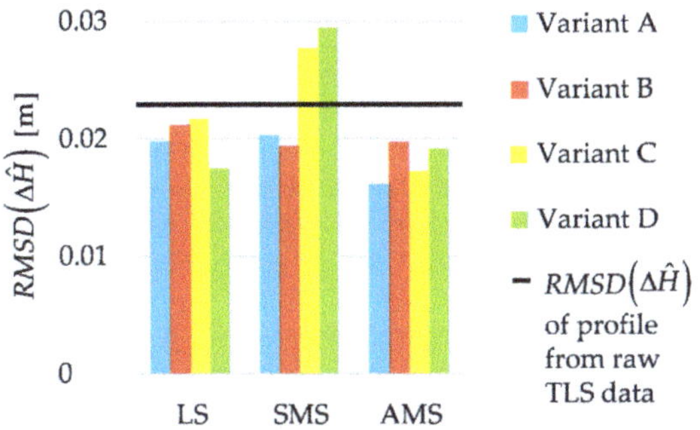

Figure 15. Lines III and IV; accuracy of the differences between the estimated profiles compared to accuracy of the differences obtained from raw TLS data.

The values of *RMSD* are usually close to one another. The smallest *RMSD* is obtained when the differences between profiles are based on the AMS estimates (Variant A). In contrast, the highest *RMSD*s are obtained by applying the SMS estimates (Variants C and D). Except for these two cases, *RMSDs* are lower than *RMSDs* obtained for the profiles determined from the raw TLS data. Moreover, they are twice as small as the respective *RMSD*s of the terrain profiles of lines III and IV. This means that during the estimation of height differences, the influence of the systematic error (shift) was significantly reduced.

Table 4 shows the maximum, mean and median values of the absolute differences $|\Delta\hat{H}_j - \Delta H_j|$ to describe the height differences obtained in the different variants.

Table 4. Maximum, mean and median of $|\Delta\hat{H}_j - \Delta H_j|$ between the terrain points of lines III and IV.

Variant	max $\lvert\Delta\hat{H}_j-\Delta H_j\rvert$ [m]				mean $\lvert\Delta\hat{H}_j-\Delta H_j\rvert$ [m]				median $\lvert\Delta\hat{H}_j-\Delta H_j\rvert$ [m]			
	LS	SMS	AMS	Raw TLS Data	LS	SMS	AMS	Raw TLS Data	LS	SMS	AMS	Raw TLS Data
A	0.0523	0.0483	0.0361	0.0595	0.0159	0.0157	0.0127	0.0174	0.0115	0.0119	0.0087	0.0138
B	0.0460	0.0438	0.0520		0.0171	0.0155	0.0154		0.0125	0.0137	0.0118	
C	0.0620	0.0900	0.0468		0.0171	0.0211	0.0131		0.0144	0.0175	0.0111	
D	0.0396	0.0803	0.0695		0.0136	0.0217	0.0143		0.0113	0.0148	0.0122	

The mean and median values confirm the conclusions drawn from the *RMSD* values. The smallest discrepancies are obtained when the profiles are estimated by applying the AMS estimation. On average, the discrepancies between the estimated and the reference profile are close to 10–15 mm. For the other methods, the discrepancies are usually bigger, about 2–3 mm.

5. Discussion

This paper presents an application of two variants of M_{split} estimation, i.e., the SMS and AMS estimation, in the context of the determination of profiles of the terrain surface based on TLS data. Let us recall that up to now, the SMS estimation was applied in several problems of the laser scanning data analysis [1,15,22,31]. It was proved that the method in question could provide satisfactory results in the context mentioned.

The empirical tests were based on two types of data: simulated or real TLS data, respectively. The processing of the simulated TLS data by applying M_{split} estimation, especially the AMS estimation, gives very promising results for all three polynomials examined. The fit of the estimated profile for the simulated data is very good. The AMS estimation does not fail even for 50% outliers in the observation set. It confirms the results presented in the previous papers, which indicate that such a method is generally less sensitive to outliers of the moderate magnitude than the SMS estimation [34,35].

In real TLS data, we chose the areas with many terrain details (i.e., shrubs and fallen tree branches) or where fallen leaves covered the terrain surface. Thus, we considered the objects for which it was not so easy to determine terrain profiles because many points in a point cloud did not refer to the real terrain surface. For the first object, where there were many shrubs and fallen tree branches, M_{split} estimation was able to provide the terrain profiles which were better fitted in the real profile (determined by geometric leveling) than the profiles provided by the LS estimation. The profiles in question are also better fitted than the profiles determined directly from the raw TLS data. The best-fitted profiles were obtained by applying the AMS estimation; the accuracy of the fit is from 0.02 m to 0.04 m, which depends on the estimation variant. One can say that such accuracy is twice as small as the accuracy of the fit of the profiles obtained by applying the LS estimation or determined directly from the raw TLS data. Unfortunately, if the obstacles "obscure" the terrain surface to a large extent and hence the number of TLS points related to the terrain was very small, then even the variants of the M_{split} estimation examined might not give us satisfactory results.

For the second object, there were only fallen leaves and fallen small tree branches (no high obstacles). The fallen leaves covered the terrain surface almost uniformly. Hence, the point cloud included only a few points related to the terrain surface. In many intervals considered, there were no such points at all. It was impossible to determine the real terrain profile from the TLS data in such a case. The shapes of estimated profiles were close to the real terrain profile; however, all determined profiles seemed to be shifted up for about 0.03–0.04 m. The accuracy of the fit for all estimated profiles was usually close to 0.04 m, which appeared to be strongly influenced by the systematic error (the shift). The accuracy was the highest when the profile was estimated by applying the AMS estimation. However, the differences between all estimation methods were much smaller than for the first object. Generally, the accuracy of fit of the estimated profile was a little bit better than the accuracy obtained for the profile determined from the raw TLS data.

As we considered applying the proposed approach in deformation analysis, we estimated differences between two terrain profiles at the second object. When estimating such differences, the systematic error was reduced in a significant way. The accuracy of the fit increased to 0.02 m, which is twice smaller than in the case of a single profile. We can expect much higher accuracy when the terrain would not be covered by a "contamination layer." In such a context, the approach presented can be applied in, e.g., assessing the mining damages and ground disasters [20,21].

Another interesting issue is the analysis of the M_{split} estimation itself. A study with a focus on profile estimation has never been done before. The results confirm the general properties of the SMS and AMS estimations. Firstly, the results are better when the difference between profiles of terrain or terrain details are more significant [34,44,45]. Secondly, the AMS estimation proved its robustness against outliers of the moderate magnitude, in contrast to the SMS estimation and, of course, the LS estimation. It is worth noting that the presented application of both M_{split} estimations is not a usual one. Here, we were especially interested in one estimated profile (the terrain profile). The second profile (the profile of the terrain details) is out of our interest. Its main objective is to attract the outlying observations (observations related to terrain details or observations disturbed with gross errors) and reduce the influence of such observations on the terrain profile estimation. Such an approach is similar to the variant of the SMS estimation proposed in [46].

Here, we consider four variants that were applied in the profile estimation. They differ from each other in the length of the distance interval considered and the polynomial degree used. Although there are some differences between the respective variants' results, it is hard to indicate the most suitable variant for profile estimation. Generally, one can suggest that polynomials of the third degree and shorter rather than longer intervals should be applied when there are many terrain details of medium height (e.g., shrubs).

Furthermore, an interesting issue to consider in the future is the application of $M_{split(q)}$ estimation [30] for more than two competitive models for processing TLS or ALS point cloud. Another option would be the application of M_{split} estimation to determine horizontal displacements from TLS data, based on an analysis of changing terrain details location.

6. Conclusions

The paper presents an application of M_{split} estimation in the determination of terrain profiles from TLS point clouds. The profiles in question are determined by applying the polynomials, of which the coefficients are estimated by the respective estimation method. The analysis presented in the paper proves that such an approach is efficient and can provide good terrain profiles even if there are outliers in an observation set. In other words, if there are many shrubs or other significant terrain details, namely if the difference between terrain heights and terrain details heights is vivid, M_{split} estimation might be applied successfully to the determination of the terrain profile. The obtained results suggest that in the application at hand, AMS estimation is a better option than SMS estimation. Regardless of the methods analyzed in this paper, too few terrain data due to terrain obstacles may lead to unacceptable results; however, other methods cannot provide satisfactory results in such a case.

The application of M_{split} estimation, especially the AMS estimation, yields profiles that are better fitted in the real terrain profile compared to profiles obtained by applying the LS estimation or computed from the raw TLS data.

Author Contributions: Conceptualization, R.D. and P.W.; methodology, R.D., P.W. and A.D.; software, P.W. and A.D.; validation, R.D. and P.W.; formal analysis, R.D.; investigation, P.W.; data acquisition, A.D. and P.W.; data curation, P.W.; writing—original draft preparation, P.W. and R.D.; writing—review and editing, R.D. and P.W.; visualization, P.W.; supervision, R.D. All authors have read and agreed to the published version of the manuscript.

Funding: This research was funded by the Department of Geodesy, University of Warmia and Mazury in Olsztyn, Poland, statutory research no. 28.610.002-110.

Institutional Review Board Statement: Not applicable.

Informed Consent Statement: Not applicable.

Data Availability Statement: Not applicable.

Conflicts of Interest: The authors declare no conflict of interest.

References

1. Błaszczak-Bąk, W.; Janowski, A.; Kamiński, W.; Rapiński, J. Application of the M_{split} method for filtering airborne laser scanning data-sets to estimate digital terrain models. *Int. J. Remote Sens.* **2015**, *36*, 2421–2437. [CrossRef]
2. Rodríguez-Gonzálvez, P.; Jiménez Fernández-Palacios, B.; Muñoz-Nieto, Á.L.; Arias-Sanchez, P.; Gonzalez-Aguilera, D. Mobile LiDAR System: New possibilities for the documentation and dissemination of large cultural heritage sites. *Remote Sens.* **2017**, *9*, 189. [CrossRef]
3. Forlani, G.; Nardinocchi, C. Adaptive filtering of aerial laser scanning data. In Proceedings of the ISPRS Workshop on Laser Scanning 2007 and SilviLaser 2007, Espoo, Finland, 12–14 September 2007; pp. 130–135.
4. Spaete, L.P.; Glenn, N.F.; Derryberry, D.R.; Sankey, T.T.; Mitchell, J.J.; Hardegree, S.P. Vegetation and slope effects on accuracy of a LiDAR-derived DEM in the sagebrush steppe. *Remote Sens. Lett.* **2011**, *2*, 317–326. [CrossRef]
5. Zhao, X.; Guo, Q.; Su, Y.; Xue, B. Improved progressive TIN densification filtering algorithm for airborne LiDAR data in forested areas. *ISPRS J. Photogramm. Remote Sens.* **2016**, *117*, 79–91. [CrossRef]
6. Crespo-Peremarch, P.; Tompalski, P.; Coops, N.C.; Ruiz, L.Á. Characterizing understory vegetation in Mediterranean forests using full-waveform airborne laser scanning data. *Remote Sens. Environ.* **2018**, *217*, 400–413. [CrossRef]
7. Chen, C.; Li, Y. A fast global interpolation method for digital terrain model generation from large LiDAR-derived data. *Remote Sens.* **2019**, *11*, 1324. [CrossRef]
8. Wang, C.; Hsu, P.-H. Building detection and structure line extraction from airborne LiDAR data. *J. Photogramm. Remote Sens.* **2007**, *12*, 365–379.

9. He, M.; Zhu, Q.; Du, Z.; Hu, H.; Ding, Y.; Chen, M. A 3D shape descriptor based on contour clusters for damaged roof detection using airborne LiDAR point clouds. *Remote Sens.* **2016**, *8*, 189. [CrossRef]
10. Zhang, Z.; Vosselman, G.; Gerke, M.; Persello, C.; Tuia, D.; Yang, M.Y. Detecting building changes between Airborne Laser Scanning and photogrammetric data. *Remote Sens.* **2019**, *11*, 2417. [CrossRef]
11. Zhao, R.; Pang, M.; Liu, C.; Zhang, Y. Robust normal estimation for 3D LiDAR point clouds in urban environments. *Sensors* **2019**, *19*, 1248. [CrossRef]
12. Błaszczak-Bąk, W.; Suchocki, C.; Janicka, J.; Dumalski, A.; Duchnowski, R.; Sobieraj-Żłobińska, A. Automatic threat detection for historic buildings in dark places based on the modified OptD method. *ISPRS Int. J. Geo Inf.* **2020**, *9*, 123. [CrossRef]
13. Yang, H.; Omidalizarandi, M.; Xu, X.; Neumann, I. Terrestrial laser scanning technology for deformation monitoring and surface modeling of arch structures. *Compos. Struct.* **2017**, *169*, 173–179. [CrossRef]
14. Zhao, X.; Kargoll, B.; Omidalizarandi, M.; Xu, X.; Alkhatib, H. Model selection for parametric surfaces approximating 3D point clouds for deformation analysis. *Remote Sens.* **2018**, *10*, 634. [CrossRef]
15. Janicka, J.; Rapiński, J.; Błaszczak-Bąk, W.; Suchocki, C. Application of the Msplit estimation method in the detection and dimensioning of the displacement of adjacent planes. *Remote Sens.* **2020**, *12*, 3203. [CrossRef]
16. Corso, J.; Roca, J.; Buill, F. Geometric analysis on stone façades with terrestrial laser scanner technology. *Geosciences* **2017**, *7*, 103. [CrossRef]
17. Nurunnabi, A.; Belton, D.; West, G. Robust statistical approaches for local planar surface fitting in 3D laser scanning data. *ISPRS J. Photogramm. Remote Sens.* **2014**, *96*, 106–122. [CrossRef]
18. Kwoczyńska, B.; Dobek, J. Elaboration of the 3D model and survey of the power lines using data from airborne laser scanning. *J. Ecol. Eng.* **2016**, *17*. [CrossRef]
19. Suchocki, C. Conception of monitoring of cliff shores. *Rep. Geod.* **2011**, *90*, 461–467.
20. Lian, X.; Hu, H. Terrestrial laser scanning monitoring and spatial analysis of ground disaster in Gaoyang coal mine in Shanxi, China: A technical note. *Environ. Earth Sci.* **2017**, *76*, 287. [CrossRef]
21. Szwarkowski, D.; Moskal, M. Assessment of deformations in mining areas using the Riegl VZ-400 terrestrial laser scanner. *E3S Web Conf.* **2018**, *36*, 02009. [CrossRef]
22. Janowski, A.; Rapiński, J. M-split estimation in laser scanning data modeling. *J. Indian Soc. Remote Sens.* **2013**, *41*, 15–19. [CrossRef]
23. Carrilho, A.C.; Galo, M.; Santos, R.C. Statistical outlier detection method for airborne LiDAR data. In Proceedings of the ISPRS—International Archives of the Photogrammetry, Remote Sensing and Spatial Information Sciences, Karlsruhe, Germany, 10–12 October 2018; Copernicus GmbH: Göttingen, Germany, 2018; Volume XLII-1, pp. 87–92.
24. Carrilho, A.C.; Galo, M. Automatic object extraction from high resolution aerial imagery with simple linear iterative clustering and convolutional neural networks. *ISPRS—Int. Arch. Photogramm. Remote Sens. Spat. Inf. Sci.* **2019**, *XLII-2/W16*, 61–66. [CrossRef]
25. Schnabel, R.; Wahl, R.; Klein, R. Efficient RANSAC for point-cloud shape detection. *Comput. Graph. Forum* **2007**, *26*, 214–226. [CrossRef]
26. Nguyen, A.; Le, B. 3D point cloud segmentation: A survey. In Proceedings of the 2013 6th IEEE Conference on Robotics, Automation and Mechatronics (RAM), Manila, Phillippines, 12–15 November 2013; pp. 225–230.
27. Li, L.; Yang, F.; Zhu, H.; Li, D.; Li, Y.; Tang, L. An improved RANSAC for 3D point cloud plane segmentation based on normal distribution transformation cells. *Remote Sens.* **2017**, *9*, 433. [CrossRef]
28. Chen, Z.; Gao, B.; Devereux, B. State-of-the-art: DTM generation using airborne LIDAR data. *Sensors* **2017**, *17*, 150. [CrossRef] [PubMed]
29. Wiśniewski, Z. Estimation of parameters in a split functional model of geodetic observations (M_{split} estimation). *J. Geod.* **2009**, *83*, 105–120. [CrossRef]
30. Wiśniewski, Z. $M_{split(q)}$ estimation: Estimation of parameters in a multi split functional model of geodetic observations. *J. Geod.* **2010**, *84*, 355–372. [CrossRef]
31. Janowski, A. The circle object detection with the use of M_{split} estimation. *E3S Web Conf.* **2018**, *26*, 00014. [CrossRef]
32. Zienkiewicz, M.H.; Baryła, R. Determination of vertical indicators of ground deformation in the Old and Main City of Gdansk area by applying unconventional method of robust estimation. *Acta Geodyn. Geomater.* **2015**, *12*, 249–257. [CrossRef]
33. Zienkiewicz, M.H.; Hejbudzka, K.; Dumalski, A. Multi split functional model of geodetic observations in deformation analyses of the Olsztyn castle. *Acta Geodyn. Geomater.* **2017**, *14*, 195–204. [CrossRef]
34. Wyszkowska, P.; Duchnowski, R. M_{split} estimation based on L_1 norm condition. *J. Surv. Eng.* **2019**, *145*, 04019006. [CrossRef]
35. Wyszkowska, P.; Duchnowski, R. Performance of M_{split} estimates in the context of vertical displacement analysis. *J. Appl. Geod.* **2020**, *14*, 149–158. [CrossRef]
36. Wyszkowska, P.; Duchnowski, R. Systematic bias of selected estimates applied in vertical displacement analysis. *J. Geod. Sci.* **2020**, *10*, 41–47. [CrossRef]
37. Li, J.; Wang, A.; Xinyuan, W. M_{split} estimate the relationship between LS and its application in gross error detection. *Mine Surv.* **2013**, *2*, 57–59. [CrossRef]
38. Janicka, J.; Rapiński, J. M_{split} transformation of coordinates. *Surv. Rev.* **2013**, *45*, 269–274. [CrossRef]
39. Nowel, K. Squared $M_{split(q)}$ S-transformation of control network deformations. *J. Geod.* **2019**, *93*, 1025–1044. [CrossRef]
40. Czaplewski, K.; Wąż, M.; Zienkiewicz, M.H. A novel approach of using selected unconventional geodesic methods of estimation on VTS areas. *Mar. Geod.* **2019**, *42*, 447–468. [CrossRef]

41. Marshall, J.; Bethel, J. Basic concepts of L_1 norm minimization for surveying applications. *J. Surv. Eng.* **1996**, *122*, 168–179. [CrossRef]
42. Marshall, J. L_1-norm pre-analysis measures for geodetic networks. *J. Geod.* **2002**, *76*, 334–344. [CrossRef]
43. Wyszkowska, P.; Duchnowski, R. Iterative process of $M_{split(q)}$ estimation. *J. Surv. Eng.* **2020**, *146*, 06020002. [CrossRef]
44. Duchnowski, R.; Wiśniewski, Z. Robustness of squared $M_{split(q)}$ estimation: Empirical analyses. *Stud. Geophys. Geod.* **2020**, *64*, 153–171. [CrossRef]
45. Wiśniewski, Z.; Duchnowski, R.; Dumalski, A. Efficacy of M_{split} estimation in displacement analysis. *Sensors* **2019**, *19*, 5047. [CrossRef] [PubMed]
46. Zienkiewicz, M.H. Application of M_{split} estimation to determine control points displacements in networks with unstable reference system. *Surv. Rev.* **2015**, *47*, 174–180. [CrossRef]

remote sensing

MDPI

Article

Analysis and Discussion on the Optimal Noise Model of Global GNSS Long-Term Coordinate Series Considering Hydrological Loading

Yuefan He [1], Guigen Nie [1,2,*], Shuguang Wu [1] and Haiyang Li [1]

[1] GNSS Research Center, Wuhan University, Wuhan 430079, China; heyuefan@whu.edu.cn (Y.H.); shgwu@whu.edu.cn (S.W.); haiyangli@whu.edu.cn (H.L.)
[2] Collaborative Innovation Center for Geospatial Information Technology, Wuhan 430079, China
* Correspondence: ggnie@whu.edu.cn

Abstract: The displacement of Global Navigation Satellite System (GNSS) station contains the information of surface elastic deformation caused by the variation of land water reserves. This paper selects the long-term coordinate series data of 671 International GNSS Service (IGS) reference stations distributed globally under the framework of World Geodetic System 1984 (WGS84) from 2000 to 2021. Different noise model combinations are used for noise analysis, and the optimal noise model for each station before and after hydrologic loading correction is calculated. The results show that the noise models of global IGS reference stations are diverse, and each component has different optimal noise model characteristics, mainly white noise + flicker noise (WN+FN), generalized Gauss–Markov noise (GGM) and white noise + power law noise (WN+PL). Through specific analysis between the optimal noise model and the time series velocity of the station, it is found that the maximum influence value of the vertical velocity can reach 1.8 mm when hydrological loading is considered. Different complex noise models also have a certain influence on the linear velocity and velocity uncertainty of the station. Among them, the influence of white noise + random walking noise is relatively obvious, and its maximum influence value in the elevation direction can reach over 2 mm/year. When studying the impact of hydrological loading correction on the periodicity of the coordinate series, it is concluded whether the hydrological loading is calculated or not, and the GNSS long-term coordinate series has obvious annual and semi-annual amplitude changes, which are most obvious in the vertical direction, up to 16.48 mm.

Keywords: hydrological loading; long-term coordinate series; noise model; velocity; amplitude

Citation: He, Y.; Nie, G.; Wu, S.; Li, H. Analysis and Discussion on the Optimal Noise Model of Global GNSS Long-Term Coordinate Series Considering Hydrological Loading. *Remote Sens.* **2021**, *13*, 431. https://doi.org/10.3390/rs13030431

Academic Editor: Alex Hay-Man Ng
Received: 14 December 2020
Accepted: 20 January 2021
Published: 26 January 2021

1. Introduction

The global IGS base station location time series accumulated in the past 20 years have provided valuable basic data for geodesy and geodynamic research. Many scholars have also launched a lot of research on it, including periodic signal analysis, stochastic model research, spatial filtering method and surface loading model research, etc., in GNSS coordinate time series [1–16]. Analyzing it can get the precise movement trend of the station, so as to further study the internal driving mechanism of the station movement. It has far-reaching theoretical significance and broad application prospects to explain plate tectonic movement, establish and maintain a dynamic earth reference frame, study post-ice rebound and sea level changes, and invert the changes in ice and snow quality.

The noise of IGS station mainly comes from some random factors. As the stations are located all over the world, the geographical environment is very different and the sources of noise influence are also different. For example, different types of observation piers, missing time series data, time series of different spans, different geographical environments, etc. may generate some noise [17]. Currently, a large number of researchers generally believe that the characteristics of GNSS coordinate series noise model are WN+FN [18–20].

However, it turns out that this statement is not entirely true. The noise model characteristics of the reference station are actually more complicated. Langbein [3] estimated the noise model of 236 continuous GPS operation stations in California and southern Nevada in 2008; the results showed that the best noise model of FN or random walk noise (RW) station accounted for 50% to 60%, 25%~30% of the station performance are FN+RW or non-integer spectral exponential powerlaw noise (PL). The remaining stations are expressed as band pass+powerlaw noise (BPPL) and first-order Gauss–Markov+random walk noise (FOGMRW). In 2019, Wang et al. [21] analyzed the coordinate time series of 260 continuous GPS stations in the Crustal Movement Observation Network of China (CMONOC). In the noise evaluation, the influence of periodic signals is considered, and the maximum likelihood estimation method is used to discuss the noise characteristics of the residual time series that remove seasonal signals and the coordinate time series that further remove other periodic signals. It is concluded that FN is the main noise related to time. If it is assumed to be pure white noise, its speed uncertainty will be underestimated by 8-10 times. In 2008, Yuan et al. [22] analyzed the noise characteristics of 12 GPS reference stations in Hong Kong. It was concluded that the noise model after the common mode error (CME) could be removed by the principal component, and spatial filtering can be described by the variable white noise (VW) plus flicker noise. In 2010, Jiang et al. analyzed the coordinate series data of the China Continuously Operating Reference Stations (CORS) network in the past 10 years. The results point out that under the framework of the China Geodetic Coordinate System 2000 (CGCS2000), the noise model of coordinate time series appears as WN, FN and RW after principal component filtering. Additionally, if only white noise is considered, the velocity error will be 2~6 times smaller than when colored noise is considered [23].

The reason why WN+FN is considered to be the noise model representing the optimal random characteristics of the station may be analyzed in two ways: (1) When analyzing the time series, a more complex noise model that sufficiently represents the noise characteristics of the reference station is not used; (2) The time scale of the accumulated coordinate series is not long enough to explain the long-period component of the noise model. With the passage of time, the GNSS reference station coordinate time series continue to grow, and the long-period components of noise (such as RW noise with spectral index=2) will become more prominent, which provides favorable conditions for detecting the existence of low-frequency noise. Therefore, it is meaningful to conduct a more comprehensive noise analysis on GNSS time series and obtain a noise model that can more accurately represent the random characteristics of the reference station.

The GNSS coordinate series contains various ground loading changes related to the elastic deformation of Earth, and is closely related to changes in the atmosphere, seabed pressure, snow, ice, surface water and groundwater [24,25]. Some scholars have deeply analyzed the spatial and temporal distribution characteristics of the vertical displacements of IGS stations caused by the above environmental loadings. Their influence on the nonlinear variation and common mode error (CME) of GPS coordinate time series are estimated. The results show that the average root mean square (RMS) of vertical displacement caused by environmental loadings is 4.0 mm. The impact on one station can reach centimeter level [26]. Collilieux et al. [27] conducted the study based on 748 stations around the world, with many stations concentrated in North America and Europe. They found that when the time series were corrected for loading, the in-phase and out-of-phase coefficients of the annual signal in GPS height coordinates were reduced by 84% and 83 %, respectively. These factors may have a profound influence on the characteristics of the generated noise [28,29]. Among them, the land hydrological loading change is an important factor that causes the periodic vertical deformation of GNSS stations [30–32]. The increase of the hydrological loading on the land will cause the ground to sink, causing the vertical position of the GNSS station to move downward. Decreasing the hydrological loading on the land will cause the surface to rebound, causing the vertical position of the GNSS station to move upward.

The main innovation of this article is to use longer GNSS coordinate time series, adopt more types of noise model combinations and more IGS reference stations, and analyze the optimal noise model characteristics of global regional coordinate time series. At the same time, the displacement of the station caused by the hydrological loading is calculated, and the change analysis of the optimal noise model before and after the hydrological loading correction is given. On this basis, the relationship between the optimal noise model and the velocity, velocity uncertainty and amplitude of the stations are established. The organization structure of this article is as follows. The second part contains GNSS, hydrological data and theoretical methods to be analyzed in this article. The third part analyzes the optimal noise model of the selected station, including the comparison of the optimal noise model before and after hydrological loading correction. In the fourth part, the velocity and velocity uncertainty of the stations and the amplitude of the long-term coordinate series are discussed with different noise models before and after the hydrological loading correction. Finally, the fifth part summarizes the main results.

2. Data and Methods

2.1. GNSS Data

The GNSS time series data used in this article are released by the Scripps Orbit and Permanent Array Center (SOPAC), a data analysis center organized by the IGS [33]. This organization not only publishes some raw GNSS time series results, but also publishes its processed (Clean) time series products. In this article, we mainly use the linearized mean GNSS time series (CleanDetrend) after removing the step (coseismic and non-coseismic) and singular values. Seasonal period term is retained. In order to analyze the impact of various noises on it and explain the results of the global optimal noise model after adding hydrological loading.

Before data analysis, this paper needs to filter the long-term coordinate series data of 671 IGS reference stations distributed globally under the framework of WGS84 from 2000 to 2021. The screening criteria are as follows: first, within the 21 years, the length of the GNSS coordinate series shall not be less than 3 years to ensure the integrity of the long time series and the reliability of the data processing results. Then, delete sites with obvious abnormal nonlinear motion, including obvious post-earthquake deformation caused by a major earthquake, local surface deformation or abnormal motion caused by other unknown causes. Through the elimination of the above steps, 671 sites around the world are finally selected for data processing. The time length distribution of the selected GNSS stations in the past 21 years are shown in Figure 1. The longest time series can reach 20.54 years, the shortest time length is 3.47 years, the average time length can reach 13.25 years, and the average data integrity rate can be up to 63.1%. In this article, the average data integrity rate equals average time series length of selected sites divided by the total length of GNSS coordinate time series (21 years).

2.2. Hydrological Loading Data

Environmental loadings including hydrological loading, atmospheric loading, non-tidal ocean loading and other environmental factors are closely related to the elastic deformation of the earth, and these factors will also have a significant impact on the noise characteristics generated in the GNSS time series [25,28]. The impact of this loadings are not considered in the GNSS coordinate time series data processing under the WGS84 framework provided by SOPAC. Among them, the change of hydrological loading is an important factor that causes periodic vertical deformation of GNSS stations, as well as an important cause of surface subsidence and rebound. Therefore, it is particularly important to study the influence of hydrological loading on the GNSS coordinates time series [30,31].

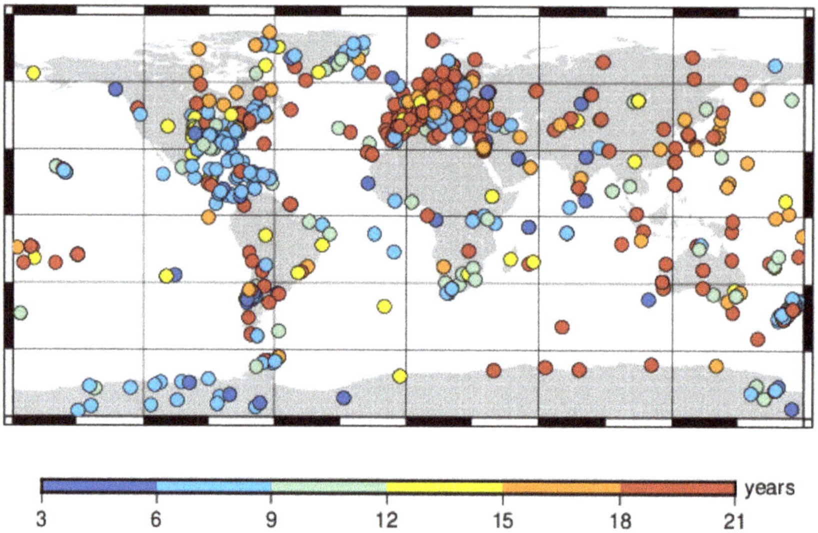

Figure 1. Distribution of 671 global GNSS reference stations and their coordinate time series length in 21 years (2000–2021).

At present, many organizations provide hydrological loading products, for example, German Research Centre for Geosciences (GFZ) (http://rz-vm115.gfz-potsdam.de:8080/repository), International Mass Loading Service (IMLS) (http://massloading.net) products released by National Aeronautics and Space Administration (NASA) and School and Observatory of Earth Sciences (EOST) (http://loading.u-strasbg.fr/) loading products from university of Strasbourg. This article first selects the hydrological loading (HYDL) data products provided by GFZ [25,34] and estimate the displacement of global GNSS stations caused by hydrological loading, and subtract the influence of hydrological loading from the GNSS coordinate series. Then, analyze the optimal noise model of the global GNSS coordinate series taking into account the hydrological loading. Finally, based on the optimal noise model combination, we will further discuss its influence on the long-term coordinate series velocity, velocity uncertainty and amplitude of the IGS stations.

The hydrological loading data selected in this paper are based on the center of figure (CF) frame, with a time resolution of 24 h and a spatial resolution of 0.5° × 0.5°. The hydrological loading data model calculated by GFZ is the Land Surface Discharge Model (LSDM) [35]. The physics and parameterization of the LSDM is based on the research of Hagemann and Düumenil (1998). LSDM includes soil moisture, snow cover, shallow groundwater, and surface water stored in rivers and lakes [36].

2.3. Theoretical Method

The GNSS coordinate sequence is represented by two parts: a deterministic model and a stochastic model [37]. The deterministic model is composed of trend, periodic terms (including anniversaries, semi-anniversaries, etc.), offset terms, etc. The stochastic model is the noise that this article will analyze. The GNSS coordinate time series is represented by the following functional model.

$$y(t_i) = y_0 + vt_i + \sum_{k=1}^{q} a_k sin(2\pi f_k t_i) + b_k cos(2\pi f_k t_i) + \sum_{j=1}^{n_g} g_j H(t_i - T_{g_j}) + r(t_i) \quad (1)$$

In the above formula, y_0 is the intercept; v is the linear trend in unit per year, where a year is defined as 365.25 days; t_i is the epoch of the GNSS time series; a_k, and b_k are the amplitude of periodic signals and f_k is the corresponding frequency; g_j and T_{g_j} is the offset and corresponding epoch, respectively; r refers to the residual time series, which can be various combinations of noise models. H is the Heaviside step function:

$$\begin{cases} H(t) = 0, t < 0 \\ H(t) = 0.5, t = 0 \\ H(t) = 1, t > 0 \end{cases} \tag{2}$$

In this article, we use the Hector software to analyze the optimal noise model of the long-term coordinate series of the IGS reference station [38]. Hector software can estimate linear trend terms, high-order polynomials, seasonal terms, periodic terms and a variety of noise model combinations in coordinate time series. It uses the Akaike Information Criterion (AIC) and the Bayesian Information Criterion (BIC) information standards to select the optimal noise model [39,40]. Both use the maximum likelihood estimate as a starting point, but in order to avoid overfitting, measures are taken to add parameters. The AIC and BIC information standards are expressed by the following formula.

$$ln(L) = -\frac{1}{2}\left[Nln(2\pi) + lndet(C) + r^T C^{-1} r\right] \tag{3}$$

In the above formula, N represents the actual number of observations, and C refers to the covariance matrix, which can be decomposed into the following formula.

$$C = \sigma^2 \overline{C} \tag{4}$$

$\overline{C}$ represents the sum of multiple noise models, and σ represents the standard deviation, which can be estimated by the residual term in the following formula.

$$\sigma = \sqrt{\frac{r^T \overline{C}^{-1} r}{N}} \tag{5}$$

Due to $\det cA = c^N detA$, Where A refers to any matrix, $detA$ refers to the determinant of matrix A. Therefore, you can get

$$ln(L) = -\frac{1}{2}[Nln(2\pi) + lndet(\overline{C}) + 2Nln(\sigma) + N] \tag{6}$$

The parameter k refers to the sum of a design matrix and the noise model parameters and the variance of the driving white noise. For example, if the white noise + power law noise model is used to estimate the linear trend term, 5 parameters are estimated, including the normal offset, the linear trend term, the power spectrum index, the variance of driving white noise and the difference between power law noise and white noise ($k = 2 + 2 + 1 = 5$).

Then,

$$AIC = 2k + 2ln(L) \tag{7}$$

$$BIC = kln(N) + 2ln(L) \tag{8}$$

AIC and BIC are commonly used optimal noise model evaluation indicators in the maximum likelihood estimation analysis of Hector software [41]. From the two calculation formulae, it can be found that the better noise model should have a smaller AIC or BIC value. Since BIC considers the number of parameters to be estimated in the model, it is more suitable for comparing noise models with different numbers of parameters to be estimated. In this article, the optimal noise model is obtained by using BIC evaluation criteria. There are many noise models to choose from in the Hector software, including PL, WN, FN, RW, GGM, first-order autogressive noise (AR1) and a combination model of these noises.

3. Results

3.1. Before Hydrological Loading Correction

As far as the global GNSS single-day solution time series are concerned, if no measures are taken to reduce the spatial correlation of the station, it can be concluded that WN + FN is the most suitable stochastic noise model reflecting global GNSS coordinates [2]. Due to the different environments of each measuring station, the sources of noise may also be different, and their noise characteristics may not be exactly the same. This paper selects six noise models, namely WN, GGM, WN+FN, WN+PL, WN+RW and WN+GGM. According to the evaluation criteria of the optimal noise model in Section 2.3 above, use Hector software to obtain the minimal BIC value corresponding to the N, E, and U directions of 671 sites around the world, then the noise model corresponding to this minimal BIC value is the optimal noise model of the selected site. Finally, statistics are made on the optimal noise models corresponding to 671 sites around the world, and the proportion of the global optimal noise models before the hydrological loading correction is obtained, as shown in Figure 2 below.

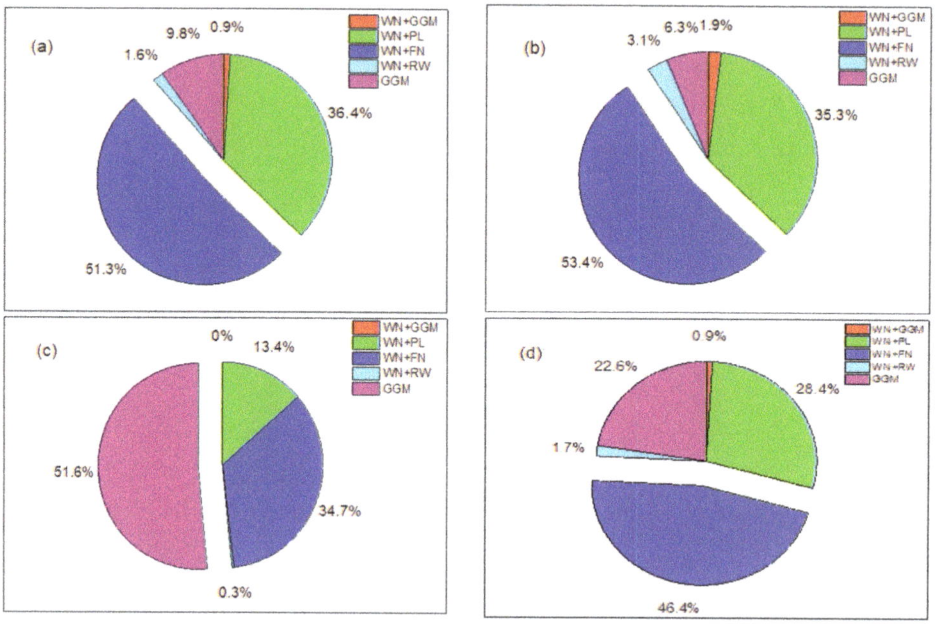

Figure 2. The optimal noise model distribution in the North (N), East (E), Up (U) directions and total components of global International Global Navigation Satellite System (GNSS) Service (IGS) stations before hydrological loading correction. (**a**), (**b**) and (**c**) respectively show the optimal noise model percentage in the N, E and U directions. The optimal noise model percentage of total components shown in (**d**).

It can be seen from Figure 2 that the noise characteristics of global IGS reference station components are diversified, mainly represented by WN+FN and GGM models. Especially in the N and E directions, the WN+FN model accounts for 51.3% and 53.4% of the world's optimal noise models, respectively. In the U direction, the GGM model has the largest proportion of the optimal noise model, which is as high as 51.6%. Next is the WN+PL noise model, and other combined noise models also occupy a certain proportion. The N, E, and U components of most stations exhibit different noise characteristics. As shown in Figure 2d above, the total components refer to the sum of the three directions of N, E, and U. Figure 2d shows the accumulation of the stations in these three directions to obtain the total optimal noise model percentage. For example, WN+FN, which accounts for the largest proportion of the optimal noise model, can reach 46.4%. It defines to the sum of the number of stations where the optimal noise model in the three directions of N, E, and U is WN+FN divided by the total number of stations in three directions (671*3). Therefore, when looking for a physical explanation for the change characteristics of time series, each component needs to be treated differently.

3.2. Hydrological Loading Displacement

This paper uses the hydrological loading products of GFZ to obtain the hydrological loading displacement values of the long-term coordinate time series of 671 IGS stations worldwide from 2000 to 2021. First, we selected three IGS stations at different latitudes, namely ACP1 (longitude 280.0501, latitude 9.3714), ALGO (longitude 281.9286, latitude 45.9558), ALRT (longitude 297.6595, latitude 82.4943). Then, we drew the displacement series diagram of their hydrological loading from 2000 to 2021. As shown in Figure 3, in the horizontal direction, the hydrological loading displacement values of these three stations are all small, and the maximum value is less than 2 mm. Especially for the station of ALRT, the maximum horizontal hydrological loading displacement is only 0.93 mm. However, it can be seen in Figure 3 that the hydrological loading influence value of these three stations is larger in the vertical direction. For the low-latitude station ACP1, the maximum vertical hydrological loading can reach 6.46 mm, and there is a periodic fluctuation in the long-term series. For the mid-latitude site ALGO, the maximum vertical hydrological loading can reach 9.48 mm, and the long-term series shows that its hydrological loading influence value has a downward trend. For the high-latitude site ALRT, the maximum vertical hydrological loading can reach 6.29 mm, but it has an upward trend in the long-term series. Regardless of whether the vertical hydrological loading series of these three stations rises, falls, or changes periodically, it may be related to the surrounding environment, including rainfall, soil moisture, etc.

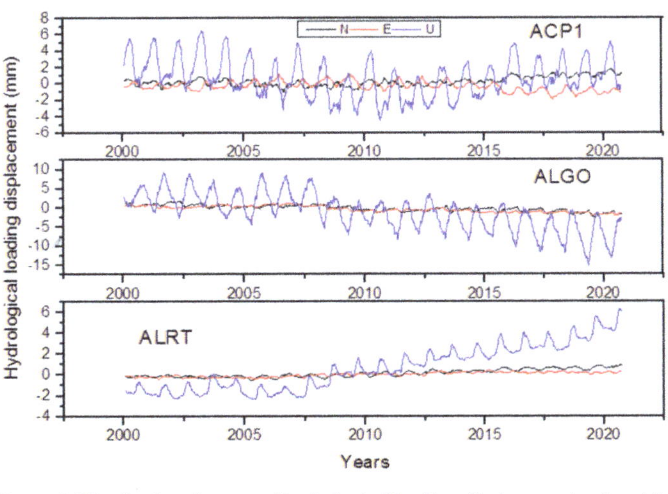

Figure 3. Distribution diagram of hydrological loading displacement series at the three IGS stations ACP1, ALGO, and ALRT from 2000 to 2021.

Since the time span is as long as 21 years, this article counts the RMS value in the N, E, and U directions of the selected 671 IGS reference stations from 2000 to 2021 due to hydrological loading. Then, the maximum RMS value of the three directions of N, E, and U is represented by a histogram. As shown in Figure 4, the magnitude of the hydrological loading in the N and E directions are relatively small. Among them, the RMS value of more than 90% of the stations are less than 1 mm, the maximum RMS value in the E direction can reach 2.07 mm, and the maximum RMS value in the N direction can reach 2.15 mm. In the vertical direction, the magnitude of the influence of hydrological loading is relatively large, and the RMS value can reach 13.48 mm. Especially in the Amazon River Basin in South America, the Mississippi River Basin in North America, the Congo River Basin in Africa, the Yangtze River Basin in China and the Indochina Peninsula, the magnitude of the hydrological load is relatively large. Therefore, when considering the influence factors of the displacement of the reference frame, the hydrological loading must be considered.

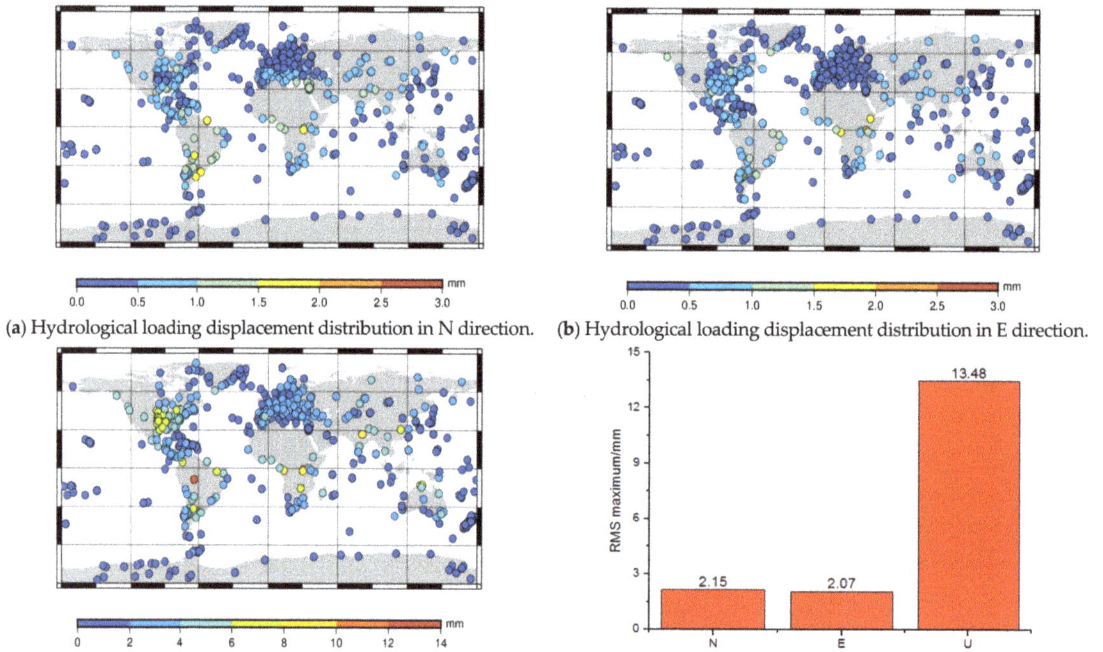

(**a**) Hydrological loading displacement distribution in N direction. (**b**) Hydrological loading displacement distribution in E direction.

(**c**) Hydrological loading displacement distribution in U direction. (**d**) Maximum RMS value in N, E and U direction.

Figure 4. A statistical diagram of the RMS value of hydrological loading displacement in the N, E, and U directions and their maximum RMS values. (**a**), (**b**) and (**c**) respectively show the RMS value of hydrological loading displacement in the N, E, and U directions and their maximum RMS values are shown in (**d**).

3.3. *After Hydrological Loading Correction*

Research shows that environmental loading, including hydrological loading, atmospheric loading and non-tidal ocean loading are the main factors that cause non-linear changes in stations [42]. Among them, the hydrological loading change is an important reason for the periodic vertical changes of GNSS stations. The GNSS time series itself contains the influence of loading. This article deducts it on the basis of it, and then analyzes the influence of hydrological loading on GNSS time series. Therefore, based on the above calculation results, this paper corrects the hydrological loading of these 671 IGS reference stations, re-analyzes the noise of the corrected coordinate time series, and obtains the optimal noise model of the global IGS reference station N, E, U components after deducting the influence of hydrological loading. Figure 5 shows the proportional distribution of the optimal noise model for each station component.

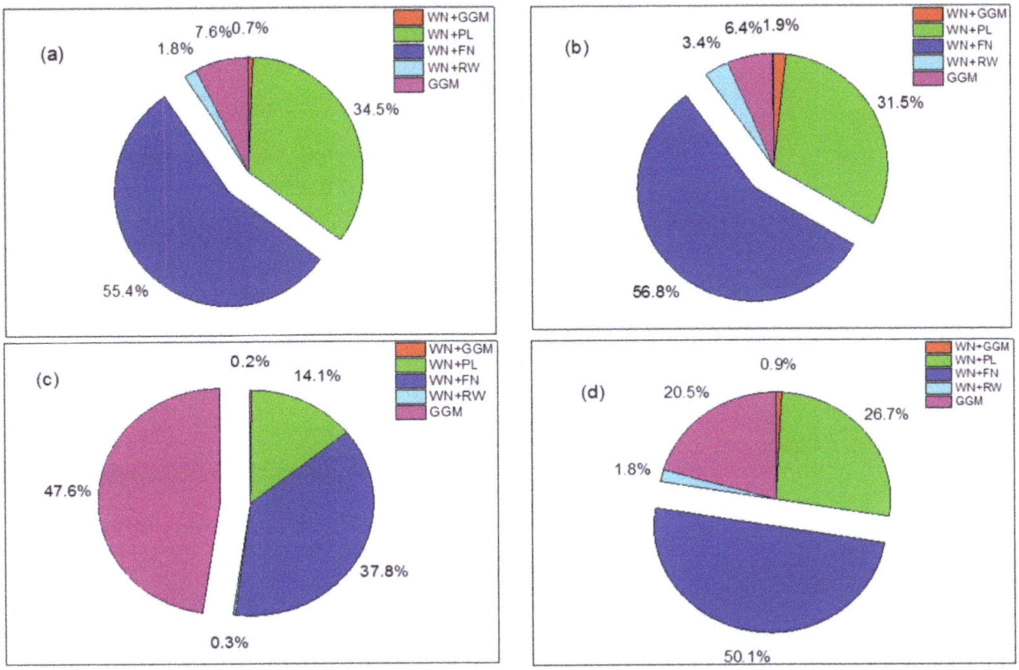

Figure 5. The optimal noise model distribution in the N, E, U directions and total components of global IGS stations after hydrological loading correction. (**a**), (**b**) and (**c**) respectively show the optimal noise model percentage in the N, E and U directions and the optimal noise model percentage of total components are shown in (**d**).

From Figure 5a–d, it can be seen that, compared with Figure 2a–d, after calculating the hydrological loading correction, the percentage of the WN+FN model in the N, E, and U directions obviously increases, and the percentage of GGM model somewhat decreases. However, in the N and E directions, the optimal noise model, WN+FN, still accounts for the largest proportion with 55.4% and 56.8%, respectively. Compared with the case without employing hydrological loading correction, the percentage increased by 4.1% and 3.4%, respectively. The second largest proportion is WN+PL. In the U direction, the GGM model still has the largest proportion of the optimal noise model, accounting for 47.6%, followed by the WN+PL model. Regardless of whether the impact of hydrological loading is considered, in the long-term coordinate series analysis of global IGS reference stations, the smaller proportion of the optimal noise model are the combination of WN+GGM and WN+RW models.

4. Discussion

4.1. Velocity and Velocity Uncertainty Analysis

Studies have shown that the estimated values of velocity and velocity uncertainty in GNSS time series will vary with the noise model used [43–45]. Therefore, in the velocity error estimation, the influence of different noise models should be considered. According to the above analysis, in the global GNSS long-term coordinate series, WN+FN is the optimal noise model. The results obtained in this article are the same as those obtained by many previous scholars [46,47]. In this article, we used the CleanDetrend data provided by SOPAC, and the trend item information has been deduced from the GNSS time series. We added the optimal noise model combination WN+FN, using Hector software to extract the velocity of the selected 671 IGS long-term coordinate series. The uncertainty of the velocity is also generated immediately, which actually represents the fluctuation of the velocity value. When considering the optimal noise model combination WN+FN, compare the difference of the velocity and its uncertainty relative to SOPAC. That is, when only WN + FN is considered, the velocity and velocity uncertainty difference distribution of each station components are shown in Figure 6 below:

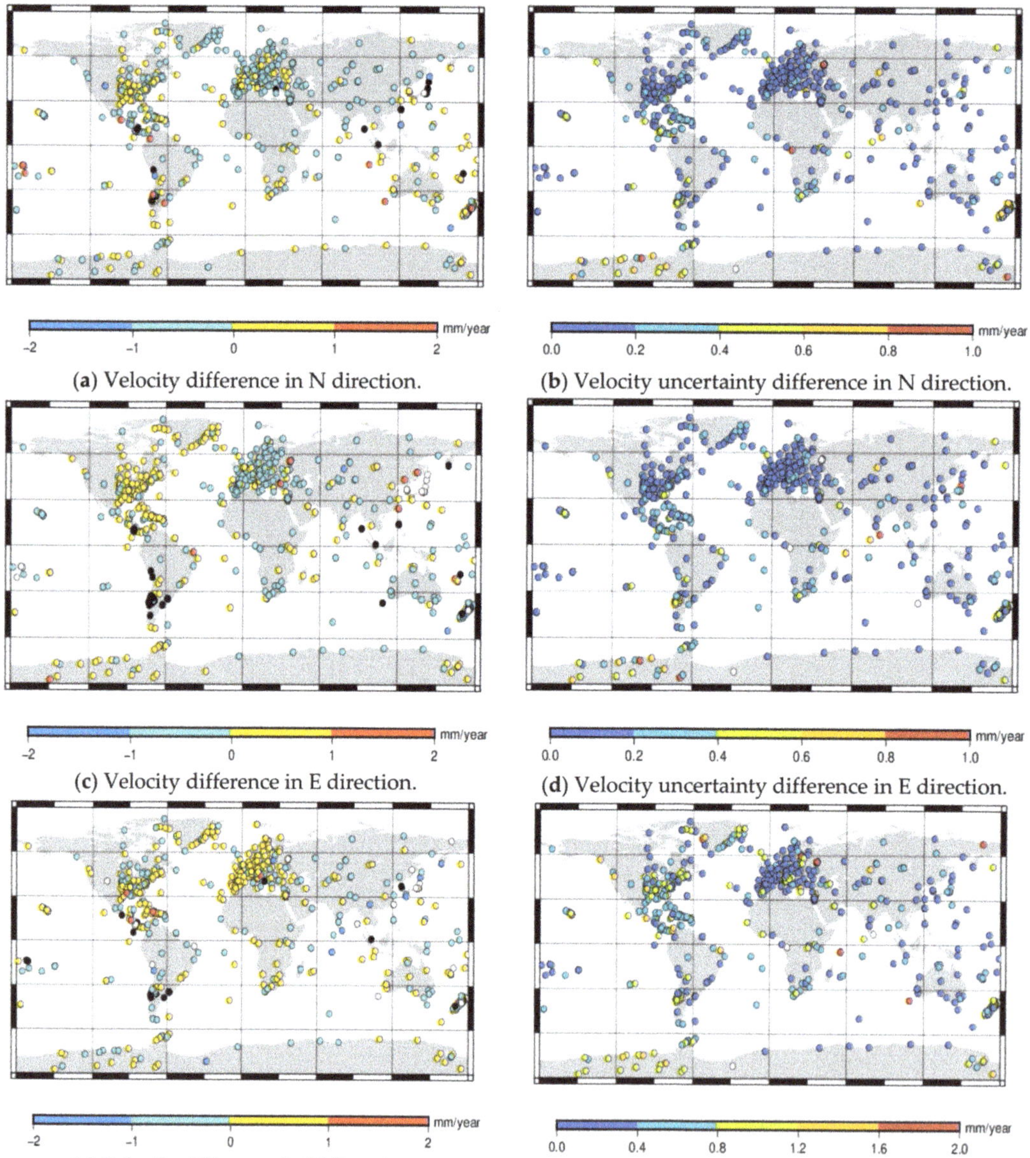

(**a**) Velocity difference in N direction.
(**b**) Velocity uncertainty difference in N direction.
(**c**) Velocity difference in E direction.
(**d**) Velocity uncertainty difference in E direction.
(**e**) Velocity difference in U direction.
(**f**) Velocity uncertainty difference in U direction.

Figure 6. The difference between the velocity and its uncertainty estimation under the optimal noise model of global IGS stations. (**a**,**c**,**e**) show the difference between the velocity under the optimal noise model of global IGS stations. (**b**,**d**,**f**) show the difference between the velocity uncertainty estimation under the optimal noise model of global IGS stations.

In order to more clearly and intuitively see the specific size of the velocity difference and the velocity uncertainty difference of each direction in each interval, this article lists the percentage of the interval of velocity and velocity uncertainty of each station. As shown in Table 1, in the direction of N, E and U, the absolute value of the velocity difference is more than 85% within 1 mm, and only less than 10% of the stations have the absolute value of the velocity difference between 1 and 2 mm, and some of them are larger than 2 mm. As shown in Figure 6a,c,e, these sites over 2 mm are marked with black and white dots. We can see that most of the black and white dots are near the coastal area. For example, ANTC (longitude 288.4680°, latitude −37.3387°), MIZU (longitude 141.1328°, latitude 39.1352°) stations, etc. The reason for analysis may be related to the imperfect simulation of noise models in these coastal areas. For the velocity uncertainty difference, in the N and E directions, 86% and 87% of the stations velocity uncertainty differences are less than 0.4 mm/year, and the maximum value is less than 1 mm. In the U direction, the influence value is slightly larger than that in the N and E directions. A total of 87% of the stations velocity uncertainty differences are less than 0.8 mm, and the maximum value is less than 2 mm. If you consider the effects of rounding errors, the calculated results can be considered reasonable.

Table 1. Distribution of station velocity and velocity uncertainty difference.

	Velocity Difference (mm/Year)				Velocity Uncertainty Difference (mm/Year)				
	(−2, −1)	**(−1, 0)**	**(0, 1)**	**(1, 2)**	**(0, 0.2)**	**(0.2, 0.4)**	**(0.4, 0.8)**	**(0.6, 0.8)**	**(0.8, 1.0)**
N	2.2%	51.1%	38.5%	2.2%	65.7%	20.1%	9.2%	4.0%	0.75%
E	1.3%	45.5%	42.5%	2.4%	63.0%	24.0%	8.6%	2.4%	1.0%
					(0, 0.4)	(0.4, 0.8)	(0.8, 1.2)	(1.2, 1.6)	(1.6, 2.0)
U	3.0%	32.6%	52.8%	2.2%	44.7%	28.8%	19.8%	4.9%	1.1%

In order to further study the influence of hydrological loading on GNSS long-time series velocity and its uncertainty, considering only the WN+FN noise, this paper analyzes the differences in the velocities and its uncertainties of the 671 IGS stations before and after the hydrological loading correction, and draws the percentage diagrams in the N, E, and U directions. The result shows that it has basically no effect on the velocity uncertainty, but it will cause varying degrees of velocity changes at the station, especially in the vertical direction. The statistical results are shown in Figure 7. It can be seen that in the N direction, after the hydrological loading correction, the velocity influence value can reach up to 0.9 mm, and 97.4% of the values are between 0 and 0.2 mm. In the E direction, the maximum influence value can reach 0.5 mm, and 89.8% of the influence values are between 0 and 0.1 mm. In the U direction, we can see that the influence value is more obvious relative to the horizontal direction. Its maximum value can reach 1.8 mm, most of the sites are between 0 and 0.4 mm, and the percentage can account for 82.6%. Therefore, when estimating the vertical velocity, the influence of hydrological loading must be considered.

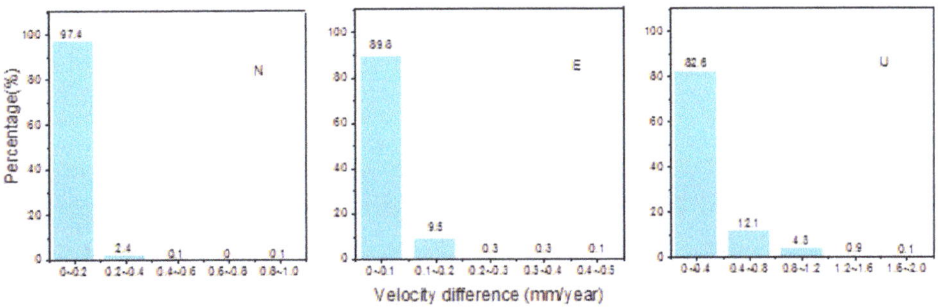

Figure 7. The percentage diagram of velocity influence value before and after hydrological loading correction.

Compared with the WN+FN model, different complex noise models will also have a certain impact on the station velocity and its uncertainty, as shown in Figure 8 below, the vertical direction is larger and the horizontal direction is smaller. Among them, regarding the influence of velocity, in the directions of N and E, the remaining four noise models are basically the same, and more than 85% of the station velocity differences are within 1 mm in absolute value. However, in the U direction, the ratio of the WN+RW velocity difference between (−1, 1) is significantly smaller than other noise models, accounting for only 68.7%, while in (−2, 1) and (1, 2), the proportion between them is slightly larger. The difference between these four noise models is mainly reflected in the velocity uncertainty. For the velocity uncertainty difference, in the directions of N, E, and U, the GGM noise model accounts for the largest proportion between (0, 0.2), accounted for 76.5%, 79.1%, 64.5%, respectively, followed by WN+GGM, WN+PL noise model combination. Among them, WN+RW has the largest impact level, and the velocity uncertainty difference of more than 85% of the stations are greater than 1 mm/year. Especially in the U direction, the velocity uncertainty difference is greater, and more than 95% of the stations are greater than 2 mm/year. It can be concluded that for the establishment of mm-level high-precision reference frames and plate motion analysis, we need to take into account the differences brought about by different noise models.

4.2. Amplitude Analysis

To study the relationship between the hydrological loading and the amplitude of the selected global IGS reference stations. In this paper, the annual and half-annual amplitudes of the stations considering the influence of hydrological loading under different noise models are estimated. The results under the optimal noise model are shown in Figure 9. It can be seen that no matter whether the hydrological loading is corrected or not, there are significant annual and semi-annual amplitudes at each station. Without considering the influence of hydrological loading, the U direction is the largest, and its maximum annual amplitude can reach 16.48 mm, and 95.7% of the stations have an annual amplitude between 0 and 8 mm. The N and E directions are next. In the N direction, the maximum annual amplitude can reach 9.87 mm, and in the E direction, the maximum annual amplitude can reach 10.07 mm. A total of 92.1% and 91.8% of the station sizes are between 0 and 2 mm. In addition to the influence of the annual amplitude, the half-annual amplitude also has a significant change, which is still more affected in the U direction, and its maximum value can reach 5.11 mm. Compared with the vertical direction, the horizontal half-annual amplitude is smaller. In the N and E directions, 93.4% and 90.6% of the stations are less than 1 mm.

After the hydrological loading is corrected, when only the WN+FN noise model is considered, it will have a certain impact on the annual and half-annual amplitude of the stations. Figure 10 plots the annual and half-annual amplitude differences before and after the hydrological loading correction in each direction. It can be seen that after the hydrological loading correction is added, it still has a relatively large impact on the U direction. The maximum annual amplitude difference can reach 8.08 mm, and the annual amplitude difference of most stations are between −3 and 3 mm, which can account for 97.5%. Compared with the annual amplitude, the half-annual amplitude difference is smaller. In the U direction, the maximum value can reach 1.71 mm, and the absolute value of the half-annual amplitude difference of 98.6% of the stations are less than 1 mm. In the horizontal direction, whether it is the annual amplitude or the half-annual amplitude, its influence value is relatively small. In the N direction, the maximum annual amplitude difference is 1.04 mm, and the absolute value of the annual amplitude difference in 99.7% of the stations are within 1 mm. The maximum half-annual amplitude difference is 0.21 mm, and the absolute value of the half-annual amplitude difference in 98.3% of the stations are less than 0.2 mm. In the E direction, the maximum annual amplitude difference is 1.01 mm, and the absolute value of the annual amplitude difference in 99.9% of the stations are within 1 mm. The maximum half-annual amplitude difference is 0.32 mm, and the absolute value of the half-annual amplitude difference in 99.4% of the stations are less than 0.2 mm. From this, it can be concluded that the calculated hydrological loading will indeed cause the annual and semi-annual movements of IGS reference stations in the global region. The impact varies from station to station and is related to the geographical environment around the stations. The annual amplitude is larger than the half-annual amplitude, but it cannot completely reduce the annual and half-annual amplitudes of the GNSS long time coordinate series, especially the half-annual amplitude.

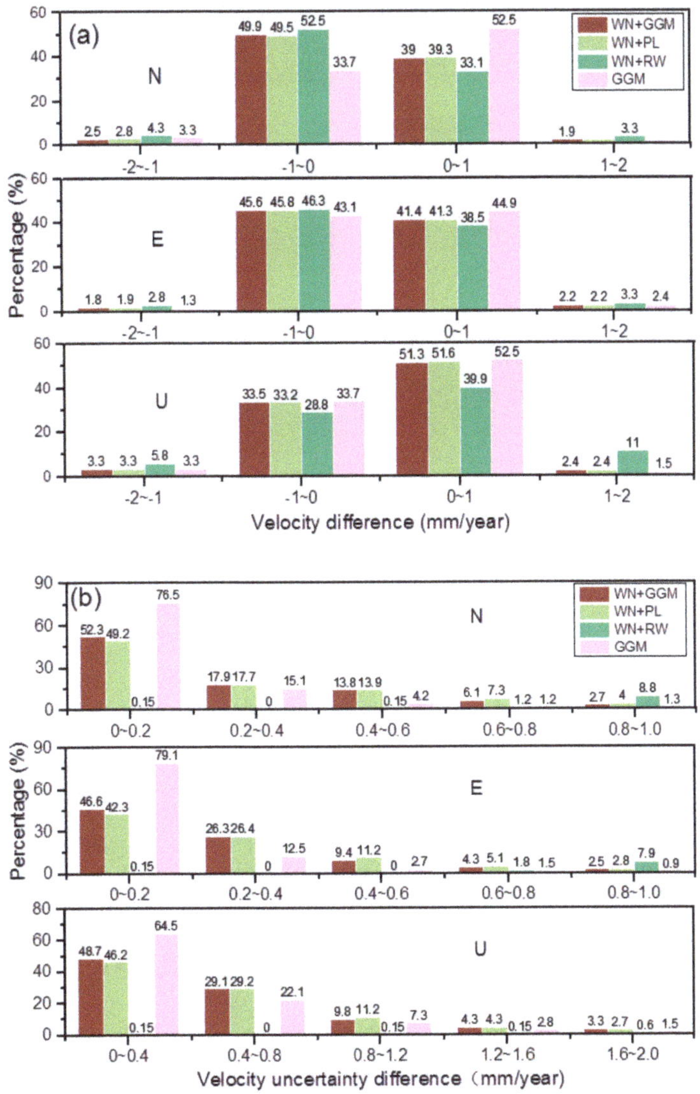

Figure 8. The distribution of velocity and velocity uncertainty differences under different noise models. (**a**) shows the distribution of velocity differences under different noise models and (**b**) shows the distribution of velocity uncertainty differences under different noise models.

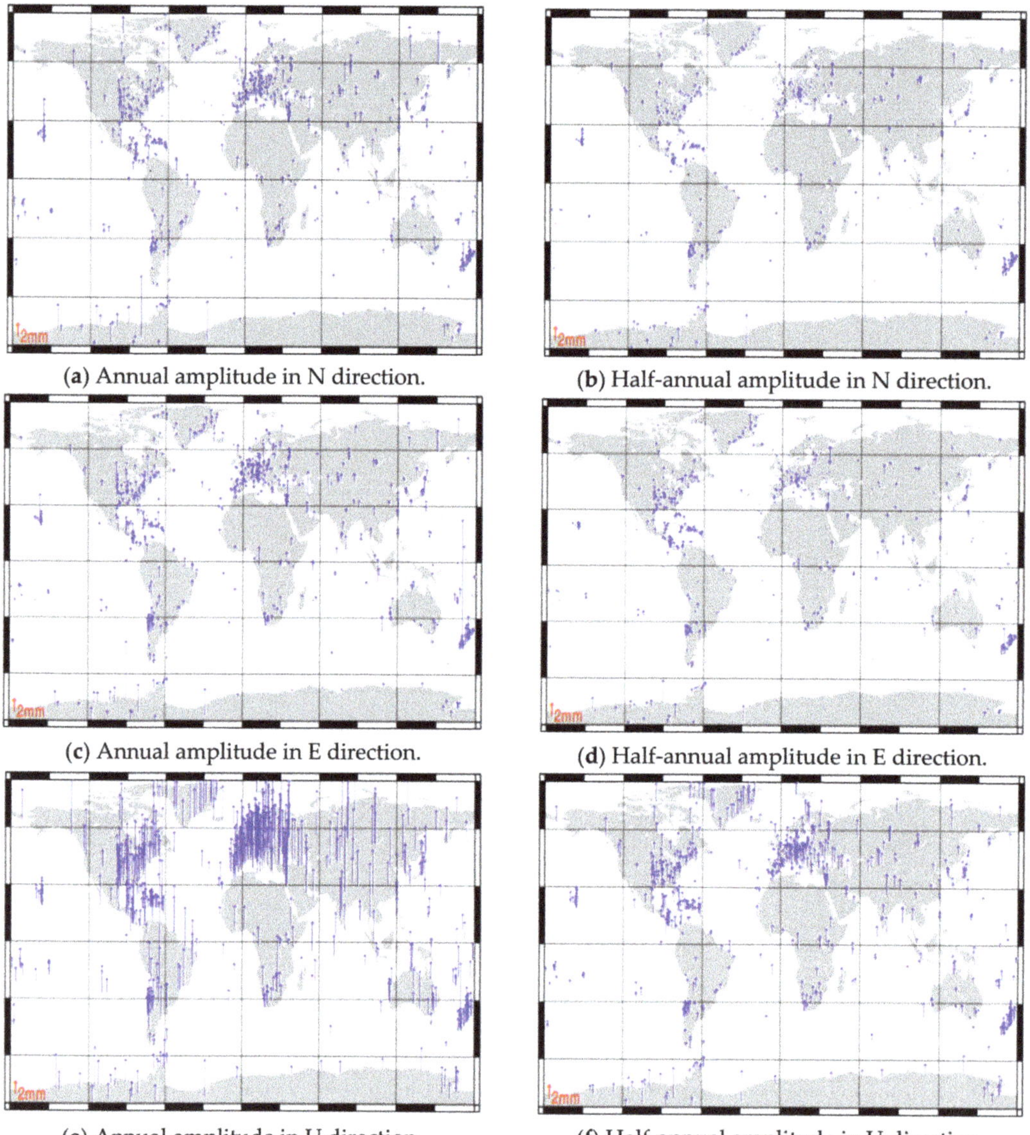

(**a**) Annual amplitude in N direction.
(**b**) Half-annual amplitude in N direction.
(**c**) Annual amplitude in E direction.
(**d**) Half-annual amplitude in E direction.
(**e**) Annual amplitude in U direction.
(**f**) Half-annual amplitude in U direction.

Figure 9. Distribution of the annual amplitudes before hydrological loading correction are shown (**a**,**c**,**e**). (**b**,**d**,**f**) show the half-annual amplitude before hydrological loading correction. (unit: mm).

In order to more clearly explain the influence of the hydrological loading correction on the annual and half-annual amplitudes, Table 2 lists the percentages of the annual and half-annual amplitude differences in each interval in the N, E, and U directions in detail.

Figure 11 shows the statistics of the annual and half-annual amplitude difference before and after the hydrological loading correction in various directions under different noise models. It can be seen, no matter which noise model is selected before and after the hydrological loading correction, that the influence of the annual amplitude and half-annual amplitude is larger in the vertical direction and smaller in the horizontal direction. Therefore, it can be concluded when estimating the influence of the hydrological loading on the annual and semi-annual amplitude difference of the IGS station that the noise models including WN+FN, WN+PL, WN+RW, GGM and other noise models have negligible influence.

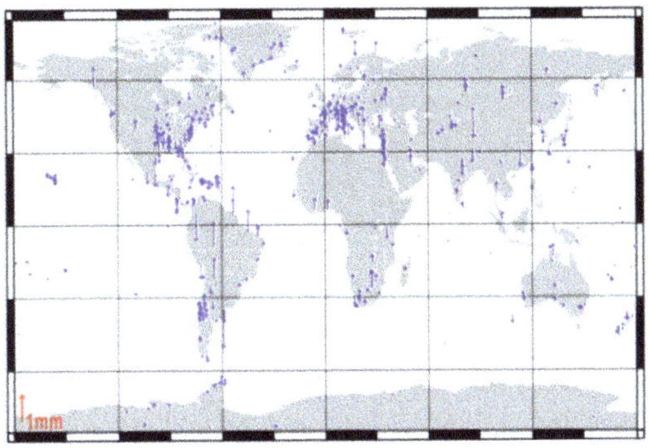

(**a**) Annual amplitude difference in N direction.

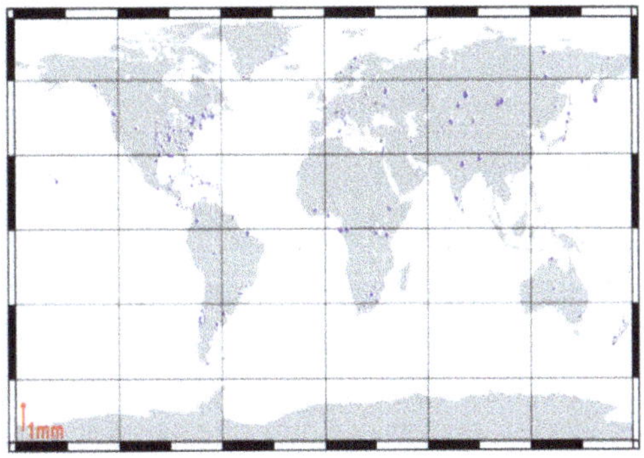

(**b**) Half-annual amplitude difference in N direction.

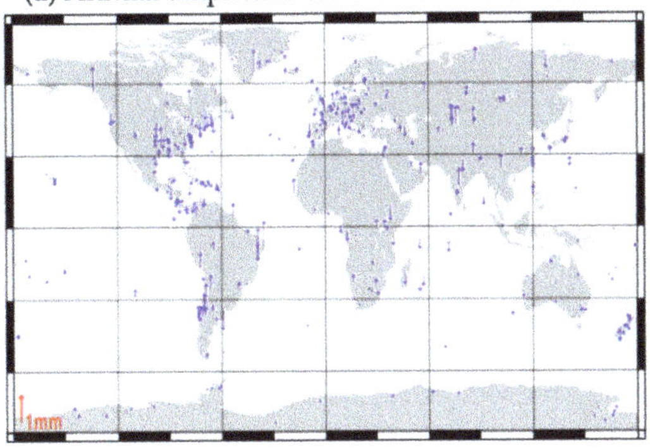

(**c**) Annual amplitude difference in E direction.

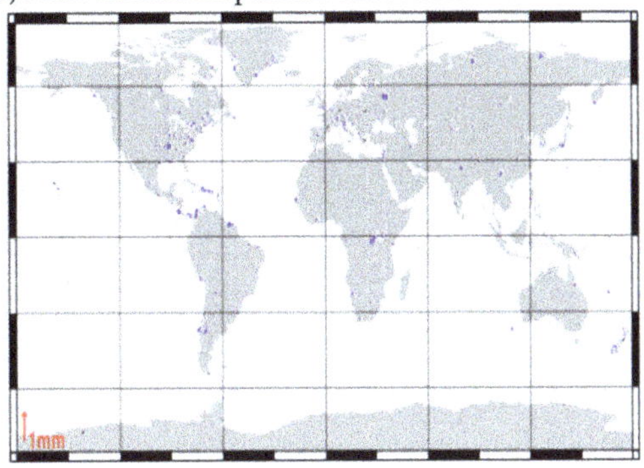

(**d**) Half-annual amplitude difference in E direction.

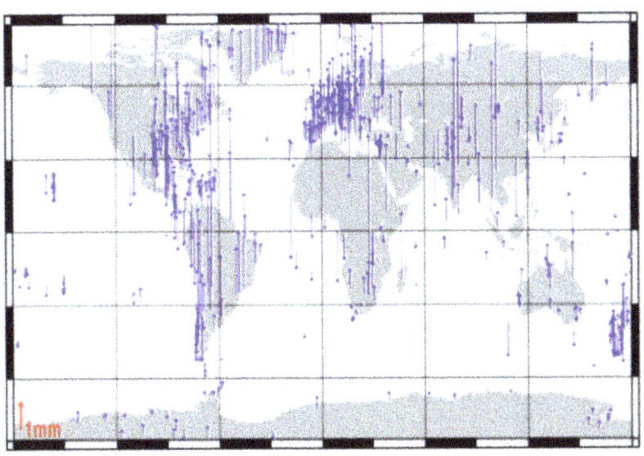

(**e**) Annual amplitude difference in U direction.

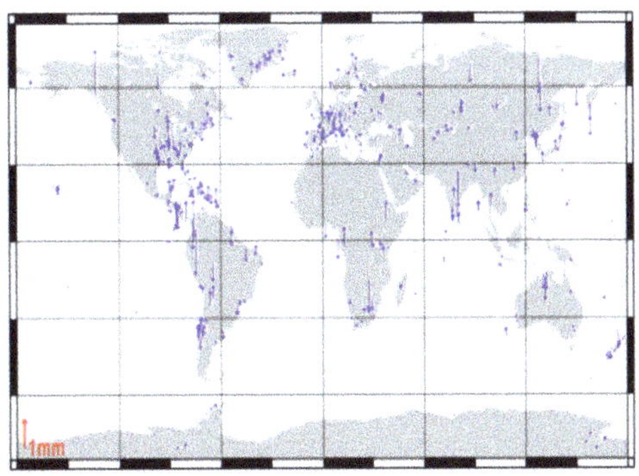

(**f**) Half-annual amplitude difference in U direction.

Figure 10. Statistics of annual and half-annual amplitude differences before and after hydrological loading correction. (**a**,**c**,**e**) show the statistics of annual amplitude differences before and after hydrological loading correction. The half-annual amplitude differences before and after hydrological loading correction are shown in (**b**,**d**,**f**). (unit: mm).

Table 2. Annual and half-annual amplitude difference percentage table before and after hydrological loading correction.

	Annual Amplitude Difference (mm)				**Half-Annual Amplitude Difference (mm)**			
	(−2, −1)	**(−1, 0)**	**(0, 1)**	**(1, 2)**	**(−0.4, −0.2)**	**(−0.2, 0)**	**(0, 0.2)**	**(0.2, 0.4)**
N	0.3%	51%	48.7%	0.1%	1.3%	50%	48.3%	0.4%
E	0%	49.6%	50.3%	0.1%	0.6%	52.1%	47.3%	0%
	(−3, 0)	(0, 3)	(3, 6)	(6, 9)	(−2, −1)	(−1, 0)	(0, 1)	(1, 2)
U	30%	67.5%	2.2%	0.4%	1.2%	47.2%	51.4%	0.2%

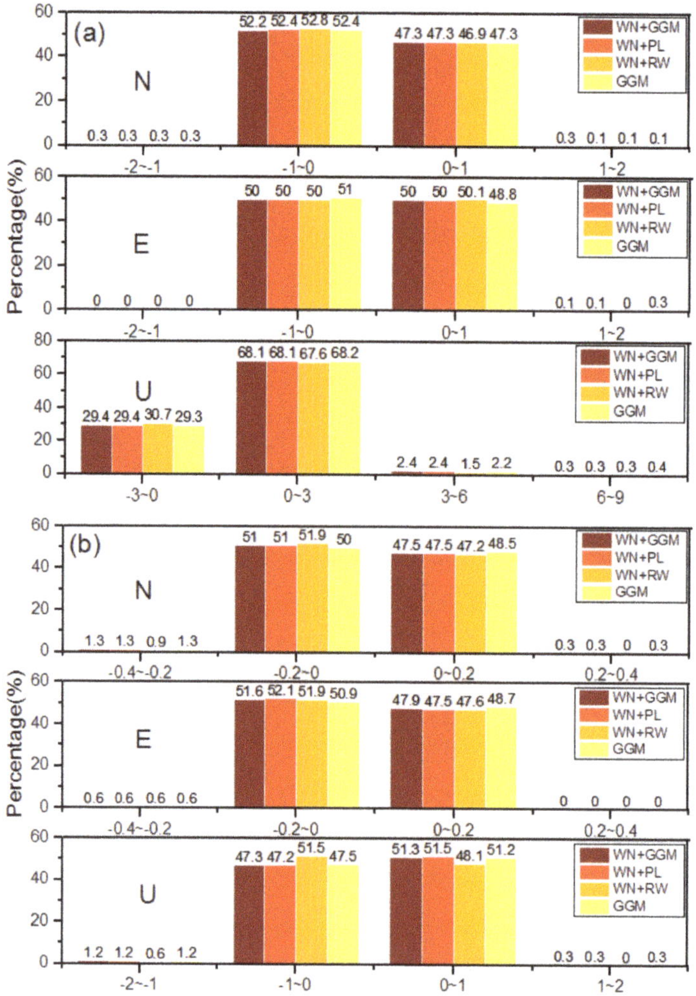

Figure 11. Statistics of annual and half-annual amplitude difference under different noise models before and after hydrological loading correction. (**a**) shows the statistics of annual amplitude difference under different noise models before and after hydrological loading correction and (**b**) shows the half-annual amplitude difference under different noise models before and after hydrological loading correction. (unit: mm).

5. Conclusions

In this paper, we used the GNSS coordinate time series of 671 IGS stations around the world from 2000 to 2021 to analyze the world's optimal noise model. More importantly, the influence of hydrological loading on the GNSS coordinates time series is discussed, and the following conclusions are obtained:

1. Pure white noise cannot represent the optimal noise model for the long-term series of global IGS reference station coordinates. The noise models of global IGS reference station coordinate time series are diverse, and the N, E, and U directions show different optimal noise model characteristics. Before the hydrological loading correction, the optimal noise model for 46.4% of the station components is WN+FN, the optimal noise model for 28.4% of the station components is WN+PL, and the 22.6% components is the GGM noise model. There is also a small number of components represented as WN+GGM and WN+RW model combinations.
2. Experiments show that the hydrological loading does cause changes in the noise characteristics of IGS stations. After calculating the hydrological loading correction, the ratio of WN+FN in the optimal noise model increased significantly (50.1%). Among them, the largest proportion of the optimal noise model in the U direction is still the GGM noise model (47.6%), followed by WN+PL.
3. When studying the influence of hydrological loading on the velocity of the stations and its uncertainty, it is found that the hydrological loading has little effect on the velocity uncertainty of the IGS long-term coordinate series, but it will affect the velocity of the stations, especially in its vertical direction, its velocity influence value can reach up to 1.8 mm. Therefore, when estimating the vertical velocity, the influence of hydrological loading must be considered.
4. Different complex noise models will affect the station velocity and velocity uncertainty. For WN+FN, 85% of the stations velocity influence value are within 1 mm. For velocity uncertainty, its influence in the vertical direction is more obvious, and the maximum value can be close to 2 mm. When analyzing the WN + RW noise model combination, its impact on the GNSS time series velocity uncertainty is quite different from other combined noise models. Therefore, when studying the influence of the world's optimal noise model on time series, this will be the next step worth pondering.
5. The hydrological loading will have a certain impact on the annual and semi-annual amplitudes of global IGS stations, which is mainly reflected in the vertical direction. The magnitude of the impact varies from station to station, mainly related to the environment around the station. The annual amplitude motion caused is greater than the half-annual amplitude motion. After adding hydrological loading correction, it cannot completely reduce the annual and half-annual amplitude movement of the station. The amplitude of some stations not only does not decrease, but shows an increasing trend. When considering the influence of different noise models on the annual and half-annual amplitude difference before and after hydrological loading correction, it is found that the influence value is very small and can be ignored.

However, there are currently many research institutions that have calculated the hydrological loading. In this article, only the hydrological loading products of GFZ are discussed, and other hydrological loading products are not explained. Therefore, in future research, multiple hydrological loading models should be compared and analyzed, and the influence of hydrological loading on the GNSS coordinates time series should be studied at a deeper level.

Author Contributions: Conceptualization, Y.H. and G.N.; methodology, Y.H. and S.W.; software, Y.H.; validation, G.N. and H.L.; formal analysis, H.L.; resources, G.N.; writing—original draft preparation, Y.H.; writing—review and editing, S.W.; visualization, H.L.; supervision, G.N.; funding acquisition, G.N. All authors have read and agreed to the published version of the manuscript.

Funding: This work is supported by the National Natural Science Foundation of China (Grant No. 41074023). National key Research and Development Scheme Strategic International Coopearation in Science and Technology Innovtion Program, and the funding number is No.2018YFE0206500.

Acknowledgments: The 671 long-time coordinate series calculated in this article from 2000 to 2021 were provided by SOPAC (ftp://garner.ucsd.edu/archive/garner/timeseries/measures/ats/). Hydrological loading data were provided by the GFZ agency from Germany (http://rz-vm115.gfz-potsdam.de:8080/ repository). The Hector software (http://segal.ubi.pt/hector/) was used to calculate and analyze GNSS long-term coordinate series. We used GMT software and Origin software to plot the calculation results. We express our heartfelt thanks to these organizations and software providers.

Conflicts of Interest: The authors declare no conflict of interest.

Abbreviations

AR1	First-order autogressive noise
AIC	Akaike Information Criterion
BIC	Bayesian Information Criterion
BPPL	band pass+powerlaw noise
CMONOC	Crustal Movement Observation Network of China
CME	common mode error
CORS	Continuously Operating Reference Stations
CGCS2000	China Geodetic Coordinate System 2000
CF	center of figure
EOST	School and Observatory of Earth Sciences
FN	flicker noise
FOGMRW	first order Gauss–Markov+random walk noise
GNSS	Global Navigation Satellite System
GGM	generalized Gauss–Markov noise
GFZ	German Research Centre for Geosciences
HYDL	hydrological loading
IGS	International GNSS Service
IMLS	International Mass Loading Service
LSDM	Land Surface Discharge Model
NASA	National Aeronautics and Space Administration
PL	power-law noise
RW	random walk noise
SOPAC	Scripps Orbit and Permanent Array Center
VW	variable white noise
WGS84	World Geodetic System 1984
WN	white noise

References

1. Blewitt, G.; Lavallée, D.; Clarke, P.; Nurutdinov, K. A new global mode of Earth deformation: Seasonal cycle detected. *Science* **2001**, *294*, 2342–2345. [CrossRef] [PubMed]
2. Williams, S.D.P.; Bock, Y.; Fang, P.; Jamason, P.; Nikolaidis, R.M.; Prawirodirdjo, L.; Miller, M.; Johnson, D.J. Error analysis of continuous GPS position time series. *J. Geophys. Res. Solid Earth* **2004**, *109*, B03412. [CrossRef]
3. Langbein, J. Noise in GPS Displacement Measurements from Southern California and Southern Nevada. *J. Geophys. Res. Solid Earth* **2008**, *113*, 1–12. [CrossRef]
4. Van Dam, T.; Altamimi, Z.; Collilieux, X.; Ray, J. Topographically induced height errors in predicted atmospheric loading effects. *J. Geophys. Res. Solid Earth* **2010**, *115*, B07415. [CrossRef]
5. Tian, Y.; Shen, Z. Extracting the regional common-mode component of GPS station position time series from dense continuous network. *J. Geophys. Res. Solid Earth* **2016**, *121*, 1080–1096. [CrossRef]
6. Gu, Y.; Yuan, L.; Fan, D.; You, W.; Su, Y. Seasonal crustal vertical deformation induced by environmental mass loading in mainland China derived from GPS, GRACE and surface loading models. *Adv. Space Res.* **2017**, *59*, 88–102. [CrossRef]
7. Chanard, K.; Fleitout, L.; Calais, E.; Rebischung, P.; Avouac, J. Toward a global horizontal and vertical elastic load deformation model derived from GRACE and GNSS station position time series. *J. Geophys. Res. Solid Earth* **2018**, *123*, 3225–3237. [CrossRef]

8. Klos, A.; Bos, M.S.; Bogusz, J. Detecting time-varying seasonal signal in gps position time series with different noise levels. *GPS Solut.* **2018**, *22*, 21. [CrossRef]
9. Klos, A.; Olivares, G.; Teferle, F.N.; Hunegnaw, A.; Bogusz, J. On the combined effect of periodic signals and colored noise on velocity uncertainties. *GPS Solut.* **2018**, *22*, 1–13. [CrossRef]
10. Klos, A.; Gruszczynska, M.; Bos, M.S.; Boy, J.-P.; Bogusz, J. Estimates of vertical errors for IGS ITRF2014 stations by applying the improved singular spectrum analysis method and environmental loading models. *Pure Appl. Geophys.* **2018**, *175*, 1823–1840. [CrossRef]
11. Yuan, P.; Li, Z.; Jiang, W.; Ma, Y.; Chen, W.; Sneeuw, N. Influences of environmental loading corrections on the nonlinear variations and velocity uncertainties for the reprocessed global positioning system height time series of the crustal movement observation network of China. *Remote Sens.* **2018**, *10*, 958. [CrossRef]
12. Ming, F.; Yang, Y.; Zeng, A.; Zhao, B. Spatiotemporal filtering for regional GPS network in China using independent component analysis. *J. Geod.* **2017**, *91*, 419–440. [CrossRef]
13. Ma, C.; Li, F.; Zhang, S.-K.; Lei, J.-T.; E, D.-C.; Hao, W.-F.; Zhang, Q.-C. The coordinate time series analysis of continuous GPS stations in the Antarctic Peninsula with consideration of common mode error. *Chin. J. Geophys.* **2016**, *59*, 2783–2795.
14. Wang, M.; Shen, Z.; Dong, D. Effects of non-tectonic crustal deformation on continuous GPS position time series and correction to them. *Chin. J. Geophys.* **2005**, *48*, 1045–1052. [CrossRef]
15. Ray, J.; Altamimi, Z.; Collilieux, X.; Van Dam, T. Anomalous harmonics in the spectra of GPS position estimates. *GPS Solut.* **2008**, *12*, 55–64. [CrossRef]
16. Beavan, R.J. Noise properties of continuous GPS data from concrete pillar geodetic monuments in New Zealand, and comparison with data from deep drilled braced monuments. *J. Geophys. Res. Solid Earth* **2005**, *110*, B8. [CrossRef]
17. Johnson, H.O.; Agnew, D.C. Monument motion and measurements of crustal velocities. *Geophys. Res. Lett.* **1995**, *22*, 2905–2908. [CrossRef]
18. Mao, A.; Harrison, C.G.A.; Dlxon, T.H. Noise in GPS Coordinate Time Series. *J. Geophys. Res. Atmos.* **1999**, *104*, 2797–2816. [CrossRef]
19. Williams, S.D.P. The Effect of Coloured Noise on the Uncertainties of Rates Estimated from Geodetic Time Series. *J. Geod.* **2003**, *76*, 483–494. [CrossRef]
20. He, X.; Bos, M.S.; MontilletiD, J.-P.; Fernandes, R. Investigation of the noise properties at low frequencies in long GNSS time series. *J. Geod.* **2019**, *93*, 1271–1282. [CrossRef]
21. Wang, W.; Qiao, X.; Wang, D.; Chen, Z.; Yu, P.; Lin, M.; Chen, W. Spatiotemporal noise in GPS position time-series from Crustal Movement Observation Network of China. *Geophys. J. Int.* **2019**, *216*, 1560–1577. [CrossRef]
22. Yuan, L.-G.; Ding, X.-L.; Chen, W.; Kwok, S.; Chan, S.-B.; Hung, P.-S.; Chau, K.-T. Characteristics of Daily Position Time Series from Hong Kong GPS Fiducial Network. *Chin. J. Geophys.* **2008**, *51*, 1372–1384. [CrossRef]
23. Jiang, Z.; Zhang, P.; Bi, J.; Liu, L. Velocity Estimation on the Colored Noise Properties of CORS Network in China Based on the CGCS2000 Frame. *Acta Geod. Cartogr. Sin.* **2010**, *39*, 355–363.
24. Farrell, W.E. Deformation of the earth by surface loads. *Rev. Geophys.* **1972**, *10*, 761–797. [CrossRef]
25. Li, C.; Huang, S.; Chen, X.; Van Dam, T.; Fok, H.S.; Zhao, Q.; Wu, W.; Wang, X. Quantitative Evaluation of Environmental Loading Induced Displacement Products for Correcting GNSS Time Series in CMONOC. *Remote Sens.* **2020**, *12*, 594. [CrossRef]
26. Gong, G.; Hua, X.; He, X.; Shu, Y.; Ma, M. Analysis of Regional Characteristics of Environment Load Effect in GPS Coordinate Time Series. *J. Geod. Geodyn.* **2017**, *37*, 961–967.
27. Collilieux, X.; Van Dam, T.; Ray, J.; Coulot, D.; Métivier, L.; Altamimi, Z. Strategies to mitigate aliasing of loading signals while estimating GPS frame parameters. *J. Geod.* **2012**, *86*, 1–14. [CrossRef]
28. Amiri-Simkooei, A.R.; Tiberius, C.C.J.M.; Teunissen, P.J.G. Assessment of noise in GPS coordinate time series: Methodology and results. *J. Geophys. Res. Atmos.* **2007**, *112*. [CrossRef]
29. Bogusz, J.; Klos, A. On the significance of periodic signals in noise analysis of GPS station coordinates time series. *GPS Solut.* **2016**, *20*, 655–664. [CrossRef]
30. Van Dam, T.M.; Wanr, J. Modeling environment loading effects: A review. *Phys. Chem. Earth* **1998**, *23*, 1077–1087. [CrossRef]
31. Liao, H.H.; Zhong, M.; Zhou, X.H. Climate driven annual vertical deformation of the solid earth calculated from GRACE. *Chin. J. Geophys.* **2010**, *53*, 321–328. [CrossRef]
32. Jiang, W.; Xia, C.; Li, Z.; Guo, Q.; Zhang, S. Analysis of environmental loading effects on regional GPS coordinate time series. *Acta Geod. Cartogr. Sin.* **2014**, *43*, 1217–1223.
33. Wu, S.; Nie, G.; Meng, X.; Liu, J.; He, Y.; Xue, C.; Li, H. Comparative Analysis of the Effect of the Loading Series from GFZ and EOST on Long-Term GPS Height Time Series. *Remote Sens.* **2020**, *12*, 2822. [CrossRef]
34. Dill, R.; Dobslaw, H. Numerical simulations of global scale high-resolution hydrological crustal deformations. *J. Geophys. Res. Solid Earth* **2013**, *118*, 5008–5017. [CrossRef]
35. Dill, R. Hydrological model LSDM for operational Earth rotation and gravity field variations. *GFZ* **2008**. [CrossRef]
36. Hagemann, S.; Dümenil, L. A parametrization of the lateral waterflow for the global scale. *Clim. Dyn.* **1998**, *14*, 17–31. [CrossRef]
37. Bevis, M.; Brown, A. Trajectory models and reference frames for crustal motion geodesy. *J. Geod.* **2014**, *88*, 283–311. [CrossRef]
38. Bos, M.S.; Fernandes, R.M.S.; Williams, S.D.P.; Bastos, L. Fast error analysis of continuous GNSS observations with missing data. *J. Geod.* **2013**, *87*, 351–360. [CrossRef]

39. Akaike, H.T. A new look at the statistical model identification. *IEEE Trans. Autom. Control* **1974**, *19*, 716–723. [CrossRef]
40. Schwarz, G. Estimating the Dimension of a Model. *Ann. Stat.* **1978**, *6*, 31–38. [CrossRef]
41. Bos, M.S.; Fernandes, R.M.S. Hector User Manual Version 1.6. 2016. Available online: http://segal.ubi.pt/hector/manual_1.7.2.pdf (accessed on 22 January 2021).
42. Sun, H.P.; Ducarme, B.; Dehant, V. Effect of the Atmospheric Pressure on Surface Displacement. *J. Geod.* **1995**, *70*, 131–139. [CrossRef]
43. Van Dam, T.; Collilieux, X.; Wuite, J.; Altamimi, Z.; Ray, J. Nontidal ocean loading: Amplitudes and potential effects in GPS height time series. *J. Geod.* **2012**, *86*, 1043–1057. [CrossRef]
44. Li, Z.; Yue, J.; Li, W.; Lu, D.; Li, X. A comparison of hydrological deformation using GPS and global hydrological model for the Eurasian plate. *Adv. Space Res.* **2017**, *60*, 587–596. [CrossRef]
45. Wu, S.; Nie, G.; Liu, J.; Xue, C.; Wang, J.; Li, H.; Peng, F. Analysis of deterministic and stochastic models of GPS stations in the crustal movement observation network of China. *Adv. Space Res.* **2019**, *64*, 335–351. [CrossRef]
46. Zhang, J.; Bock, Y.; Johnson, H.; Fang, P.; Williams, S.; Genrich, J.; Wdowinski, S.; Behr, J. Southern Californie Permanent GPS Geodetic Array: Error Analysis of Daily Position Estimates and Site Velocities. *J. Geophys. Res. Solid Earth* **1997**, *102*, 18035–18055. [CrossRef]
47. Huang, L. Noise properties in time series of coordinate component at GPS fiducial stations. *J. Geod. Geodyn.* **2006**, *26*, 31–33.

remote sensing

MDPI

Article

An Accurate Geocoding Method for GB-SAR Images Based on Solution Space Search and Its Application in Landslide Monitoring

Jialun Cai [1], Hongguo Jia [1,2,*], Guoxiang Liu [1,2], Bo Zhang [1], Qiao Liu [3], Yin Fu [1], Xiaowen Wang [1,2] and Rui Zhang [1,2]

1 Faculty of Geosciences and Environmental Engineering, Southwest Jiaotong University, Chengdu 611756, China; caijialun@my.swjtu.edu.cn (J.C.); rsgxliu@swjtu.edu.cn (G.L.); rsbozh@my.swjtu.edu.cn (B.Z.); rsyinfu@my.swjtu.edu.cn (Y.F.); insarwxw@swjtu.edu.cn (X.W.); zhangrui@swjtu.edu.cn (R.Z.)
2 State-Province Joint Engineering Laboratory of Spatial Information Technology of High-Speed Rail Safety, Southwest Jiaotong University, Chengdu 611756, China
3 Institute of Mountain Hazards and Environment, Chinese Academy of Sciences, Chengdu 610041, China; liuqiao@imde.ac.cn
* Correspondence: rsjia@swjtu.edu.cn

Citation: Cai, J.; Jia, H.; Liu, G.; Zhang, B.; Liu, Q.; Fu, Y.; Wang, X.; Zhang, R. An Accurate Geocoding Method for GB-SAR Images Based on Solution Space Search and Its Application in Landslide Monitoring. *Remote Sens.* **2021**, *13*, 832. https://doi.org/10.3390/rs13050832

Academic Editors: Alex Hay-Man Ng, Linlin Ge, Hsing-Chung Chang and Zheyuan Du

Received: 1 February 2021
Accepted: 22 February 2021
Published: 24 February 2021

Abstract: Although ground-based synthetic aperture radar (GB-SAR) interferometry has a very high precision with respect to deformation monitoring, it is difficult to match the fan-shaped grid coordinates with the local topography in the geographical space because of the slant range projection imaging mode of the radar. To accurately identify the deformation target and its position, high-accuracy geocoding of the GB-SAR images must be performed to transform them from the two-dimensional plane coordinate system to the three-dimensional (3D) local coordinate system. To overcome difficulties of traditional methods with respect to the selection of control points in GB-SAR images in a complex scattering environment, a high-resolution digital surface model obtained by unmanned aerial vehicle (UAV) aerial photogrammetry was used to establish a high-accuracy GB-SAR coordinate transformation model. An accurate GB-SAR image geocoding method based on solution space search was proposed. Based on this method, three modules are used for geocoding: framework for the unification of coordinate elements, transformation model, and solution space search of the minimum Euclidean distance. By applying this method to the Laoguanjingtai landslide monitoring experiment on Hailuogou Glacier, a subpixel geocoding accuracy was realized. The effectiveness and accuracy of the proposed method were verified by contrastive analysis and error assessment. The method proposed in this study can be applied for accurate 3D interpretation and analysis of the spatiotemporal characteristic in GB-SAR deformation monitoring and should be popularized.

Keywords: ground-based SAR; UAV aerial photogrammetry; geocoding; solution space search; landslide monitoring

1. Introduction

Ground-based synthetic aperture radar (GB-SAR) interferometry is an active microwave remote sensing technology that was developed in the past two decades. GB-SAR interferometry can be used to image the radiation field of radar antenna with high spatiotemporal resolution, yielding single look complex (SLC) images of the target area in the two-dimensional (2D) plane radar coordinate system [1,2]. The emergence of GB-SAR interferometry technology is an important complementarity with spaceborne or airborne interferometric SAR (InSAR) technology, which greatly promotes the application of InSAR technology in the field of deformation monitoring [3]. Owing to the satellite revisit period and side looking angle of the sensor, monitoring with high spatiotemporal resolution on the surface of the local small area is difficult to achieve using traditional spaceborne SAR [3,4]. Compared with spaceborne SAR, GB-SAR can be used to establish specific geometry observation scene in accordance with the monitoring target. A minute- and submeter-level spatial–temporal monitoring resolution can be achieved by using a flexible baseline based on the monitoring needs [5–7]. In addition, a submillimeter-level precision can be reached with respect to the deformation measurement by interferometry, which compensates for the monitoring deficiencies of spaceborne InSAR and traditional geodetic monitoring technology, such as deficiencies in the spatiotemporal resolution and accuracy [2,8–10]. After years of development and practice, GB-SAR has been widely used for monitoring the motion of landslides [11–17], dams [18,19], side slopes [8,20,21], open-pit mining areas [22,23], and glaciers [24–28]. This method has good application prospects and research value in the industry.

As a new technology based on microwave interferometry, GB-SAR is mainly based on the stepped-frequency continuous wave (SFCW) [5,24], synthetic aperture and differential interferometry to obtain 2D high-precision displacement maps and high-precision deformation results. Specifically, GB-SAR uses SFCW technology to synthesize signals to improve range resolution and keep a high average transmit power at the same time. It uses synthetic aperture technology to improve azimuth resolution, and uses InSAR technology to compare the phase information of the target at different times, so as to calculate a small displacement change and obtain high accuracy deformation results. The data collected by GB-SAR system is a 2D image including range and azimuth resolution. Each pixel includes amplitude and phase information. The GB-SAR system can distinguish different monitoring targets mainly based on the slant range from the target to the radar center and the deflection angle from the target to the center of the radar antenna beam [29]. In contrast to the orthographic projection of traditional topographic maps and central projection used in photogrammetry, slant range projection in the radar 2D plane coordinate system is used as imaging mode of the GB-SAR. The spaceborne SAR satellite satisfies the far-field approximation conditions of the radar antenna, which can be regarded as SAR image data in a regular grid [30]. Due to the installation mode and antenna size limitations, the imaging distance of GB-SAR is relatively short (maximum distance: 5–10 km), and the target area is mainly in the near field of the radar antenna radiation field, which does not satisfy the far-field approximation conditions [3,11]. Therefore, a special fan-shaped grid coordinate system is used for GB-SAR images.

As the imaging methods and spatial geometric relations vary, the method used for the transformation of the GB-SAR image coordinate system significantly differs from traditional methods. Therefore, the most important task in GB-SAR image data processing and interpretation is to determine the actual position of the deformation according to the 2D plane coordinates of the radar. To address the gap with the real three-dimensional (3D) deformable body, researchers should directly interpret the 2D radar image or interferogram in the imaging mode based on their experience. Otherwise, recognition errors with respect to the deformation target or area could occur [19]. To correctly identify the deformation target and position and compare them with the results of other monitoring technologies or carry out fusion analysis, it is necessary to establish a high-accuracy coordinate system transformation model to accurately geocode the radar images in the 3D space system.

Methods that are commonly used for the transformation from the 2D plane coordinate system to the 3D local coordinate system of the GB-SAR image mainly include parameter transformation and coordinate coding [31–34]. Parameter transformation mainly includes direct transformation and similar transformation. However, owing to the specific geometric projection method of GB-SAR imaging, direct transformation will cause a large plane coordinate offset error when the elevation angle from the target to the radar center is large [29]. Based on similar transformation, the transformation parameters are calculated by using a limited number of reference points. A certain transformation accuracy can be achieved in the area with a small change in the terrain slope, but it is difficult to achieve a holistic high accuracy because of the imprecise transformation model [34]. The coordinate coding method is relatively rigorous, but it also has several deficiencies. On one hand, it strongly relies on the accuracy and integrity of external auxiliary data such as the digital surface model (DSM) and digital elevation model (DEM). If 3D auxiliary data in the monitoring area are missing, it is difficult to match the 3D information based on coordinate coding using the corresponding pixels [35]. On the other hand, the accuracy of the coordinate coding method depends on the selection of control points and the quality of the positioning. In practical work, artificial corner reflectors (CRs) are generally installed as control points and their positions in the 3D coordinate system are measured [9]. However, it is difficult to distinguish these control points in areas with dense rocks or buildings with strong and complex scattering of ground features, such as glaciers covered by moraines, landslides with surface debris, and mining areas with large ore accumulation. In addition, if the control point is far from the radar center, the azimuth resolution of the pixel increases, and also the side lobe plays a role. Therefore, it is difficult to accurately locate the positions of the CRs in the radar image. Thus, accurate geocoding based on GB-SAR cannot be realized in practice because of the shortcomings of the coordinate transformation methods.

Therefore, in this study, unmanned aerial vehicle (UAV) photogrammetry data, which are widely used in the surveying and mapping industry, were utilized for GB-SAR coordinate transformation as the first consideration to realize high-accuracy geocoding. The digital orthophoto map (DOM) and DSM that were obtained based on the use of UAV aerial photogrammetry data have a high precision and integrity, and thus, effectively compensate for the lack of 3D data based on visual field limitations with respect to ground-based 3D laser scanning and close-range photogrammetry. In addition, the ground photo control points (PCPs) laid out for UAV photogrammetry can be combined with the GB-SAR deployment location, and the coordinates of the points and radar rail azimuth angle can be accurately measured by using the global navigation satellite system (GNSS) or total station. This method can effectively avoid the defects of inaccurate coordinates measurement by using CRs as ground control points (GCPs). With the help of external auxiliary data, an accurate GB-SAR geocoding model was established in this study based on searching the solution space. This model was applied for the monitoring of the Laoguanjingtai (LGJT) landslide that occurred on the Hailuogou Glacier, China. A subpixel level geocoding accuracy was achieved, and the effectiveness and accuracy of the model were verified. The method proposed in this paper can be applied for accurate 3D interpretation and spatiotemporal analysis in GB-SAR deformation monitoring and should be popularized.

2. Materials and Methods

2.1. Dataset

The following data were used in this study.

1. GB-SAR original observation dataset. Obtained by GB-SAR by continuous monitoring under specific observation geometry conditions. The azimuth resolution was 2–5 mrad, the maximum range resolution was 0.5 m, and the deformation monitoring accuracy was at the submillimeter level. This was the main dataset used for geocoding in this paper.
2. Ground control survey dataset. The azimuth angle of the GB-SAR guide rail was measured by the GNSS or total station and the radar center was combined with ground PCPs in the survey area. The coordinate and angle measurement accuracies reached the centimeter and 0.5 s level, respectively, fully meeting the requirements of GB-SAR coordinate transformation [33,36]. After the measurement, the location point of GB-SAR became the link between the radar and 3D coordinate systems. This dataset was mainly used for the transformation between the two coordinate systems.
3. External ancillary 3D information. These data were obtained from aerial photogrammetry or light detection and ranging (LiDAR) carried on UAV. Due to the data gap caused by the shadow of sight, several transformation defects occur during ground-based 3D laser scanning. In this study, UAV aerial photogrammetry technology was used to collect high-resolution images of the monitored area. The PCPs were

arranged on the ground to obtain external auxiliary data such as DSM, DOM, and 3D real-scene model (five-lens oblique photogrammetry) using a local unified coordinate system. These data were mainly used to realize the solution space search for geocoding transformation.

The related parameters of the above datasets are shown in Table 1.

Table 1. Available equipment and related indicators of datasets.

Dataset	Available Equipment	Outputs	Resolutions	Characteristics
GB-SAR observation dataset	GB-SAR	SLC/deformation	Azimuth: 2–5 mrad; Range: 0.5 m	The deformation accuracy can reach submillimeter level
Ground control survey dataset	GNSS	Coordinates	10 mm ± 1 ppm	Convenient measurement and uniform global accuracy
	Total station	Coordinates/azimuth	Range: 2 mm + 2 ppm Angle: 0.5″–2″	High precision, measurement needs intervisibility
External ancillary 3D information	UAV aerial photogrammetry	DSM/DEM/DOM	≥5 cm	With high precision and rich ground texture
	Airborne LiDAR	DSM/DEM	≥5 cm	With high precision and less ground texture
	Ground 3D laser scan	DSM/DEM	≥5 cm	With shadows and inconvenience of interpretation

2.2. Solution Space Search Geocoding Model

For accurate geocoding from the 2D radar plane coordinate system of the GB-SAR to the 3D local coordinate system, accurate deployment data of the GB-SAR are required, including the center coordinates of the radar sensor and azimuth angle of the radar guide rail, which are used as a link to realize the unification of the two coordinate systems. For geocoding, with the help of high-precision external elevation data such as DSM, DEM, and 3D terrain point cloud data, the Euclidean distance between the radar center and the ground point is calculated by using MATLAB software programming, and the minimum Euclidean distance is searched and matched with the corresponding azimuth distance in the radar image coordinate system. The geocoding method used in this study mainly included the following parts.

2.2.1. Unified Coordinate System Frame of Each Element

To realize high-accuracy GB-SAR image geocoding, the position and orientation of each coordinate element in the target coordinate system should be accurately determined using the GNSS or total station. The element survey mainly includes the measurement of PCPs and the precise measurement of the GB-SAR location point. Based on the specific requirements of actual work, the target coordinate system also has a certain particularity and independence. By measuring the PCPs, the UAV photogrammetry results can be unified to the target coordinate system and can be used as the basic supporting data for subsequent coordinate transformation. Further, based on the joint measurement with GCPs, the coordinates of the GB-SAR center point and azimuth angle of the radar guide rail can be accurately measured. The azimuth angle can also be calculated by measuring the center coordinates of the positioning screw of the observation pier when the GB-SAR system is set up. After the above-mentioned elements are unified in the target coordinate system, the coordinate transformation model can be established.

2.2.2. Coordinate Transformation Model

Owing to the fan-shaped projection mode of GB-SAR imaging, its range resolution is fixed and the azimuth resolution decreases with increasing observation distance. In addition, the GB-SAR projection mode is slant range projection, therefore, there is a certain angle between the imaging projection plane and horizontal plane, which leads to a scale inconsistency of the same ground feature in the two coordinate systems. If the terrain elevation significantly changes, the coordinate offset will be aggravated [1]. As shown in Figure 1, in the radar coordinate system G-XY, "a," "b," and "c" are three ground feature points, which have the same distance from the radar center. However, in the 3D local coordinate system O-ENZ, the radar line of sight (LOS) angle and ground feature elevation differ and the coordinate projection can be Na, Nb, and Nc, respectively. The figure shows that the LOS to the horizontal direction is closer and the coordinate offset is smaller. The larger the elevation difference between the ground object and radar center is, the larger is the projection deviation. This is the error source of the parameter transformation method, which explains why it is difficult to accurately transform the coordinates.

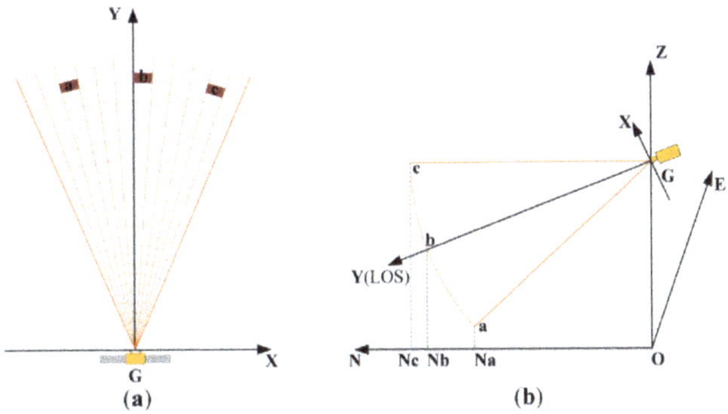

Figure 1. Schematic diagram of ground-based synthetic aperture radar (GB-SAR) coordinate offset. (**a**) Three feature points with equal distance from the center of GB-SAR in the radar coordinate system; (**b**) the north direction projection offset of the three pixels in the three-dimensional (3D) local coordinate system.

After determining the error source, the basic principle of radar imaging is used as the starting point, that is, independent of the coordinate system, the Euclidean distance from the feature point to the GB-SAR center is equal. Based on the assumption that the coordinate of feature point P in the radar coordinate system G-XY is $P(X_P, Y_P)$, the distance can be expressed as follows:

$$R_{GP} = \sqrt{(X_P - X_G)^2 + (Y_P - Y_G)^2}, \tag{1}$$

$$\theta = \arctan\frac{X_P - X_G}{Y_P - Y_G}, \tag{2}$$

where R_{GP} represents the distance from point P to the radar center point G and θ is the deflection angle from point P to the radar center line direction. Due to the limited scanning angle of GB-SAR, the angle is generally within the range of $\pm 50°$. Under real conditions, the θ range can be limited to reduce the calculation amount and improve the transformation efficiency (e.g., $\theta \in [-50°, 50°]$).

Based on the assumption that the coordinate of the ground feature point P in the 3D local coordinate system O-ENZ is $P(E_P, N_P, Z_P)\theta$ and the coordinate of the GB-SAR center point G is $G(E_G, N_G, Z_G)$, the feature point P can be expressed as follows:

$$S_{GP} = \sqrt{(E_P - E_G)^2 + (N_P - N_G)^2 + (Z_P - Z_G)^2}, \tag{3}$$

$$\alpha = \arctan\frac{E_P - E_G}{N_P - N_G}, \tag{4}$$

where S_{GP} is the Euclidean distance from the ground feature point P to the radar center point G and α is the horizontal plane projection quadrant angle of GP.

After establishing the corresponding distance and angle information of the ground feature points in two coordinate systems, it is necessary to further match the angle relationship based on the equal distance, establish a set of connection equations, and provide a mathematical foundation for the searching and matching of the distance solution space. As the orientation of the radar LOS (Y-axis direction) differs from the north direction (N-axis direction) of the 3D local coordinate system, it is necessary to determine the relationships among the coordinate axes when setting up the GB-SAR system. Figure 2 shows the relationships between the 2D image coordinate system and local plane coordinate system of GB-SAR for different orientations. The parameter β is the horizontal rotation angle of the two coordinate systems, which can be calculated by using common points. During actual monitoring, the CRs are usually arranged and measured in the 3D coordinate system and the corresponding points in the radar images are identified and calculated [19]. Although the method is easy to operate, it has several disadvantages such as a poor recognition of the reflected echo, low resolution, and strong dependence on manual interpretation. Therefore, it is difficult to accurately calculate the horizontal angle.

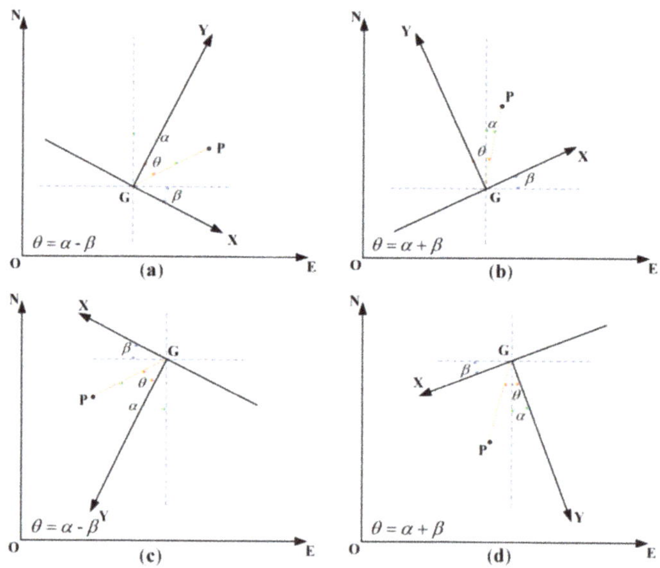

Figure 2. Diagram of the relationship between the angle of the coordinate axis and orientation of the GB-SAR. The Y-axis represents the line of sight direction of the GB-SAR. (**a**,**c**) Relational expression $\theta = \alpha - \beta$ when the line of sight of the GB-SAR is directed toward the first and third quadrant, respectively. (**b**,**d**) Relational expression $\theta = \alpha + \beta$ when the line of sight of the GB-SAR is directed toward the second and fourth quadrant, respectively. The horizontal rotation angle of axes β is scalar, θ and α are vectors, and the values should be positive and negative.

In this paper, precise engineering measurement was used as global control measurement to measure the GB-SAR center and azimuth of the guide rail. The results were combined with the UAV photogrammetry PCPs or regional control network in the whole region. The coordinates of the GB-SAR center in the 3D local coordinate system and the horizontal angle β of the two coordinate axes can be accurately solved. The coordinate precision reached the mm level and the angle measurement precision was 0.5″, meeting the requirements of an accurate coordinate transformation of GB-SAR.

Based on determining the above-mentioned axes relationships, the following coordinate transformation equations can be obtained.

$$\begin{cases} R_{GP} = S_{GP} \\ \theta = \alpha \pm \beta \end{cases}, \quad (5)$$

That is,

$$\begin{cases} \sqrt{(X_P - X_G)^2 + (Y_P - Y_G)^2} = \sqrt{(E_P - E_G)^2 + (N_P - N_G)^2 + (Z_P - Z_G)^2} \\ \arctan\frac{X_P - X_G}{Y_P - Y_G} = \arctan\frac{E_P - E_G}{N_P - N_G} \pm \beta \end{cases}, \quad (6)$$

The overall relationship is changed into the process of obtaining the corresponding pixel $P(E_P, N_P, Z_P)$ in the local 3D coordinate system according to pixel $P(X_P, Y_P)$ in the 2D radar coordinate system. The sign "±" in the coordinate axis angle relationship equation is based on the GB-SAR orientation quadrant in the 3D local coordinate system, as shown in Figure 2.

2.2.3. Geocoding Based on Solution Space Search

It is not difficult to identify the rank defect of the above-mentioned equations, but it is difficult to solve massive transformation data using traditional methods (e.g., the full search and iterative method). With the help of computing power, we can identify the radar coordinate pixel P corresponding to local 3D coordinates by searching the solution space using the minimum Euclidean distance criterion.

The main ideas are as follows: first, the azimuth angle and range resolutions of the transformation model are set according to the actual monitoring resolution requirements of the project. As the GB-SAR resolution (range resolution: ≈0.5 m, azimuth resolution: ≈4 mrad) is lower than the DSM grid resolution (reaching the cm level) of the UAV photogrammetry results, GB-SAR is generally oversampled or the DSM is downsampled to adapt to the fixed grid resolution. After the processing, speckle noise interferences can be reduced to a certain extent, and the subsequent matching is smoother. The range set $\{S_{GP}\}$ and quadrant angle set $\{\alpha\}$ from each pixel center to the radar center in the DSM and the range set $\{R_{GP}\}$ and deflection angle set $\{\theta\}$ from each pixel to the radar origin in GB-SAR coordinate system are calculated according to the grid size. Finally, based on azimuth angle matching, the minimum Euclidian distance principle is adopted to match the distances in this direction and the pixels of the range matching in the 2D radar coordinate system are assigned to the corresponding coordinates in the local 3D coordinate system for geocoding.

In addition, to reduce the search workload and improve the matching accuracy, the search scope and range elimination threshold can be set according to the actual GB-SAR observation system and local geographic conditions. If there is no effective distance matching value in the azimuth direction, the range solution is further searched within the scope of the search threshold set in the azimuth direction of the point. The principle of the coordinate search and match method is shown in Figure 3. To accurately calculate the coordinate value $P(E_P, N_P, Z_P)$ in the local 3D coordinate system corresponding to the pixel $P(X_P, Y_P)$ in the radar image coordinate system, we can determine the 3D topographic profile of the DSM corresponding to the pixel P of the radar based on the unified coordinate frame and through the matching of angular geometric relations among β, θ, and α. The distance between the radar sensor center and DSM pixel in the topographic profile is matched sequentially, and the minimum value of the distance within the search threshold is obtained as the optimal solution.

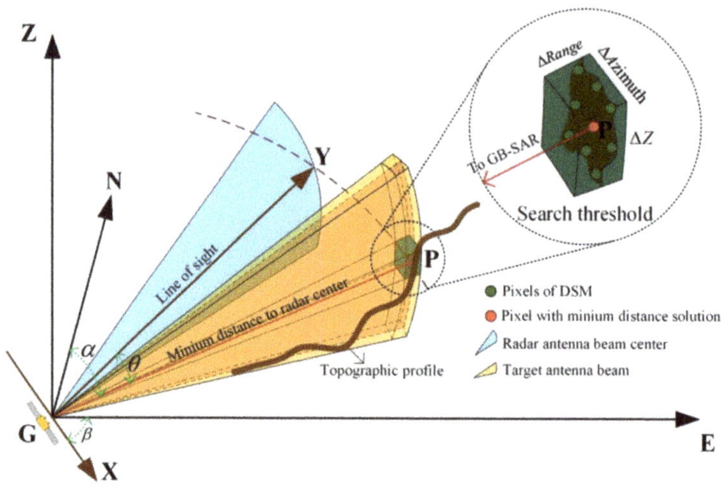

Figure 3. Schematic diagram of the solution space search. The main figure shows the geometric positional relationship of the GB-SAR in the radar 2D plane coordinate system (XY) and local 3D coordinate system (ENZ). The red line in the figure indicates the minimum Euclidean distance matching solution between the ground target point P and radar sensor center in the topographic profile in the beam direction. The inset in the upper right corner (enlarged picture in the circle) shows the schematic diagram of the solution space search threshold. The $\Delta Range$; $\Delta Azimuth$, and ΔZ are the range, azimuth, and elevation search thresholds, respectively. The search threshold can be customized according to the actual monitoring needs and the distance matching solutions beyond this scope will be eliminated.

The setting of the search threshold is shown in the inset in Figure 3. The $\Delta Range$, $\Delta Azimuth$, and ΔZ are the range, azimuth, and elevation search thresholds, respectively. The search threshold can be customized according to the actual monitoring needs and distance matching solutions beyond the scope will be eliminated. In general monitoring cases, $\Delta Range$ and $\Delta Azimuth$ should be compatible with the grid resolution of the radar or DSM to unify the order of magnitude. To achieve a comprehensive spatial search of the distance solution, ΔZ can be defaulted to be the entire topographic profile elevation range, that is, to search and match all DSM elevation points in the beam direction of the radar antenna. The pixel points corresponding to the minimum Euclidian distance from the center of the GB-SAR sensor are selected as the optimal matching points. Based on this method, all radar image pixels are searched and matched sequentially to realize the highly accurate geocoding of the GB-SAR images. The flowchart of the GB-SAR geocoding method based on solution space search is shown in Figure 4.

2.3. Method of Geocoding Accuracy Assessment

For the effectiveness of geocoding results, we used the method of contrastive interpretation to analyze and evaluate. Combined with the high-precision DOM and 3D model produced by UAV photogrammetry, we interpreted and analyzed the geocoding situation of pixels in uncoded region, obvious deformation region and invisible region, and compared it with the actual geographical environment to determine whether it is consistent with the radar imaging mechanism and the actual situation, so as to evaluate the correctness of the geocoding results and verify whether the method is effective.

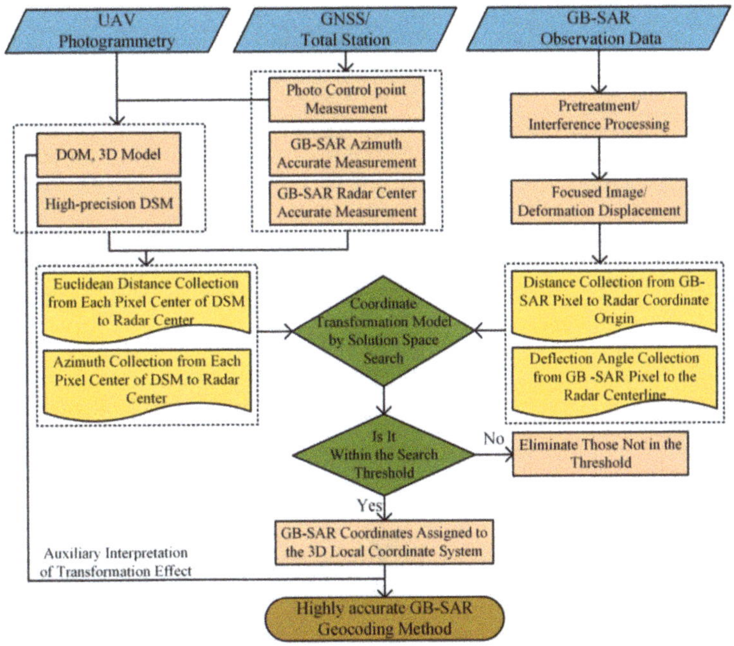

Figure 4. Flowchart of the GB-SAR geocoding method based on solution space search.

On the basis of the effectiveness of the method, we further assessed the logical accuracy of the geocoding results. The coordinate components of pixels geocoded in the 3D local coordinate system O-ENZ were reprojected into the corresponding GB-SAR image coordinate system G-XY to obtain the distribution of E, N and Z coordinate components, and check whether the plane component E and N are smoothly distributed, and whether the elevation component Z is consistent with the DSM distribution. If the plane components have a mutation or the elevation component does not match the actual DSM distribution, it means that there is a logical error in the process of solution space search and matching, otherwise, the geocoding result is logically accurate.

Finally, we calculated the geocoding error and assessed its accuracy quantitatively. The accuracy index of geocoding includes range error and azimuth error. The basic method is to make differential between the corresponding pixel P_{geo} after geocoding in the local 3D coordinate system and the pixel P in the original GB-SAR coordinate system, and calculate the range error and azimuth error of each pixel, so as to realize the quantitative assessment of the accuracy index of each pixel. Then, the accuracy indicators of all pixels were counted to assess the overall accuracy of geocoding. The error can be expressed as follows:

$$\begin{cases} Error_{Range} = S_{GP_{geo}} - R_{GP} = \sqrt{(E_{P_{geo}} - E_G)^2 + (N_{P_{geo}} - N_G)^2 + (Z_{P_{geo}} - Z_G)^2} - \sqrt{(X_P - X_G)^2 + (Y_P - Y_G)^2} \\ Error_{Azimuth} = (\alpha_{geo} \pm \beta) - \theta = \left\{ \arctan\frac{E_{P_{geo}} - E_G}{N_{P_{geo}} - N_G} \pm \beta \right\} - \arctan\frac{X_P - X_G}{Y_P - Y_G} \end{cases} \tag{7}$$

where $Error_{Range}$ is the range error, $Error_{Azimuth}$ is the azimuth error, $S_{GP_{geo}}$ is the distance from the pixel P_{geo} in the local 3D coordinate system to the radar center G, and α_{geo} is the corresponding quadrant angle of pixel P_{geo} in the 3D coordinate system.

3. Experiments

3.1. Overview of the Study Area

To verify the effectiveness and accuracy of the proposed method, we used it to monitor the LGJT landslide in the Hailuogou Glacier area, Mountain Gongga, Sichuan Province, China. Mountain Gongga is located in the southeastern Qinghai–Tibet Plateau. It is the highest peak of the Hengduan Mountains, with an elevation of 7556 m. There are 74 glaciers around the main peak. The largest glacier is the Hailuogou Glacier on the eastern slope, which is ≈13 km long, 250–1200 m wide, and covers an area of 24.7 km^2 [37]. It is the most famous glacier in this area, and also a national scenic spot in China, which is visited by many tourists throughout the year. The Hailuogou Glacier is a typical monsoon temperate glacier. Due to the influence of the monsoon and topography, the annual rainfall at the end of glacier tongue above an elevation of 3000 m is 1956 mm and the average number of annual precipitation days is 251 d. Recent observations showed that the glaciers in Hailuogou Valley are notably shrinking. From 1966 to 2008, the glacier receded by ≈728 m (average rate of ≈17.3 m/a). The average surface thinning of the glacier ablation area is 33.9 m (average thinning rate of ≈1.8 m/a). The area near the glacier tongue exhibits the fastest thinning rate of up to 4.3 m/a [24]. The study area is shown in Figure 5.

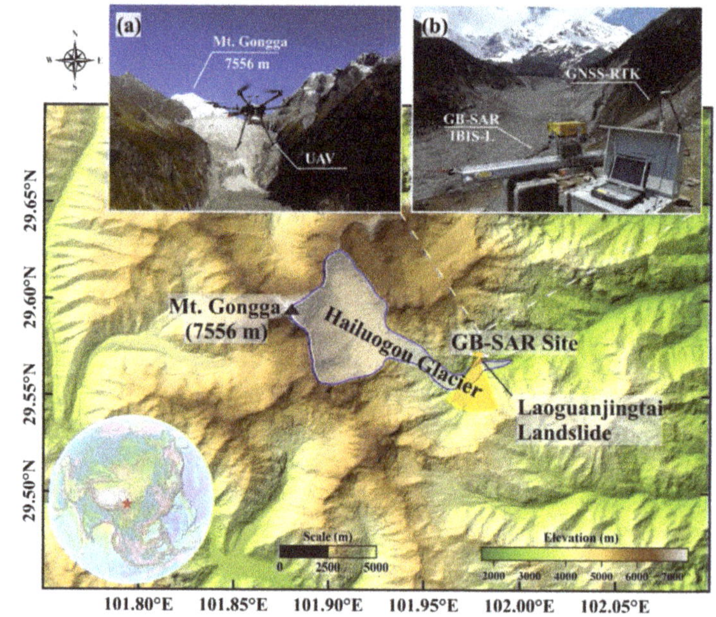

Figure 5. Main map of the study area shows the relationship between the GB-SAR monitoring station and Hailuogou Glacier and the red spot in the picture indicates the Laoguanjingtai (LGJT) landslide. (**a**) Five-lens oblique photogrammetric unmanned aerial vehicle (UAV) used to obtain external ancillary data such as the digital surface model (DSM) and digital orthophoto map (DOM). (**b**) Location of the GB-SAR monitoring station. The LGJT landslide is directly below the station and the Hailuogou Glacier can be seen in the foreground. This station has a good visibility and jointly monitors the landslide and glacier.

The degradation and thinning of the Hailuogou Glacier caused a large number of secondary landslide hazards in the surrounding mountains among which the LGJT landslide was the most representative one. The LGJT landslide occurred beneath the observation platform of the terminal moraine dam of the Hailuogou Glacier, with an area of ≈1.07 × 10^5 m^2, an elevation distribution of ≈3050–3250 m. A tourist footpath leads from the observation platform to the glacier surface. The average slope of the landslide was as high as 80%, and the drop was large. In recent years, as the thinning movement of the glacier intensified, several large slips occurred. The LGJT landslide posed a great risk to the lives of tourists in the scenic area, as shown in Figure 6. Therefore, continuous and effective landslide monitoring should be carried out to ensure the tourists' safety and normal operation of the Hailuogou scenic spot.

Figure 6. LGJT landslide on the Hailuogou Glacier. (**a**) Overview of LGJT landslide. The main scarp that formed by the separation of the rear edge of the landslide's main body from the mountain mass is very prominent. The front edge of the landslide body is tangent to the Hailuogou Glacier and local collapse is notable. (**b**) Detailed landslide surface structures can be observed such as partially damaged footpaths, the rockfall protection net, and ground fissures. (**c**) Details of the main scarp and rear edge of the main body. The altitude difference of the main scarp is >50 m, which causes the trees below to seriously incline and appearance of the unique phenomenon of a drunken forest in the landslide movement. (**d**) Ground fissures on the surface of the landslide; the fissure width generally ranges from 15 to 30 cm.

To accurately determine the movement and mechanism of the LGJT landslide, both the landslide and tangent glacier must be monitored. After the field investigation, we finally emplaced the GB-SAR system in a stable area in the northeast of the landslide, as shown in Figure 5b. The site conditions were mainly based on the following considerations: elevation of 3195 m, that is, higher than the top of the main body of the landslide; good visibility; joint monitoring of the landslide mass and glacier at its lower edge. The observation platform service station was ≈20 m away from the GB-SAR station, providing stable power and network coverage for long-term monitoring.

3.2. Dataset Acquisition

To implement the method proposed in this study, a high-accuracy and high-resolution DSM for the whole monitoring area should be obtained to accurately geocode the pixels in the GB-SAR radar coordinate system into the 3D local coordinate system. In this paper, the DJI M600 Pro six-rotor flight platform equipped with a Rainpoo DG3 aerial camera (Chengdu Rainpoo Technology Co., Ltd.; Chengdu, China; www.rainpoo.com, accessed on 1 February 2021) was used to collect aerial images covering the middle and lower parts of the Hailuogou Glacier and the slopes around the glacier, as shown in Figure 5a. The longitudinal overlap of the collected image set was 80%, the lateral overlap was 60%. The coverage area was ≈4.5 km^2, and the average ground resolution of the images was 3 cm.

Prior to the photogrammetry, we set up several PCPs on the surface of the landslide and glacier and four GCPs as datum points in the stable areas next to the GB-SAR and far away from the landslide, respectively. The 3D coordinates of the PCPs were collected using GNSS-RTK; the coordinate acquisition accuracy reached the cm level [38,39]. Finally, the PCPs and photogrammetry image set were joint calculated by GodWork v.1.0 software (www.uavmap.cn, accessed on 20 January 2021). After image preprocessing, aerial triangulation, point cloud editing, and other processes, the high-accuracy and high-resolution DSM and DOM of the target 3D coordinate system were obtained. The DSM grid resolution of this experiment was 5 cm.

In terms of GB-SAR data acquisition, the IBIS-L GB-SAR system developed by the Ingegneria Dei Sistemi (IDS) corporation of Italy was adopted as the monitoring instrument in this study. The radar system was composed of a sensor unit, linear slide rail, control unit, and power supply unit. Based on the principle of spaceborne SAR interferometry, SFCW technology was introduced to overcome low resolution of spaceborne SAR [40]. The system platform was installed on the ground and the sensor periodically obtained monitoring data based on a nearly zero spatial baseline on a fixed linear sliding rail. The sampling frequency was less than 2 min, which solved the problem of temporal and spatial decorrelation that may be caused by spaceborne SAR [7,15,41,42]. Based on the above-mentioned characteristics and the advantages of a small volume, flexible layout, and noncontact measurement method, personnel did not need to enter the target monitoring area. This ensured the safety of the personnel and prevented any disturbance of the deformation body. In conclusion, the IBIS-L GB-SAR system has a great advantage in landslide monitoring.

Before the continuous GB-SAR observation, we used the total station Leica TS02 Plus (www.leica-geosystems.com, accessed on 1 February 2021) to accurately measure the 3D coordinates of the radar guide rail and its azimuth angle, installed a reflection prism at the fixed bolts of the guide rail, and measured the 3D coordinates of the prism center. The angle measurement accuracy was 2″ and the ranging accuracy was 1.5 mm + 2 ppm, with a high measurement accuracy. The GNSS-RTK was used to measure the PCPs and GCPs on the surface of the landslide, glacier, and in the stable zone. The coordinates of the radar center point G in the 3D local coordinate system O-ENZ (independent rectangular coordinate system) were obtained based on the arithmetic average of repeated measurements. The coordinates of the radar center point G in this experiment were 498,358.612 m, 3,272,392.383 m, and 3195.448 m and those of the azimuth were 285°42′15.47″ (from left to right of the rail) and 105°42′15.47″ (from right to left of the rail). The horizontal rotation angle of two axes is 15°42′15.47″ (angle β in Figure 2).

After the center coordinates of the GB-SAR were determined, the LGJT landslide could be continuously monitored using the GB-SAR system. At a sampling interval of 10 min, a total of 4288 GB-SAR images were acquired from May 20 to June 20, 2018, with a range resolution of 0.75 m and azimuth resolution of 4.38 mrad.

3.3. Data Processing

3.3.1. DSM and DOM Processing

The magnitudes of the GB-SAR radar image and DSM resolutions significantly differ. Combined with the actual requirement of monitoring, we downsampled the DSM and resampled the grid resolution from 5 cm to 0.5 m, representing the same magnitude level as the GB-SAR image resolution. The DSM obtained after downsampling is shown in Figure 7a. The boundary of the LGJT landslide is marked by a red line. Its elevation is ≈3050–3250 m, with a large drop height. In addition, we overlayed the DOM on the DSM to determine the land coverage and obtain information about the geographic environment. After the coordinate frame was unified, the GB-SAR center point and its orientation were marked. The relative position of the radar coordinate system in the 3D local coordinate system is shown in Figure 7b.

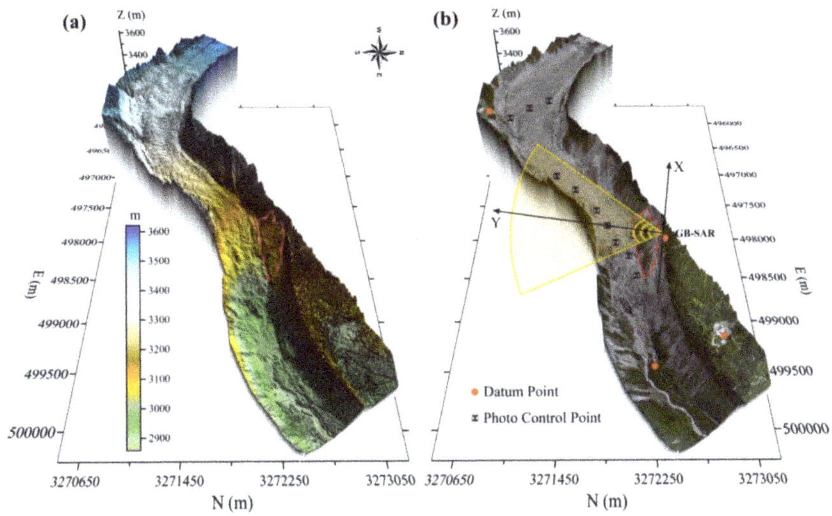

Figure 7. (**a**) DSM of the Hailuogou Glacier and its surrounding side slopes. The red dot indicates the LGJT landslide with an elevation of 3050–3250 m. (**b**) DOM superimposed on the DSM. The positional relationship between the radar 2D plane coordinate system and 3D local coordinate system is indicated.

3.3.2. GB-SAR Image Processing

Each IBIS-L acquisition produces a focused image made of an N×M complex matrix, where N and M values depend on the radar configuration and on focusing parameters specified by the user in the IBISDV v.03.04.005 software (www.idsgeoradar.com, accessed on 1 February 2021). In the data processing of this experiment, N and M correspond to the field distance of 500 and 2000 m, respectively. The average power, coherence coefficient, phase stability, and estimated signal-to-noise ratio (SNR) distribution of the radar image in the study region after processing are shown in Figure 8. The interpretation and analysis of the various quality indicators of the radar image show that the surface of the glacier has a high overall reflection intensity due to the rocks and moraines (Figure 8a). The sampling interval of 10 min ensures the high coherence in the whole region (Figure 8b). Figure 8c shows that the phase stability is better in the area with higher reflection intensity. Based on the DOM ancillary identification in this area, this stability is related to the backscattering characteristics of moraines (rocks) covered by the surface of the Hailuogou Glacier. Almost no movement can be detected at the exposed bedrock (glacial polish) on the south side of the U-shaped glacier valley and the radar backscattering is stable; therefore, the SNR is the highest, as shown in Figure 8d.

After focusing, we carried out GB-SAR data processing on the 4288 SLC images using the interference stacking method by IBISDV software. Considering the high humidity in the valley of Hailuogou Glacier terminal, especially the rainy and foggy in summer, in order to improve the accuracy of radar interferometry, the atmospheric delay was estimated and corrected by selecting the stable target point as the GCPs. This processing relied on the stability of the GCPs. If the scatterer with a large deformation was selected, a part of the low-frequency component in the monitored displacement signal would be regarded as the atmosphere and filtered. In this experiment, the stable bedrock on the south side of Hailuogou U-shaped valley with high SNR was selected as the control point to participate in the correction of atmospheric phase [43,44], and obtained the cumulative displacement map of the entire study area, as shown in Figure 9. Figure 9 shows that the area near the radar center point shows a positive value (i.e., the movement direction is far away from the radar LOS) and the glacier long-range perspective area shows a negative value (i.e., the movement direction is close to the radar

LOS) during the monitoring period, which is consistent with the downward flow of the glacier and the landslide movement away from the GB-SAR station.

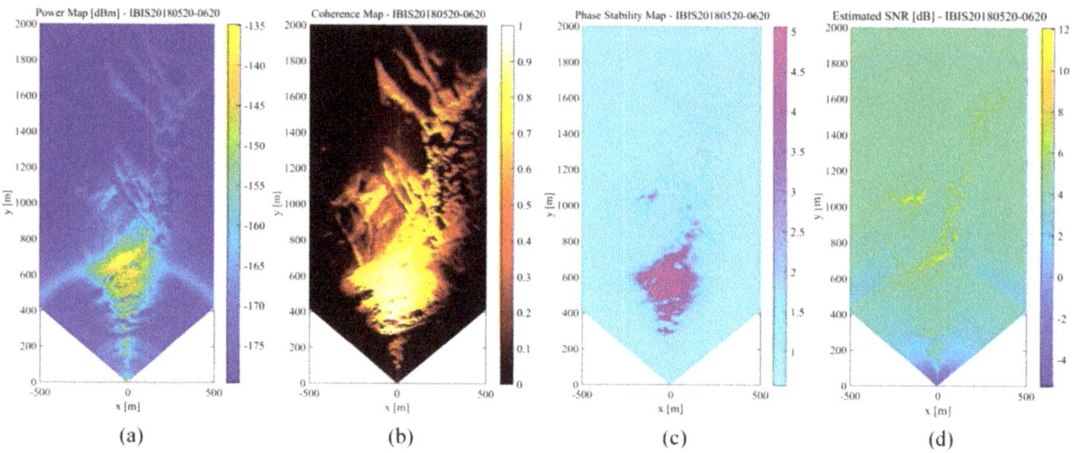

Figure 8. GB-SAR imaging quality index. (**a**) Average power map. (**b**) Coherence coefficient map. (**c**) Phase stability map. (**d**) Estimated signal-to-noise ratio (SNR) map.

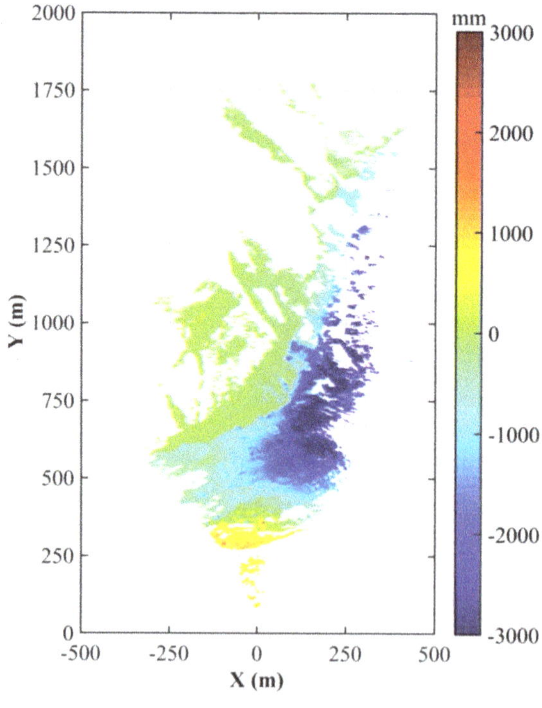

Figure 9. GB-SAR cumulative displacement distribution from 20 May 2018, 12:07:38 p.m. to 20 June 2018, 12:01:49 p.m. The negative displacement reflects movement towards the sensor along the line of sight (LOS) direction and the positive displacement represents movement away from the sensor.

4. Results

Based on the above-mentioned preliminary processing, we can roughly judge the movement trends of the landslide and glacier at the lower edge. However, we cannot precisely analyze the monitoring area, that is, we cannot effectively determine the specific point position corresponding to each pixel in the radar image. It is difficult to distinguish the boundary between the landslide and glacier and analyze their motion coupling relationship. Therefore, accurate geocoding based on the method proposed in this study is required.

As mentioned above, using MATLAB software programming based on the Formula (1)–(4), we calculated the range set $\{S_{GP}\}$ and quadrant angle set $\{\alpha\}$ from each pixel center to the radar center in the DSM and the range set $\{R_{GP}\}$ and deflection angle set $\{\theta\}$ from each pixel to the radar origin in the GB-SAR coordinate system according to the grid size. Based on azimuth angle matching, the minimum Euclidian distance principle was adopted to sequentially match the distances in this direction. The pixels of the range matching in the radar 2D coordinate system were assigned to the corresponding coordinates in the local 3D coordinate system for geocoding. In addition, we reduced the workload of the solution space search and improved the matching accuracy by setting search range and distance elimination thresholds. In this study, the DSM resolution was 0.5 m. Therefore, we set the angle matching accuracy to $\pm$ 0.05° (i.e., $\approx$1 km from the GB-SAR center point, the azimuth resolution was consistent with the DSM grid size). The range elimination threshold was set to $\pm$0.5 m. The pixels exceeding the distance threshold were not geocoded in the 3D local coordinate system to ensure that the geocoding accuracy of the entire region was at the subpixel level.

The cumulative displacement map of GB-SAR after searching the solution space is shown in Figure 10. Figure 10 shows that the distribution of cumulative displacement map before and after geocoding is consistent, and there is no mutation. Combined with the ancillary interpretation of the 3D terrain model, the zero boundary of the accumulative displacement coincides with the boundary of the glacier bedrock's polished surface with high SNR. The boundary between the bedrock and glacier is notable, which is consistent with the local geographic environment. A good geocoding effect was achieved, as shown in Figure 10a. The pixels in the area outside the DSM were eliminated because the minimum Euclidian distance solution space search exceeded the set threshold (Figure 10b). Thus, this method is only applicable if the target monitoring area has an external ancillary data coverage. Based on the different accuracies of the geographic dataset, external ancillary data that can be used to calculate the distance and azimuth values include but are not limited to the DSM, DEM, digital terrain model (DTM), digital line graphic (DLG) with contour, and point cloud generated by 3D laser scanning or a LiDAR system.

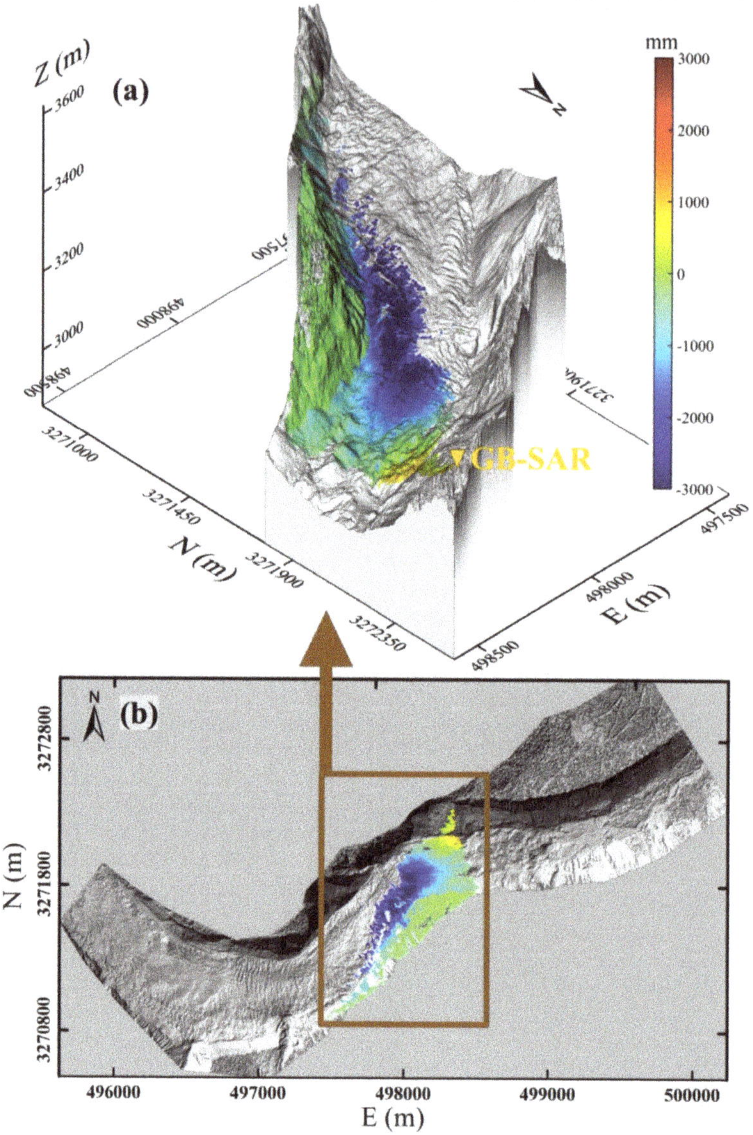

Figure 10. Geocoding result of the GB-SAR cumulative displacement. (**a**) Based on the superimposition on the GB-SAR cumulative displacement geocoding result of the 3D terrain model, the displacement value of the polished surface of the bedrock on the southern slope of the glacier is almost zero and the boundary is notable, which is consistent with the local geographic environment. (**b**) Distribution map of the GB-SAR result geocoded to the DSM. The area outside the DSM cannot be effectively geocoded because the distance cannot be calculated. The method proposed in this paper must be supported by 3D ancillary data in the target monitoring area, such as DSM and DEM data.

5. Discussion

5.1. Discussion on GB-SAR Geocoding Results

The surface of Hailuogou Glacier is covered by massive moraine rocks. It has strong backscattering characteristics and a high average reflection intensity. It is difficult to effectively identify and measure the radar images using the traditional method of placing CRs on the surface of the glacier as GCPs. Thus, accurate geocoding is difficult. The geocoding method used in this paper, that is, searching the solution space, can be realized by reducing the GB-SAR deployment position to the local 3D coordinate system frame using the control survey. It is not necessary to deploy CRs in the monitoring area and a highly accurate GB-SAR geocoding result can be realized.

Based on the analysis of the details of the geocoding results in the experiment (Figure 11), the area including the lower part of the landslide body, which is tangent to the glacier, significantly collapses. The GB-SAR geocoding results show notable boundaries in this area (Figure 11c). Based on the interpretation of the superimposed 3D terrain model, a backslope terrace has formed in the collapsed area at the lower edge of the landslide, which is invisible to GB-SAR due to the melting of the glacier. Therefore, this area cannot be effectively geocoded, which is consistent with the actual geographic condition of the area (Figure 11e). Similarly, there is no effective backscattering of radar waves on the water surface in the area including the glacial lake under the GB-SAR station because of the radar imaging mechanism, the GB-SAR system cannot effectively image in this area, resulting in data holes. Thus, there is no effective geocoding coverage, and the geocoding results around the glacial lake have notable boundaries (Figure 11f). Based on the combination of these results with the auxiliary analysis of the 3D terrain model of this area, the other uncoded regions of the block are also the areas that cannot be reached by the GB-SAR LOS due to terrain undulations. The geocoding results of this area conform to the GB-SAR imaging mechanism and local terrain.

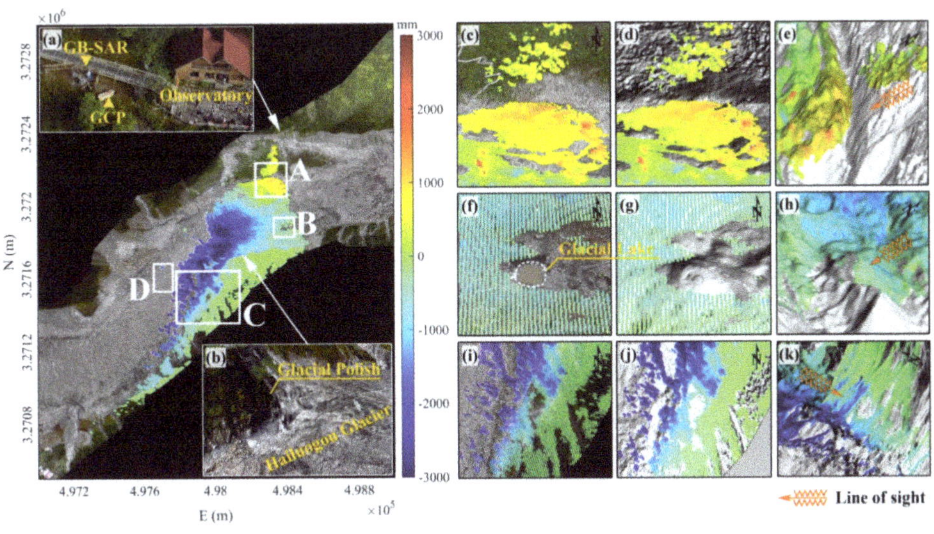

Figure 11. Map of the GB-SAR geocoding results including superimposed DOM image data. (**a**) The GB-SAR location and information about its environment. (**b**) Geographic environment information near the dividing line with a cumulative displacement of 0 mm. The area with a cumulative displacement of 0 mm is mainly the glacial polish. The lower part includes the Hailuogou Glacier covered by a large amount of surface moraine. (**c–e**) Local details of the area marked by the rectangular box A in the main figure, that is, the geocoding results of the superimposed DOM, superimposed DSM, and 3D terrain model, respectively. Similarly, (**f–h**) and (**i–k**) represent the geocoding results of the area marked by the rectangular boxes B and C, respectively. Based on the comparison and interpretation of the local enlarged details of the DOM, DSM, and 3D terrain model superimposed in the three regions A, B, and C, the uncoded areas are mainly concentrated in the backslope, glacial lake, and invisible valley, which conforms to the GB-SAR imaging mechanism and local geographic conditions. The geocoding effect is satisfactory.

The geocoding results of the glacier and its southern side slope far away from the LOS of GB-SAR show that the geocoding results have no effective surface deformation coverage in valley areas that were formed by the glacier and the side slope. Based on the deformation geocoding results and the 3D terrain model, the main reason is the lack of effective backscattering due to the invisible LOS of GB-SAR. However, the deformation area near the center of the glacier is not effectively geocoded or there are few geocoding points, which may be related to the spatial decorrelation of the phase caused by the rapid movement of the middle part of the glacier, as shown in area "D" in Figure 11. In addition, we identified several areas with negative displacement values on the side slope body, such as area "C," which move toward to the radar sensor and slide downward. The field survey image (Figure 11b) indicates a downward sliding surface debris flow in the side slope area, which is consistent with the actual geographic environment.

Based on the above-mentioned analysis, the GB-SAR geocoding results are highly consistent with the average power map and cumulative displacement distribution map of GB-SAR. Uncoded areas are mainly concentrated in the invisible backslopes, glacial lakes without effective backscattering, and invisible valley terrain. The results of geocoding are in accordance with the radar imaging mechanism and local geographic conditions, which verifies the effectiveness and universality of the solution space search geocoding method proposed in this paper.

5.2. Assessment of the Geocoding Accuracy

As the pixels beyond the scope of DSM cannot be used to solve the 3D spatial distance and effective distance solution space matching cannot be carried out, the values outside the monitoring range should be discarded, and only GB-SAR image pixels within the DSM range should be geocoded.

To verify the accuracy of the solution space search geocoding method, we decomposed the geocoded 3D local coordinates into the GB-SAR image coordinate system and obtained the distribution of the 3D local coordinate system coordinate components of E, N, Z in the radar coordinate system, as shown in Figure 12. Figure 12b,c shows that the coordinate values of each component of E and N are uniformly distributed, smooth, and lack a saltus step in the radar coordinate system G-XY, indicating that the GB-SAR geocoding plane position is accurate. The components of the Z coordinate axis in Figure 12d are consistent with the DSM; that is, the geocoding results have a high logical accuracy in elevation. The GB-SAR pixels correspond to the pixels in the 3D local coordinate system and no logic error occurs, confirming the accuracy of the solution space search geocoding method proposed in this paper.

The azimuth and range errors of each pixel in the GB-SAR image during geocoding were calculated. Their distributions are shown in Figure 13. The geocoding error of the GB-SAR image is a standard normal distribution in the radar azimuth direction and uniform distribution in the range direction, which is related to the imaging mechanism of the radar image. This implies that the range resolution in the radar coordinate system is a fixed value and the azimuth resolution is radian. With the increase in the distance, the radar pixel size gradually increases in the azimuth direction and the pixel resolution decreases. The basic logic of the solution space search method proposed in this study is azimuth angle matching, followed by minimum Euclidean distance matching of the radar pixels in the range direction. If the range matching in the azimuth direction in ineffective, a distance solution space search is carried out within the angle threshold range set in the azimuth direction. Therefore, most of the matching pixels are close to the azimuth angle center line near the 0° value of the azimuth error. As a result, the geocoding error of GB-SAR presents a normal distribution in the azimuth direction, as shown in Figure 13b. As the range resolution of the GB-SAR is fixed and the ground surface of the monitoring area is continuous but uneven, the range solutions in the range direction are matched within the set threshold range, forming a uniform distribution, as shown in Figure 13d. Based on the error distribution, the geocoding pixel position errors in this paper are all smaller than the single-pixel resolution and the matching accuracy is at the subpixel level. This proves that the solution space search geocoding method has a high geocoding accuracy.

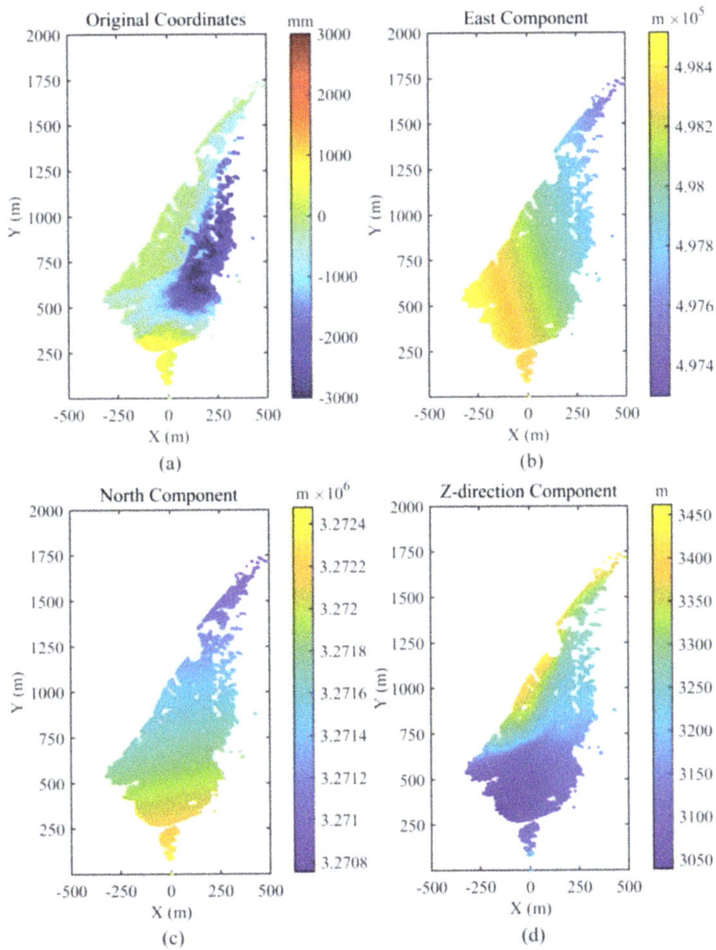

Figure 12. Distribution of the 3D local coordinate system components in the radar coordinate system. (**a**) Original GB-SAR monitoring results. (**b**) Distribution of the east component of the 3D coordinate system in the radar coordinate system. (**c**) Distribution of the north component of the 3D coordinate system in the radar coordinate system. (**d**) Distribution of the Z-direction component of the 3D coordinate system in radar coordinate system.

5.3. Landslide Migration Mechanism

Based on the combination of the chart data obtained from continuous GB-SAR monitoring and the field investigation of the landslide area, the movement of the LGJT landslide on the Hailuogou Glacier can be inferred. On one hand, owing to the loose soil and rocks on the lateral moraine of Hailuogou Glacier, the thinning of the glacier will lead to the decrease of the sliding resistance section at the lower edge of the landslide mass in the beginning of annual ablation period. The lower part of the landslide significantly collapses, forming a notable collapse terrace, which leads to the increase of the slope free height and the decrease of the overall anti sliding force of landslide mass [24] (Figure 11c,e). On the other hand, the eastward tangential movement of the glacier at the lower edge of the landslide has a strong traction effect on the erosion of the lower part of the landslide and the abundant rainfall in the area promotes the aggravation of the landslide movement [45–47]. Based on this information, we can conclude that the migration mechanism of the LGJT landslide is retrogressive movement [39,48], that is, the lower part of the landslide first slides under the traction of the glacier, causing the upper part to lose support, deform, and slide. While sliding down, the landslide's main body further squeezes the edge of the glacier under the action of gravity, changing the flow direction of the glacier in this area. This leads to the positive accumulated displacement value of GB-SAR in the glacier area at the lower edge of the landslide. However, when it reaches the middle of the glacier, the stress gradually decreases and positive value gradually changes to a negative value in the way of moving close to the GB-SAR sensor center.

This monitoring experiment shows that the high spatiotemporal resolution of the GB-SAR monitoring system directly reflects the cumulative displacement distribution between the landslide mass and its tangent glacier. Based on the combination of the monitoring chart data and information about the local geographic environment, the migration mechanism of the landslide body can be well explained.

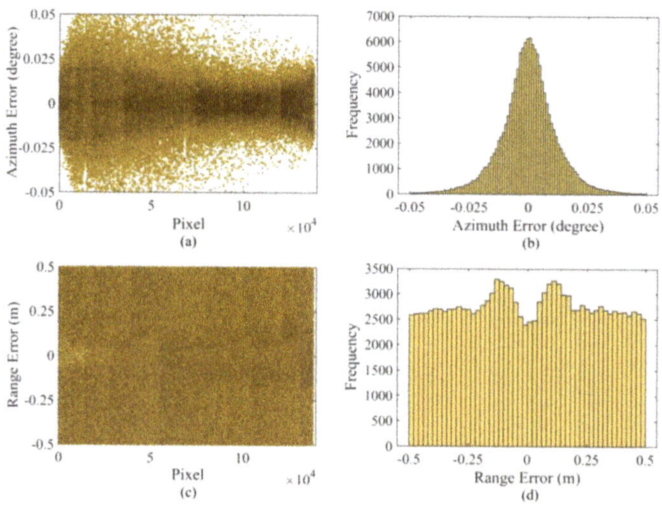

Figure 13. Distribution charts of geocoding error: (**a**) distribution of azimuth error, (**b**) histogram of azimuth error distribution, (**c**) distribution of range error, and (**d**) histogram of range error distribution.

6. Conclusions

In this study, the high-resolution DSM obtained by UAV aerial photogrammetry was used to establish a high-accuracy coordinate transformation model of GB-SAR. A high-accuracy geocoding method for GB-SAR based on searching the solution space was proposed, and the coordinate conversion of radar image from 2D plane coordinate system to the local 3D coordinate system was realized. By applying this method to LGJT landslide monitoring experiment on Hailuogou Glacier, the effectiveness was verified by contrastive interpretation and the logical accuracy and accuracy indicators were assessed by quantitative analysis. Finally, a subpixel geocoding accuracy was realized. Based on the geocoding method proposed in this paper, the GB-SAR monitoring system can accurately monitor and interpret the movement of landslides, providing important data and decision support for the monitoring and early warning of high-risk landslides and related geological disasters.

The geocoding method proposed in this study has several limitations. For the geocoding of the target monitoring area, external 3D ancillary data coverage is required such as DSM data (the radar center can be outside of the DSM, only its coordinates and azimuth angle must be measured). In addition, the GB-SAR center and DSM must be in a unified coordinate system to effectively search and match the distance solution space. Therefore, the geocoding method proposed in this study should be supported by spatial data acquired by UAV aerial photogrammetry or LiDAR surveys to obtain the highest geocoding accuracy and best interpretation of the GB-SAR monitoring results.

Author Contributions: Conceptualization, G.L. and H.J.; methodology, J.C.; validation, B.Z. and Y.F.; formal analysis, Q.L. and X.W.; investigation, B.Z. and J.C.; data curation, J.C., B.Z. and R.Z.; writing—original draft preparation, J.C.; writing—review and editing, H.J. and Q.L.; supervision, G.L.; project administration, G.L. All authors have read and agreed to the published version of the manuscript.

Funding: This research was jointly funded by the National Key Research and Development Program of China (Grant No. 2017YFB0502700); the National Natural Science Foundation of China (Grant Nos. 41771402, 41701535, 41871069, and 41804009); the Sichuan Science and Technology Program (Grant Nos. 2019YJ0224, 2019ZDZX0042, 2020JDTD0003, and 2020YJ0322); and the Major Special Airborne Observation System Project for High-resolution Earth Observations (Grant No. 30-H30C01-9004-19/21).

Institutional Review Board Statement: Not applicable.

Informed Consent Statement: Not applicable.

Data Availability Statement: Samples of data are available from the authors.

Acknowledgments: The authors would like to thank the Hailuogou Scenic Spot Management Bureau for providing UAV photogrammetry and GB-SAR observation data acquisition permission and traffic convenience. The authors are also grateful to Chengdu Beidou Tiandi Co., Ltd. (www.compassgnss.com, 20 January 2021) for providing the GodWork software support.

Conflicts of Interest: The authors declare no conflict of interest.

References

1. Wang, Y.; Hong, W.; Zhang, Y.; Lin, Y.; Li, Y.; Bai, Z.; Zhang, Q.; Lv, S.; Liu, H.; Song, Y. Ground-Based Differential Interferometry SAR: A Review. *IEEE Geosci. Remote Sens. Mag.* **2020**, *8*, 43–70. [CrossRef]
2. Caduff, R.; Schlunegger, F.; Kos, A.; Wiesmann, A. A Review of Terrestrial Radar Interferometry for Measuring Surface Change in the Geosciences. *Earth Surf. Process. Landf.* **2015**, *40*, 208–228. [CrossRef]
3. Pieraccini, M.; Miccinesi, L. Ground-based Radar Interferometry: A Bibliographic Review. *Remote Sens.* **2019**, *11*, 1029. [CrossRef]
4. Lingua, A.; Piatti, D.; Rinaudo, F. Remote Monitoring of a Landslide Using an Integration of GB-INSAR and LIDAR Techniques. *Int. Arch. Photogramm. Remote Sens. Spat. Inf. Sci.* **2008**, *37*, 133–139.
5. Monserrat, O.; Crosetto, M.; Luzi, G. A Review of Ground-based SAR Interferometry for Deformation Measurement. *ISPRS-J. Photogramm. Remote Sens.* **2014**, *93*, 40–48. [CrossRef]
6. Lowry, B.; Gomez, F.; Zhou, W.; Mooney, M.A.; Held, B.; Grasmick, J. High Resolution Displacement Monitoring of a Slow Velocity Landslide Using Ground Based Radar Interferometry. *Eng. Geol.* **2013**, *166*, 160–169. [CrossRef]
7. Wang, Z.; Li, Z.; Mills, J. A New Approach to Selecting Coherent Pixels for Ground-based SAR Deformation Monitoring. *ISPRS-J. Photogramm. Remote Sens.* **2018**, *144*, 412–422. [CrossRef]
8. Atzeni, C.; Barla, M.; Pieraccini, M.; Antolini, F. Early Warning Monitoring of Natural and Engineered Slopes with Ground-Based Synthetic-Aperture Radar. *Rock Mech. Rock Eng.* **2015**, *48*, 235–246. [CrossRef]
9. Zheng, X.; Yang, X.; Ma, H.; Ren, G.; Zhang, K.; Yang, F.; Li, C. Integrated Ground-based SAR Interferometry, Terrestrial Laser Scanner, and Corner Reflector Deformation Experiments. *Sensors* **2018**, *18*, 4401. [CrossRef] [PubMed]
10. Cossu, R.; Schoepfer, E.; Bally, P.; Fusco, L. Near real-time SAR-based processing to support flood monitoring. *J. Real-Time Image Process.* **2009**, *4*, 205–218. [CrossRef]
11. Li, Y.; Jiao, Q.; Hu, X.; Li, Z.; Li, B.; Zhang, J.; Jiang, W.; Luo, Y.; Li, Q.; Ba, R. Detecting the Slope Movement After the 2018 Baige Landslides Based on Ground-based and Space-borne Radar Observations. *Int. J. Appl. Earth Obs. Geoinf.* **2020**, *84*, 1–12. [CrossRef]
12. Casagli, N.; Catani, F.; Ventisette, C.D.; Luzi, G. Monitoring, prediction, and early warning using ground-based radar interferometry. *Landslides* **2010**, *7*, 291–301. [CrossRef]
13. Pieraccini, M.; Casagli, N.; Luzi, G.; Tarchi, D.; Mecatti, D.; Noferini, L.; Atzeni, C. Landslide Monitoring by Ground-based Radar Interferometry: A Field Test in Valdarno (Italy). *Int. J. Remote Sens.* **2003**, *24*, 1385–1391. [CrossRef]
14. Zhang, R.; Song, Y.; Liu, G.; Zhang, H. Landslide Monitoring Based on GB-SAR Interferometry for the Instable Slopes of Yaoji Reservoir. In Proceedings of the IET International Radar Conference, Hangzhou, China, 14–16 October 2015.
15. Noferini, L.; Pieraccini, M.; Mecatti, D.; Macaluso, G.; Atzeni, C.; Mantovani, M.; Marcato, G.; Pasuto, A.; Slivano, S.; Tagliavini, F. Using GB-SAR Technique to Monitor Slow Moving Landslide. *Eng. Geol.* **2007**, *95*, 88–98. [CrossRef]

16. Tarchi, D.; Casagli, N.; Fanti, R.; Leva, D.D.; Luzi, G.; Pasuto, A.; Pieraccini, M.; Slivano, S. Landslide Monitoring by Using Ground-based SAR Interferometry: An Example of Application to the Tessina Landslide in Italy. *Eng. Geol.* **2003**, *68*, 15–30. [CrossRef]
17. Bozzano, F.; Cipriani, I.; Mazzanti, P.; Prestininzi, A. Displacement Patterns of a Landslide Affected by Human Activities: Insights from Ground-based InSAR Monitoring. *Nat. Hazards* **2011**, *59*, 1377–1396. [CrossRef]
18. Xing, C.; Huang, J.; Han, X. Research on the Environmental Effects of GB-SAR for Dam Monitoring. *Adv. Mater. Res.* **2014**, *919*, 392–397. [CrossRef]
19. Wang, P.; Xing, C. Research on Coordinate Transformation Method of GB-SAR Image Supported by 3D Laser Scanning Technology. In Proceedings of the ISPRS TC III Mid-term Symposium, Beijing, China, 7–10 May 2018; pp. 1757–1763.
20. Rouyet, L.; Kristensen, L.; Derron, M.H.; Michoud, C.; Blikra, L.H.; Jaboyedoff, M.; Lauknes, T.R. Evidence of Rock Slope Breathing Using Ground-based InSAR. *Geomorphology* **2017**, *289*, 152–169. [CrossRef]
21. Frodella, W.; Ciampalini, A.; Gigli, G.; Lombardi, L.; Raspini, F.; Nocentini, M.; Scardigli, C.; Casagli, N. Synergic Use of Satellite and Ground Based Remote Sensing Methods for Monitoring the San Leo Rock Cliff (Northern Italy). *Geomorphology* **2016**, *264*, 80–94. [CrossRef]
22. Liu, B.; Ge, D.; Li, M.; Zhang, L.; Wang, Y.; Zhang, X. Using GB-SAR Technique to Monitor Displacement of Open Pit Slope. In Proceedings of the 2018 IEEE International Geoscience and Remote Sensing Symposium, Beijing, China, 10–15 July 2016; pp. 5986–5989.
23. Pipia, L.; Fabregas, X.; Aguasca, A.; López-Martínez, C. Polarimetric Temporal Analysis of Urban Environments with a Ground-Based SAR. *IEEE Trans. Geosci. Remote Sens.* **2013**, *51*, 2343–2360. [CrossRef]
24. Liu, G.; Zhang, B.; Zhang, R.; Cai, J.; Fu, Y.; Liu, Q.; Yu, B.; Li, Z. Monitoring Dynamics of Hailuogou Glacier and the Secondary Landslide Disasters Based on Combination of Satellite SAR and Ground-based SAR. *Geomat. Inf. Sci. Wuhan Univ.* **2019**, *44*, 980–995.
25. Dematteis, N.; Luzi, G.; Giordan, D.; Zucca, F.; Allasia, P. Monitoring Alpine glacier surface deformations with GB-SAR. *Remote Sens. Lett.* **2017**, *8*, 947–956. [CrossRef]
26. Noferini, L.; Mecatti, D.; Macaluso, G.; Pieraccini, M.; Atzeni, C. Monitoring of Belvedere Glacier Using a Wide Angle GB-SAR Interferometer. *J. Appl. Geophys.* **2009**, *68*, 289–293. [CrossRef]
27. Luzi, G.; Pieraccini, M.; Mecatti, D.; Noferini, L.; Macaluso, G.; Tamburini, A.; Atzeni, C. Monitoring of an Alpine Glacier by Means of Ground-Based SAR Interferometry. *IEEE Geosci. Remote Sens. Lett.* **2007**, *4*, 495–499. [CrossRef]
28. Xie, S.; Dixon, T.H.; Voytenko, D.; Deng, F.; Holland, D.M. Grounding Line Migration Through the Calving Season at Jakobshavn Isbræ, Greenland, Observed with Terrestrial Radar Interferometry. *Cryosphere* **2018**, *12*, 1–29. [CrossRef]
29. Wang, P.; Xing, C. A Method of Transforming GB-SAR Image Coordinate to 3D Terrain Coordinate. *J. Yangtze River Sci. Res. Inst.* **2018**, *35*, 122–127.
30. Bardi, F.; Raspini, F.; Ciampalini, A.; Kristensen, L.; Rouyet, L.; Lauknes, T.R.; Frauenfelder, R.; Casagli, N. Space-Borne and Ground-Based InSAR Data Integration: The Aknes Test Site. *Remote Sens.* **2016**, *8*, 237. [CrossRef]
31. Zhou, J.; Tian, J.; Chen, Y.; Mao, Q.; Li, Q. Research on Data Fusion Method of Ground-Based SAR and 3D Laser Scanning. *J. Geomat.* **2015**, *40*, 26–30.
32. Tapete, D.; Casagli, N.; Luzi, G.; Fanti, R.; Gigli, G.; Leva, D. Integrating Radar and Laser-based Remote Sensing Techniques for Monitoring Structural Deformation of Archaeological Monuments. *J. Arch. Sci.* **2013**, *40*, 176–189. [CrossRef]
33. Dematteis, N.; Giordan, D.; Zucca, F.; Luzi, G.; Allasia, P. 4D Surface Kinematics Monitoring Through Terrestrial Radar Interferometry and Image Cross-correlation Coupling. *ISPRS-J. Photogramm. Remote Sens.* **2018**, *142*, 38–50. [CrossRef]
34. Monserrat, O.; Crosetto, M. Deformation Measurement Using Terrestrial Laser Scanning Data and Least Squares 3D Surface Matching. *ISPRS-J. Photogramm. Remote Sens.* **2008**, *63*, 142–154. [CrossRef]
35. Tian, W.; Zhao, Z.; Hu, C.; Wang, J.; Zeng, T. GB-InSAR-Based DEM Generation Method and Precision Analysis. *Remote Sens.* **2019**, *11*, 997. [CrossRef]
36. Kuras, P.; Ortyl, Ł.; Owerko, T.; Salamak, M.; Łaziński, P. GB-SAR in the Diagnosis of Critical City Infrastructure—A Case Study of a Load Test on the Long Tram Extradosed Bridge. *Remote Sens.* **2020**, *12*, 3361. [CrossRef]
37. Liu, Q.; Liu, S.; Cao, W. Seasonal Variation of Drainage System in the Lower Ablation Area of a Monsoonal Temperate Debris-covered Glacier in Mt. Gongga, South-eastern Tibet. *Water* **2018**, *10*, 1050. [CrossRef]
38. Hassan, A.; Xu, J.; Zhang, L.; Liu, G.; Schmit, A.; Xing, C.; Xu, Y.; Ouyang, C.; Schwieger, V. Towards Integration of GNSS and GB-SAR Measurements: Exemplary Monitoring of a Rock Fall at the Yangtze River in China. In Proceedings of the Embracing Our Smart World Where the Continents Connect: Enhancing the Geospatial Maturity of Societies, Proceedings of the FIG Congress 2018, Istanbul, Turkey, 6–11 May 2018.
39. Cai, J.; Zhang, W.; Li, Y.; Ren, S. Landslide Stability Analysis Based on Law of Surface Displacement. *Sci. Surv. Mapp.* **2016**, *41*, 96–99.
40. Liu, G.; Chen, Q.; Luo, X.; Cai, G. *Principle and Application of InSAR*; Science Press: Chengdu, China, 2019; pp. 237–242.
41. Yang, H.; Cai, J.; Peng, J.; Wang, J.; Jiang, Q. A Correcting Method About GB-SAR Rail Displacement. *Int. J. Remote Sens.* **2017**, *38*, 1483–1493. [CrossRef]
42. Jiang, M.; Monti-Guarnieri, A. Distributed Scatterer Interferometry with the Refinement of Spatiotemporal Coherence. *IEEE Trans. Geosci. Remote Sens.* **2020**, *58*, 3977–3987. [CrossRef]

43. Jehle, M.; Perler, D.; Small, D.; Schubert, A.; Meier, E. Estimation of Atmospheric Path Delays in TerraSAR-X Data Using Models vs. Measurements. *Sensors* **2008**, *8*, 8479–8491. [CrossRef]
44. Noferini, L.; Pieraccini, M.; Mecatti, D.; Luzi, G.; Atzeni, C.; Tamburini, A.; Broccolato, M. Permanent Scatterers Analysis for Atmospheric Correction in Ground-based SAR Interferometry. *IEEE Trans. Geosci. Remote Sens.* **2005**, *43*, 1459–1471. [CrossRef]
45. Nie, Y.; Pritchard, H.D.; Liu, Q.; Hennig, T.; Wang, W.; Wang, X.; Liu, S.; Nepal, S.; Samyn, D.; Hewitt, K.; et al. Glacial change and hydrological implications in the Himalaya and Karakoram. *Nat. Rev. Earth Environ.* **2021**, *2*, 91–106. [CrossRef]
46. Huggel, C. Recent extreme slope failures in glacial environments: Effects of thermal perturbation. *Quat. Sci. Rev.* **2009**, *28*, 1119–1130. [CrossRef]
47. Huggel, C.; Clague, J.J.; Korup, O. Is climate change responsible for changing landslide activity in high mountains. *Earth Surf. Process. Landf.* **2012**, *37*, 77–91. [CrossRef]
48. Wang, X.; Liu, Q.; Liu, S.; He, G. Manifestations and mechanisms of mountain glacier-related hazards. *Sci. Cold Arid Reg.* **2020**, *12*, 436–446.

remote sensing

Article

A Semi-Automatic Method for Extracting Small Ground Fissures from Loess Areas Using Unmanned Aerial Vehicle Images

Hongguo Jia [1], Bowen Wei [2,*], Guoxiang Liu [1], Rui Zhang [1], Bing Yu [3] and Shuaiying Wu [1]

1 Faculty of Geosciences and Environmental Engineering, Southwest Jiaotong University, Chengdu 611756, China; rsjia@swjtu.edu.cn (H.J.); rsgxliu@swjtu.edu.cn (G.L.); zhangry@my.swjtu.edu.cn (R.Z.); shining@my.swjtu.edu.cn (S.W.)
2 Guangxi Natural Resources Survey and Monitoring Institute, Nanning 530022, China
3 School of Civil Engineering and Geomatics, Southwest Petroleum University, Chengdu 610500, China; 201599010140@swpu.edu.cn
* Correspondence: rswbw@my.swjtu.edu.cn

Abstract: Remote sensing-based ground fissure extraction techniques (e.g., image classification, image segmentation, feature extraction) are widely used to monitor geological hazards and large-scale artificial engineering projects such as bridges, dams, highways, and tunnels. However, conventional technologies cannot be applied in loess areas due to their complex terrain, diverse textural information, and diffuse ground target boundaries, leading to the extraction of many false ground fissure targets. To rapidly and accurately acquire ground fissures in the loess areas, this study proposes a data processing scheme to detect loess ground fissure spatial distributions using unmanned aerial vehicle (UAV) images. Firstly, the matched filter (MF) algorithm and the first-order derivative of the Gaussian (FDOG) algorithm were used for image convolution. A new method was then developed to generate the response matrices of the convolution with normalization, instead of the sensitivity correction parameter, which can effectively extract initial ground fissure candidates. Directions, the number of MF/FDOG templates, and the efficiency of the algorithm are comprehensively considerate to conclude the suitable scheme of parameters. The random forest (RF) algorithm was employed for the step of the image classification to create mask files for removing non-ground-fissure features. In the next step, the hit-or-miss transform algorithm and filtering algorithm in mathematical morphology is used to connect discontinuous ground fissures and remove pixel sets with areas much smaller than those of the ground fissures, resulting in a final binary ground fissure image. The experimental results demonstrate that the proposed scheme can adequately address the inability of conventional methods to accurately extract ground fissures due to plentiful edge information and diverse textures, thereby obtaining precise results of small ground fissures from high-resolution images of loess areas.

Keywords: ground fissure extraction; loess landform; modified MF-FDOG algorithm; RF algorithm; UAV image

Citation: Jia, H.; Wei, B.; Liu, G.; Zhang, R.; Yu, B.; Wu, S. A Semi-Automatic Method for Extracting Small Ground Fissures from Loess Areas Using Unmanned Aerial Vehicle Images. *Remote Sens.* **2021**, *13*, 1784. https://doi.org/10.3390/rs13091784

Academic Editors: Alex Hay-Man Ng, Linlin Ge, Hsing-Chung Chang and Zheyuan Du

Received: 30 March 2021
Accepted: 27 April 2021
Published: 3 May 2021

1. Introduction

Loess is a yellow silt quaternary deposit transported by the wind; the most typical loess area in the world is the Loess Plateau in China. Loess is well developed in the He'nan, Shanxi, and Gansu provinces near the middle basin of the Yellow River, which has been gradually transformed into representative loess landform areas by the power of water, gravity, and the wind [1]. Loess landforms are typically divided into gully, erosion, and valley landscapes, the latter of which are prone to major geohazards such as floods, collapses, and landslides, which threaten human life [1,2]. Ground fissures are the most typical representative of the instability of loess sediments. It may be caused by natural or human-made factors, which will cause damage to buildings and other structures on the ground and endanger people's lives [3]. In addition, it also seriously threatens the safe operation of high-speed railways, highways and other transportation networks [4,5]. Therefore, it is necessary to monitor these geological dynamics effectively. As mentioned in previous studies [6], a useful way to monitor such geohazards is to extract ground fissure information from loess areas using unmanned aerial vehicle (UAV) images.

Ground fissures, which exhibit dark and linear characteristics in aerial images, are a typical feature of loess areas. Many computer vision methods are used to extract linear targets, such as pixel-oriented edge extraction [7,8], object-oriented feature extraction [9], and LiDAR [10]. Edge extraction operators based on computer vision are particularly popular. Methods based on edge extraction operators are typically divided into two categories: first-order derivative methods and second-order derivative methods [11]. The edge in the images is the pixel sets covering sharp grayscale changes, whose trend can be represented by the first-order derivative function. The greater the value of the first-order derivative function, the more likely the pixel is to be marked as an edge indicator. Frequently used first-order derivative functions include the Roberts operator [12], Sobel operator [13], Prewitt operator [14], and Canny operator [11]. With all these operators, an appropriate threshold is selected after the convolution to the segment the image and extract the edge information; however, the arbitrary threshold used in image segmentation has a substantial impact on the final results of fissure extraction.

Therefore, second-order derivative methods were proposed to address this problem. The essence of second-order derivative methods is to find the maximum position of local gradient values calculated by the first-order derivative function (i.e., the local peak position of the first-order derivative function); the value of this position in the second-order derivative function is zero. This position is therefore considered the ground fissure location of interest. The Laplacian operator and LOG operator [15] are two typical second-order derivative strategies. Recently, the aforementioned edge detection operators, as well as other improved operators [16,17], have been widely studied and applied to highway extraction, bridge recognition, and tunnel and pipeline fissure detection [18–21].

Although edge extraction operators based on pixel grayscale change detection can extract ground fissures from high-resolution UAV images, one primary limitation is their applicability to different types of ground fissure. Firstly, these operators only detect the outer edges of the ground fissure and not the fissure itself when the fissure is treated as a planar target. Secondly, they are not sensitive to pixels with a small grayscale gradient. Moreover, broken soil, pits, withered grass, snow, and ice (in winter) are often widely distributed in loess areas [22], whose edges are commonly recognized as ground fissures by the above methods because of their rich textural information.

In the medical field, blood vessels in retina images scanned by computed tomography (CT) present a similar shape and hue to ground fissures. Using the spatial attributes of blood vessels in CT images, Chaudhuri et al. [23] successfully recognized blood vessels with the matched filter (MF) algorithm. When the vertical profile of a blood vessel is drawn as a graph, its distribution curve is similar to that of an inverted Gaussian distribution. Therefore, the use of a convolution operation with the available inverted Gaussian template can enhance the contrast between the blood vessels and the background, enabling the CT image to be segmented by setting an appropriate grayscale threshold. However, the MF algorithm cannot accurately distinguish blood vessel targets and blood vessel edges. After matched filtering, the response signals of blood vessels and their edges showed similar patterns but different strengths. To address this issue, Zhang et al. [24] proposed the combined matched filter and first-order derivative of the Gaussian (MF-FDOG) algorithm to improve the accuracy of blood vessel extraction. In 2013, Stumpf et al. [25] successfully applied the MF-FDOG algorithm to investigate ground fissures before landslides due to the similarity of their vertical profile compared to blood vessels. By analyzing the results of ground fissure extracted for different periods, they were able to predict the orientation and extent of the landslide.

Nevertheless, the scheme proposed by Stumpf et al. [25] has some limitations regarding its practical application. In loess areas, vegetation often appears as linear targets in images; i.e., its spectral curve does not always satisfy the normal spectral curve pattern of green vegetation. As a result, traditional vegetation index methods are not capable of removing vegetation that is incorrectly detected as ground fissures. Moreover, the sensitivity correction parameter in the MF-FDOG algorithm is indeterminate and not in the range from 3 to 4 for different UAV images. In addition, the optimal direction number of the convolution template is not considered and it can reduce the accuracy and efficiency of the algorithm. Therefore, to overcome these shortcomings of the MF-FDOG algorithm, we propose an advanced processing scheme for ground fissure extraction using different UAV images.

2. Methodology of Ground Fissure Extraction

2.1. Modified MF-FDOG Algorithm for Ground Fissure Extraction

The MF-FDOG algorithm comprises the MF algorithm and the FDOG algorithm. The MF [26] algorithm was initially applied in the digital signal processing field to detect whether a complex signal contains a simple known signal [27,28]. As a digital image is one type of digital signal; thus, methods used in the digital signal field can be used for digital image processing.

In terms of the structure (e.g., width, length, shape) of ground fissures in different images, we can expand a one-dimensional template to two-dimensional space. Then, using the two-dimensional template. The function of the template is as follows:

$$\mathrm{MF} = \mathrm{g}(\mathrm{x},\mathrm{y};\sigma) = \begin{cases} y_1 = -\frac{1}{\sqrt{2\pi}\sigma}\exp\left(-\frac{|x|^2}{2\sigma^2}\right) - m, \\ \cdots\cdots \\ y_n = -\frac{1}{\sqrt{2\pi}\sigma}\exp\left(-\frac{|x|^2}{2\sigma^2}\right) - m; \end{cases} \quad |x| \leq 3\sigma, |y| \leq L/2 \tag{1}$$

where x and y denote variables along the horizontal and vertical directions of the two-dimensional template, respectively; and σ denotes the standard deviation of the inverted Gaussian distribution. In a Gaussian distribution, the probability that variable x lies in the range of $[-3\sigma, 3\sigma]$ is up to 99.7%, and the vertical profile of ground fissures is similar to an inverted Gaussian distribution (Figure 1). Therefore, it is reasonable to select this interval as the range of variable x. In other words, the width of the ground fissure is 6σ. Moreover, L denotes the minimum length of the ground fissure and m denotes the mean template. In computer memory, continuous digital image signals have a regular digital matrix format. Therefore, m is represented by

$$m = \sum_{i \in N} K_i(x,y)/N, \tag{2}$$

where K_i represents the response after MF processing at position (x,y) and N represents the size of the two-dimensional template.

(a) The photograph of ground fissure in loess areas

(b) Pixel value and fitting curve of fissure

Figure 1. (**a**) Photograph and (**b**) pixel value distribution (red line shows the fitting curve) of ground fissures in loess areas.

FDOG [24] refers to the first-order derivative function of the Gaussian function, then expands it to a two-dimensional filter template. The template is established in the same way as that of the MF template. The FDOG function is represented by

$$\mathrm{FDOG} = g(x, y; \sigma) = \begin{cases} y_1 = \frac{x}{\sqrt{2\pi\sigma^3}} \exp\left(-\frac{|x|^2}{2\sigma^2}\right), \\ \cdots\cdots \\ y_n = \frac{x}{\sqrt{2\pi\sigma^3}} \exp\left(-\frac{|x|^2}{2\sigma^2}\right); \end{cases} \quad |x| \leq 3\sigma, |y| \leq L/2, \tag{3}$$

where all parameters are the same as those in Equation (1). Figure 2a,b show examples of the three-dimensional diagram about the MF template and FDOG template, respectively. The parameters of both templates are set to $\sigma = 1.5$, $L = 9.0$, and $\theta = 90°$ (i.e., the template is along the vertical direction).

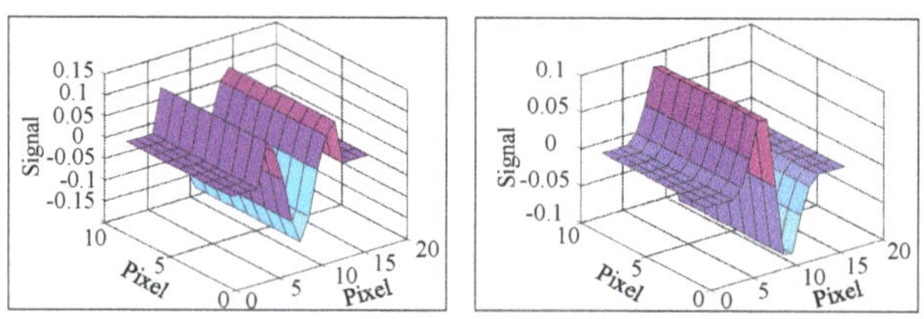

(a) 3D template of MF filter

(b) 3D template of FDOG filter

Figure 2. (**a**,**b**) are templates of MF and FODG filters, respectively (parameters: $\sigma = 1.5$, $L = 9.0$, $\theta = 90°$).

The orientation and distribution of ground fissures are arbitrary. Thus, we must process images with the optimal direction number of templates to ensure that all ground fissures are detected. In experiments, the template direction number will have a strong effect on the sensitivity and accuracy of the modified MF-FDOG algorithm; however, ground fissures are a linear target, which means that they can be regarded as a straight line with an anti-parallel characteristic in a local area [29]. Therefore, we only need to create a template with directions of 0°–180° in two-dimensional space. To create these templates with different directions, the central pixel of the templates is set as the origin (0, 0) of the coordinate system. We must decide how many directions, N, we want in the 180° space. Then, the angular interval of direction θ_{int} is calculated using Equation (4). The rotation angle, θ_i, between the direction of the template and the x axis (see Figure 3) and the rotation matrix r_i are represented by Equations (5) and (6), respectively:

$$\theta_{int} = 180^\circ \div N, \tag{4}$$

$$\theta_i = n \times \theta_{int}, \tag{5}$$

$$r_i = \begin{bmatrix} \cos\theta_i & -\sin\theta_i \\ \sin\theta_i & \cos\theta_i \end{bmatrix}, \tag{6}$$

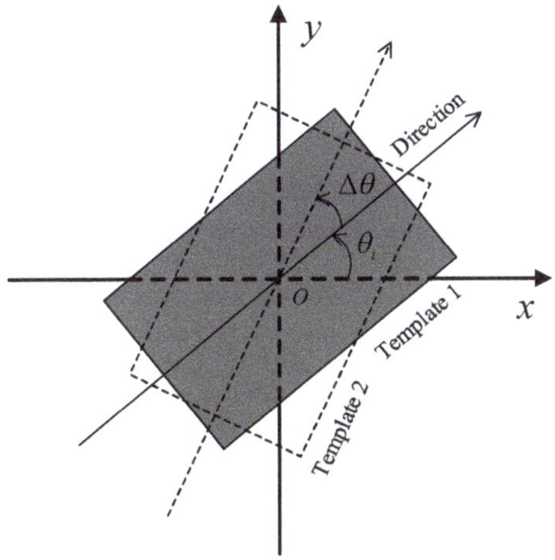

Figure 3. Illustration of the template direction.

Through rotation, we can obtain a group of MF and FDOG templates. The template direction corresponding to the maximum response is taken as the direction of ground fissures in the processed images. The derived maximum response represents the result of the MF operation at the current pixel location. The angle corresponding to this response represents the orientation angle of the ground fissures. The function for selecting the template angle is defined as

$$\theta_{max(x,y)} = argmax\left(I_{(x,y)} \bigotimes MF_{\theta_i}\right),\ 0 < \theta_i \le \pi \tag{7}$$

where $\theta_{max(x,y)}$ denotes the angle corresponding to the maximum response obtained by convolving an image with n MF templates processing different directions; $I_{(x,y)}$ represents the UAV image; MF_{θ_i} denotes the template with the rotation angle θ_i; and $\otimes$ denotes convolution operation.

For ground fissure extraction, the difference between the MF response matrix and the FDOG response matrix must be calculated. Before this step, to reduce the probability of false ground fissure extraction, [25] suggested implementing some special procedures: (1) after convolution by the MF template, all negative responses should be set to zero; (2) for the FDOG response matrix, a mean filter should be performed before calculating the absolute value for all responses. The two steps are represented by

$$R_{(x,y)} = \left(I_{(x,y)} \bigotimes MF_{\theta_{max(x,y)}}\right) > 0 \tag{8}$$

$$D_{(x,y)} = \left| I_{(x,y)} \bigotimes FDOG_{\theta_{max(x,y)}} \bigotimes M \right| \tag{9}$$

where $R_{(x,y)}$ and $D_{(x,y)}$ denote the response matrices obtained by convolution of the MF and FDOG templates, respectively; $MF_{\theta_{max(x,y)}}$ and $FDOG_{\theta_{max(x,y)}}$ denote the template corresponding to the maximum response of the convolution operation in n different directions, and M denotes the template of the mean filter, whose value is set to 6σ [24]. Additionally, the mean filter not only reduces noise and smooths the image, but also expands the width of $D_{(x,y)}$. To do this, the differential signals are enhanced.

The results indicate that it is difficult to directly differentiate between $R_{(x,y)}$ and $D_{(x,y)}$ matrices because their ranges are substantially different. Therefore, a sensitivity correction parameter should be used to correct their ranges. In [25], the empirical ranges of the parameter are between 3 and 4. However, the ranges of $R_{(x,y)}$ and $D_{(x,y)}$, do not satisfy the linear relationship shown by ground fissure extraction experiments using different images. Therefore, there is a limitation of using the sensitivity correction parameter to correct their ranges. In this study, we propose a solution to normalize the range of $R_{(x,y)}$ and $D_{(x,y)}$ by a linear stretching method. Their difference is then calculated as

$$\overline{R_{(x,y)}} = R'_{(x,y)} - D'_{(x,y)} \tag{10}$$

where $R'_{(x,y)}$ and $D'_{(x,y)}$ indicate the response matrices of $R_{(x,y)}$ and $D_{(x,y)}$ stretched into [0,1], respectively.

Figure 4 is the frequency histogram of the $\overline{R_{(x,y)}}$ matrix, which exhibits a Gaussian distribution. Generally, ground fissures are small targets in UAV images, whose area is far less than that of the loess, vegetation, and other ground targets. Ground fissure signals show a strong response after the modified MF-FDOG convolution operation. It is reasonable to choose a threshold T, according to the mean and standard deviation of $\overline{R_{(x,y)}}$ to segment the image using (11) and (12). Then, pixels with values greater than T can be considered ground fissure locations. Finally, ground fissure candidates can be extracted from the images; i.e., locations with $F_{map} = 1$:

$$T = \mu_{\overline{R}} + 2\sigma_{\overline{R}} \tag{11}$$

$$\begin{cases} \overline{R_{(x,y)}} \geq T, \ F_{map} = 1 \\ \overline{R_{(x,y)}} < T, \ F_{map} = 0 \end{cases} \tag{12}$$

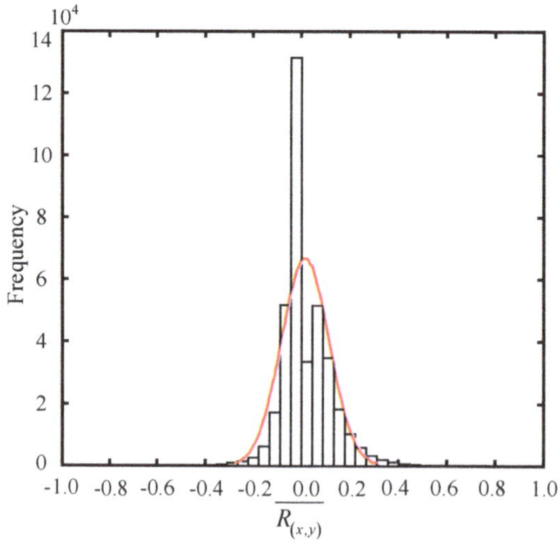

Figure 4. Frequency histogram of $\overline{R_{(x,y)}}$.

In Equations (11) and (12), $\mu_{\overline{R}}$ denotes the mean of $\overline{R_{(x,y)}}$; $\sigma_{\overline{R}}$ denotes the standard deviation of $\overline{R_{(x,y)}}$; T denotes the threshold selected for image segmentation; and F_{map} denotes the final result the ground fissure extraction.

For a better description of the modified MF-FDOG algorithm, we present a simulation experiment. The simulated signals of the ground fissure and edges, as well as the results of the modified MF-FDOG algorithm, are shown in Figure 5. Both the ground fissure and edge signals are dramatically enhanced by the MF convolution operation. However, the ground fissure information obtained by FDOG convolution shows the exact opposite response from that of MF convolution. As the response signals of edges are stronger than those of ground fissures, these two targets can be clearly distinguished by mean filter processing.

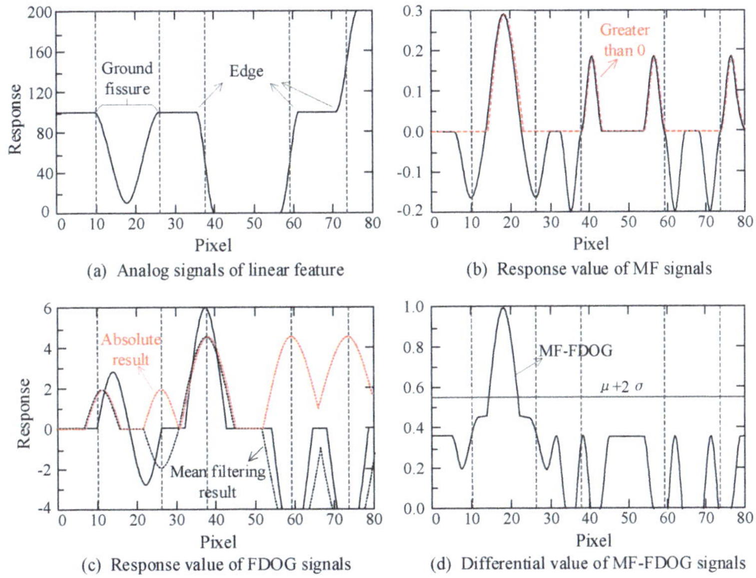

Figure 5. Analog signals of ground fissures and edges (modified from Stumpf et al. 2013): (**a**) analog signals of linear features; (**b**) response value of MF signals; (**c**) response value of FDOG signals; and (**d**) differential value of MF-FDOG signals.

High-resolution UAV images contain copious features, representing complex and diverse textures; this increases the difficulty of ground fissure identification, as these textures can be mistaken for ground fissures. Therefore, it is inevitable that multiple false ground fissure results will result from modified MF-FDOG processing. These false results must be removed. In this study, a "non-ground fissure removal" method based on random forest (RF) classification was introduced to improve the ground fissure results obtained using the modified MF-FDOG algorithm.

2.2. RF Classification in Loess Landform Areas

In this study, the RF algorithm was utilized to identify different ground targets (e.g., withered vegetation, snow, and bare soil). Vegetation indexes are common methods for extracting green vegetation from true color images [30,31]; however, they cannot extract withered vegetation [32]. Feng et al. [33] proved that the RF algorithm using both gray and texture information is better than other classification methods such as vegetation indexes, the nearest distance method, and the object-oriented method in urban vegetation cartography using true color images. In our experiment, the RF algorithm was used to remove "non-ground fissure" targets, such as snow and vegetation.

3. Post-Processing of Ground Fissure Extraction

3.1. Ground Fissure Connection

The modified MF-FDOG algorithm and the RF algorithm are operations based on pixels instead of objects. The grayscale values of ground fissures in the image often change unevenly at different positions. Linear targets (e.g., ground fissures) will inevitably become discontinuous after template convolution, resulting in small gaps along the linear targets. To eliminate small gaps distributed along the ground fissure candidates, the hit-or-miss transform [34] was adopted in our experiment. This method can connect broken targets within a small gap distance according to complementary templates M_1 $M_1$and M_2, which are termed the "hit template" and "miss template,", respectively. The central pixel of a patch that satisfies the conditions of both the "hit template" and "miss template" is used for further processing.

For the UAV images used in our study, we propose that gaps of no more than one-pixel length in a four-connected field (first sub-figure in Figure 6) should be connected. Therefore, the 3 × 3 template is built in the hit-or-miss transform. The pixel value for the foreground is set to 1, whereas the pixel value for the background is set to 0 (third sub-figure in Figure 7). Figure 7a (1)–(12) show cases with a pixel gap and (13)–(16) show cases that only satisfy the diagonal condition in an eight-connected field (second sub-figure in Figure 6). All "miss templates" are shown in Figure 7b. It should be noted that all "hit templates" and "miss templates" follow a corresponding one-to-one relationship.

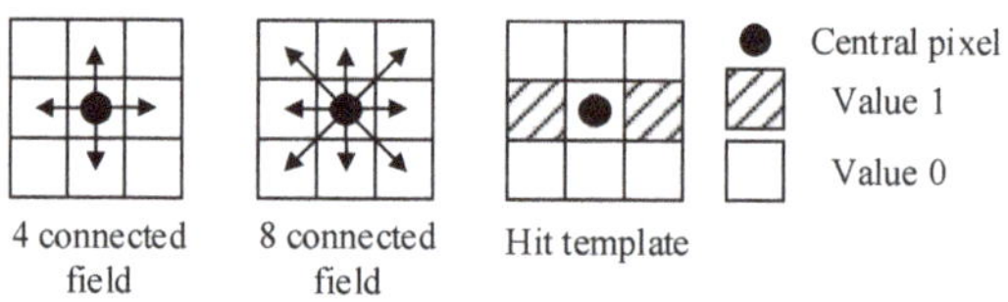

Figure 6. Illustration of connected fields and hit templates.

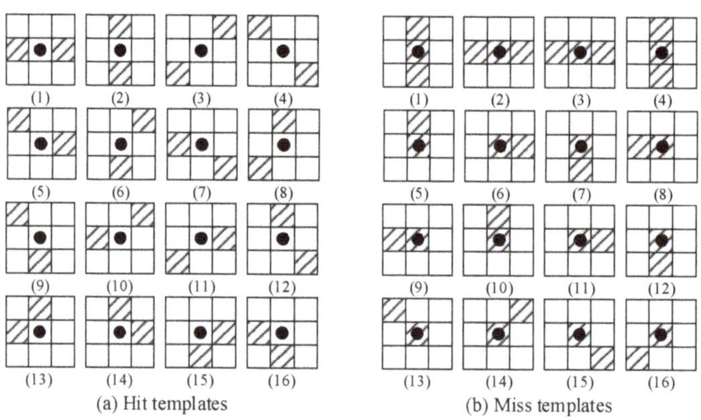

Figure 7. Hit templates and miss templates((**a**) and (**b**) represent all cases of hit template and miss template respectively).

In the process of gap elimination, the "hit template" is used to match the images pixel by pixel. A case is determined as a "hit" if the distribution of pixels within the coverage of the "hit template" is the same as the "hit template" itself in the binary image. Similarly, a case is termed a "miss" when the distribution of pixels within the coverage of the "miss template" is the same as the "miss template." "Hit" and "miss," which detect the position of the pixels of interest, are the key points of this method. To improve the efficiency of the calculation, the procedures of hit-or-miss transform are changed according to the mathematical morphology. The original binary image, I_1, is eroded by the "hit" template. Meanwhile, the complementary binary image I_2 caused by I_1 is eroded by the "miss template." The pixel intersection of I_2 and I_1 is the final result. The flow chart is shown in Figure 8. The results from the hit-or-miss transform are shown in Figure 9a,b.

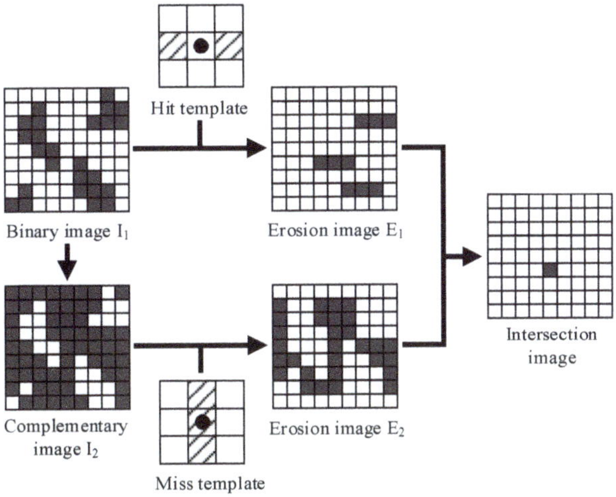

Figure 8. Flow chart of hit-or-miss transform.

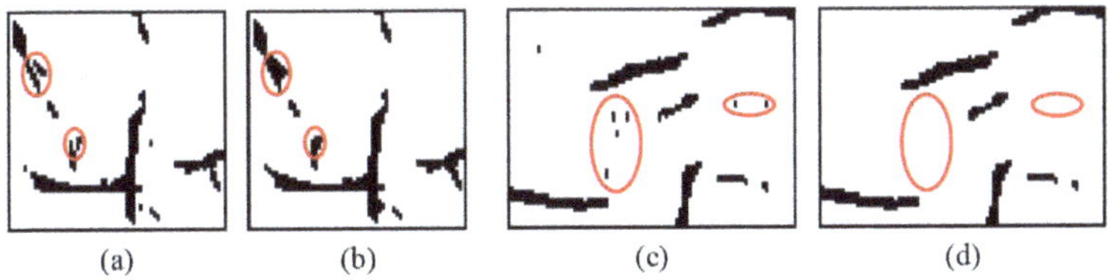

Figure 9. Results of hit-or-miss transform and fragment removal.

It is noted that the length of the used template is typically odd; thus, the central pixel can be chosen as the target pixel. However, a non-central pixel can also be chosen as the target pixel. In some special cases, the length of the template can be set as even. Moreover, it is unnecessary that the "hit" template and "miss" template are completely complementary for each pixel when they are built (see Figure 7a,b), provided that the pixels set to one do not overlap each other.

3.2. Fragment Removal from the Ground Fissure Candidates

Among the ground fissure candidate sets, there are some discrete pixels (or small sets of pixels), whose area is far less than the area of the ground fissure. To improve the results, these small patches must be removed. After connecting the disconnected ground fissure sections, these patches should be deleted if they contain three or fewer pixels that satisfy the relationship of the eight-connected field in the range of 3×3 (Figure 9c,d).

4. Application of the Ground Fissure Extraction Scheme to the Study Area

4.1. Introduction of the Ground Fissure Extraction Scheme

In this study, we propose an advanced scheme of ground fissure extraction based on UAV images according to typical feature information in loess areas, which involves image acquisition, image preprocessing, ground fissure candidate extraction, removing "non-ground fissures" and accuracy assessment of linear targets. The principle procedures are summarized in the following steps:

Step 1: Image acquisition. According to the survey results, the appropriate UAV is selected to obtain the images of the study area. Before flight, the terrain, wind direction and other factors should be considered, and the appropriate flight parameter and the position of the control points should be set. After flight, RTK equipment is used to obtain the coordinates of control points. If image quality is low in some areas, much flight time is added in order to improve the image quality.

Step 2: Image pre-processing. Firstly, the lens parameters will be imported into PixelGrid software to correct the distortion of the original image, in order to reduce the systematic error caused by the lens. After importing the control point data, the geometric correction of images will be processed by PixelGrid software, and the resampling method of three convolutions is selected to ensure the image accuracy. Then, the accuracy evaluation method of PixelGrid will be automatically carried out. Finally, the appropriate filtering and histogram matching method is used to enhance images for conveniently extracting ground fissure.

Step 3: Extraction of ground fissure candidates with the modified MF-FDOG algorithm. According to information obtained from surveying and image analysis, the program is run with suitable parameter settings (i.e., the optimal σ, L, and θ values) to obtain ground fissure candidates.

Step 4: "Non-ground fissure" target removal. Banded vegetation, edges of remaining snow and shadows from micro-topography will result in several "non-ground fissure" targets in the candidates extracted by step 2. Additionally, UAV images only include red, green, and blue bands. It is difficult to completely remove vegetation targets with such limited spectral information from the candidates. However, the inherent textural features of vegetation are valuable for vegetation information extraction, which is used to eliminate vegetation-induced false ground fissures. Therefore, the RF algorithm is used to classify the images based on both textural and spectral information. Then, the different categories, such as vegetation, are used to create mask files. Applying the mask files to the ground fissure candidate image can exclude non-ground fissure targets and improve the accuracy of ground fissure extraction. Finally, a binary image of the ground fissure is derived.

Step 5: Post-processing. The modified MF-FDOG algorithm is an image processing method based on single pixels. Therefore, the obtained binary image of ground fissures inevitably contains many small gaps and pixel sets whose areas are far less than that of the corresponding true ground fissures. To achieve more accurate results, mathematical morphology was employed to connect broken ground fissures and remove these scattered pixels.

Step 6: Accuracy assessment. In our study, an image classification method was used to identify non-ground fissure targets. The accuracy of image classification significantly affects the final results; thus, the Kappa coefficient and confusion matrix were used to evaluate RF classification accuracy in this study. In addition, the receiver operating characteristic curve (ROC) was used to evaluate the results of ground fissure extraction [35].

4.2. Study Area and Data Source

A mountain over a tunnel in Qinghai province and a mining area in Gansu province were selected as the study areas to represent typical developing loess landforms (red five-pointed stars in Figure 10). Due to the instability of loess units, as well as other natural and anthropogenic factors, ground fissures are a very common geological phenomenon. These two areas predominantly comprise very complex mountainous terrain. The exposed loess surface is unstable because of its typical semi-arid climate and small forest coverage. To prevent soil erosion, the local inhabitants have planted banded vegetation and trees on the mountain.

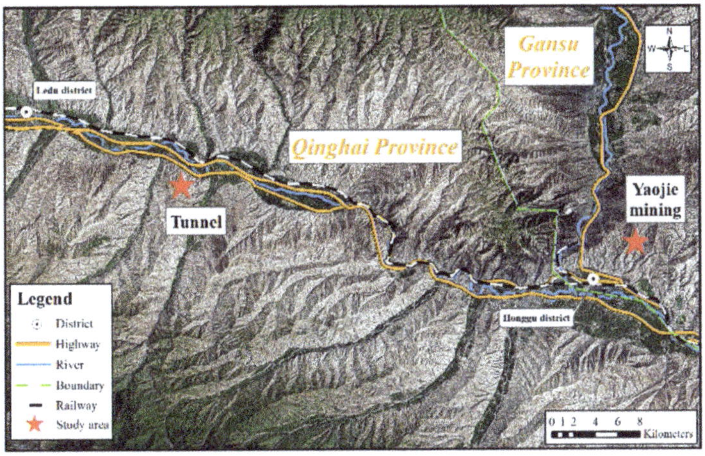

Figure 10. Image of the locations of the two study areas, accessed on 7 May 2018 (http://www.91weitu.com/).

True color images of the tunnel area and mining area were obtained at approximately 13:00 on 25 March 2015, and 09:00 on 18 November 2016, respectively. The former was acquired by a small fixed-wing UAV oblique photography platform (Figure 11a, come from Third Aerial Survey and Remote Sensing Institute of the Ministry of Natural Resources, Chengdu, China) and covers a larger area around the tunnel. The latter was acquired by a small four-rotor UAV platform (Figure 11b) and covers a small area around the mining area. The detailed attributes of the two platforms are listed in Table 1.

Figure 11. Equipment used for image acquisition ((**a**) is a homemade UAV that comes from the Third Aerial Survey and Remote Sensing Institute of the Ministry of Natural Resources; (**b**) is Phantom 4 Pro comes from SZ DJI Technology Co., Ltd.).

Table 1. Parameters of the UAV equipment.

		Tunnel Area	Mining Area
Properties of sensor	Model	SONY DSC-RX1	PHANTOM4
	Resolution (dpi)	6000 × 4000	4000 × 2250
	Effective focal length (mm)	35	20
	Photoreceptor size (mm)	35.8 × 23.9	25.4 × 25.4
UAV platform	Flight height (m)	3000	150
	Flight speed (km/h)	90–120	72
	Payload (kg)	3–5	-

In our experiment, the computer with a CPU model Intel (R) Core (TM) i5-2450M was used for data processing. The frequency of the CPU is 2.5 GHz, and the memory of the computer is 4 GB. The modified MF-FDOG algorithm was developed by C/C++

programming language based on Visual Studio 2013. PixelGrid software was used to preprocess the raw images. For the RF algorithm, the EnMAP-Box2.1.1 software (Earth Observation Center of DLR. German: EnMap Box) was used.

For the tunnel area (Figure 12), the coverage of the image was 5 km × 2 km, with an elevation of 1930–2500 m. According to the field investigation, there are a large number of ground fissures, loess craters, remaining snow, and withered grasses in the mountain areas. The maximum width of ground fissures is no more than 30 cm: however, the depth is up to 120 cm. Some ground fissures are covered by broken soil after a period of weathering and snow melting. The coverage of the mining area (Figure 13), which is less than that of the tunnel area, is predominantly composed of bare soil, withered grass, ground fissures, and artificial structures. As shown in Table 1, the flight altitude of PHANTOM4 is approximately 150 m. Therefore, we can obtain high-resolution images in which the ground fissures can be clearly recognized.

Figure 12. Experimental images of mountains covering the tunnel (the two local areas marked by the rectangles I–II were selected for ground fissure extraction).

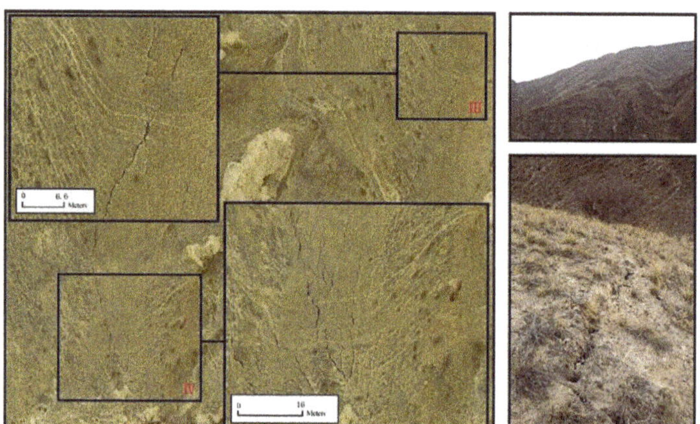

Figure 13. Experimental images of the mining area (the two local areas marked by the rectangles III–IV were selected for ground fissure extraction).

To improve the efficiency of the data processing, we selected and cropped small subsections from the complete images (I,II,III, and IV in Figures 12 and 13). All previously described typical objects were found in these patches; thus, they were deemed highly representative for testing the proposed ground fissure extraction scheme.

5. Ground Fissure Extraction Results and Analysis in a Loess Area

5.1. Selection of Parameter σ and Direction Number in the Template

As described in Section 2.1, the key aspect of the modified MF-FDOG algorithm is the convolution operation using specific templates that are similar to the vertical profile of ground fissures. After convolution, the contrast between ground fissures and other targets will be enhanced. Thus, it is crucial to select an appropriate parameter, σ, for fitting the vertical profile of ground fissures. A previous study [25] noted that: (1) the parameter σ should increase as the image resolution is enhanced; (2) for a ground fissure in a single image, the value of the detectable width will be larger as σ increases. Additionally, σ maintains good sensitivity to five times the optimal width of a ground fissure. Therefore, parameter σ can be improved using the image resolution and the width of ground fissures in the real scene, which is selected according to Table 2.

Table 2. Relationship between the image resolution and parameter σ.

Resolution (m)	0.1	0.08	0.5
σ (px)	0.67	0.75	1.34

Additionally, the template of the convolution operation is related to parameter σ and the minimum length of the ground fissure, L. The two parameters define the width and length of the template, respectively. An appropriate template can achieve a better effect when the minimum length of the ground fissure is greater than or equal to L. Therefore, it is necessary to select an optimal value of L according to the actual minimum length of the ground fissure and effectively reduce the possibility of "non-ground fissure" targets in the ground fissure extraction results.

Another important parameter of the modified MF-FDOG algorithm is the template direction number. The direction of the ground fissure is arbitrary; thus, it is necessary to process the entire image using a group of templates with different directions for detecting ground fissures with unknown positions and directions. Multiple template directions can be obtained by setting different angular intervals. For example, if six is selected as the template direction number in the algorithm, the angular interval will be computed by Equation (4). Generally, the larger the template direction number, the more accurate the results.

The template direction number, N, is a crucial parameter that affects the performance of the modified MF-FDOG algorithm in ground fissure extraction from two major aspects: (1) the template direction number, which is related to the sensitivity of direction detection, and the accuracy of ground fissure extraction. A larger direction number not only does fail to improve the accuracy of ground fissure extraction, but also increases the redundancy. Increasing N can even reduce the accuracy of ground fissure extraction. (2) Time consumed in the modified MF-FDOG algorithm has a positive correlation with an increase of N. If we reduce the template direction number, the efficiency of the algorithm will increase linearly. However, the performance of sensitivity in the direction detection will decrease. Therefore, to maintain high accuracy in ground fissure extraction, the template direction number should be reduced as much as possible to improve the efficiency of the modified MF-FDOG algorithm.

To determine the relationship between the accuracy of the modified MF-FDOG algorithm and the direction number of the template, we designed a specific experiment, which can provide important insights into selecting the template direction number. Firstly, only one group of templates is used to extract ground fissures with different widths from the UAV image. To determine whether the template has the same directionality characteristics as the ground fissures of different width, it is necessary to simulate two ground fissures with different widths covering the full direction of 360° (Figure 14). The length and width of the simulated images are 295 pixels. Secondly, we created eight groups of templates with different direction numbers, which have the same values of parameters σ and L. Finally, the modified MF-FDOG convolution operation results are shown in Figures 15 and 16.

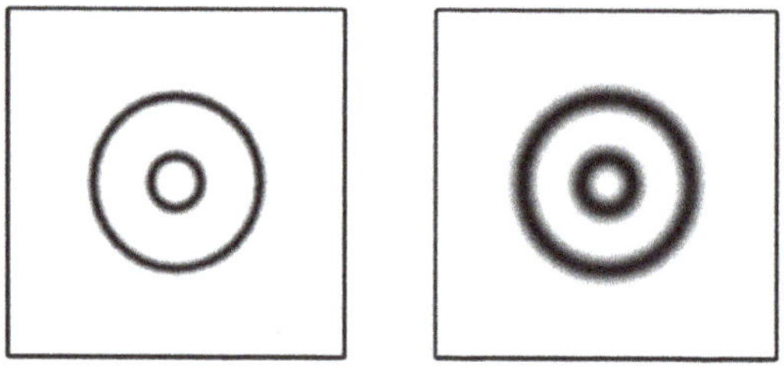

(a) Ground fissure with smaller width (b) Ground fissure with larger width

Figure 14. Illustration of analog images: (**a**) small-width ground fissure; and (**b**) large-width ground fissure.

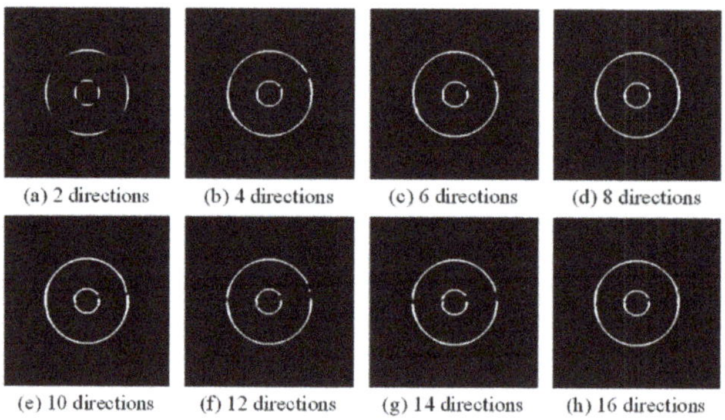

Figure 15. Modified MF-FDOG convolution results for the small-width ground fissure.

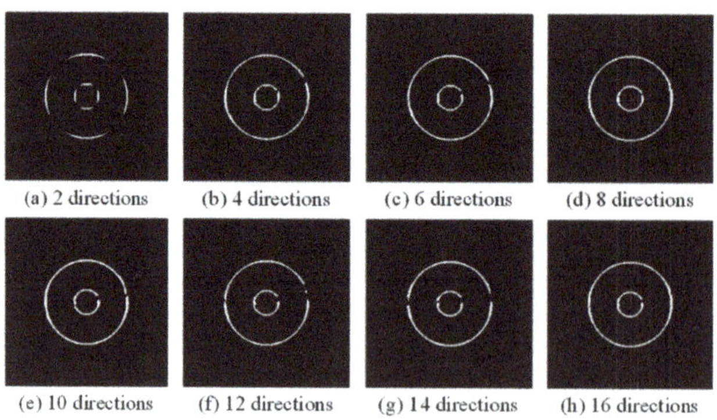

Figure 16. Modified MF-FDOG convolution results for the large-width ground fissure.

To evaluate the accuracy of the results of the modified MF-FDOG algorithm, the results of the visual interpretation were considered as accurate results of the ground fissures (i.e., expert maps). Following a comparison of the results of ground fissure extraction and the expert maps, the number of ground fissure pixels that were correctly extracted is shown in Table 3, where TP indicates true pixels, i.e., correctly extracted pixels. The statistical results indicate that TP reaches a maximum when the direction number is 10. Furthermore, data fitting was conducted using the two result sets in Table 3 to demonstrate the relationship between the accuracy of ground fissure extraction and the direction number of the template (Figure 17a,b). The two fitted curves exhibit the same trend and achieve a peak value around the position associated with direction number 10. It can be inferred that ground fissures with different widths have no effect on the selection of template direction numbers. In addition, the templates with 10 directions can achieve optimal results. It should be noted that the direction in the modified MF-FDOG algorithm only uses even numbers; therefore, 10 is selected to be the optimal direction number in our study. The relevant angular interval is 18°.

Table 3. Statistics of true pixels for analog images.

	Direction							
TP	2	4	6	8	10	12	14	16
Small	908	1667	1708	1854	1979	1679	1788	1880
Large	1154	1938	2155	2267	2331	2054	2186	2132

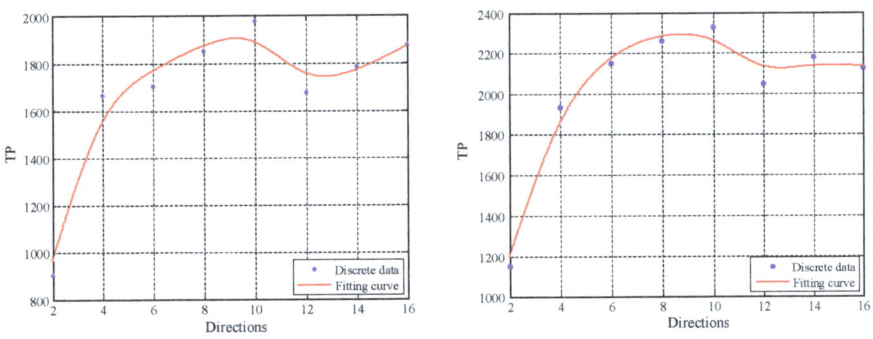

Figure 17. Fitted curve representing the relationship between template direction numbers and true ground fissure pixel numbers obtained for smaller ground fissures (Figure 14a) and larger ground fissures (Figure 14b).

5.2. Results of Ground Fissure Extraction Based on Modified MF-FDOG Algorithm

Figure 12 (I and II) and Figure 13 (III and IV) show the original experimental images of this study. After processing using the modified MF-FDOG algorithm, the results from each step are shown in Figure 18a–c. The ground fissure candidates are presented as bright white against a black background (Figure 18a). The candidate sets include "non-ground fissure" targets, e.g., the edges of snow cover and withered grass.

The RF algorithm was used to identify these targets. The spectral information of snow is most obvious due to its bright white spectral signature. However, it is difficult for the RF algorithm to distinguish ground fissures and the boundaries between snow and loess because they have similar vertical profiles. The use of a dilation operation in the mathematical morphology for the boundaries can eliminate the effect of this phenomenon. Specifically, we combined six types of textural information (mean, variance, homogeneity, entropy, and second moment), which were created by a 7 × 7 co-occurrence matrix, as well as the spectral information, to calculate the required parameters of the RF algorithm by sample training. The samples should be a small polygon shape [36]. The results of extracted vegetation and snow by the RF algorithm are shown in Figure 19 I–IV. Then, the vegetation and snow-induced "false ground fissures" were excluded by mask files. The RF algorithm was integrated into the EnMAP-Box software.

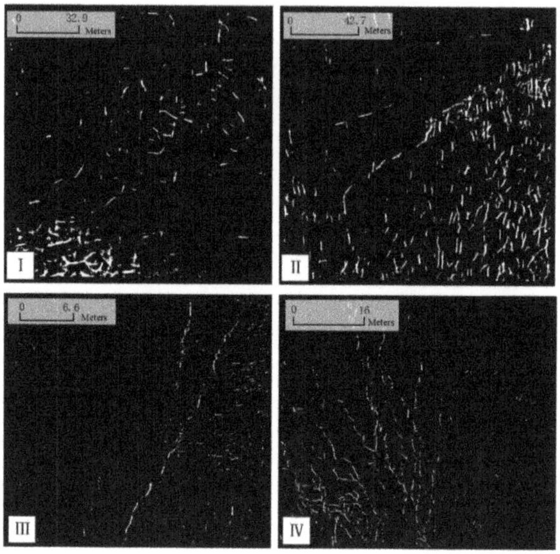

(a) Results of ground fissure extraction based on the modified MF-FDOG algorithm.

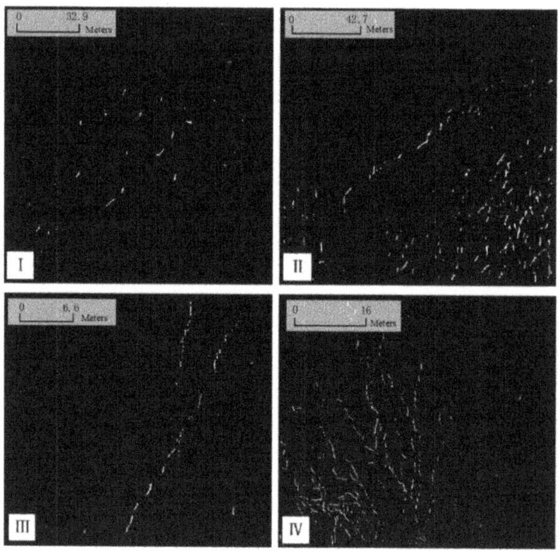

(b) Results of ground fissure extraction.

Figure 18. *Cont.*

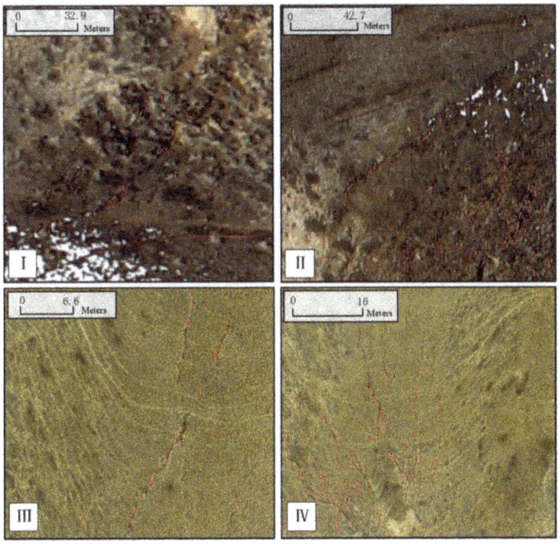

(c) Results of ground fissures superposed on the original experimental images.

Figure 18. Results of experimental images of ground fissure extraction.

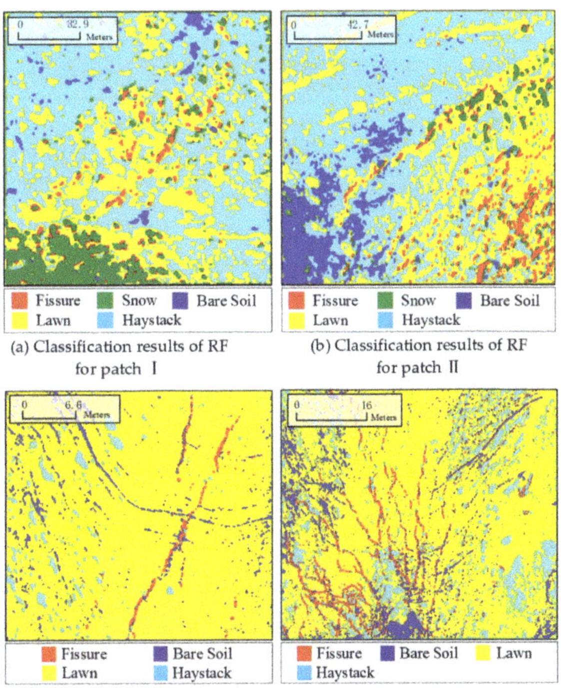

(a) Classification results of RF for patch I

(b) Classification results of RF for patch II

(c) Classification results of RF for patch III

(d) Classification results of RF for patch IV

Figure 19. Results of the RF algorithm: (**a**–**d**) are the classification results of RF for patch I–IV, respectively.

For the RF classification results, overall accuracy [37] and Kappa coefficient [38] are typically employed for the accuracy assessment. The assessment results for image patches I, II, III, and IV are listed in Table 4. According to the accuracy indices, all four groups of RF classification results exhibit good precision. Moreover, the Kappa coefficients for image patches II and III are higher than 90%, which further demonstrates the accuracy and reliability of the RF algorithm. The accuracy of image patch I is lower because of the more complex spectral and textural information in this area.

Table 4. User accuracy and producer accuracy.

	I	II	III	IV
Overall accuracy/%	79.81	93.10	94.98	97.22
Kappa accuracy/%	73.55	90.89	89.57	96.01

Finally, the post-processing procedure introduced in Section 3 was conducted, and the final results of ground fissure extraction were derived (Figure 18c I–IV). According to the original image, the ground fissures in Figure 18a I do not satisfy the characteristics of a linear distribution due to their complex and broken shape. Therefore, only a small fraction of ground fissures with a discrete distribution were extracted. After applying the RF algorithm, many error targets caused by shadows remain in the lower right corner of Figure 18b II, along with correct ground fissures. In comparison, Figure 18b III,IV clearly shows the distribution and shape of ground fissures. In order to better demonstrate the effect of the modified MF-FDOG algorithm, the results of ground fissure extraction are superposed on the original experimental image in Figure 18c.

5.3. Accuracy Analysis of the Modified MF-FDOG Algorithm

To evaluate the effectiveness and accuracy of the modified MF-FDOG algorithm used in ground fissure extraction based on UAV images in a loess area, expert images (Figure 20) of ground fissures were obtained by visual interpretation according to surveys and previous experience. These expert images have sufficient precision for use as reference images. Furthermore, three other traditional algorithms (e.g., Canny operator, image segmentation, and SVM (support vector machine) algorithm) were used to extract ground fissures. The results of these algorithms are shown in Figures 21–24, in which (a), (b), (c) and (d) represent the results of the Canny operator, image segmentation, and the SVM algorithm, respectively. The accuracy of the shapes marked A–H was analyzed in this study.

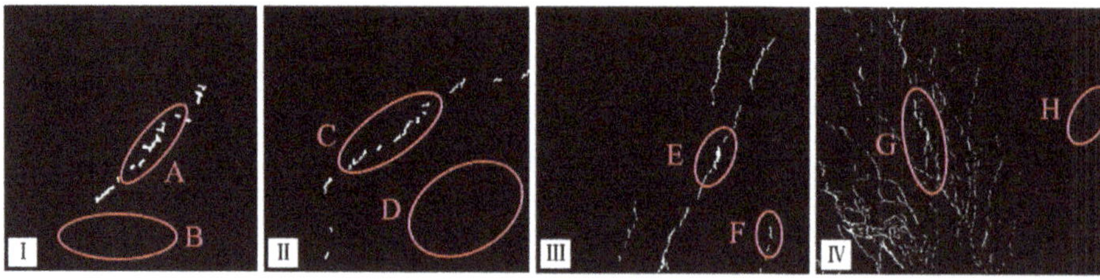

Figure 20. Results of visual interpretation.

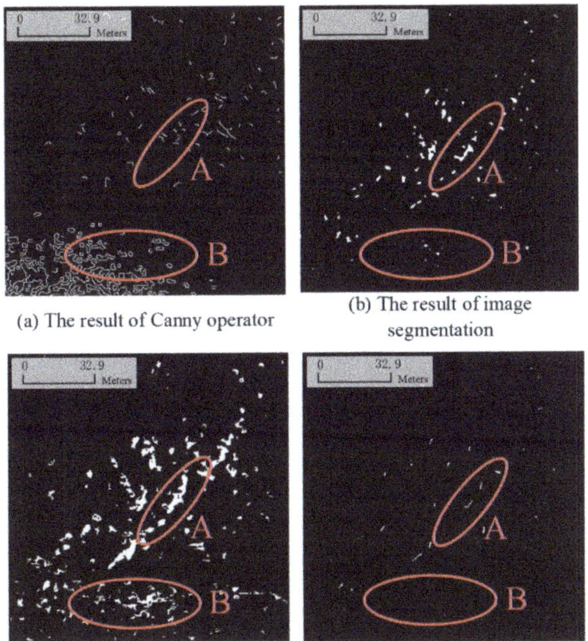

(a) The result of Canny operator

(b) The result of image segmentation

(c) The result of SVM algorithm

(d) The result of modified MF-FDOG algorithm

Figure 21. Comparison of ground fissure extraction in image patch I.

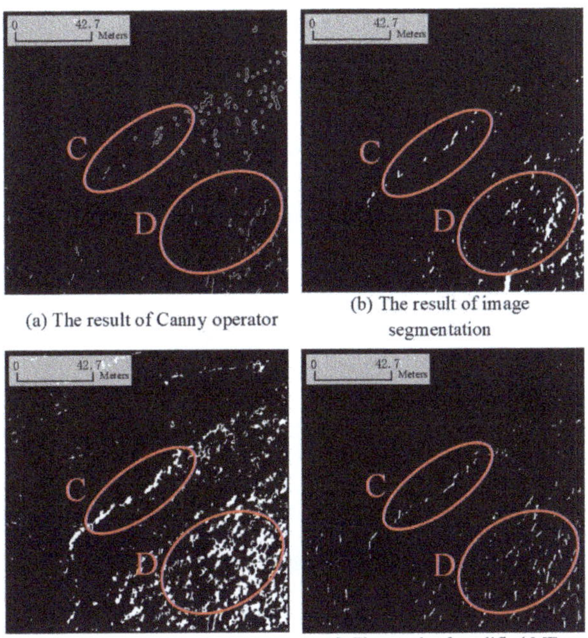

(a) The result of Canny operator

(b) The result of image segmentation

(c) The result of SVM algorithm

(d) The result of modified MF-FDOG algorithm

Figure 22. Comparison of ground fissure extraction in image patch II.

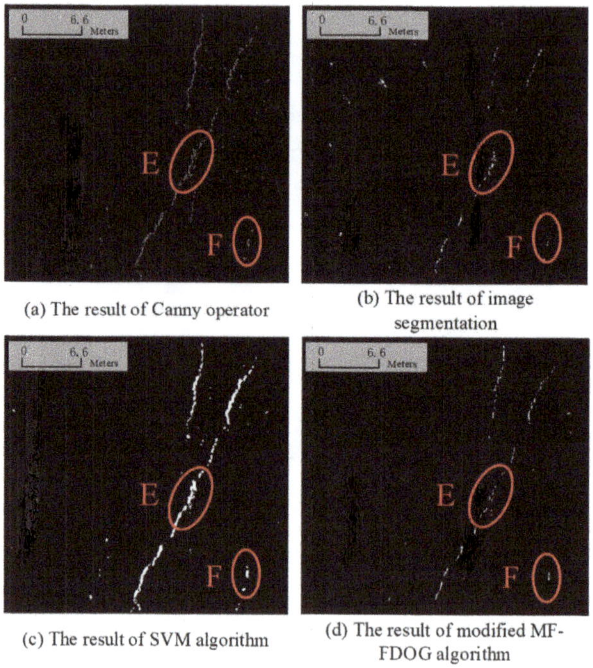

(a) The result of Canny operator
(b) The result of image segmentation
(c) The result of SVM algorithm
(d) The result of modified MF-FDOG algorithm

Figure 23. Comparison of ground fissure extraction in image patch III.

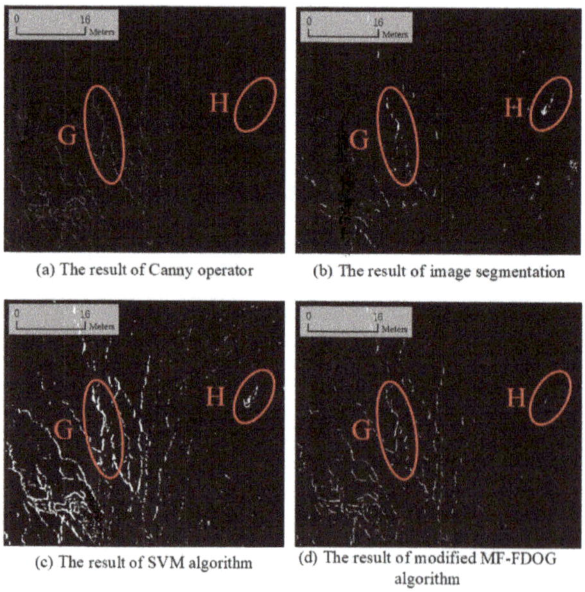

(a) The result of Canny operator
(b) The result of image segmentation
(c) The result of SVM algorithm
(d) The result of modified MF-FDOG algorithm

Figure 24. Comparison of ground fissure extraction in image patch IV.

The Canny operator, which is only sensitive to edge information in the image, can extract many edge targets. However, it also contains many "non-ground fissures" (e.g., the snow marked by shape B in Figure 21a and the shadow marked by shape D in Figure 22a. In addition, the Canny operator cannot efficiently extract ground fissures with blurred edges (e.g., the ground fissures marked by shapes A in Figure 21a and C in Figure 22a). A key characteristic of the Canny operator for ground fissure extraction is that the majority of ground fissure edges are not completely closed; therefore, it is difficult to extract complete ground fissures.

A crucial aspect of the image segmentation method is dividing the image grayscale into many levels, which can be set to the same range or not. To do this, some objects in the same grayscale level, such as shadows, are considered as ground fissures (as shown by the area marked by shape B in Figure 21b and shape D in Figure 22b). Moreover, the ground fissure results do not retain their clear linear features (see Figure 24b).

Grayscale information and textural information are the main classification bases of the SVM algorithm, which exhibits a good ground fissure extraction result (as shown by the area marked by shapes A–H in Figure 21c to Figure 24c). However, withered grass (see shape B in Figure 21c), shadows caused by micro topography (see shape D in Figure 22c), loess pits (see shape H in Figure 24c), and even loess close to ground fissures (see shape A in Figure 21c) are wrongly extracted by the SVM algorithm without considering the geometric information of ground fissures. For ground fissures with large variations in width, the SVM algorithm exhibits good adaptability during extraction (as shown by the areas marked by shape E in Figure 23c and shape G in Figure 24c).

In contrast to the above algorithms, the modified MF-FDOG algorithm is sensitive to ground fissures with small and narrow features. The extraction results retain good ground fissure geometry (as shown by shape A in Figure 21d, shape C in Figure 22d, and shape G in Figure 24d). However, the modified MF-FDOG algorithm does not obtain a good result if the ground fissures do not exhibit a linear distribution in the local area (see shape E in Figure 23d).

5.4. Accuracy Assessment of the Different Ground Fissure Extraction Algorithms

5.4.1. Self-Assessment of the Modified MF-FDOG Algorithm

To evaluate the results of ground fissure extraction, the ROC was used for the accuracy assessment. The ROC was initially applied in the medical field to evaluate the performance of feature extraction methods. The first application of the ROC in machine learning was in 1889 [35]. For the accuracy assessment of linear targets extractions, Tveite et al. [39] suggested that the buffer zone of the result of the modified MF-FDOG algorithm should be increased step by step and overlaid on an expert map. Accordingly, we selected the horizontal distance between the central point of two adjacent pixels as the single unit of the buffer zone. and then increased the buffer zone from one to ten units. Subsequently, the different buffer zones of each ground fissure were used for the overlay analysis.

By searching each ground fissure pixel after buffer processing, ground fissure pixels were deemed to be correctly extracted if the pixels of the expert map in the same position are also ground fissure pixels. On the contrary, pixels are incorrectly extracted if the pixels of the expert map in the same position are not ground fissure pixels. If the pixels in the raw image corresponding to ground fissure pixels in the expert map are not recognized by the algorithm, the ground fissure pixels are missed. We count the number of instances of the above three cases and calculate the TPR (true positive rate) and FPR (false positive rate) according to Equations (13) and (14), which express the ratio of correct and incorrect extraction, respectively. The results are shown in Table 5:

$$TPR = \frac{TN}{M_{total}} \tag{13}$$

$$FPR = \frac{FN}{N_{total} - M_{total}} \tag{14}$$

Table 5. Buffer analysis of the tunnel area and mining area.

-	Tunnel Area				Mining Area			
	Subset I		Subset II		Subset III		Subset IV	
	TPR	**FPR**	**TPR**	**FPR**	**TPR**	**FPR**	**TPR**	**FPR**
Buffer 1	0.1195	0.0022	0.2938	0.0080	0.2490	0.0005	0.3387	0.0027
Buffer 2	0.2489	0.0058	0.4276	0.0177	0.4081	0.0017	0.5165	0.0075
Buffer 3	0.4323	0.0144	0.6472	0.0357	0.5638	0.0046	0.6537	0.0184
Buffer 4	0.5338	0.0223	0.7253	0.0513	0.6312	0.0074	0.7090	0.0282
Buffer 5	0.6150	0.0331	0.7821	0.0710	0.6853	0.0108	0.7547	0.0405
Buffer 6	0.6767	0.0468	0.8181	0.0936	0.7270	0.0144	0.7872	0.0526
Buffer 7	0.7135	0.0574	0.8399	0.1104	0.7579	0.0182	0.8141	0.0652
Buffer 8	0.7534	0.0725	0.8635	0.1330	0.7823	0.0223	0.8384	0.0795
Buffer 9	0.7895	0.0895	0.8744	0.1444	0.8038	0.0270	0.8581	0.0947
Buffer 10	0.8218	0.1063	0.8957	0.1803	0.8180	0.0315	0.8750	0.1087

Here, *TN* and *FN* denote the correctly extracted pixel number and incorrectly extracted pixel number, respectively; N_{total} represents the total pixel number of an image; and M_{total} represents the pixel number of the expert maps derived by visual interpretation. The range of *TPR* and *FPR* is [0, 1].

The ROCs were drawn using Table 5 and are shown in Figure 25. The vertical axis represents the TPR, and the horizontal axis represents the FPR. As the FPRs for part I, II, and IV are close to 10%, their TPRs all exceed 80%. The FPR for patch III is only 2.6% when the TPR exceeds 80%, demonstrating that the ground fissure extraction accuracy for patch III is the highest among all four image patches.

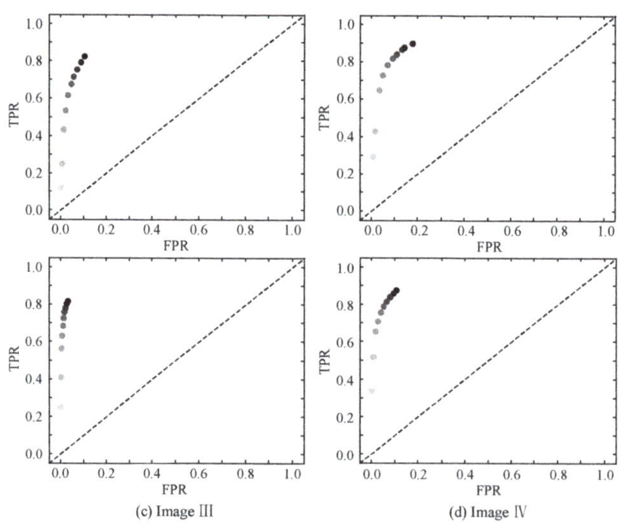

Figure 25. ROCs of the ground fissure extraction results.

5.4.2. Accuracy Comparison of the Different Ground Fissure Extractions Algorithms

Ground fissures can be extracted by the Canny algorithm in the form of edges, which cannot be compared with the SVM algorithm, grayscale threshold algorithm, or modified MF-FDOG algorithm. Therefore, we compared and analyzed the latter three algorithms. The overall accuracy [37] is calculated using Equation (15), and results are shown in Table 6:

$$\varepsilon = \frac{TN + \overline{TN}}{N_{sum}} \quad (15)$$

where ε denotes the overall accuracy and $\overline{TN}$ denotes the number of correctly extracted background pixels. As shown by the results (Table 6 and Figure 26), the modified MF-FDOG algorithm is better than both the SVM algorithm and the grayscale threshold segmentation algorithm.

Table 6. Comparison of the overall accuracy of three ground fissure extraction algorithms.

Subset	SVM Algorithm	Grayscale Threshold Algorithm	The Modified MF-FDOG Algorithm
I	0.9462	0.9782	0.9836
II	0.9217	0.9758	0.9806
III	0.9826	0.9869	0.9871
IV	0.9587	0.9754	0.9749

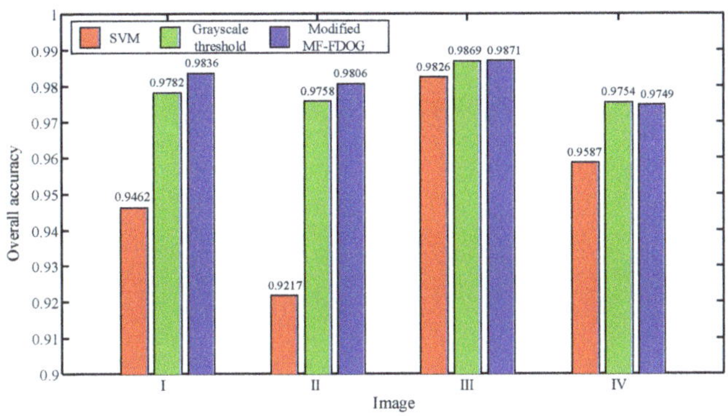

Figure 26. Overall accuracy of the ground fissure extraction results.

6. Conclusions and Future Works

The image of the loess region contains rich texture information, which makes it difficult for traditional edge extraction operators to accurately distinguish ground cracks and other linear objects, resulting in limited extraction accuracy. In order to improve the accuracy of ground fissure extraction, this study proposes an improved MF-FDOG algorithm. Experimental results show that this method has high accuracy and robustness.

In this study, the normalization was used to generate the response matrices of the convolution, which solves the instability of the sensitivity correction parameter and effectively extract initial ground fissure candidates. At the same time, the automatic evaluation of optimal directions and the number of the MF/FDOG template can effectively improve the efficiency of the algorithm. In addition, using the sum of the mean value and two times the mean square error in the Gaussian curve as the threshold instead of an arbitrary threshold can achieve an adaptive threshold parameter, which effectively avoids the influence of uncertainty in image segmentation.

In future work, our research will focus on the improvement of the efficiency of data processing and the accuracy of ground fissure extraction. Firstly, the parallel processing technology based on graphics processing unit (GPU) [40] calculations and the development of multiple programming languages (such as CUDA and OpenCL [41]) can greatly improve the efficiency of data processing; secondly, although the vertical photography image has the advantage that the image data can be processed quickly, for areas with large slope fluctuations, the ground crack target will be deformed. Therefore, the images of different spatial resolutions or different types of images can be combined to improve the accuracy of fracture extraction, thereby further ensuring the robustness of the proposed method.

Author Contributions: Conceptualization, G.L. and H.J.; methodology, B.W.; validation, R.Z.; formal analysis, B.Y.; investigation, S.W.; data curation, R.Z.; writing—original draft preparation, B.W.; writing—review and editing, H.J. and G.L.; supervision, G.L.; project administration, G.L. All authors have read and agreed to the published version of the manuscript.

Funding: This research was funded by the National Key R&D Program of China under Grant 2017YFB0502700, the National Natural Science Foundation of China under Grant 41701535 and Grant 41771402 and the Sichuan Science and Technology Program under Grant 2019YJ0224.

Institutional Review Board Statement: Not applicable.

Informed Consent Statement: Not applicable.

Data Availability Statement: Data are available on request.

Acknowledgments: The authors would like to thank the Third Aerial Survey and Remote Sensing Institute of the Ministry of Natural Resources for providing the UAV images.

Conflicts of Interest: The authors declare no conflict of interest.

References

1. Wang, S.; Dong, Y.G.L. *Geology and Geomorphology*; China Agricultural University Press: Beijin, China, 2013; pp. 257–267.
2. Chai, H.X.; Cheng, W.M.; Qiao, Y.L. Classification system of 1:1000,000 digital loess geomorphology in China Geo-information. *Science* **2006**, *8*, 6–13.
3. Wang, X.Y.; Mao, S.H.; Zhang, Y.J. Differential Settlement due to Ground Fissures in Xi'an. *Adv. Mater. Res.* **2015**, *1065–1069*, 410–413. [CrossRef]
4. Li, Q.C.; Ye, P.; Shao, G.Z.; Luo, W.B. Integrated Geophysical Detection on Ground Fissure in Shanxi Jingyang Seismostation of National Highway 211. *Chin. J. Eng. Geophys.* **2014**, *11*, 106–111.
5. Wang, Z.F.; Shen, S.L.; Cheng, W.C.; Xu, Y.S. Ground fissures in Xi'an and measures to prevent damage to the Metro tunnel system due to geohazards. *Environ. Earth Sci.* **2016**, *75*, 511. [CrossRef]
6. Stumpf, A.; Niethhammer, U.; Rothmund, S.; Mathieu, A.; Malet, J.P.; Kerle, N.; Joswing, M. Advanced image analysis for automated mapping of landslide surface fissures. *Landslide Sci. Pract.* **2013**, *2*, 357–363.
7. Giuseppe, P.; Nicolai, P. Edge and line oriented contour detection: State of the art. *Image Vis. Comput.* **2011**, *29*, 79–103.
8. Quackenbush, L.J. A review of techniques for extracting linear features from imagery. *Photogramm. Eng. Remote Sens.* **2004**, *70*, 1383–1392. [CrossRef]
9. Shruthi, B.V.; Kerle, N.; Jetten, V.G. Object-based gully feature extraction using high spatial resolution imagery. *Geomorphology* **2011**, *134*, 260–268. [CrossRef]
10. Baruch, A.; Filin, S. Detection of gullies in roughly textured terrain using airborne laser scanning data. *J. Photogramm. Remote Sens.* **2011**, *66*, 564–578. [CrossRef]
11. Jain, R.; Kasturi, R.; Schunck, B.G. *Machine Vision*; McGraw-Hill: New York, NY, USA, 1995.
12. Lawrence, G.R. *Machine Perception of Three-Dimensional Solids*; Institute of Technology: Boston, MA, USA, 1963.
13. Sobel, I. *Camera Models and Machine Perception*; California Department of Computer Science, Stanford University: Stanford, CA, USA, 1970.
14. Prewitt, J.M.S. *Object Enhancement and Extraction. Picture Processing and Psychopictorics*; Academic Press: Cambridge, MA, USA, 1970.
15. Koenderink, J.J. Theory of "Edge-Detection". *Anal. Sci. Eng. Beyond* **2012**, *6*, 35–49.
16. Gonzalez, C.I.; Melin, P.; Castro, J.R.; Mendoza, O.; Castillo, O. An improved sobel edge detection method based on generalized type-2 fuzzy logic. Soft Computing—A Fusion of Foundations. *Methodol. Appl.* **2016**, *20*, 773–784.
17. Manasa, N.; Mounica, G. Brain tumor detection based on canny edge detection algorithm and its area calculation. *Int. Comput. Math. Sci.* **2016**, *5*, 10–13.
18. Hu, X. *Research on Subway Tunnel Crack Detection Technology Based on Image Processing*; Beijing Jiaotong University: Beijing, China, 2014.
19. Alam, M.A.; Ali, M.M.N.; Syed, M.A.A.; Sorif, N.; Rahaman, M.A. An algorithm to detect and identify defects of industrial pipes using image processing. In Proceedings of the 8th International Conference on Software, Knowledge, Information Management and Applications (SKIMA), Dhaka, Bangladesh, 18–20 December 2014; pp. 1–6.
20. Khalifa, I.; Aboutabl, A.E.; Barakat, G.S.A. A New Image-Based Model for Predicting Cracks in Sewer Pipes. *Int. J. Adv. Comput. Sci. Appl.* **2013**, *4*, 65–71. [CrossRef]
21. Kirstein, S.; Muller, K.; Walecki, M.N.; Desemo, T.M. Robust adaptive flow line detection in sewer pipes. *Autom. Constr.* **2012**, *21*, 24–31. [CrossRef]
22. Li, Z.; Liu, X.L. Geomorphic regionalization and agricultural evaluation of geomorphic featuresin Gansu. *J. Gansu Agric. Univ.* **1994**, *30*, 444–449.

23. Chaudhuri, S.; Chatterjee, S.; Katz, N.; Nelson, M.; Goldbaum, M. Detection of blood vessels in retinal images using two-dimensional matched filters. *IEEE Trans. Med Imaging* **1989**, *8*, 263–269.
24. Zhang, B.; Zhang, L.; Zhang, L.; Karray, F. Retinal vessel extraction by matched filter with first-order derivative Gaussian. *Comput. Biol. Med.* **2010**, *40*, 438–445. [CrossRef] [PubMed]
25. Stumpf, A.; Malet, J.P.; Kerle, N.; Niethammer, U.; Rothmund, S. Image-based mapping of surface fissures for the investigation of landslide dynamics. *Geomorphology* **2013**, *186*, 12–27. [CrossRef]
26. Breimal, L. *Random Forests. Machine Learning*; Springer: Berlin/Heidelberg, Germany, 2001; Volume 45, pp. 5–32.
27. Turin, G. An introduction to matched filters. *IRE Trans. Inf. Theory* **1960**, *6*, 311–329. [CrossRef]
28. Woodward, P.M. *Probability and Information Theory with Applications to Radar*; Pergamon Press Ltd.: New York, NY, USA, 1953; Volume 41, pp. 59–68.
29. Nevatla, R.; Babu, K.R. Linear feature extraction and description. *Comput. Graph. Image Process.* **1980**, *13*, 257–269. [CrossRef]
30. Torres, S.J.; Peńa, J.M.; Castro, A.I.D.; López, G.F. Multi-temporal mapping of the vegetation fraction in early-season wheat fields using images from UAV. *Comput. Electron. Agric.* **2014**, *103*, 104–113. [CrossRef]
31. Rasmussen, J.; Ntakos, G.; Nielsen, J.; Svensgaard, J.; Poulsen, R.N.; Christensen, S. Are vegetation indices derived from consumer-grade cameras mounted on UAVs sufficiently reliable for assessing experimental plots. *Eur. J. Agron.* **2016**, *74*, 75–92. [CrossRef]
32. Hung, C.; Xu, Z.; Sukkarieh, S. Feature learning based approach for weed classification using high resolution aerial images from a digital camera mounted on a UAV. *Remote Sens.* **2014**, *6*, 12037–12054. [CrossRef]
33. Feng, Q.; Liu, J.; Gong, J. UAV remote sensing for urban vegetation mapping using random forest and texture analysis. *Remote Sens.* **2015**, *7*, 1074–1094. [CrossRef]
34. Serra, J. *Image Analysis and Mathematical Morphology. Orlando*; Academic Press: Cambridge, MA, USA, 1982.
35. Fawcett, T. An introduction to ROC analysis. *Pattern Recognit. Lett.* **2006**, *27*, 861–874. [CrossRef]
36. Chen, D.M.; Stow, D. The effect of training strategies on supervised classification at different spatial resolutions. *Photogramm. Eng. Remote Sens.* **2002**, *68*, 1155–1161.
37. Emami, H.; Mojaradi, B. A New Method for Accuracy Assessment of Sub-Pixel Classification Results. *Am. J Eng. Appl. Sci.* **2009**, *2*, 456–465. [CrossRef]
38. Janssen, L.L.F.; Vanderwel, F.J.W. Accuracy assessment of satellite-derived land-cover data: A review. *Photogramm. Eng. Remote Sens.* **1994**, *60*, 419–426.
39. Tveite, H.; Langaas, S. An accuracy assessment method for geographical line data sets based on buffering. *Int. J. Geogr. Inf.* **1999**, *13*, 27–47. [CrossRef]
40. Owens, J.D.; Houston, M.; Luebke, D.; Green, S.; Stone, J.E.; Phillips, J.C. GPU Computing. *IEEE* **2008**, *96*, 879–899. [CrossRef]
41. Wei, B.W.; Li, T.; Li, G.Y.; Wang, Z.H.; He, M.; Shi, Y.L.; Liu, L.Y.; Zhang, R. Applied analysis of image accelerating distortion correction of OpenCL technology on heterogeneous platform. *Comput. Sci.* **2016**, *43*, 167–169.

remote sensing

Article

Retrieve Ice Velocities and Invert Spatial Rigidity of the Larsen C Ice Shelf Based on Sentinel-1 Interferometric Data

Faming Gong, Kui Zhang * and Shujun Liu

School of Microelectronics and Communication Engineering, Chongqing University, Chongqing 400044, China; 20134576@cqu.edu.cn (F.G.); liusj@cqu.edu.cn (S.L.)
* Correspondence: zk@cqu.edu.cn

Abstract: The Larsen C Ice Shelf (LCIS) is the largest ice shelf in the Antarctica Peninsula, and its state can be considered to be an indicator of local climate change. The goal of this paper is to invert the rigidity of the LCIS based on the interferometric synthetic aperture radar (InSAR) technique using Sentinel-1 images. A targeted processing chain is first used to obtain reliable interferometric phase measurements under the circumstance of rapid ice flow. Unfortunately, only the descending data are available, which disallows the corresponding 2-D velocity field to be directly obtained from such measurements. A new approach is thus proposed to estimate the interferometric phase-based 2-D velocity field with the assistance of speckle tracking offsets. This approach establishes an implicit relationship between range and azimuth displacements based on speckle tracking observations. By taking advantage of such a relationship, the equivalent interferometric signals in the azimuth direction are estimated, thereby recovering the interferometric phase-based 2-D ice velocity field of the LCIS. To further investigate the state of the LCIS, the recovered 2-D velocity field is utilized to invert the ice rigidity. The shallow-shelf approximation (SSA) is the core of the reverse model, which is closely dependent on boundary conditions, including kinematic and dynamic conditions. The experimental results demonstrate that the spatial distribution of the rigidity varies approximately from 70 MPa·s$^{1/3}$ to 300 MPa·s$^{1/3}$. This rigidity distribution can reproduce a similar ice flow pattern to the observations.

Keywords: InSAR; ice velocity field; Sentinel-1 images; missing data recovering; ice rigidity inversion

Citation: Gong, F.; Zhang, K.; Liu S. Retrieve Ice Velocities and Invert Spatial Rigidity of the Larsen C Ice Shelf Based on Sentinel-1 Interferometric Data. *Remote Sens.* **2021**, *13*, 2361. https://doi.org/10.3390/rs13122361

Academic Editors: Alex Hay-Man Ng, Linlin Ge, Hsing-Chung Chang and Zheyuan Du

Received: 19 April 2021
Accepted: 12 June 2021
Published: 17 June 2021

1. Introduction

Ice shelves account for 75% of the Antarctic coastline, covering an area over 1.561 million km^2 [1]. They retreat/thin and collapse when the climate warms [2], which reduces the buttressing effect regulating the contribution of the continental ice sheet to the sea level. These changes of ice shelves influence the ocean circulation as they are generally accompanied by a plethora of melting water [3]. To a certain degree, ice shelves can be considered to be a potential signal of the environmental change. The Larsen C Ice Shelf (LCIS), approximately 51,000 km^2 [4], is the largest ice shelf in the Antarctica Peninsula [5] and is sensitive to atmospheric and oceanic forcing [6]. In the past few years, as warm air/ocean water drives more melting, the LCIS has continuously thinned at a rate of 0.38 m/a [7]. Especially, it calved on 12 July 2017, along with a greater than 200 km crack, forming a huge tabular iceberg. Because of discharges, the corresponding resistance decreases, which leads to an increase in inland glacier flow velocity. Such events render the LCIS minimum observation size since the birth of the remote sensing technique.

In this paper, a velocity field of the LCIS is first mapped by the differential interferometric synthetic aperture radar (D-InSAR) method [8]. It retrieves the interferometric phase to map a glacier velocity field with a higher accuracy and a finer spatial resolution [9], which is beneficial to obtaining the spatial rigidity of the LCIS more reliably. However, the interferometric observations can be only derived from one direction (descending orbit) [10], which leads to a failure in resolving the complete horizontal velocity components. Theoretically, the correlation between the displacements in the range and the azimuth directions is fixed at a given time. That is to say, as long as a velocity map is retrieved from the same SAR images, it is possible to recover the 2-D horizontal velocity vectors from the available displacements in the line-of-sight (LOS) direction. It is well known that the speckle tracking technique can obtain a 2-D velocity field with a sub-pixel measurement accuracy [11], which can be used to explore the implicit interrelation between the azimuth and the range displacements. Due to the large size and the high nonlinearity of the displacement data, the relationship is extremely complicated. In other words, it is impossible to express it as an explicit mathematical function. It has been repeatedly proofed that the artificial neural network (ANN) is capable of accurately fitting a function with arbitrary forms [12]. Noting this, a targeted ANN is constructed to depict the implicit relationship between the displacements in the azimuth and the range directions. The speckle tracking observations are used to train this network in a supervised manner. With the use of the model, an implicit function correlating the range and the azimuth displacements is generated, thereby estimating the equivalent interferometric signals in the azimuth direction based on the single direction phase observations. The proposed method has two advantages: (1) Because the ice motion is associated with several parameters (e.g., ice thickness [13], surface slope [14], and ice bed conditions [15]), it is necessary to take these factors into account when computing the function. The proposed method can extend the input to multiple variables, allowing the introduction of a series of parameters. It enables the relationship to be depicted more comprehensively, leading to an improvement in estimation reliability. (2) The proposed method is immune to outlying speckle tracking estimates to some extent as the network is defined in a global sense.

Next, the inverse control model of the ice sheet and sea-level system model (ISSM) [16] is applied to the interferometric phase-based ice velocity field, thereby deriving the spatial rigidity of the LCIS. The model consists of the forward and the inverse parts [17]. The foundation of the forward part is actually established by a set of diagnostic equations of ice shelf flow [18], which connects the surface velocity with the depth-averaged viscosity. Especially, the viscosity is a function of effective strain rate and ice rigidity [18]. In order to avoid the interpretation trouble, the rigidity is directly substituted to calculate rather than deduced through viscosity [19]. On the other hand, the inverse part is responsible for seeking an appropriate rigidity distribution to match the observed velocity field as precisely as possible. A cost function is constructed in this part to calculate the misfit between the observed field and the modeled counterpart. The steepest-descent algorithm is introduced to minimize the cost function by adjusting the ice rigidity along with the direction of the gradient of the cost function. The strengths of the inverse control model are enumerated as follows: (1) It is unpractical to find the exact solution of the inverse parameter. By iteratively optimizing the cost function, the model can achieve the best match between the observations and the model outputs, thereby providing an optimal solution of the rigidity. (2) The model is easily automated with a little help of the programming technique [20].

2. Data and InSAR Data Processing

2.1. Input Data

The following datasets are employed in this paper.

1. The ice velocity map of the LCIS is retrieved from six standard Sentinel 1A/B terrain observation with progressive scan (TOPS) Level-1 single look complex (SLC) images acquired in interferometric wide (IW) swath mode. The images were acquired on 15 May 2020 (orbit number: 21589) and 21 May 2020 (orbit number: 32660). These SLC products preserving phase information have been referenced in radar range observation coordinate using the orbit and attitude data from the satellite [21]. The heading angle of the acquired images is around −61.27 degrees, whereas the incidence angle varies approximately from 29 degrees to 42 degrees. Note that the horizontal transmit and horizontal receive (HH) polarization possesses higher signal-to-noise ra-

tio and supplies better amplitude characteristics compared to the horizontal transmit and vertical receive (HV) polarization, implying that the former is more suitable for mapping the ice motion [22]. Therefore, the HH channel is chosen in this paper.

2. The MEaSUREs BedMachine Antarctica (Version 2) [23] dataset provides the necessary ice thickness (Figure 1a) and the surface topography (Figure 1b) for the modeling manipulation. The bed parameter is derived from the difference between the surface elevation and the ice thickness. The ice thickness and the surface elevation are presented in ice equivalent, which avoids the trouble of correcting for the firn density [24]. These data have a spatial resolution of 500 m and are geolocated in a polar stereographic projection with the standard longitude of 0°E and latitude of 71°S.

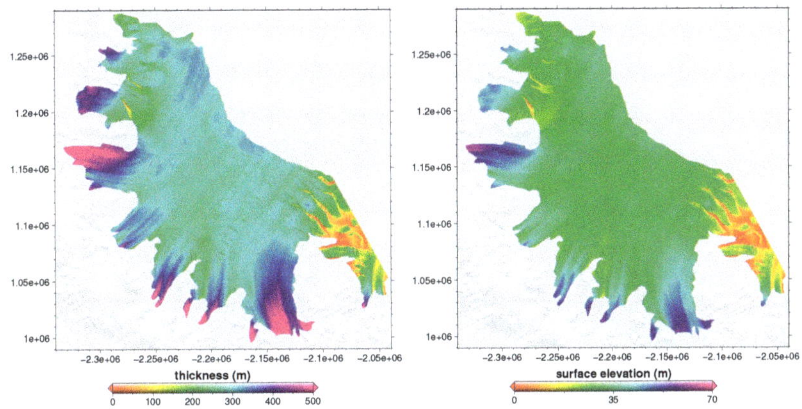

(a) (b)

Figure 1. The ice thickness (**a**) and the ice surface elevation (**b**) of the LCIS, respectively.

2.2. InSAR Processing

Due to the rapid flow of the glacier, the traditional procedure of the D-InSAR technique cannot be directly utilized to generate an interferogram with respect to the glacier velocity [25]. A targeted InSAR processing chain is used to deal with the input data (see Figure 2), which is roughly described as follows.

(1) A standard Sentinel-1 SAR image in IW mode generally consists of three sub-swaths, and each sub-swath usually comprises nine bursts. Fractional areas of neighboring bursts overlap in azimuth. Similarly, neighboring sub-swaths cover a certain common area in range. Consequently, burst alignment operations are first applied to each sub-swath of the TOPS acquisitions according to the time parameters. Following this, the corresponding sub-swaths are merged to generate the consecutive master and slave magnitude images.

(2) With the use of the external digital elevation model (DEM) and the precise orbit information generated by Copernicus precise orbit determination (POD) service, the relative offset of each measurement point is estimated using the geometric coregistration. As a result, an initial rough look-up table is generated.

(3) According to the look-up table, the slave magnitude image is resampled with respect to the master magnitude image.

(4) The offsets between the master and the resampled slave images are obtained by the speckle tracking technique with a window size of 32 × 128, which utilizes the classic cross-correlation function to determine the range and the azimuth offsets within a sub-pixel measurement accuracy. The speckle tracking result also plays another crucial role in this paper. It is used to supervise the training process of the artificial neural network depicting the relationship between the range and the azimuth displacements (see Section 3.1).

(5) The look-up table is updated based on the speckle tracking estimates, and the slave image is resampled again according to it.

(6) As the accuracy of the speckle tracking based look-up table cannot meet the requirement of TOPS image registration, inevitable phase jumps will be presented at burst/swath edges. Following the idea proposed in [26], phase jumps are estimated and compensated.

(7) Eventually, the interferogram is generated based on the result in (6). Note that the interferometric phase components induced by topography are estimated based on the external DEM and removed during the process of interferogram generation.

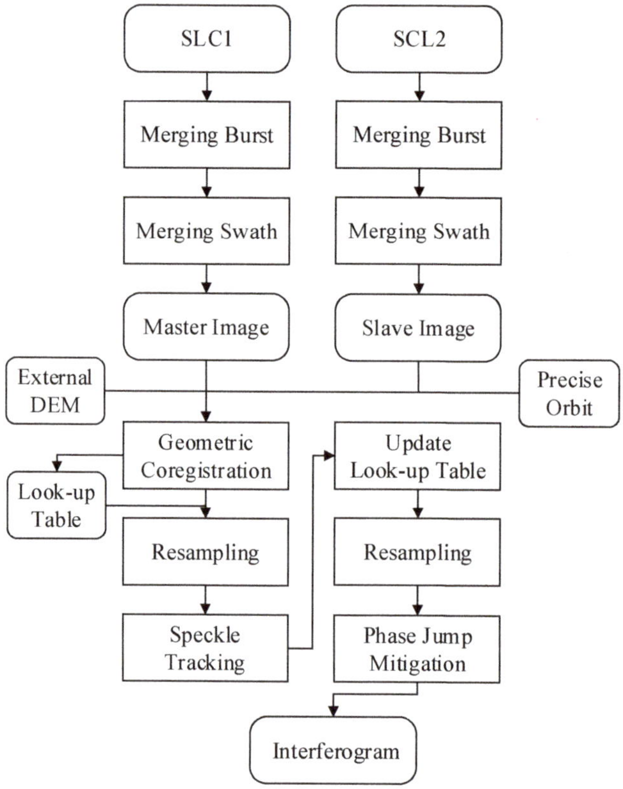

Figure 2. The interferometric synthetic aperture radar (InSAR) processing chain used in this paper.

3. Mathematical Model of Rigidity

3.1. Ice Velocity Field Recovering

Based on the speckle tracking technique, the 2-D ice surface velocity field of the LCIS can be produced with the use of the offsets in the range and the azimuth directions. Theoretically, there exists an interrelation between the displacements in the two directions. However, it cannot be expressed as a determinate function (e.g., a sine function or an exponential function). This study describes this implicit function as follows:

$$v_{azi} = f(v_{ran}, \mathbf{x}) \tag{1}$$

where v_{azi} and v_{ran} refer to the azimuth and the range displacements, respectively. $f(\cdot)$ is the function delineating the correlation between v_{azi} and v_{ran}. The vector $\mathbf{x}$ is a series of physical parameters associated with v_{azi} and v_{ran}, including spatial coordinates, surface slopes, ice thickness, etc. Frankly, it is hard to find a function $f(\cdot)$ by the traditional fitting method, as the large dimensions of the input-output data bring about severe nonlinearity and unacceptable computational time. In recent years, the artificial neural network widely attracts the focus of researchers [27,28], because it has made several breakthroughs in science and society [29,30]. It is an information processing system which imitates the human's brain. It plays an important role in highly complex, nonlinear, and parallel computing.

The architecture of the network used in this study is illustrated in Figure 3. In this network, the range displacements obtained from the speckle tracking technique, the coordinates of the points in the ice displacement field, the surface slopes, and the ice thickness are regarded as the inputs. As the input layer is desired to have as many independent data dimensions as possible, surface slopes are decomposed into the partial derivatives of the surface elevations. The azimuth displacements gained from the speckle tracking technique are assigned as the labels associated with the output layer. The depth of the hidden layers is set to 4, and the size of each hidden layer is uniformly set to 100. Tanh is set as the activation function of each hidden layer. After the construction of the network, we commence on the hyperparameters related to the training (including the learning rate, the batch number) and the optimization strategies (including the optimization algorithm and the regulation). The learning rate and the batch number are set to 0.001 and 32, respectively. Adam is chosen to be the optimization algorithm. It combines the advantages of the AdaGrad and RMSProp, which is helpful for handling sparse gradients on noisy problems. In order to avoid model overfitting, L2 regulation is carried out.

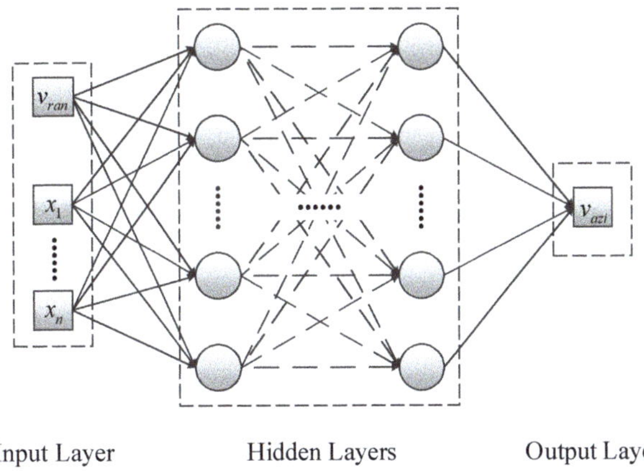

Figure 3. Architecture of the established artificial neural network (ANN).

3.2. Inversion for Rigidity

As ice can be characterized as fluid, the Stokes equations is widely used to depict the glacier ice flow [16]:

$$\nabla \cdot \mathbb{T} + \rho \mathbf{g} = 0 \tag{2}$$

$$\nabla \cdot \mathbf{u}=0 \tag{3}$$

Equations (2) and (3) represent the stress balance (neglecting the acceleration and the Coriolis forces) and the incompressibility, respectively, where ρ is the ice density, $\mathbf{u}$ is the ice velocity vector, $\mathbb{T}$ is the Cauchy stress tensor, $p = -\frac{1}{3}\sum_i \mathbb{T}_{ii}$ is the pressure, $\mathbf{g}$ is the gravity acceleration, and $\nabla\cdot$ denotes the divergence operator. The glacier ice is considered as viscous density-preserving fluid, and its constitutive relation can be expressed as

$$\mathbb{T}^D = 2\mu\mathbb{D} \tag{4}$$

where $\mathbb{T}^D$ refers to the stress deviator. It is split from the Cauchy stress tensor:

$$\mathbb{T}^D = \mathbb{T} + p\mathbb{I} \tag{5}$$

where $\mathbb{I}$ is the identity matrix in mathematics. $\mathbb{D}$ is the strain rate tensor, which can be further expressed as

$$\mathbb{D}_{ij} = \frac{1}{2}\left(\frac{\partial u_i}{\partial x_j} + \frac{\partial u_j}{\partial x_i}\right) \tag{6}$$

Note that the trace of the $\mathbb{D}$ is equal to the divergence of $\mathbf{u}$, thus the value of $tr(\mathbb{D})$ is also zero. μ is the ice effective viscosity. Due to the glacier ice belonging to the non-Newtonian fluid, μ follows the Glen's flow law [31]:

$$\mu = \frac{B}{\mathbb{D}_e^{\frac{n-1}{n}}} \tag{7}$$

where B is the ice rigidity (i.e., the objective of this inversion), n the Glen's flow law exponent (typically chosen as 3). $\mathbb{D}_e$ is the effective strain rate and defined as

$$\mathbb{D}_e = \sqrt{\frac{1}{2}\sum_{i,j}\mathbb{D}_{ij}^2} \tag{8}$$

Let x and y be horizontal coordinates, z be the vertical coordinate (positive upward). With the use of Equations (4)–(6), Equations (2) and (3) can be rewritten in the form of the velocity components in the Cartesian coordinates [16]:

$$\frac{\partial}{\partial x}\left(2\mu\frac{\partial u}{\partial x}\right) + \frac{\partial}{\partial y}\left(\mu\frac{\partial u}{\partial y} + \mu\frac{\partial v}{\partial x}\right) + \frac{\partial}{\partial z}\left(\mu\frac{\partial u}{\partial z} + \mu\frac{\partial w}{\partial x}\right) - \frac{\partial p}{\partial x} = 0 \tag{9}$$

$$\frac{\partial}{\partial x}\left(\mu\frac{\partial u}{\partial y} + \mu\frac{\partial v}{\partial x}\right) + \frac{\partial}{\partial y}\left(2\mu\frac{\partial u}{\partial y}\right) + \frac{\partial}{\partial z}\left(\mu\frac{\partial v}{\partial z} + \mu\frac{\partial w}{\partial y}\right) - \frac{\partial p}{\partial y} = 0 \tag{10}$$

$$\frac{\partial}{\partial x}\left(\mu\frac{\partial u}{\partial z} + \mu\frac{\partial w}{\partial x}\right) + \frac{\partial}{\partial y}\left(\mu\frac{\partial v}{\partial z} + \mu\frac{\partial w}{\partial y}\right) + \frac{\partial}{\partial z}\left(2\mu\frac{\partial u}{\partial z}\right) - \frac{\partial p}{\partial z} - \rho g = 0 \tag{11}$$

$$\frac{\partial u}{\partial x} + \frac{\partial v}{\partial y} + \frac{\partial w}{\partial z} = 0 \tag{12}$$

where u, v, and w are the components of the velocity vector $\mathbf{u}$ in x, y, and z directions, respectively. These equations are the so-called full-Stokes model, which contains four unknowns (u, v, w, and p). It is computationally time-consuming to find the solutions. In practice, ice shelves are relatively flat and thin. In addition, the viscosity of sea water is negligible compared to that of glacier ice. Therefore, it is reasonable to assume that the horizontal velocities and strain rate are independent of z. Based on the assumption, the full-Stokes model can be rearranged. The shallow-shelf approximation (SSA) is thus obtained [18]:

$$\frac{\partial}{\partial x}\left[2H\overline{\mu}\left(2\frac{\partial u}{\partial x} + \frac{\partial v}{\partial y}\right)\right] + \frac{\partial}{\partial y}\left[H\overline{\mu}\left(\frac{\partial u}{\partial y} + \frac{\partial v}{\partial x}\right)\right] = \rho g H\frac{\partial z_s}{\partial x} \tag{13}$$

$$\frac{\partial}{\partial y}\left[2H\overline{\mu}\left(2\frac{\partial v}{\partial y} + \frac{\partial u}{\partial x}\right)\right] + \frac{\partial}{\partial x}\left[H\overline{\mu}\left(\frac{\partial u}{\partial y} + \frac{\partial v}{\partial x}\right)\right] = \rho g H\frac{\partial z_s}{\partial y} \tag{14}$$

Additionally, Equation (7) can be simplified as

$$\overline{\mu} = \frac{\overline{B}}{2\left[\left(\frac{\partial u}{\partial x}\right)^2 + \left(\frac{\partial v}{\partial y}\right)^2 + \frac{1}{4}\left(\frac{\partial u}{\partial y} + \frac{\partial v}{\partial x}\right)^2 + \frac{\partial u}{\partial x}\frac{\partial v}{\partial y}\right]^{\frac{1}{3}}} \tag{15}$$

where z_s and H are the elevation of an ice shelf and the local ice thickness, respectively. Obviously, the SSA model has only two unknowns (u and v), the computational complexity is therefore reduced. It must be noted that the viscosity in the Equations (13)–(15) refers to the depth-averaged effective viscosity due to the temperature dependence of ice rigidity.

To solve the SSA model that consists of two partial differential equations, the appropriate boundary conditions are required. In this paper, both the kinematic and the dynamic conditions are introduced. The kinematic conditions specify the velocities (usually the observed surface velocities) at locations neighboring the grounding line of an ice shelf. On the other hand, the dynamic conditions apply the stress constraints to the ice front, which is the depth-integrated balance:

$$\int_{z_b}^{z_s} \mathbb{T} \cdot \mathbf{n} dz = -\frac{\rho_w g}{2}\left(\frac{\rho}{\rho_w} H\right)^2 \tag{16}$$

where $\mathbf{n}$ is the outward-pointing normal vector of the ice front, ρ_w is the density of sea water.

Inversely, the SSA model can be also considered as a direct approach to solve the depth-averaged effective viscosity when the velocities, the local ice thickness, and the ice surface elevation are known. However, it may not be applicable in practice mainly due to the errors in the observed velocities [20]. To deal with the problem, a control method is introduced to estimate the distribution of viscosity [20]. The method treats the SSA equations as the "forward model" to compute the modeled velocity (u_m, v_m) beginning with an initial viscosity value. Moreover, it adjusts the viscosity iteratively by gradient descent algorithm to achieve the best match between the modeled velocities (u_m, v_m) and the observed velocities (u_d, v_d) under the constraints given by Equations (13) and (14). The misfit between the model outputs and the observations is evaluated by a least-squares cost function with Lagrange- multipliers:

$$\begin{aligned} J = &\iint_\Omega \frac{1}{2}\left[(u_m - u_d)^2 + (v_m - v_d)^2\right] dxdy + \\ &\iint_\Omega \alpha\left\{\frac{\partial}{\partial x}\left[2H\overline{\mu}\left(2\frac{\partial u_m}{\partial x} + \frac{\partial v_m}{\partial y}\right)\right] + \frac{\partial}{\partial y}\left[H\overline{\mu}\left(\frac{\partial u_m}{\partial y} + \frac{\partial v_m}{\partial x}\right)\right] - \rho g H\frac{\partial z_s}{\partial x}\right\} dxdy + \\ &\iint_\Omega \beta\left\{\frac{\partial}{\partial y}\left[2H\overline{\mu}\left(2\frac{\partial v_m}{\partial y} + \frac{\partial u_m}{\partial x}\right)\right] + \frac{\partial}{\partial x}\left[H\overline{\mu}\left(\frac{\partial u_m}{\partial y} + \frac{\partial v_m}{\partial x}\right)\right] - \rho g H\frac{\partial z_s}{\partial y}\right\} dxdy \end{aligned} \tag{17}$$

where Ω is the study domain and (α, β) is the Lagrange-multiplier vector. In mathematics, obtaining an extremum of the cost function necessitates that all the variations of the function with respect to the free variables are equal to zero. Note that the variations of α and β are naturally equal to zero due to (u_m, v_m) being obtained from the "forward model". The variation of J with respect to $\overline{\mu}$ is written as [32]

$$\delta J = \iint_\Omega \delta\overline{\mu} \cdot H \begin{bmatrix} 2\left(2\frac{\partial u_m}{\partial x} + \frac{\partial v_m}{\partial y}\right)\frac{\partial \alpha}{\partial x} + \left(\frac{\partial u_m}{\partial y} + \frac{\partial v_m}{\partial x}\right)\left(\frac{\partial \alpha}{\partial y} + \frac{\partial \beta}{\partial x}\right) \\ +2\left(2\frac{\partial v_m}{\partial y} + \frac{\partial u_m}{\partial x}\right)\frac{\partial \beta}{\partial y} \end{bmatrix} dxdy \tag{18}$$

in this equation, (α, β) is produced by the following equations that ensure the variations of u_m and v_m to be zero:

$$\frac{\partial}{\partial x}\left[2H\overline{\mu}\left(2\frac{\partial \alpha}{\partial x} + \frac{\partial \beta}{\partial y}\right)\right] + \frac{\partial}{\partial y}\left[H\overline{\mu}\left(\frac{\partial \alpha}{\partial y} + \frac{\partial \beta}{\partial x}\right)\right] = u_d - u_m \tag{19}$$

$$\frac{\partial}{\partial y}\left[2H\overline{\mu}\left(2\frac{\partial \beta}{\partial y} + \frac{\partial \alpha}{\partial x}\right)\right] + \frac{\partial}{\partial x}\left[H\overline{\mu}\left(\frac{\partial \alpha}{\partial y} + \frac{\partial \beta}{\partial x}\right)\right] = v_d - v_m \tag{20}$$

To invert the ice rigidity directly, Equation (15) is used to deduce the variation of J with respect to $\overline{B}$. Due to the noise presented in the observed velocities, a Tikhonov regularization term is introduced to the cost function to stabilize the inversion [16].

The partial differential equations are discretized by the continuous Galerkin finite element method [33], and implemented with the use of triangular Lagrange P1 elements in the ISSM, which helps us improve the computational efficiency by constructing meshes with different resolution on the whole study region. In addition, it is beneficial to preserving small-scale features in the area where the ice dynamics are of great concern.

4. Results

Note that the results obtained from both the speckle tracking and the D-InSAR techniques are projected to the polar stereographic coordinate from the radar coordinate. As the observations in the non-glacier regions are not associated with glacier activities, they may impose negative effect on the ANN performance and the inversion of the ice rigidity. Therefore, the displacement estimates outside the LCIS were masked out. Figure 4a,b represents the speckle tracking displacement components of the LCIS in the range and the azimuth directions, respectively. It can be easily observed that there is a primary signal leap extending from the top left to the bottom right. Moreover, a few small distinct inconsecutive signals are presented in Figure 4b. This phenomenon should stem from the influence of the inaccurate radar timing information. Figure 4c demonstrates the LOS displacement map transformed from the interferometric phase. Essentially, Figure 4a reflects an identical motion information with Figure 4c. They both exhibit that the range motion varies from −6 m to −2 m. The negative values indicate that the ice moves along the opposite direction of the LOS. The minimum motion is observed around the promontories, whereas the maximum motion is located in the neighboring region of the ice front. Compared to Figure 4a, Figure 4c is much smoother (without evident jumps), primarily due to the higher spatial resolution and measurement accuracy of the interferometric phase based estimates. With the use of the ANN, Figure 4d maintains an almost same distribution pattern of the displacement as that in Figure 4b. Moreover, it gets rid of abrupt signals, which mainly because of two reasons. First, the introduction of the surface slope information and the ice thickness to the ANN forces the model to match the mechanism of ice flow, which mitigates the influence of outlying estimates. Second, the ANN is trained with regulations, the high-frequency noise is therefore eliminated. In this study, the results in Figure 4c,d are used as the observations for the inverse model. Figure 4b,d reveals that the maximum azimuth displacement is located at the northern part of the shelf. Further, both the positive and the negative azimuth displacements are observed, which implies the significant change of the ice flow direction. Figure 4e,f shows the displacement magnitudes of Figure 4a,b and Figure 4c,d, respectively, which demonstrate that the ice displacement speeds up from the inlet areas of the shelf to the ice front and reaches a maximum value (nearly 14 m) at the middle of the ice front.

Due to the phenomenon of interferometric decorrelation, there are several no data points in the ice displacement fields represented by Figure 4c,d, which may exert negative influences on the model inversion process. Noting this, these no data points were recovered using the method proposed in [34]. Next, the complete ice displacement fields were decomposed and transformed into two mean annual velocity components in the Cartesian coordinate(Figure 5a,b) to satisfy the requirement of the model inversion. Evidently, the x-velocity component approaches zero in the northernmost and the southernmost regions. On the other hand, the y-velocity component is nearly zero in the northernmost areas. The maximum values of the two velocity components are both located in the middle of the ice front. The study domain is adaptively discretized into 260,190 finite elements in the inverse model. Figure 5c,d represents the magnitudes of observed mean annual velocities and modeled counterparts, respectively. The black arrows indicate the ice velocity vectors. In general, the ice moves along a uniform direction from the inland to the sea, and it speeds up in the same direction. Figure 5e demonstrates the misfit between the observations and the model outputs. It indicates that the disagreement between them is small. Most discrepancies of the two velocity fields are distributed within 12 m/a.

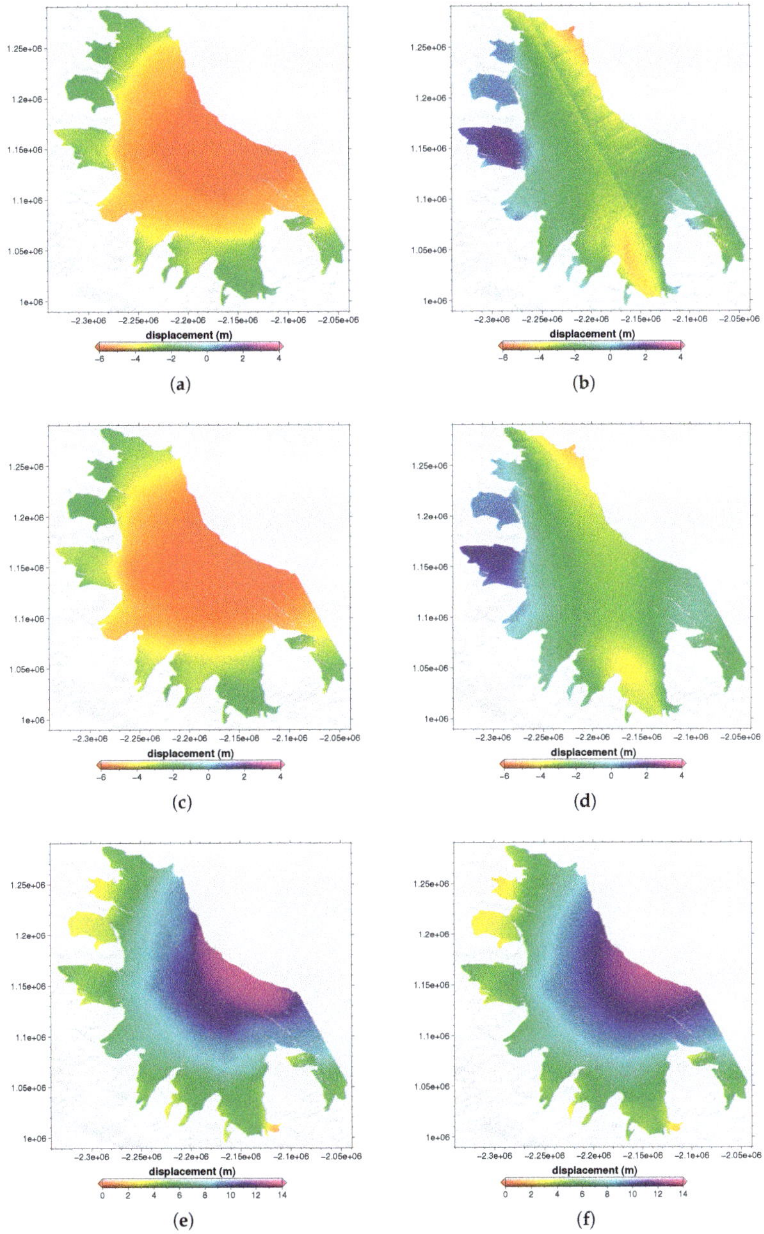

Figure 4. (**a**,**b**) The speckle tracking ice displacements of the LCIS along the range and the azimuth directions. (**c**) The interferometric phase-based displacements along the range direction. (**d**) The azimuth displacements recovered by the artificial neural network. (**e**) The displacement magnitudes calculated from panels (a,b). (**f**) The displacement magnitudes calculated from panels (c,d).

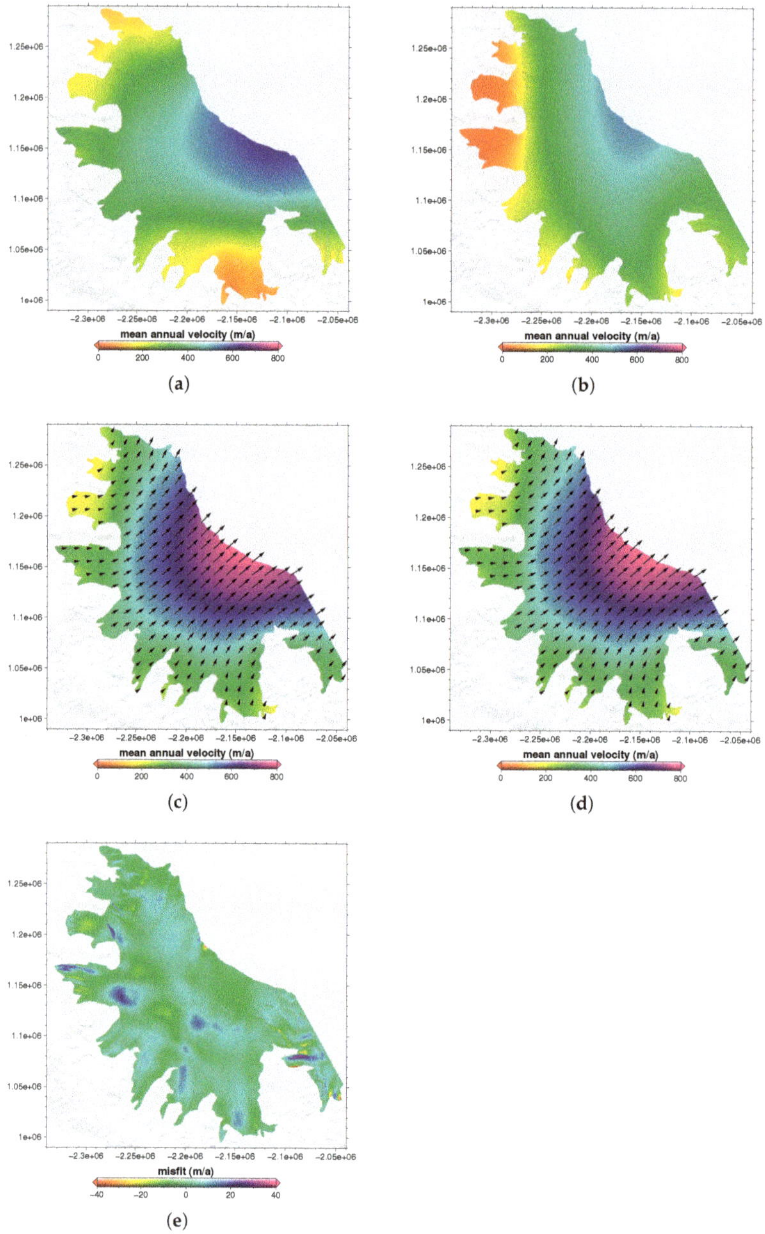

Figure 5. (**a**,**b**) The x and the y components of observed mean annual velocities of the LCIS in the polar stereographic projection, respectively. The magnitudes of observed mean annual velocities and modeled counterparts are presented in panels (**c**,**d**), respectively. The black arrows indicate the velocity vectors. (**e**) The misfit between panels (c,d).

Figure 6 exhibits the depth-averaged spatial distribution of the ice rigidity inverted from the ice velocity fields (Figure 5a,b). The rigidity varies in a huge span, approximately from 70 MPa·s$^{1/3}$ to 300 MPa·s$^{1/3}$, and its distribution pattern is not random. For the convenience of description, five typical areas are circled in purple dash line, and they are marked as A–E, respectively. The ice in area A is relatively stiff (with a rigidity of 250 MPa·$s^{1/3}$), which is corresponding to an isothermal ice at approximately −28 °C (see Table 1). As the inverted rigidity is depth-averaged, the estimated temperature according to Table 1 is logically between the base and the surface temperatures. The soft ice is presented in areas B and E, which generally has a rigidity within 100 MPa·$s^{1/3}$. Correspondingly, the temperature in these areas is no less than −7 °C. Note that the soft ice is observed along the three distinct rifts in area E. Area C is of great interest, as strong bipolar singularities of the rigidity are presented in a relatively small area.

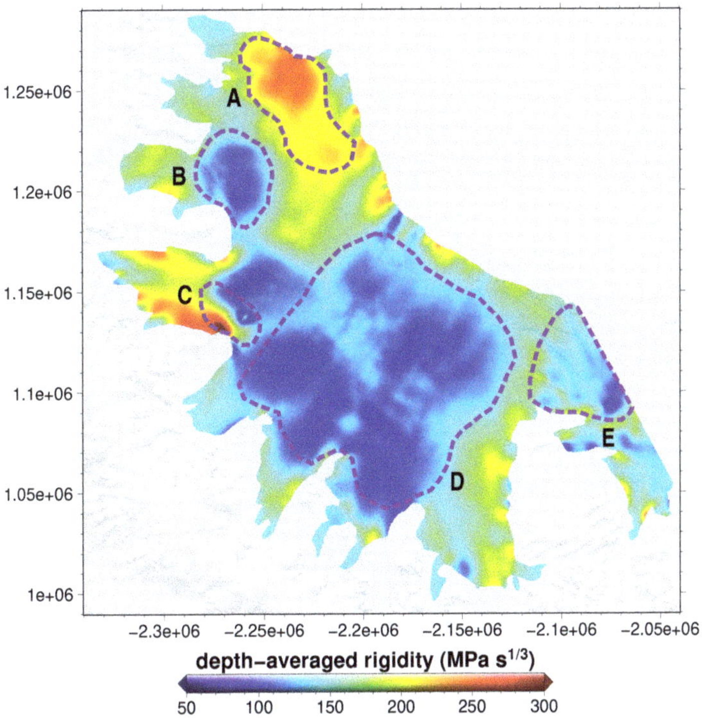

Figure 6. The spatial distribution of the depth-averaged rigidity of the LCIS.

Table 1. Recommended values of ice rigidity B corresponding to different temperature T, following Paterson [35].

T (°C)	0	−2	−5	−10	−15	−20	−25	−30	−35	−40	−45	−50
B (MPa·$s^{1/3}$)	53	75	86	127	151	180	220	270	333	415	523	630

5. Discussion

5.1. Analysis of Ice Rigidity Distribution

Rigidity refers to the property of an object to resist deformation when forces are exerted on it. On the basis of Glen's law [31], the proportional factor between deviatoric stress and strain rate is the so-called glacier ice rigidity. Its value is dependent on temperature, water content, grain size, impurities, and preferred-orientation fabric [35]. Frankly, it is hard to obtain the variables like impurities for most researchers. However, temperature is relatively within reach, as glaciologists can obtain temperature data from laboratory experiments or public meteorology archives. Consequently, the relationship between temperature and rigidity is best understood. Due to the lack of adequate theoretical formulations about rigidity, researchers calculate it only through the temperature distribution over interested areas in most ice shelf modeling applications [32]. Obviously, the value of rigidity obtained by this method is not accurate.

In this paper, the rigidity distribution of the LCIS is inverted based on the 2-D velocity field, which provides us with an opportunity to investigate the state of the LCIS further. Several rifts and crevasses can be observed in the area B, D, and E from the base map of the topography, where the ice rigidity is low. To some extent, the effectiveness of the inversion model used in this paper is therefore confirmed. Figure 7 is the driving stress deduced from the modeling results. It can be observed that higher driving stress usually accompanies with the rifts and crevasses, which brings about higher strain rates and lower ice rigidity. Note that no information with respect to rifts and crevasses is involved in the model inversion process, because the field measurements have related the fractures to higher strain rates and larger velocities. The ice in the regions closed to the grounding line is relatively stiff, and its rigidity is between 150 MPa·s$^{1/3}$ and 250 MPa·s$^{1/3}$. This is possibly due to three factors. First, the basal melting is dramatic around the grounding line, which causes a thinning of the ice shelf and a fall of the temperature, the rigidity thus arises. Second, the advection of cold ice from tributary glaciers reduces the temperature and enlarges the ice rigidity. Third, abutting the grounding line, the sea water beneath the ice shelf is comparatively shallow. Consequently, vigorous tidal mixing can easily transfer heat to the ice shelf, which triggers a great deal of melting, thereby increasing the rigidity. In area A, where the ice is relatively stiff, the accretion of marine ice might be extremely plentiful. As a result, it may isolate the ice shelf from warm sea water to a large extent, which might cool the temperature-depth profile. The existence of the strong bipolar singularities in area C is a little difficult to interpret. As there is no evident stagnant ice, the ice over this area should be very stiff. Such an abnormity is never mentioned in previous research works. It is suspected that this may result in a systematic bias, due to the non-local characteristic of the "forward model" [36].

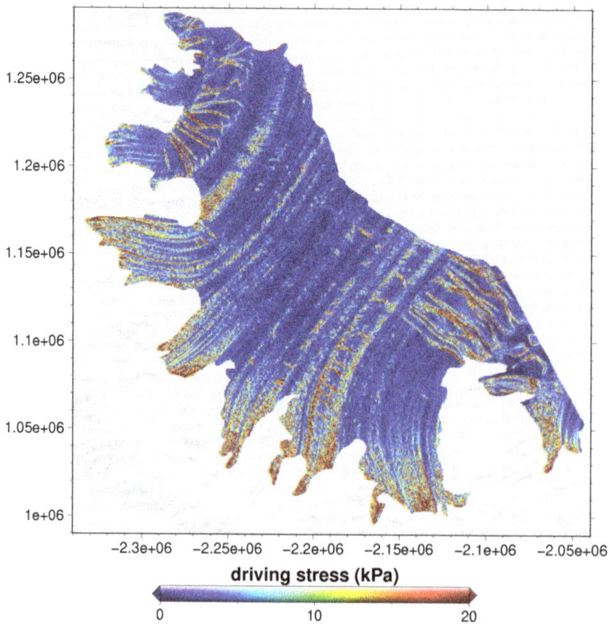

Figure 7. The driving stress of the LCIS.

5.2. *Importance of Control Method*

Traditionally, the ice rigidity parameter $\overline{B}$ is calculated based on the temperature. Unfortunately, it is hard to obtain the temperature data with a high spatial resolution. In most cases, $\overline{B}$ is usually inferred to be a uniform value over an extensive coverage, which is not capable of depicting the real state of the ice rheology and leads to inaccurate results in many glacier modeling applications. Owing to the use of the control method, a rigidity field with significant spatial variability can be generated. The velocities (Figure 5d) computed based on the inversion result reproduce the pattern of the observations accurately. To examine the effectiveness of the control method for inferring the ice rigidity parameter, the mean annual velocities were directly calculated by the "forward model" with a uniform $\overline{B}$ (Figure 8a). The value of $\overline{B}$ is set to 175 MPa·s$^{1/3}$ according to the recommended temperature (−19 °C) in the LCIS [37]. Figure 8b demonstrates that the misfit between the modeled velocities and the observations. Obviously, the disagreements are approximately 35 m/a in most areas. These areas are mainly covered by soft ice.

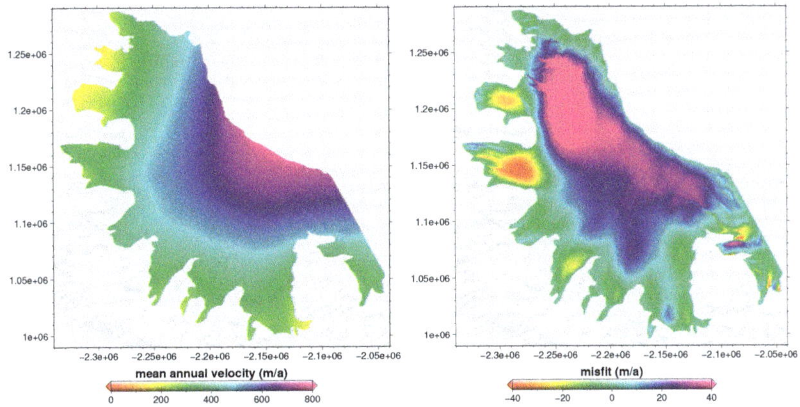

(**a**) (**b**)

Figure 8. (**a**) The mean annual velocities obtained from the "forward model" based on a uniform rigidity. (**b**) The misfit between panel (a) and Figure 5c.

6. Conclusions

Because only descending data are available over the LCIS, it is directly impossible to retrieve the corresponding 2-D velocity field using the D-InSAR technique. In this paper, a new method is proposed to recover the azimuth displacement. This basic idea of this method is stimulated by the theory of ANN. A targeted network is constructed to fit the implicit relationship between range and azimuth displacements based on speckle tracking results. It introduces physical parameters related to the ice motion as the inputs, which constrains the inputs and outputs to the mechanism of glacier flow to some extent. It is beneficial in reducing the influence of the outlying estimates and noise. Next, the flow mechanism of ice shelves is discussed. The corresponding inversion implementation in the ISSM is used to infer the ice rigidity. It seeks the best rigidity distribution by minimizing the misfit between the interferometric phase-based 2-D velocities and the modeled counterparts. Compared to the uniform ice rigidity calculated from a fixed temperature, the inverted variable rigidity can reproduce the observed flow patterns more accurately.

Author Contributions: Conceptualization, F.G. and K.Z.; methodology, F.G. and K.Z.; software, K.Z. and S.L.; validation, F.G.; formal analysis, F.G.; investigation, F.G. and K.Z.; resources, K.Z.; data curation, F.G., K.Z., and S.L.; writing—original draft preparation, F.G.; writing—review and editing, K.Z.; visualization, F.G.; supervision, K.Z.; project administration, K.Z.; funding acquisition, K.Z. All authors have read and agreed to the published version of the manuscript.

Funding: This research was jointly funded by the National Natural Science Foundation of China (Grant No. 42074034) and Chongqing Science and Technology Bureau (Grant No. cstc2019jxj100007).

Institutional Review Board Statement: Not applicable.

Informed Consent Statement: Not applicable.

Data Availability Statement: MEaSUREs BedMachine Antarctica (Version 2) presented in this study are openly available in National Snow and Ice Data Center at https://doi.org/10.5067/E1QL9HFQ7A8M (accessed on 24 February 2021), reference number [23]. The Sentinel-1 data were analyzed in this study. This data can be found here: https://asf.alaska.edu/ (accessed on 24 February 2021).

Acknowledgments: The authors are very grateful to the European Space Agency for providing Sentinel-1 data. The authors are very grateful to National Snow and Ice Data Center for providing MEaSUREs BedMachine Antarctica (Version 2) dataset.

Conflicts of Interest: The authors declare no conflict of interest.

References

1. Rignot, E.; Jacobs, S.; Mouginot, J.; Scheuchl, B. Ice-shelf melting around Antarctica. *Science* **2013**, *341*, 266–270. [CrossRef]
2. Ingels, J.; Aronson, R.B.; Smith, C.R.; Baco, A.; Bik, H.M.; Blake, J.A.; Brandt, A.; Cape, M.; Demaster, D.; Dolan, E.; et al. Antarctic ecosystem responses following ice-shelf collapse and iceberg calving: Science review and future research. *Wiley Interdiscip. Rev. Clim. Chang.* **2021**, *12*, e682. [CrossRef]
3. Hogg, A.E.; Gilbert, L.; Shepherd, A.; Muir, A.S.; McMillan, M. Extending the record of Antarctic ice shelf thickness change, from 1992 to 2017. *Adv. Space Res.* **2020**, *68*. [CrossRef]
4. Jansen, D.; Luckman, A.; Kulessa, B.; Holland, P.R.; King, E.C. Marine ice formation in a suture zone on the Larsen C Ice Shelf and its influence on ice shelf dynamics. *J. Geophys. Res. Earth Surf.* **2013**, *118*, 1628–1640. [CrossRef]
5. Cook, A.J.; Vaughan, D.G. Overview of areal changes of the ice shelves on the Antarctic Peninsula over the past 50 years. *Cryosphere* **2010**, *4*, 77–98. [CrossRef]
6. Jansen, D.; Luckman, A.J.; Cook, A.; Bevan, S.; Kulessa, B.; Hubbard, B.; Holland, P. Brief Communication: Newly developing rift in Larsen C Ice Shelf presents significant risk to stability. *Cryosphere* **2015**, *9*, 1223–1227. [CrossRef]
7. Hogg, A.E.; Gudmundsson, G.H. Impacts of the Larsen-C Ice Shelf calving event. *Nat. Clim. Chang.* **2017**, *7*, 540–542. [CrossRef]
8. Bamler, R.; Hartl, P. Synthetic aperture radar interferometry. *Inverse Probl.* **1998**, *14*, R1. [CrossRef]
9. Joughin, I. Ice-sheet velocity mapping: A combined interferometric and speckle-tracking approach. *Ann. Glaciol.* **2002**, *34*, 195–201. [CrossRef]
10. Wright, T.J.; Parsons, B.E.; Lu, Z. Toward mapping surface deformation in three dimensions using InSAR. *Geophys. Res. Lett.* **2004**, *31*. [CrossRef]
11. Hu, J.; Li, Z.; Ding, X.; Zhu, J.; Zhang, L.; Sun, Q. Resolving three-dimensional surface displacements from InSAR measurements: A review. *Earth-Sci. Rev.* **2014**, *133*, 1–17. [CrossRef]
12. Zainuddin, Z.; Pauline, O. Function approximation using artificial neural networks. *WSEAS Trans. Math.* **2008**, *7*, 333–338.
13. Layberry, R.; Bamber, J. A new ice thickness and bed data set for the Greenland ice sheet: 2. Relationship between dynamics and basal topography. *J. Geophys. Res. Atmos.* **2001**, *106*, 33781–33788. [CrossRef]
14. Kamb, B.; Echelmeyer, K.A. Stress-gradient coupling in glacier flow: I. Longitudinal averaging of the influence of ice thickness and surface slope. *J. Glaciol.* **1986**, *32*, 267–284. [CrossRef]
15. Piotrowski, J.A.; Mickelson, D.M.; Tulaczyk, S.; Krzyszkowski, D.; Junge, F.W. Were deforming subglacial beds beneath past ice sheets really widespread? *Quat. Int.* **2001**, *86*, 139–150. [CrossRef]
16. Larour, E.; Seroussi, H.; Morlighem, M.; Rignot, E. Continental scale, high order, high spatial resolution, ice sheet modeling using the Ice Sheet System Model (ISSM). *J. Geophys. Res. Earth Surf.* **2012**, *117*. [CrossRef]
17. Khazendar, A.; Rignot, E.; Larour, E. Acceleration and spatial rheology of Larsen C ice shelf, Antarctic Peninsula. *Geophys. Res. Lett.* **2011**, *38*. [CrossRef]
18. MacAyeal, D.R. Large-scale ice flow over a viscous basal sediment: Theory and application to ice stream B, Antarctica. *J. Geophys. Res. Solid Earth* **1989**, *94*, 4071–4087. [CrossRef]
19. Larour, E.; Rignot, E.; Joughin, I.; Aubry, D. Rheology of the Ronne Ice Shelf, Antarctica, inferred from satellite radar interferometry data using an inverse control method. *Geophys. Res. Lett.* **2005**, *32*, doi:10.1029/2004GL021693. [CrossRef]
20. MacAyeal, D.R. A tutorial on the use of control methods in ice-sheet modeling. *J. Glaciol.* **1993**, *39*, 91–98. [CrossRef]
21. Nagler, T.; Rott, H.; Hetzenecker, M.; Wuite, J.; Potin, P. The Sentinel-1 mission: New opportunities for ice sheet observations. *Remote Sens.* **2015**, *7*, 9371–9389. [CrossRef]
22. Sánchez-Gámez, P.; Navarro, F.J. Glacier surface velocity retrieval using D-InSAR and offset tracking techniques applied to ascending and descending passes of Sentinel-1 data for southern Ellesmere ice caps, Canadian Arctic. *Remote Sens.* **2017**, *9*, 442. [CrossRef]

23. Morlighem, M.; Rignot, E.; Binder, T.; Blankenship, D.; Drews, R.; Eagles, G.; Eisen, O.; Ferraccioli, F.; Forsberg, R.; Fretwell, P.; et al. Deep glacial troughs and stabilizing ridges unveiled beneath the margins of the Antarctic ice sheet. *Nat. Geosci.* **2020**, *13*, 132–137. [CrossRef]
24. Cuffey, K.M. A matter of firn. *Science* **2008**, *320*, 1596–1597. [CrossRef] [PubMed]
25. Scheiber, R.; Jäger, M.; Prats-Iraola, P.; De Zan, F.; Geudtner, D. Speckle tracking and interferometric processing of TerraSAR-X TOPS data for mapping nonstationary scenarios. *IEEE J. Sel. Top. Appl. Earth Obs. Remote Sens.* **2014**, *8*, 1709–1720. [CrossRef]
26. Wang, K.; Xu, X.; Fialko, Y. Improving burst alignment in TOPS interferometry with bivariate enhanced spectral diversity. *IEEE Geosci. Remote Sens. Lett.* **2017**, *14*, 2423–2427. [CrossRef]
27. Joshi, R.; Kumar, K.; Adhikari, V.P.S. Modelling suspended sediment concentration using artificial neural networks for Gangotri glacier. *Hydrol. Process.* **2016**, *30*, 1354–1366. [CrossRef]
28. Haq, M.A.; Azam, M.F.; Vincent, C. Efficiency of artificial neural networks for glacier ice-thickness estimation: A case study in western Himalaya, India. *J. Glaciol.* **2021**, 1–14. [CrossRef]
29. Huesken, D.; Lange, J.; Mickanin, C.; Weiler, J.; Asselbergs, F.; Warner, J.; Meloon, B.; Engel, S.; Rosenberg, A.; Cohen, D.; et al. Design of a genome-wide siRNA library using an artificial neural network. *Nat. Biotechnol.* **2005**, *23*, 995–1001. [CrossRef] [PubMed]
30. Rem, B.S.; Käming, N.; Tarnowski, M.; Asteria, L.; Fläschner, N.; Becker, C.; Sengstock, K.; Weitenberg, C. Identifying quantum phase transitions using artificial neural networks on experimental data. *Nat. Phys.* **2019**, *15*, 917–920. [CrossRef]
31. Glen, J.W. The creep of polycrystalline ice. *Proc. R. Soc. Lond. Ser. A Math. Phys. Sci.* **1955**, *228*, 519–538.
32. Rommelaere, V.; MacAyeal, D.R. Large-scale rheology of the Ross Ice Shelf, Antarctica, computed by a control method. *Ann. Glaciol.* **1997**, *24*, 43–48. [CrossRef]
33. Karakashian, O.; Makridakis, C. A space-time finite element method for the nonlinear Schrödinger equation: The continuous Galerkin method. *SIAM J. Numer. Anal.* **1999**, *36*, 1779–1807. [CrossRef]
34. Zhang, K.; Gong, F.; Li, Z.; Liu, S.; Shen, Y. Recover Glacier Velocity Fields Derived From the SAR Speckle Tracking Technique Using Artificial Neural Network. *IEEE Geosci. Remote Sens. Lett.* **2019**, *16*, 1250–1253. [CrossRef]
35. Cuffey, K.M.; Paterson, W.S.B. *The Physics of Glaciers*; Butterworth-Heinemann: Burlington, MA, USA, 2010.
36. MacAyeal, D.R.; Thomas, R.H. The effects of basal melting on the present flow of the Ross Ice Shelf, Antarctica. *J. Glaciol.* **1986**, *32*, 72–86. [CrossRef]
37. Thomas, E.R.; Tetzner, D.R. The climate of the Antarctic Peninsula during the twentieth century: Evidence from ice cores. In *Antarctica-A Key To Global Change*; IntechOpen: London, UK, 2018.

remote sensing

MDPI

Article

A Loading Correction Model for GPS Measurements Derived from Multiple-Data Combined Monthly Gravity

Jiesi Luo [1], Wei Chen [1,*], Jim Ray [2,†], Tonie van Dam [3] and Jiancheng Li [1]

1 Key Laboratory of Geospace Environment and Geodesy, School of Geodesy and Geomatics, Wuhan University, Wuhan 430079, China; jiesiluo@whu.edu.cn (J.L.); jcli@sgg.whu.edu.cn (J.L.)
2 National Oceanic and Atmospheric Administration, Silver Spring, MD 20910, USA; jimr.ray@gmail.com
3 Interdisciplinary Center for Security and Trust, Campus Kirchberg, University of Luxembourg, 1359 Luxembourg, Luxembourg; tonie.vandam@uni.lu
* Correspondence: wchen@sgg.whu.edu.cn
† Retired.

Abstract: Time-dependent loading deformations of the Earth's surface, due to nontidal changes in the atmosphere, ocean, land water/ice, etc., contribute significantly to the seasonal and secular Global Positioning System (GPS) site displacements, especially for the up component. While loading deformations derived from general circulation model (GCM) outputs are usually used to correct loading signals in the GPS site displacements, this study aims to provide a loading correction model based on the multiple-data combined monthly gravity products LDCmgm90. We have adopted GPS measurements from 249 IGS reference frame stations and 3 different GCM-based loading models to test the reliability of the LDCmgm90 model. Compared to the GCM-based models, the LDCmgm90 loading correction is more effective in attenuating seasonal (especially annual) loading signals and can bring more significant improvements to most stations for both the data-trend-removed and the data-trend-retained cases. Thus, we have validated the LDCmgm90 model from the loading aspect and proved it to be a reliable loading-correction model for GPS displacements. The relatively better secular loading signals provided by the LDCmgm90 loading model may provide us a chance to study the long-term, nonloading signals in GPS data.

Keywords: GPS; GRACE; LDCmgm90; time-variable gravity; loading deformation

Citation: Luo, J.; Chen, W.; Ray, J.; van Dam, T.; Li, J. A Loading Correction Model for GPS Measurements Derived from Multiple-Data Combined Monthly Gravity. *Remote Sens.* **2021**, *13*, 4408. https://doi.org/10.3390/rs13214408

Academic Editors: Alex Hay-Man Ng, Linlin Ge, Hsing-Chung Chang and Zheyuan Du

Received: 7 October 2021
Accepted: 29 October 2021
Published: 2 November 2021

1. Introduction

Despite successful applications in various fields, Global Positioning System (GPS) measurements suffer from scatters in GPS position time series, which reduce GPS measurement accuracy significantly. Ray [1] listed many sources of measurement error and other systematic effects, such as draconitic errors, local site errors, temperature cycle errors and errors in IERS models [2] (also see Ray et al. [3,4]; more details can be found in Appendix B). Besides these causes, time-dependent loading deformations, due to mass fluctuations in the Earth's surficial geophysical fluids, i.e., atmosphere, ocean and land water/ice, lead to variable loads on the crust and thus variable crustal deformations, thus also contributing greatly to GPS position data scatters, especially at seasonal frequency bands [5–16]. Therefore, accurate loading deformation corrections are urgently needed in GPS data analyses, or else the explorations of the geophysical causes of the remaining crustal deformations will be affected. Some studies on loading are reviewed below.

Van Dam and Wahr [17] studied the displacement of the Earth's surface due to atmospheric loading. Mangiarotti and Cazenave [18] compared loading results derived from 16 Doppler orbitography and radiopositioning integrated by satellite (DORIS) stations and those from annual surface mass redistributions (atmosphere and ocean mass, soil moisture and snow loadings). Van Dam et al. [5] compared vertical displacements modeled by general circulation models (GCMs) and GPS heights observed by 147 globally distributed sites (Δr_M and Δr_O) and found that the root mean square (RMS) of seasonal, especially annual, Δr_O adjusted by Δr_M was significantly reduced. Dong et al. [8] conducted a comprehensive study on the effectiveness of loading corrections and showed that ~40% of the power of the annual vertical variations in continuous GPS time series can be explained by the joint contribution of seasonal surficial mass redistributions. Blewitt [19] discussed the self-consistency in reference frames, geocenter definition and surface loading of the solid Earth. Since these studies, GCM-derived loadings were widely used to correct the GPS measurements. However, with longer and more accurate time series of GPS measures and GCM outputs available, it became obvious that GCM-derived loading corrections do not always improve the GPS data, but sometimes lead to increased noises, partly due to limitations in the GCMs, such as hydrostatic approximation in atmospheric and oceanic models, and violation of the conservation of global mass when summing atmospheric, oceanic and hydrological masses (see Chen et al. [20,21] for more details). Therefore, the GCM-based loadings may have biases in both phase and amplitude and sometimes lead to even more excess seasonal signals (worse results) in GPS time series (in Section 4, one can see that seasonal noises in GPS data are sometimes amplified rather than attenuated after correction by GCM-derived loadings).

The 2002 launch of the Gravity Recovery and Climate Experiment (GRACE) twin satellites has made available several time series of geopotential coefficients [22], which can provide information for surficial mass redistributions and therefore loading deformations meeting the conservation of global mass. Notable discrepancies are found between GCM-based and GRACE-based loading deformation models. Davis et al. [23] compared the annual deformations measured by GPS stations in the Amazon River Basin with predictions calculated from GRACE measurements and found high correlation. They also confirmed the superiority of GRACE-measured surface water variations over some hydrology models, indicating that the adoption of GRACE measurements can be a better approach to study loading deformation compared with GCMs. Tregoning et al. [11] computed elastic deformation using continental water loading estimates derived from GRACE, which agreed well with GPS observations. Yan et al. [24] proved that using either GRACE or GCM 36 degree/order (d/o) spherical harmonic coefficients can explain 90% of the variance of total crustal vertical deformation. Fu and Freymueller [25] and Chanard et al. [26] confirmed that surficial mass redistributions observed by GRACE, combined with an elastic spherical and layered Earth model, can be used to provide first-order corrections for loading deformation observed in both horizontal and vertical components of GNSS station position time series.

Despite the advantages of GRACE-derived loading, there are also some significant defects in the GRACE-based loading corrections. Due to the longitudinal striped errors and coarse spatial resolution of traditional GRACE spherical harmonic products, additional corrections (destriping and smoothing) are required to reveal valuable signals [27,28]. Consequently, with different GRACE data processing strategies adopted, the loading deformation at the same GPS station derived by different researchers can differ from each other. Further, destriping would cause distortions of signals of interest while smoothing would attenuate them (signal leakages). These problems, together with limited spatial and temporal resolutions, make the GRACE-based results not quite reliable. The newly released Mascon products are free of the longitudinal striped errors but also contain some problems. For example, atmospheric, oceanic and hydrological model outputs (used to generate the RL06 AOD1B) violate the conservation of global mass, so a sea-level mass term is introduced to conserve the mass. However, only the ocean model outputs are used to produce the RL06 GAB (namely, parts of oceanic contributions are not modeled). In addition, like traditional GRACE spherical harmonic products, there are notable differences among Mascon products released by different institutes (more details can be found in Chen et al. [29]). Thus, further improvements are needed for GRACE products.

By inverting GPS displacement series measured at roughly 450 continuously tracking sites and ocean bottom pressure estimates of a data assimilated ocean circulation model, Wu et al. [30] obtained monthly global surface mass distributions in the spherical harmonic domain with a complete spectrum up to degree and order 50 and then combined them with GRACE gravity data to provide enhanced spectral and geographic coverage. Chen et al. [29] developed a multiple-data-based monthly geopotential model set LDCmgm90 (available at https://doi.org/10.6084/m9.figshare.7874384.v4 (accessed on 20 September 2021)) by assimilating various versions of GRACE (including Mascon products); satellite laser ranging (SLR) monthly gravity data [31,32]; and outputs from various global atmospheric, oceanic and hydrological circulation models by using the least difference combination method [20,21]. The LDCmgm90 dataset provides more reliable data trends (see [21,29] for some validations) and has no stripes as seen in most GRACE data, and it was shown to have better performance in explaining atmospheric, oceanic and hydrological long-term and seasonal time-scale mass redistributions compared with most meteorological models and time-varying gravity products, including Mascon solutions.

This study aims to provide a particular choice of improved surficial loading models by using the LDCmgm90 dataset and, in the meantime, to validate the LDCmgm90 from the loading aspect. We test our LDCmgm90 loading model at 249 IGb14 GPS reference frame stations and compare them with three GCM-based loading models respectively provided by the International Mass Loading Service (IMLS, http://massloading.net/, accessed on 27 June 2020), the School and Observatory of Earth Sciences (EOST) Loading Service (http://loading.u-strasbg.fr/index.php, accessed on 8 December 2020) and the Earth System Modeling group at the Deutsches GeoForschungsZentrum (ESMGFZ) (http://rz-vm115.gfz-potsdam.de:8080/repository, accessed on 6 January 2021). The results show that the LDCmgm90-based loading model explains more vertical crustal deformation caused by surficial loading at most GPS stations than other loading models involved in this research.

This paper is arranged as follows: Datasets and methods used in this research are described in Section 2. LDCmgm90-based load deformation is derived in Section 3 and compared with GCM-based load deformations at 249 IGS GPS stations in Section 4. Conclusions are presented in Section 5.

2. Materials and Methods

2.1. GPS Data

The IGb14 reference frame is an update to the previous IGS14 reference frame in which a large fraction of the IGS14 GPS station coordinates have been "repaired" after previously being rendered unusable due to the discontinuities caused by unpredictable events such as earthquakes or equipment changes. The transformation parameters between IGb14 and IGS14/ITRF2014 are therefore all zero because IGb14 is aligned in origin, scale and orientation to IGS14, hence ITRF2014 (please refer to ftp://igs-rf.ign.fr/pub/IGb14/IGb14.ssc (accessed on 20 September 2021) for more details about IGb14 reference frame and the 261 GPS stations involved). However, only 259 of 261 IGb14 GPS stations' 24-h-sampled final solutions are released on the website of Nevada Geodetic Laboratory (NGL).

NGL-processed GPS data contain time series of east-, north- and up-components of nonlinear displacements in the IGS14 reference frame, where multiple measurement models and corrections have been applied [33–40], including ocean tidal loading correction. However, nontidal loading correction of atmosphere, ocean and hydrology is not applied. The NGL modeling of tropospheric delays is particularly notable and effective at retaining the full nontidal atmospheric load signal in the observed station coordinate time series [41]. Please refer to NGL GPS Data Analysis Strategy and Products Summary (http://geodesy.unr.edu/gps/ngl.acn.txt, accessed on 6 December 2020) for more details.

We first abandoned 10 GPS stations whose durations are less than 2 years (730 samples) within the LDCmgm90 timespan due to the unreliability of their annual fluctuations. Then, we preprocessed the 249 remaining time series to remove their gross errors (larger than 3 times standard deviation) and used a first-order difference method to correct jumps in

GPS time series, where iteration and decreasing thresholds are required. The distribution of the 249 selected GPS stations is shown in Figure 1.

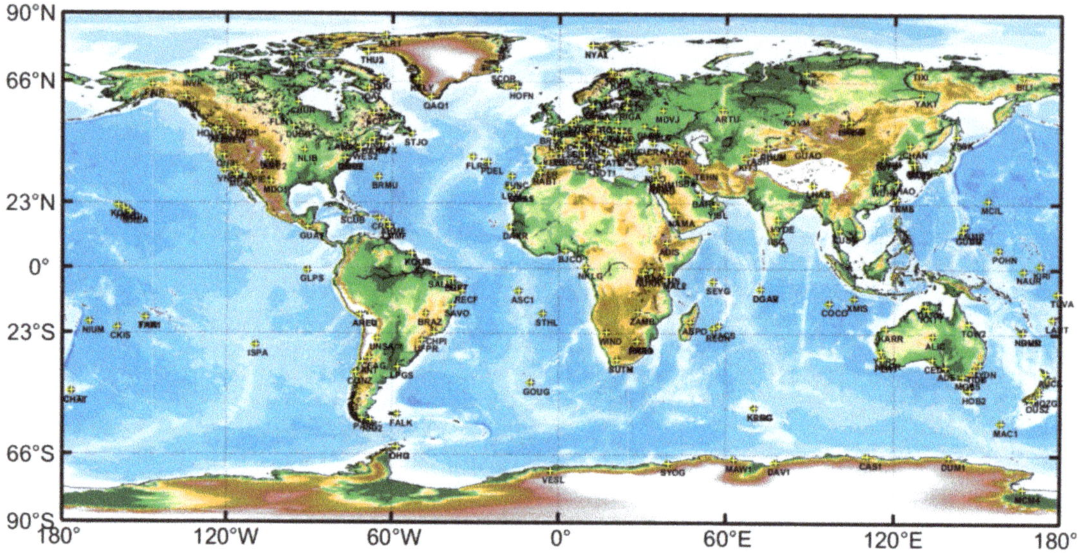

Figure 1. Distribution of 249 IGb14 GPS stations used in this research. Locations of these GPS stations are marked with yellow "+" with corresponding station name given below. Station names may overlap due to their dense distributions in several areas.

2.2. GCM-Based Loading Data

The IMLS [42,43] provides 3-h-sampled nontidal loading deformation models for 1272 stations, containing GEOS-FPIT atmospheric and hydrological models developed by Global Modeling and Assimilation Office at NASA Goddard Space Flight Center and MPIOM06 oceanic model developed at ESMGFZ. This loading model set is termed as the IMLS loading for short.

The EOST Loading Service [42] provides 1-h-sampled loading deformation datasets (in the form of Stokes coefficients up to degree and order 120; available at http://loading.u-strasbg.fr/listdata.php?dirn=st, accessed on 8 December 2020) based on the European Centre for Medium-Range Weather Forecasts (ECMWF) fifth reanalysis (ERA5) atmospheric surface pressure fields, induced TUGO-m ocean model [44] and GLDAS/Noah continental hydrology model [45]. However, unlike other classical ocean general circulation models, TUGO-m model is forced not only by atmospheric pressure but also by surface winds (please notice that both atmospheric and oceanic loadings are included in their ERA5_TUGO product while ERA5_IB contains only atmospheric loading). This loading model set is termed as the EOST loading for short.

Both the IMLS and EOST loading data are provided in the center of mass (CM) frame without corresponding data for geocenter motion corrections. We converted them to the center of figure frame to match the NGL GPS data, using geocenter motion derived from the LDCmgm90 model as described in Section 2.3 and Appendix A.

The ESMGFZ provides grid data for nontidal elastic surface loading deformation in the center of figure (CF) frame with $0.5^\circ \times 0.5^\circ$ spatial resolution [46]. ESMGFZ nontidal atmospheric loading is 3-h-sampled and based on the operational ECMWF model which is typically updated about twice a year to incorporate advanced algorithms and physical models into the system, which affects its long-term stability but improves near real-time and forecasting quality, leading to substantially higher spatial resolution compared with alternatively available atmospheric reanalysis models. ESMGFZ nontidal oceanic loading is 3-h-sampled and based on Max Planck Institute for Meteorology Ocean Model (MPIOM) ocean model and forced by the ECMWF atmospheric pressure. ESMGFZ hydrological loading is 24-h-sampled and based on the version-2 Land Surface Discharge Model (LSDM) hydrological model, in which soil moisture, snow, surface water and water in rivers and lakes are all considered with mean mass during 2013~2014 subtracted. High-resolution geographic information system (GIS)-based river network is used in this model for higher accuracy of mass. Unlike other loading datasets, ESMGFZ also provides a special 24-h-sampled sea-level loading for the conservation of global mass. Therefore, when ESMGFZ loading is used, the sum of the above four components represents the complete surface loading deformation. This loading model set is termed as the ESMGFZ loading for short.

The above-mentioned model sets are briefly summarized in Table 1. In this study, all the GCM-based loading data were downsampled to monthly time series (with signals with frequencies not exceeding 6 cycles per year (cpy)) to match the monthly LDCmgm90 loading.

Table 1. Atmospheric, oceanic and hydrological nontidal loading based on different meteorological models or products involved in four loading datasets used in this research.

Dataset	Atmosphere Loading	Ocean Loading	Hydrology Loading	Sea-Level Loading
IMLS	GEOS-FPIT	MPIOM06	GEOS-FPIT	\
EOST	ECMWF reanalysis	TUGO-m	GLDAS/Noah	\
ESMGFZ	ECMWF operational	MPIOM	LSDM	ESMGFZ SLEL
LDCmgm90	LDCmgm90 GAA	LDCmgm90 GAB	LDCmgm90 GSM	\

2.3. LDCmgm90-Based Loading Data and Geocenter Motion Correction

In this research, LDCmgm90 time-variable gravity [29] was adopted to derive LDCmgm90 global deformation model. The LDCmgm90 GSM dataset provides Stokes coefficients for degree/order 2~90, so we needed to determine the corresponding degree-1 coefficients ΔC_{10}, ΔC_{11} and ΔS_{11} in this study.

The GRACE monthly gravity products are usually released together with the GRACE AOD1B products, which provide a model-based dataset (including GAA, GAB, GAC and GAD) that describes the time variations of the gravity potential at satellite altitudes that are caused by nontidal mass variability in the atmosphere and oceans [47]. The GAA product describes the monthly nontidal atmospheric mass anomalies simulated by the operational run of the atmosphere model ECMWF. GAB refers to monthly nontidal oceanic mass anomalies simulated by the operational run of the (unconstrained) ocean model MPIOM. GAC is the sum of GAA and GAB. GAD can be regarded as a revised version of GAC with nontidal atmospheric and oceanic mass anomalies only over ocean areas. GSM is just the gravity residual after GAA and GAB are removed from the GRACE observations. Like the monthly GRACE gravity models, the LDCmgm90 dataset also has five components, namely the GSM, GAA, GAB, GAC and GAD products [29]. However, unlike common GRACE solutions, destripe correction is not used in processing LDCmgm90 solutions, which do not contain the stripe errors.

Recognizing the limited accuracies of glacial isostatic adjustment (GIA) models, Chen et al. [29] did not include the GIA correction into the LDCmgm90 dataset. The GIA effects reflect the viscoelastic behavior of the Earth's mantle, while loading deformations are only relevant with the Earth's elastic responses. Thus, we needed to remove the GIA contributions from the LDCmgm90 monthly gravity, and in this research, we chose the ICE6Gavg-GPS model (available at https://www.mdpi.com/2072-4292/12/7/1209/s1, accessed on 2 February 2020) [48], which combines multiple GIA models and global GPS observations, to remove GIA signals in the LDCmgm90 dataset. The loading model set derived from the LDCmgm90 dataset, after GIA correction, is termed as the LDCmgm90 loading for short.

The 3D NGL GPS displacements are provided in the CF frame, while the LDCmgm90 and all the relevant GRACE products are given in the CM frame. Therefore, we needed to convert the LDCmgm90-derived loading results from the CM frame to the CF frame. The CM and CF move relative to each other due to the degree-1 surficial mass loading. Let $\Delta X(t)$, $\Delta Y(t)$ and $\Delta Z(t)$ be three components of geocenter variations, reflecting differences between the Earth's CM and CF. We have [30,49]

$$\mathbf{r}_{\rm cm}-\mathbf{r}_{\rm cf}=\begin{bmatrix}\Delta X(t)\\ \Delta Y(t)\\ \Delta Z(t)\end{bmatrix}=\frac{\sqrt{3}(1+k_1)}{\rho_{ave}}\begin{bmatrix}\Delta C_{11}(t)\\ \Delta S_{11}(t)\\ \Delta C_{10}(t)\end{bmatrix} \tag{1}$$

where

$$\begin{aligned}\Delta C_{10}(t)&=\tfrac{1}{4\pi a^2}\int P_{10}(\cos\theta)\cdot F_O(\theta,\phi)\Delta\sigma(\theta,\phi,t)dS\\ \Delta C_{11}(t)&=\tfrac{1}{4\pi a^2}\int P_{10}(\cos\theta)\cdot\cos\theta\cdot F_O(\theta,\phi)\Delta\sigma(\theta,\phi,t)dS\\ \Delta S_{11}(t)&=\tfrac{1}{4\pi a^2}\int P_{10}(\cos\theta)\cdot\sin\theta\cdot F_O(\theta,\phi)\Delta\sigma(\theta,\phi,t)dS\end{aligned} \tag{2}$$

In Equations (1) and (2), $\rho_{ave} = 5517\ \mathrm{kg/m^3}$ is the average density of the Earth, $a = 6378$ km is the semimajor axis of the Earth, $\Delta\sigma$ is the incremental surface density and can be derived from Equation (14) of Wahr et al. [27] and k_l is the degree-l elastic loading Love number for geopotential variation. In this study, we used degree-1 loading number $k_1 = 0.027$ derived by Han and Wahr [50] since the origin of the coordinate system is the center of figure (also see Wahr et al. [27] for more details).

To derive ΔC_{10}, ΔC_{11} and ΔS_{11} corresponding to a global time-dependent gravity product, one possible approach is using an independent ocean bottom pressure model that provides $F_O(\theta,\phi)\Delta\sigma(\theta,\phi,t)$, where $F_O(\theta,\phi)$ is the ocean function that equals 1 in ocean areas and 0 in land areas. Another way, which was utilized in this research, is using the corresponding GRACE GAD product (please notice that GRACE GAA, GAB, GAC and GAD products are derived from meteorological model instead of GRACE observation). According to the definition, degree-1 Stokes coefficients derived from the GAD product are exactly those defined by Equation (2). Please notice that in Equation (3), GAC and GSM products should be both considered and transformed into mass spherical harmonic coefficients to present the complete global effect.

2.4. Method to Determine Loading Deformations

Loading deformations at colatitude θ and longitude ϕ caused by mass redistribution on the surface of the Earth presented by fully normalized Stokes coefficients can be derived as [51]

$$\begin{aligned}\Delta R(\theta,\phi,t)&=\tfrac{1}{g_{ave}}\sum_{l=1}^{\infty}\tfrac{h_l}{1+k_l}V_l(\theta,\phi,t)\\ \Delta E(\theta,\phi,t)&=\tfrac{1}{g_{ave}\cos(\pi/2-\theta)}\sum_{l=1}^{\infty}\tfrac{L_l}{1+k_l}\tfrac{\partial V_l(\theta,\phi,t)}{\partial\phi}=\tfrac{1}{g_{ave}\sin\theta}\sum_{l=1}^{\infty}\tfrac{L_l}{1+k_l}\tfrac{\partial V_l(\theta,\phi,t)}{\partial\phi}\\ \Delta N(\theta,\phi,t)&=\tfrac{1}{g_{ave}}\sum_{l=1}^{\infty}\tfrac{L_l}{1+k_l}\tfrac{\partial V_l(\theta,\phi,t)}{\partial(\pi/2-\theta)}=-\tfrac{1}{g_{ave}}\sum_{l=1}^{\infty}\tfrac{L_l}{1+k_l}\tfrac{\partial V_l(\theta,\phi,t)}{\partial\theta}\end{aligned} \tag{3}$$

where the $\Delta R(\theta,\phi,t)$, $\Delta E(\theta,\phi,t)$ and $\Delta N(\theta,\phi,t)$ are respectively the radial, eastward and northward loading deformations; $g_{ave} = 9.826057\ \mathrm{m/s^2}$ is the average gravitational acceleration on the surface of the Earth; and h_l and L_l are the degree-l vertical and horizontal elastic loading Love numbers, respectively. For degree > 2, we adopt k_l, h_l and L_l derived from the MATLAB code developed by Chen et al. [52]. $V_l(\theta,\phi,t)$ is degree-l gravitational potential on the surface of the Earth and can be expressed as

$$V_l(\theta,\phi,t)=\frac{GM}{a}\sum_{m=1}^{l}P_{lm}(\cos\theta)[\Delta C_{lm}(t)\cos(m\phi)+\Delta S_{lm}(t)\sin(m\phi)] \tag{4}$$

where $G = 6.67259\times10^{-11}\ \mathrm{N\cdot m^2/kg^2}$ is the gravitational constant and $M = 5.975562\times10^{24}$ kg is the total mass of the Earth.

3. Results

By using the derived degree-1 Stokes coefficients (please refer to Appendix A for more details) and degree > 1 Stokes coefficients of the LDCmgm90 dataset, we derive the corresponding horizontal and radial loading deformation separately after removing GIA effects. Obvious annual fluctuation is observed in degree-1 radial loading deformation in 2010 (Figure 2), which is similar in other years.

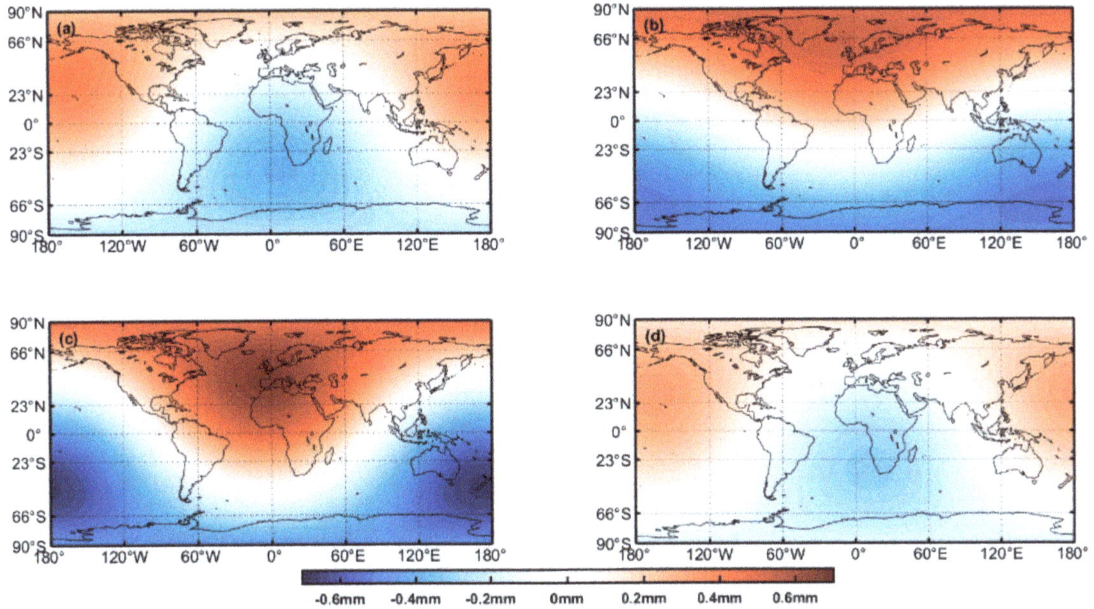

Figure 2. Global distribution of degree-1 radial loading deformation of LDCmgm90 spherical harmonic solution. Four samples in 2010 are shown: (**a**) March, 2010; (**b**) June, 2010; (**c**) September, 2010; (**d**) December, 2010.

Global distribution of degree > 1 loading deformation is significantly more complex due to the contributions of time-dependent precipitation, discharge of hydrology and atmospheric and oceanic mass redistribution.

Radial loading deformation is significantly larger than horizontal loading deformation. Globally, in 2010, absolute values of radial loading deformation in middle and northern high latitudes remain less than 15 mm, while Antarctica and the tropical continents of South America, central Africa, southern Asia and northern Australia have significant annual loading deformation variations with amplitudes larger than 15 mm (see Figure 3, line 1, GAC + GSM).

To separate the contributions of different sources, we derive the global radial loading deformation of LDCmgm90 GSM and GAC products separately. We find that continental anomalies along the equator are only observed in GSM loading deformation (Figure 3, Line 2, GSM). These prominent annual variations reflect the changes of terrestrial water storage in tropical continents of both hemispheres since precipitation brings a large amount of water in their respective rainy seasons, leading to downward loading deformation, while high temperature and drought during dry seasons vaporize water, leading to upward loading deformation. In middle- and high-latitude areas, annual variations of GSM loading deformation are not so obvious due to less precipitation and lower temperature compared with the tropical area. Seasonal melting and freezing in high latitudes also contribute a little to GSM loading deformation, among which the Greenland coastline is the most prominent. However, annual loading deformation variations in middle and high latitudes are mainly caused by changes in atmospheric pressure and ocean bottom pressure, especially in Antarctica.

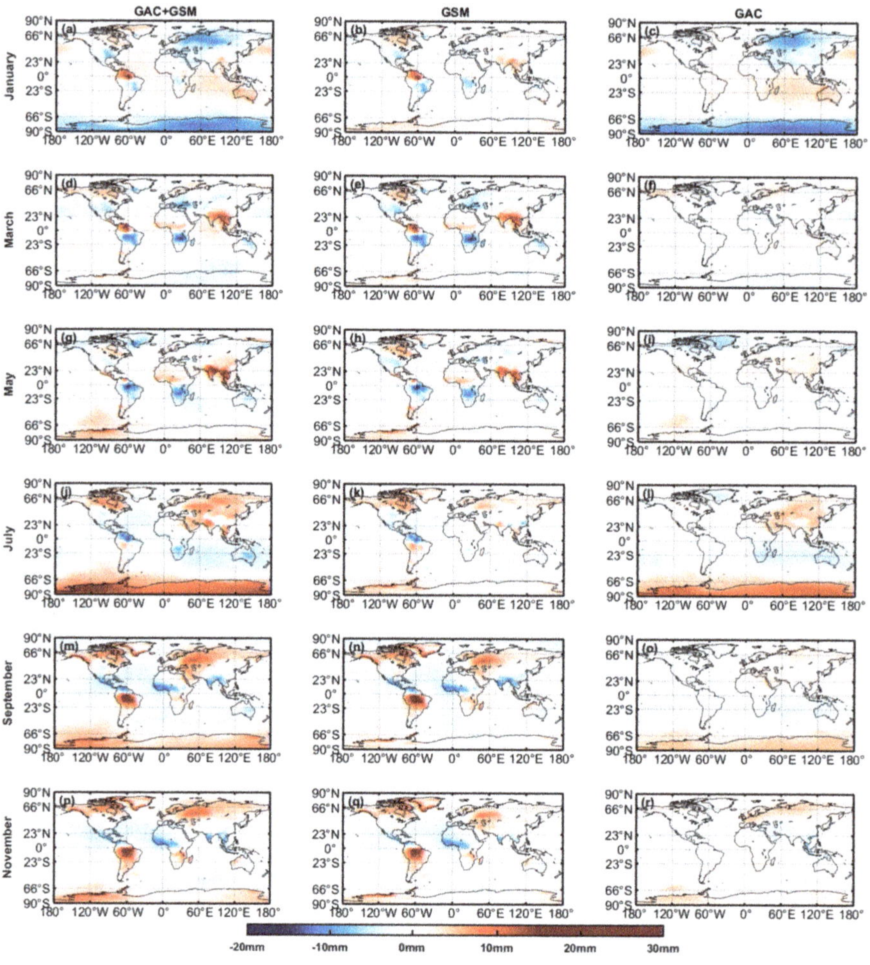

Figure 3. Global distribution of degree >1 vertical loading deformation of LDCmgm90 spherical harmonic solutions in odd months of 2010. Different sources are shown in different columns: Column 1, GAC + GSM (**a**,**d**,**g**,**j**,**m**,**p**); Column 2, GSM (**b**,**e**,**h**,**k**,**n**,**q**); Column 3, GAC (**c**,**f**,**i**,**l**,**o**,**r**); In 2010, absolute values of loading deformation of the Earth seldom reach 15 mm except in several places such as the Amazon region, where about 28 mm at most appears. Therefore, the color representing radial loading deformation >15 mm or <−15 mm remains the same to reveal smaller details.

Periodical variation of horizontal loading deformation also supports our inferences for the vertical component. Besides, due to the property of Equation (3), latitudinal and longitudinal dipole-like horizontal anomalies form near the center of radial anomalies. Eastward and northward loading deformations always appear on the east and the north of a positive radial loading deformation and on the west and the south of a negative radial loading deformation (Figures 4 and 5). On the whole, loading deformation diverges from a positive anomaly and converges to a negative anomaly. The direction of horizontal loading deformation is the same as the gradient of the vertical loading deformation field.

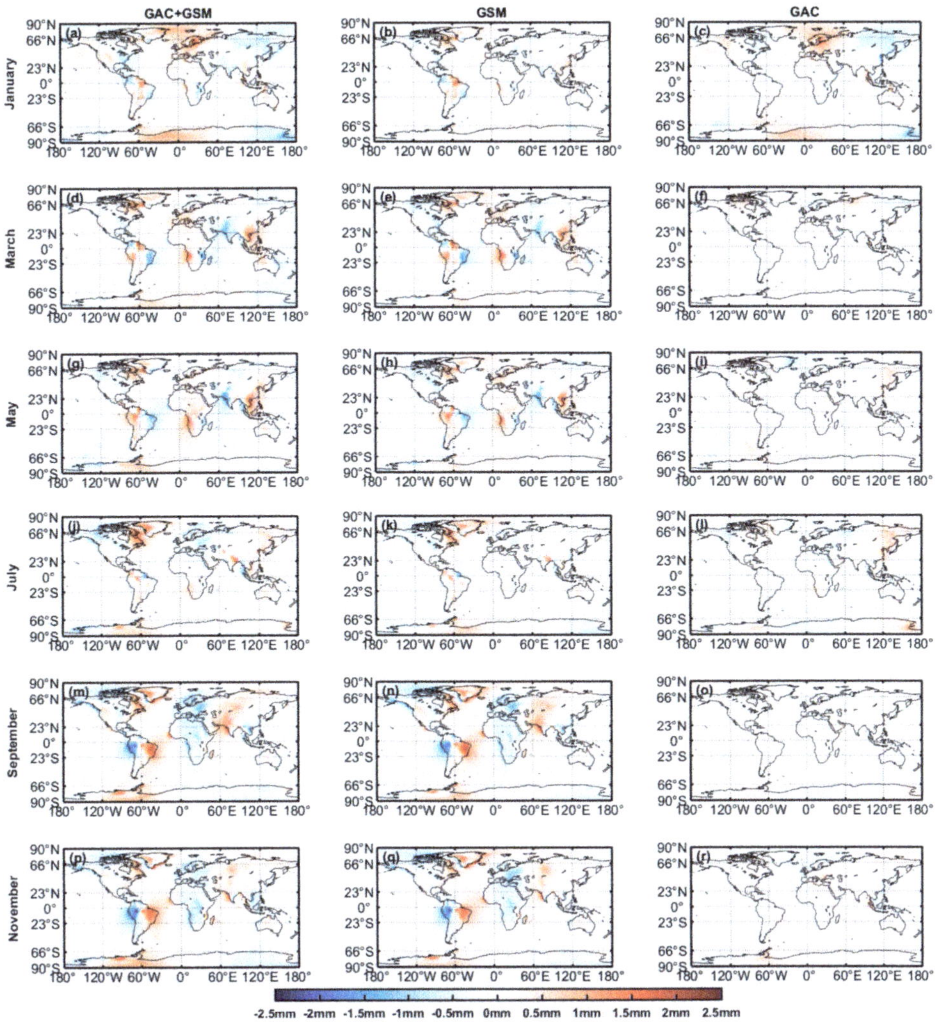

Figure 4. Global distribution of degree >1 eastward loading deformation of LDCmgm90 spherical harmonic solutions in odd months of 2010. Different sources are shown in different columns: Column 1, GAC + GSM (**a**,**d**,**g**,**j**,**m**,**p**); Column 2, GSM (**b**,**e**,**h**,**k**,**n**,**q**); Column 3, GAC (**c**,**f**,**i**,**l**,**o**,**r**).

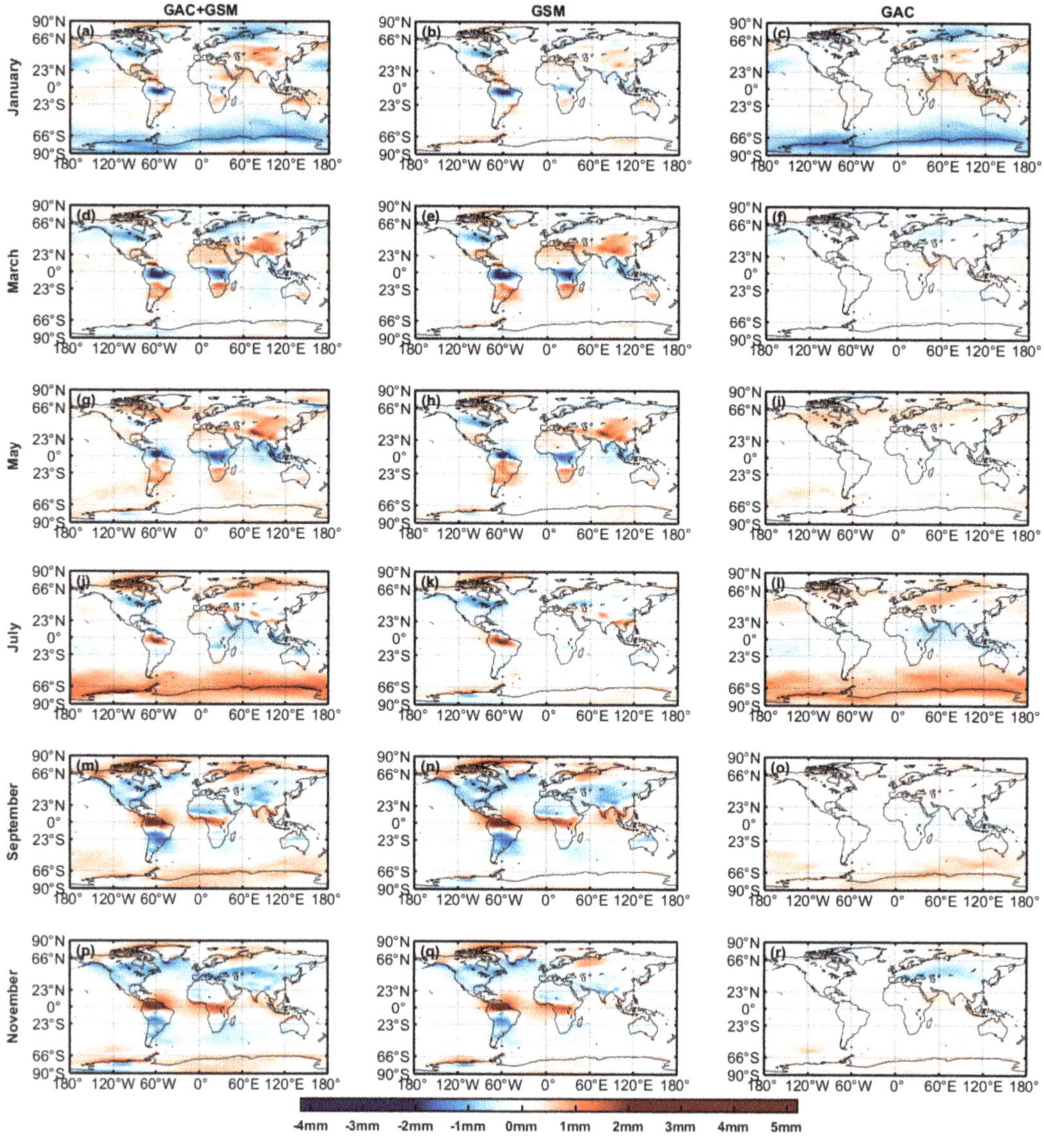

Figure 5. Global distribution of degree >1 northward loading deformation of LDCmgm90 spherical harmonic solutions in odd months of 2010. Different sources are shown in different columns: Column 1, GAC + GSM (**a**,**d**,**g**,**j**,**m**,**p**); Column 2, GSM (**b**,**e**,**h**,**k**,**n**,**q**); Column 3, GAC (**c**,**f**,**i**,**l**,**o**,**r**); In 2010, only South America northward loading deformation >2 mm in several months, reaching 5.2 mm at most. Therefore, color representing northward loading deformation >2 mm or <−2 mm remains the same to reveal smaller details.

4. Discussion

We applied our LDCmgm90 GAC- and GSM-derived loading products to 249 IGb14 GPS time series processed by NGL to correct nontidal loading effects. Likewise, EOST and ESMGFZ atmospheric, oceanic and hydrological loading products were applied for comparison. Since these stations are scattered globally with diverse data characteristics and qualities, the results of loading corrections will reflect the reliability of load deformation models adopted. We also selected 12 of 249 GPS stations as examples, in which IMLS loading products were also applied.

4.1. Comparison of Annual Residual

The up-component displacements of 12 IGb14 GPS stations are compared in Figure 6, in which loading effects have been removed by the four different loading products mentioned above (only in Figure 6 the trends are removed for easier comparisons of the seasonal signals). In most cases in the time domain, according to the remaining residuals, we find that more loading effects are removed by EOST, ESMGFZ and LDCmgm90 loadings compared with IMLS loading, especially at NRIL and WSRT. According to the corresponding residuals' power spectrum density (PSD) shown in Figure 7 and annual amplitudes of time series listed in Table 2, we find that, among these 12 stations, significantly more annual loading is removed by LDCmgm90 datasets at 6 stations: ADE1, BREW, CHPI, MOBS, TNML and WSRT. The EOST loading is better at three stations: AMC2, NKLG and TIXI. ESMGFZ loading is better at three stations: HOLM, LHAZ and NRIL. No advantage of IMLS loading is found among these stations. In fact, applying IMLS loading often increases the annual amplitudes of the 12 residuals.

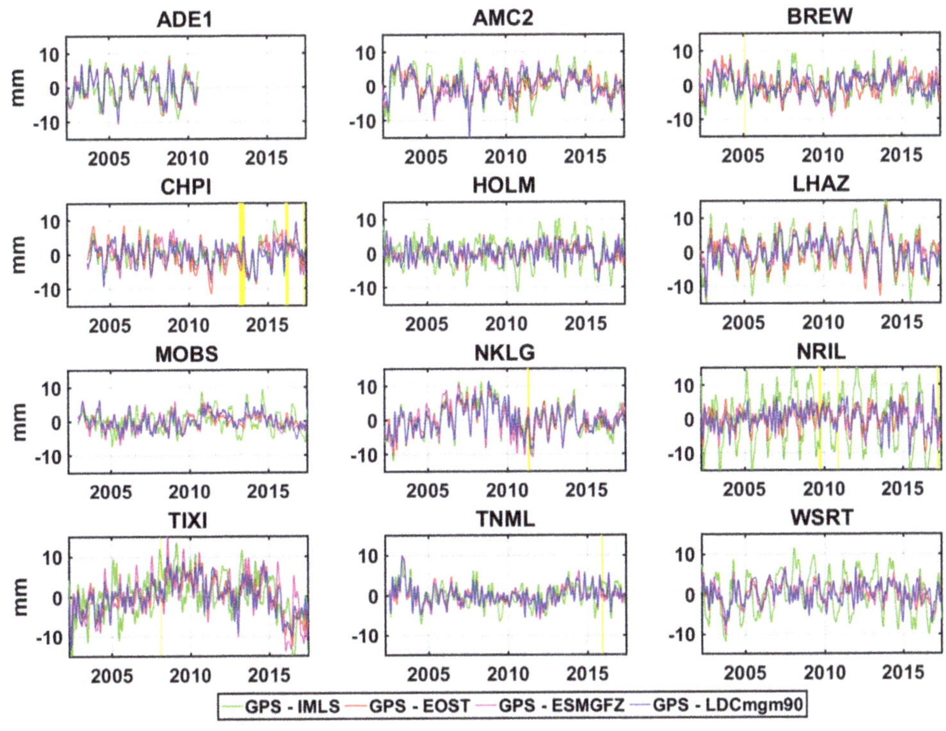

Figure 6. Detrended up-component residuals of 12 IGb14 GPS stations in which IMLS (green), EOST datasets (red), ESMGFZ (magenta) and LDCmgm90 (blue) loadings have been applied (unit: mm). Atmospheric, oceanic and hydrological loadings are all considered. Yellow perpendicular lines are marks for durations where >30 adjacent daily samples are unavailable in the original GPS time series. Signals with frequencies >6 cpy are removed.

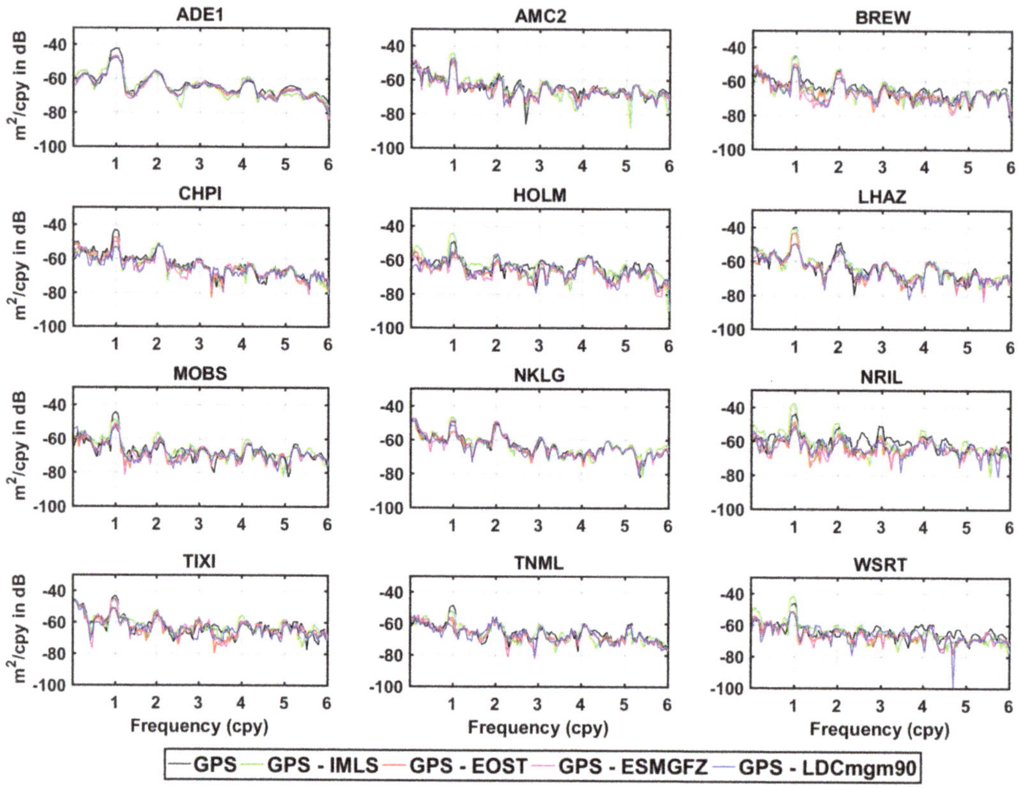

Figure 7. Power spectrum density (PSD) of detrended up-component residuals of 12 IGb14 GPS stations in which IMLS (green), EOST datasets (red), ESMGFZ (magenta) and LDCmgm90 (blue) loadings have been applied (unit: m^2/cpy in dB). Atmospheric, oceanic, and hydrological loadings are all considered. PSD of processed GPS time series is given as reference. Less PSD indicates smaller amplitudes at the corresponding frequencies.

We also notice that at several stations, applying loading corrections leads to excess annual amplitudes instead of attenuating them, which indicates that worse results are induced by loading models. According to Table 2, compared with original GPS time series, annual loading amplitudes are successfully reduced at 6 stations with IMLS loading applied, at all 12 stations with EOST loading applied, at 10 stations with ESMGFZ loading applied and at 11 stations with LDCmgm90 loading models applied. Numerically, among these 12 stations, the most prominent improvement is brought by ESMGFZ loading at HOLM, where 91% up-component annual fluctuation in the original GPS signal is removed (only 0.21 mm remains).

The accuracy of hydrological loading is not as reliable as that of atmospheric and oceanic loadings due to the uneven distribution of continent water storage and much less accurate monitoring generally. Different hydrological models diverge from each other and may enlarge the residuals instead of attenuating them further. In Table 2, we notice that at eight stations (ADE1, AMC2, HOLM, MOBS, NKLG, NRIL, TIXI and WSRT), applying IMLS hydrological loading leads to larger annual residuals than applying IMLS atmospheric and oceanic loadings only, which is significantly improved by EOST, ESMGFZ and LDCmgm90 loading corrections. Therefore, we infer that EOST, ESMGFZ and LDCmgm90 models contain better hydrological loading information compared with IMLS loading.

Divergences between different loading models are smaller in other frequency bands at these 12 stations (such as semiannual loading, see frequency = 2 cpy in Figure 7 and Table 3). However, since loading effects are mainly annual fluctuations [5,8], discrepancies among loading models are mainly introduced by their annual signals, which should represent the quality of loading models. Therefore, due to the smaller time- and frequency-domain annual residual and better hydrological loading information involved, we believe that EOST, ESMGFZ and LDCmgm90 loadings are generally better than IMLS loading.

Table 2. Effects of applying loading corrections: the case of annual signals (unit: mm).

Station	GPS	GPS—(A + O) [1]				GPS—(A + O + H) [2]			
		IMLS	EOST	ESMGFZ	LDC [3]	IMLS	EOST	ESMGFZ	LDC
ADE1	7.61	4.29	5.02	4.55	4.52	4.31	4.46	4.30	3.93
AMC2	2.62	1.18	2.38	2.09	2.34	4.54	2.37	2.72	3.28
BREW	4.28	4.52	4.32	4.46	3.80	4.37	2.46	2.47	1.95
CHPI	5.16	4.09	4.78	4.52	4.38	1.03	2.88	1.75	0.86
HOLM	2.33	1.44	1.43	1.44	0.64	4.36	0.52	0.21	0.78
LHAZ	7.29	8.23	6.79	7.08	6.79	6.96	5.00	2.16	2.25
MOBS	4.32	1.95	2.64	2.33	1.91	2.47	1.44	1.65	1.30
NKLG	2.51	2.47	1.71	1.89	1.51	3.37	1.13	2.59	1.87
NRIL	4.36	5.39	3.82	4.25	3.29	9.48	2.55	1.19	1.61
TIXI	4.75	2.64	3.08	2.62	2.07	3.75	1.93	4.24	1.96
TNML	2.79	1.93	1.70	1.52	0.81	1.86	0.94	1.15	0.51
WSRT	3.44	1.35	3.57	2.75	2.35	6.14	1.85	1.94	1.58

[1] "GPS—(A + O)" means that atmospheric loading and oceanic loading are applied. [2] "GPS—(A + O + H)" means that atmospheric loading, oceanic loading and hydrological loading are applied. [3] "LDC" denotes LDCmgm90.

Table 3. Effects of applying loading corrections: the case of semiannual signals (unit: mm).

Station	GPS	GPS—(A + O) [1]				GPS—(A + O + H) [2]			
		IMLS	EOST	ESMGFZ	LDC [3]	IMLS	EOST	ESMGFZ	LDC
ADE1	1.36	1.55	1.57	1.56	1.48	1.18	1.50	1.43	1.51
AMC2	0.83	0.50	0.41	0.40	0.40	0.86	0.30	0.53	0.44
BREW	1.51	1.62	1.20	1.31	1.04	1.67	1.20	1.45	0.75
CHPI	1.16	1.04	1.05	1.02	0.99	1.50	0.85	0.46	0.81
HOLM	0.96	0.80	0.93	1.01	0.95	0.77	0.45	0.50	0.94
LHAZ	2.35	2.47	2.36	2.44	2.34	1.34	1.53	1.07	1.46
MOBS	0.18	0.37	0.43	0.38	0.32	0.86	0.48	0.36	0.25
NKLG	0.88	0.68	0.61	0.62	0.70	1.42	1.10	1.73	1.60
NRIL	1.61	1.80	1.88	1.82	1.71	2.51	0.87	0.67	0.43
TIXI	1.32	1.55	1.59	1.48	1.30	1.71	1.16	1.34	0.90
TNML	0.91	1.24	0.98	1.18	1.18	1.09	0.86	0.94	1.13
WSRT	0.15	0.26	0.41	0.40	0.32	0.72	0.42	0.49	0.53

[1] "GPS—(A + O)" means that atmospheric loading and oceanic loading are applied. [2] "GPS—(A + O + H)" means that atmospheric loading, oceanic loading and hydrological loading are applied. [3] "LDC" denotes LDCmgm90.

4.2. Loading Due to Seasonal Mass Redistributions

We removed the means and trends of all the relevant data in order to estimate the improvement of seasonal loading correction brought by our LDCmgm90 loading models compared with EOST and ESMGFZ datasets. First of all, we calculated the mean PSD of 249 IGb14 GPS time series with different loading correction models applied (see Figure 8). One can see the LDCmgm90 is the most efficient in attenuating the annual signal in GPS data, especially in the up-component. Besides, long-period variations of up-component GPS time series are also improved by LDCmgm90 compared with EOST and ESMGFZ. However, there are also strong peaks of draconitic year error (with period 351.2 day or frequency 1.04 cpy; also see Ray et al. [3]) and its higher harmonics (peaks near the 2, 3, 4 and 5 cpy) as shown by Figure 8. These draconitic year errors in GPS data have nothing to do with loading but also contribute greatly to the scatter of GPS position residual, which can be well corrected by least-square fitting (see three bottom subplots in Figure 8). On the other hand, both the EOST and ESMGFZ have better performances than LDCmgm90 for up-component variations between draconitic peaks, the cause of which is not quite certain but could be a limitation of the frequency and spatial resolutions (~6 cpy and ~2 degrees, respectively) of the LDCmgm90 data (it is rather common that there is no data point at the frequency of interest and thus the plots are impacted by its neighboring frequency points, due to the limited frequency domain sampling rate caused by the finite length of the time series of interest).

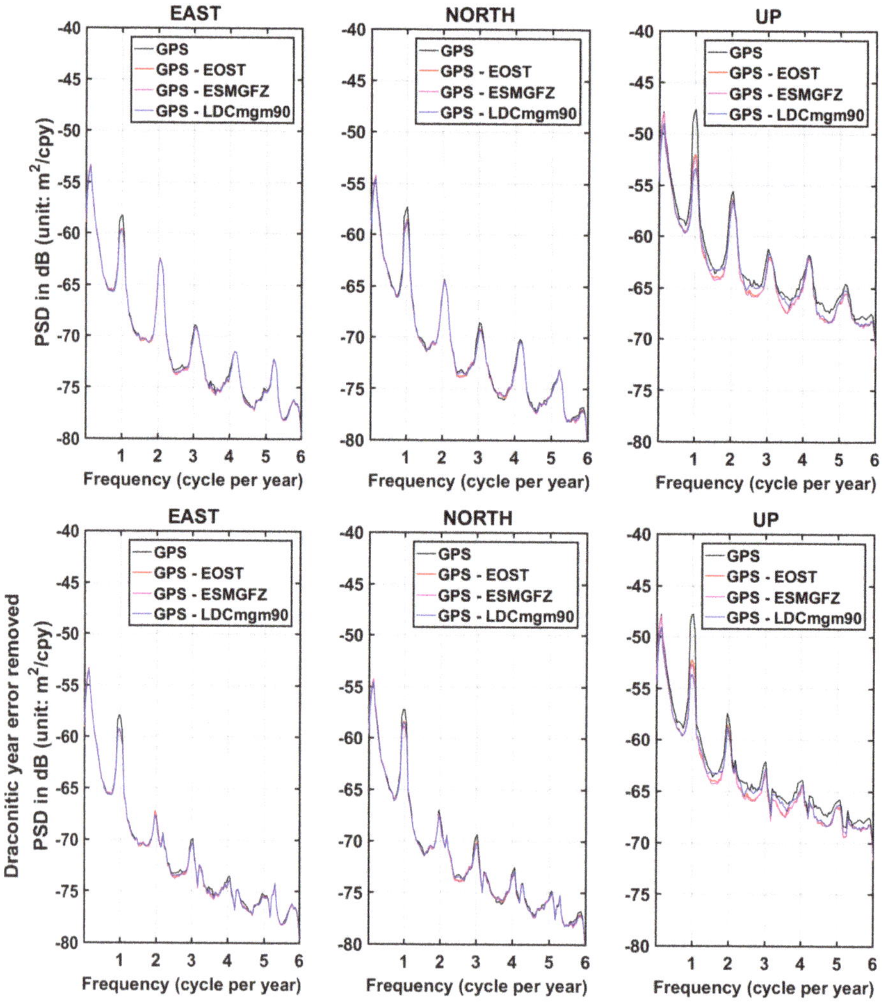

Figure 8. Mean PSD of 249 IGb14 GPS time series (black) and GPS residuals with EOST loading applied (red), with ESMGFZ loading applied (magenta) and with LDCmgm90 loading applied (blue). The 0~6 cpy east-, north- and up-components are all given. Please notice that the draconitic year errors in the 3 subplots above are removed by least-square fitting in the bottom row of subplots.

Root mean square (RMS) of detrended GPS residual is a vital index to estimate the quality of a loading model (the mean value of each time series is removed before time-domain analyses). Globally, compared with applying EOST and ESMGFZ loadings, applying LDCmgm90 loading improves the residual RMS at more of the 249 IGb14 GPS stations involved in this research. Up-component RMS variation (ΔRMS) brought by LDCmgm90 loading compared with applying either EOST loading or ESMGFZ loading is shown in Figures 9 and 10, respectively. Besides, statistical information about RMS improvement ratio is listed in Tables 4 and 5.

Compared with original GPS signal (Table 4), applying LDCmgm90 loading successfully reduces up-component RMS at 218 (equaling 87.55%) GPS stations, which is slightly more than applying EOST or ESMGFZ loading. When significantly improved (RMS improvement ratio > 5%) stations are considered only, LDCmgm90 loading is about 4% better than EOST and matches ESMGFZ loading.

In horizontal components, LDCmgm90 roughly matches EOST and ESMGFZ; differences between RMS with either loading model applied at most stations are within 5%. Numerically, according to Table 5, applying LDCmgm90 loading models instead of EOST loading models reduces 128 (equaling 51.41%) stations' east-component RMS and 136 (equaling 54.62%) stations' north-component RMS, which is slightly better when compared with ESMGFZ, in which 122 (equaling 49.00%) stations' east-component RMS and 115 (equaling 46.18%) stations' north-component RMS are reduced by applying LDCmgm90.

However, as for the up-component loading, where signals are larger than the horizontal-component loading, applying LDCmgm90 loading significantly decreases residuals' RMS at 155 (equaling 62.25%) globally distributed GPS stations compared with applying EOST loading models, among which improvements at 81 (equaling 32.53%) stations are larger than 5%. Compared with ESMGFZ loading, applying LDCmgm90 loading also reduces 142 (equaling 57.03%) stations' up-component residuals' RMS, among which improvements at 72 (equaling 28.92%) stations are larger than 5%. The most obvious RMS improvements are found in high-latitude regions of the northern hemisphere, especially at stations located along Greenland's coastline. The largest RMS decrease is found at KELY (66.987°N, 50.945°W; vs. EOST: −3.40 mm; vs. ESMGFZ: −3.78 mm), while the largest RMS increases are found at NURK (1.945°S, 30.09°S; vs. EOST: −1.02 mm) and UFPR (25.448°S, 49.231°W; vs. ESMGFZ: −1.07 mm). We infer that the better annual and long-period signals contribute to the up-component superiority of LDCmgm90 compared to EOST and ESMGFZ.

Table 4. Statistical information of GPS stations with different RMS improvement ratios: loading-removed GPS signals vs. original GPS signal [1].

Loading-Removed GPS Signal	RMS Improvement Ratio				Station Improved [2]
	>5%	0~5%	−5~0%	<−5%	
EAST					
GPS—EOST	50 (20.08%)	111 (44.58%)	72 (28.92%)	16 (6.43%)	161 (64.66%)
GPS—ESMGFZ	58 (23.29%)	94 (37.75%)	66 (26.51%)	31 (12.45%)	152 (61.04%)
GPS—LDCmgm90	54 (21.69%)	108 (43.37%)	73 (29.32%)	14 (5.62%)	162 (65.06%)
NORTH					
GPS—EOST	76 (30.52%)	117 (46.99%)	48 (19.28%)	8 (3.21%)	193 (77.51%)
GPS—ESMGFZ	92 (36.95%)	88 (35.34%)	59 (23.69%)	10 (4.02%)	180 (72.29%)
GPS—LDCmgm90	89 (35.74%)	102 (40.96%)	52 (20.88%)	6 (2.41%)	191 (76.71%)
UP					
GPS—EOST	161 (64.66%)	53 (21.29%)	28 (11.24%)	7 (2.81%)	214 (85.94%)
GPS—ESMGFZ	172 (69.08%)	44 (17.67%)	16 (6.43%)	17 (6.83%)	216 (86.75%)
GPS—LDCmgm90	171 (68.67%)	47 (18.88%)	25 (10.04%)	6 (2.41%)	218 (87.55%)

[1] Positive RMS improvement ratio means that RMS of loading-removed GPS signal is smaller than original GPS signal, indicating that loading is successfully removed, while a negative value means that applying loading model increases the RMS of the GPS signal. RMS improvement ratio compared with original GPS signal is calculated by 100%—RMS (loading-removed GPS)/RMS (original GPS). [2] "Station Improved" column shows the total number and ratio of GPS stations whose loading effects were successfully removed by EOST, ESMGFZ or LDCmgm90 loading, equaling ">5%" column plus "0~5%" column. For each component, the total number of improved stations is listed outside the brackets while the corresponding ratio is given in the brackets.

Table 5. Statistical information of GPS stations with different RMS improvement ratios: comparison between GPS signals with different loading models applied [1].

Loading-Removed GPS Signal	RMS Improvement Ratio				Station Improved [2]
	>5%	0~5%	−5~0%	<−5%	
GPS—LDCmgm90 vs. GPS—EOST [3]					
East	17 (6.83%)	111 (44.58%)	99 (39.76%)	22 (8.84%)	128 (51.41%)
North	25 (10.04%)	111 (44. 58%)	97 (38.96%)	16 (6.43%)	136 (54.62%)
Up	81 (32.53%)	74 (29.72%)	59 (23.69%)	35 (14.06%)	155 (62.25%)
GPS—LDCmgm90 vs. GPS—ESMGFZ					
East	27 (10.84%)	95 (38.15%)	106 (42.57%)	21 (8.43%)	122 (49.00%)
North	21 (8.43%)	94 (37.75%)	104 (41.77%)	30 (12.05%)	115 (46.18%)
Up	72 (28.92%)	70 (28.11%)	62 (24.90%)	45 (18.07%)	142 (57.03%)

[1] Positive RMS improvement ratio means that RMS of LDCmgm90 residual is smaller than that of EOST or ESMGFZ residual, indicating that LDCmgm90 is better, while a positive value means that EOST or ESMGFZ is better. [2] "Station Improved" column shows the total number and ratio of GPS stations whose loading effects are better removed by LDCmgm90 compared with ERA5 or ESMGFZ, equaling ">5%" column plus "0~5%" column. For each component, the total number of improved stations is listed outside the brackets while the corresponding ratio is given in the brackets. [3] RMS improvement ratio brought by LDCmgm90 compared with EOST is calculated by 100%—RMS (GPS—LDCmgm90)/RMS (GPS—EOST), which is similar to that compared with ESMGFZ.

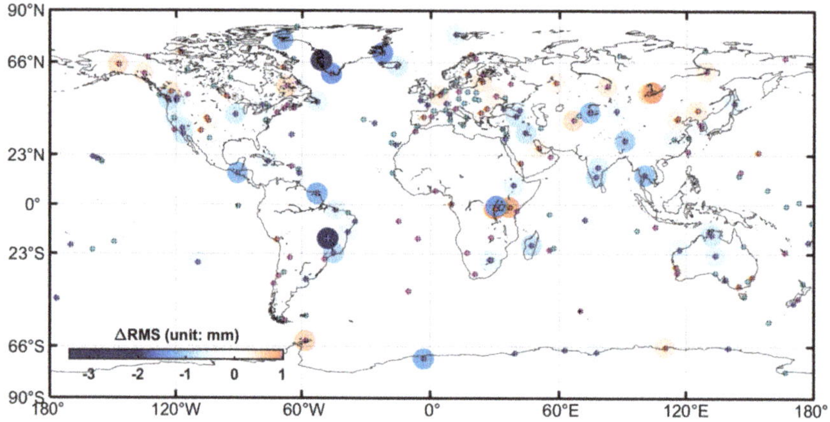

Figure 9. Changes of root mean square (ΔRMS) of 249 IGb14 GPS up-component time series: LDCmgm90 vs. EOST. Decrease in RMS means that applying LDCmgm90 loading reduces the residual's RMS compared with applying EOST loading, indicating that LDCmgm90 loading is better than EOST at these GPS stations, while increase in RMS means that EOST is better. Besides, locations of 249 stations are marked with "+" (blue "+": >5% RMS is removed; cyan "+": 0~5% RMS is removed; magenta "+": 0~5% RMS is induced; red "+": >5% RMS is induced).

To better understand these three loading models, we also compared EOST, ESMGFZ and LDCmgm90 together, finding the best loading model for each station involved in this research and calculating the RMS improvement ratio of GPS residuals brought by LDCmgm90 compared with the better result of applying EOST and ESMGFZ loadings (Figure 11). To reach the smallest RMS of GPS residuals, applying EOST is the best choice for 54 (21.69%) stations, applying ESMGFZ is the best choice for 85 (34.14%) stations and applying LDCmgm90 is the best choice for 110 (44.18%) stations. Regionally, EOST-best stations lie mainly in the middle-latitude area of North America and Asia. ESMGFZ-best stations are mainly in western Europe. LDCmgm90-best stations are widely distributed around the world, especially along the coastline of Greenland.

Therefore, according to the comparison of 249 globally distributed IGb14 GPS stations, especially the improvement in removing up-component effects, we conclude that LDCmgm90 loading is reliable and generally better than EOST loading and ESMGFZ loading.

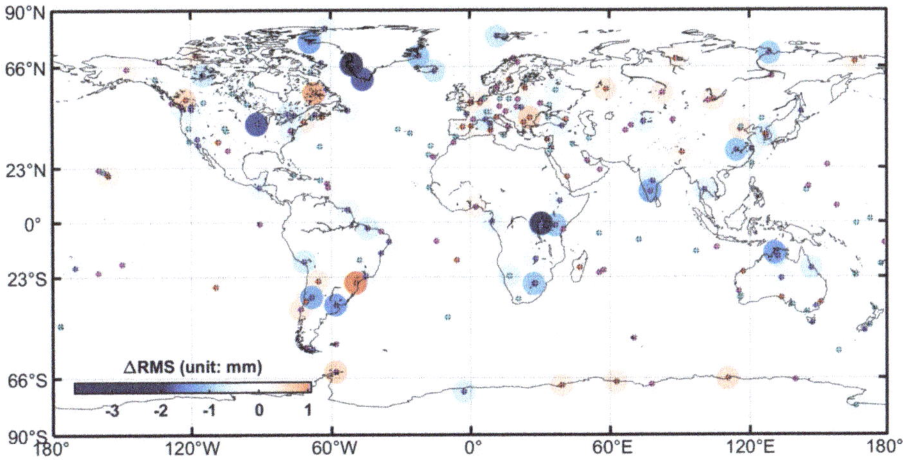

Figure 10. Changes of root mean square (ΔRMS) of 249 IGb14 GPS up-component time series: LDCmgm90 vs. ESMGFZ. Decrease in RMS means that applying LDCmgm90 loading reduces the residual's RMS compared with applying ESMGFZ loading, indicating that LDCmgm90 loading is better than EOST at these GPS stations, while increase in RMS means that ESMGFZ is better. Besides, locations of 249 stations are marked with "+" (blue "+": >5% RMS is removed; cyan "+": 0~5% RMS is removed; magenta "+": 0~5% RMS is induced; red "+": >5% RMS is induced).

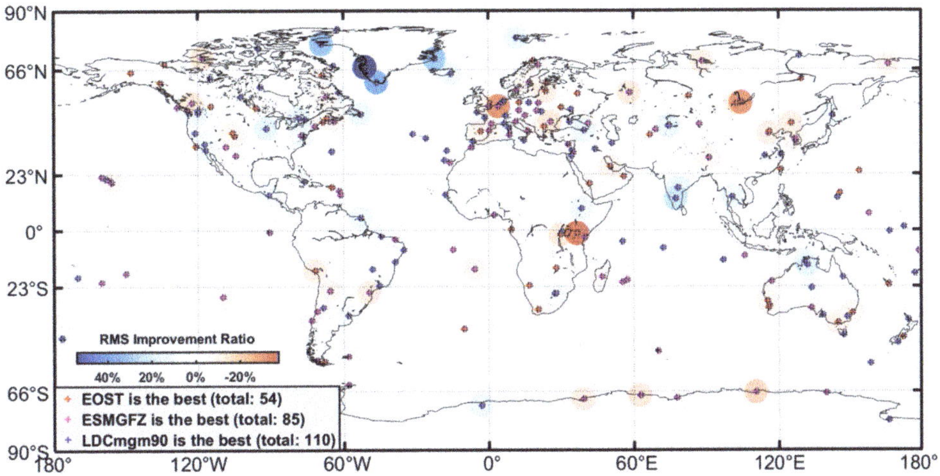

Figure 11. RMS improvement ratio of 249 IGb14 GPS Up-component time series: LDCmgm90 vs. the better of EOST and ESMGFZ. Besides, locations of 249 stations are marked with "+". Unlike Figures 9 and 10, for each station, red "+" means EOST loading is the best, magenta "+" means ESMGFZ is the best and blue "+" means LDCmgm90 is the best.

4.3. Contributions of Secular or Long-Term Mass Redistributions

Loading affects not only seasonal variations but also trends in GPS measurements, which should be appropriately considered by a complete loading model (note that many long-period geophysical signals, e.g., the ~18.6-year tide, may look like trends for the relatively short GPS time series; thus, "trends" here may contain contributions from these long-period geophysical signals). Therefore, to test the secular or long-term signals of LDCmgm90 loading, we only removed the means but retained the trends of all data, and we applied GIA corrections (ICE6Gavg-GPS model) to GPS data and LDCmgm90 loading (termed "trend-retained" for short hereafter). The analyses are similar to those in Section 4.2. Comparisons are listed in Tables 6 and 7.

According to the comparison of 249 GPS stations in Table 6, when trends are considered, the number of stations significantly improved (RMS improvement ratio > 5%) by applying EOST and ESMGFZ loadings decreases by 15~20%, while the number of stations significantly improved by applying LDCmgm90 loading varies only 2%, which is 16% better than EOST loading and 18% better than ESMGFZ loading.

Applying LDCmgm90 loading successfully reduces up-component RMS at 201 (equaling 80.72%) GPS stations, which is similar to applying EOST loading and 12% more than applying ESMGFZ loading. On the other hand, the ~14% larger improvement of the EOST model with respect to the ESMGFZ model is not found in trend-removed comparison. This may be due to the facts that the former adopts the ECMWF reanalysis (ERA5) of atmospheric data and the associated oceanic and hydrological models are forced by both surface pressure and winds (more practical), which induces better trend signals, while the ESMGFZ only uses ECMWF operational data and its associated ocean model is forced only by surface pressure (less practical; operational data usually have poorer quality than reanalysis data).

Table 6. Comparison of up-component loading corrections affected by trend in loading models [1].

Loading-Removed GPS Signal	RMS Improvement Ratio				Station Improved
	>5%	0~5%	−5~0%	<−5%	
Trend-removed [2]					
GPS—EOST	161 (64.66%)	53 (21.29%)	28 (11.24%)	7 (2.81%)	214 (85.94%)
GPS—ESMGFZ	172 (69.08%)	44 (17.67%)	16 (6.43%)	17 (6.83%)	216 (86.75%)
GPS—LDCmgm90	171 (68.67%)	47 (18.88%)	25 (10.04%)	6 (2.41%)	218 (87.55%)
Trend-retained					
GPS—EOST	125 (50.20%)	80 (32.13%)	37 (14.86%)	7 (2.81%)	205 (82.33%)
GPS—ESMGFZ	120 (48.19%)	51 (20.48%)	50 (20.08%)	28 (11.24%)	171 (68.67%)
GPS—LDCmgm90	165 (66.27%)	36 (14.46%)	28 (11.24%)	20 (8.03%)	201 (80.72%)

[1] The structure of this table is similar to Table 4. [2] Results of "Trend-removed" are the same as those of "UP" in Table 4, which are given here for comparison.

Improvement brought by the trend of LDCmgm90 loading can also be observed in Table 7, where significantly improved (RMS improvement ratio > 5%) stations rise 11% and 15% when compared with EOST and ESMGFZ results, respectively. Total numbers of improved stations also slightly increase. We also repeat the three-model comparison similar to Figure 11, the result of which is given in Figure 12. We find that the largest RMS improvement ratio is about 80%, which is nearly double that in Figure 11. Besides, when trends are considered, applying LDCmgm90 is the best choice for 116 (46.59%) stations, slightly more than trend-removed condition.

Table 7. Up-component improvement of LDCmgm90 loading compared with EOST and ESMGFZ loadings: comparison between trend-removed and trend-retained conditions [1].

Loading-Removed GPS Signal	RMS Improvement Ratio				Station Improved
	>5%	0~5%	−5~0%	<−5%	
GPS—LDCmgm90 vs. GPS—EOST					
Trend-removed [2]	81 (32.53%)	74 (29.72%)	59 (23.69%)	35 (14.06%)	155 (62.25%)
Trend-retained	108 (43.37%)	52 (20.88%)	46 (18.47%)	43 (17.27%)	160 (64.26%)
GPS—LDCmgm90 vs. GPS—ESMGFZ					
Trend-removed	72 (28.92%)	70 (28.11%)	62 (24.90%)	45 (18.07%)	142 (57.03%)
Trend-retained	111 (44.58%)	41 (16.47%)	40 (16.06%)	57 (22.89%)	152 (61.04%)

[1] The structure of this table is similar to Table 5. [2] Results of "Trend-removed" are the same as those of "UP" in Table 5, which are given here for comparison.

In order to show the advantages of long-term and annual signals in LDCmgm90 loading more clearly, we calculate the improvement of LDCmgm90 loading in long-term (with periods > 5 years) and annual frequency bands (with periods between 338 and 410 days) compared with EOST and ESMGFZ loadings (Table 8). We find that 49% EOST and 49.8% ESMGFZ long-term residuals and 57.83% EOST and 55.02% ESMGFZ annual residuals are significantly improved, larger than 43.37% and 44.58% in Table 7 where all <6 cpy signals (period > 2 months) are considered, supporting the contributions of LDCmgm90 loading in long-term and annual bands. In the meanwhile, total numbers of improved stations in Table 8 remain similar compared with Table 7, indicating that (1) general advantages of LDCmgm90 loading still exist and (2) EOST and ESMGFZ loading models also perform well in other seasonal variations, which corresponds with PSD shown in Figure 8.

Therefore, according to the trend-retained comparison in this section and band comparison in Table 8, we confirm that not only seasonal variations but also long-term signals in LDCmgm90 loading are better than those in EOST and ESMGFZ loadings.

Table 8. Up-component improvement of LDCmgm90 loading compared with EOST and ESMGFZ loadings in long-term and annual frequency bands.

Loading-Removed GPS Signal	RMS Improvement Ratio				Station Improved
	>5%	0~5%	−5~0%	<−5%	
Long-term band [1]					
LDC vs. EOST [2]	122 (49.00%)	34 (13.65%)	31 (12.45%)	62 (24.90%)	156 (62.65%)
LDC vs. ESMGFZ	124 (49.80%)	28 (11.24%)	22 (8.84%)	75 (30.12%)	152 (61.04%)
Annual band [3]					
LDC vs. EOST	144 (57.83%)	13 (5.22%)	16 (6.43%)	76 (30.52%)	157 (63.05%)
LDC vs. ESMGFZ	137 (55.02%)	24 (9.64%)	22 (8.84%)	66 (26.51%)	161 (64.66%)

[1] In long-term band, signals with periods > 5 years are involved. [2] In this table, "LDC" denotes LDCmgm90. [3] In annual band, signals with periods between 338 and 410 days are involved.

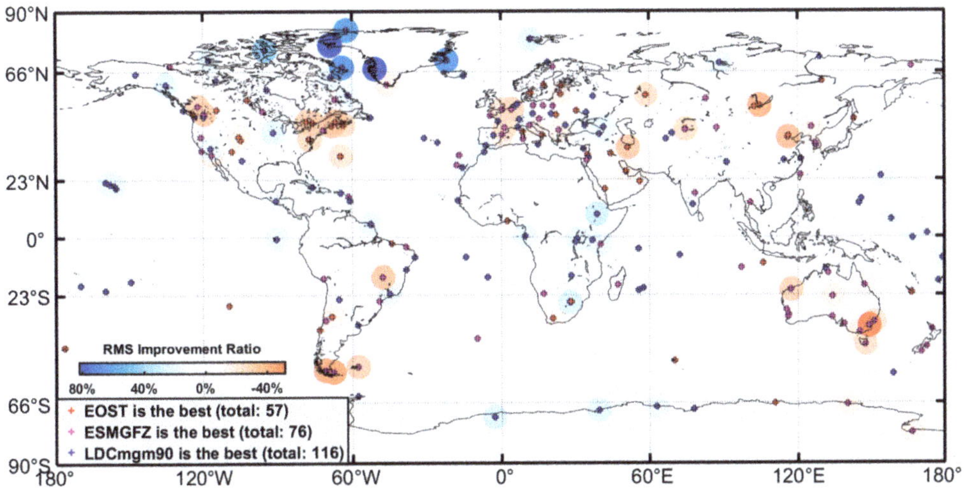

Figure 12. RMS improvement ratio of 249 IGb14 GPS up-component time series: Similar to Figure 11, but for trend-retained comparison.

5. Conclusions

By using LDCmgm90 multiple-data-based monthly gravity dataset and estimating its degree-1 Stokes coefficients, we have obtained the complete LDCmgm90 atmospheric, oceanic and hydrological loading models. Globally, we compared LDCmgm90 loading model with GCM-based loading models provided by the EOST Loading Service and ESMGFZ by applying them to 249 globally distributed IGb14 GPS stations processed by NGL to remove their loading effects. Loading models provided by IMLS were also applied to 12 GPS stations.

Our research shows that the LDCmgm90 loading model can reliably remove loading effects in GPS time series and is likely to achieve similar or even better results compared with other latest loading models. In general, according to our numerical results, we infer that the EOST and ESMGFZ loading models are better than the IMLS loading, but the LDCmgm90 loading is even better, not only for GPS sites on the continents but also for those on the islands surrounded by the oceans. The better annual, trend and long-period information, especially that of continent water, contribute to the superiority of LDCmgm90 loading compared with other loading products and therefore lead to significant improvements in the LDCmgm90 model with respect to the EOST and ESMGFZ models.

When all the data trends are removed, applying LDCmgm90 loading successfully attenuates up-component loading effects at 218 (87.55%) GPS stations. In addition, time- and frequency-domain analyses show that LDCmgm90 has better performance than EOST at 155 (62.25%) GPS stations and ESMGFZ at 142 (57.03%) GPS stations. These values increase to 160 (64.26%) for EOST and 152 (61.04%) for ESMGFZ if all the data trends are retained, implying that LDCmgm90 can provide better secular or long-term global mass redistributions. Therefore, the LDCmgm90 loading model is the best choice to remove not only seasonal but also secular or long-term loading signals in GPS site displacement data, providing a chance to explore other geophysical causes buried in the residual secular or long-term signals in the GPS data. The LDCmgm90 loading model is available upon request.

It should be mentioned that there are also some limitations in the LDCmgm90 loading model, such as limited spatial (~2-degree) and frequency (~6-cpy) resolutions and availability limited to the GRACE mission period (~2002–2017, but the LDCmgm90 will be extended to include the GRACE Follow-On data in the future). While the ~6-cpy frequency resolution should be enough for global mass loading, the ~2-degree spatial resolution could be insufficient for some small regions with large local hydrological signals, such as California. Thus, no load model is perfect; each contains certain defects and even errors. Users should beware of applying load corrections to GPS time series data blindly assuming that no new errors will be introduced.

Author Contributions: Conceptualization, J.R., J.L. (Jiancheng Li) and W.C.; methodology, J.R., W.C. and T.v.D.; software, J.L. (Jiesi Luo); validation, J.R. and W.C.; formal analysis, J.L. (Jiesi Luo); investigation, J.L. (Jiesi Luo); resources, J.L. (Jiesi Luo); data curation, J.L. (Jiesi Luo) and W.C.; writing—original draft preparation, J.L. (Jiesi Luo) and W.C.; writing—review and editing, J.R., W.C. and T.v.D.; visualization, J.L. (Jiesi Luo); supervision, J.L. (Jiancheng Li); project administration, W.C.; funding acquisition, W.C. All authors have read and agreed to the published version of the manuscript.

Funding: This research was funded by the National Natural Science Foundation of China, grant numbers 41874025 and 41474022. The APC was funded by the National Natural Science Foundation of China, grant number 41874025.

Institutional Review Board Statement: Not applicable.

Informed Consent Statement: Not applicable.

Data Availability Statement: The EOST, ESMGFZ and IMLS loading data are respectively available at http://massloading.net/, http://loading.u-strasbg.fr/index.php and http://rz-vm115.gfz-potsdam.de:8080/repository (accessed on 20 September 2021). The LDCmgm90 model is available at https://doi.org/10.6084/m9.figshare.7874384.v4 (accessed on 20 September 2021). The LDCmgm90 loading model is available upon request to W.C. and J.L. (Jiesi Luo).

Acknowledgments: We thank the EOST, ESMGFZ and IMLS teams for making their loading models publicly available.

Conflicts of Interest: The authors declare no conflict of interest.

Appendix A. Estimations of Degree-1 Stokes Coefficients and Geocenter Variation of LDCmgm90

We estimate the LDCmgm90 degree-1 Stokes coefficients ($\Delta C_{10}^{LDCmgm90}$, $\Delta C_{11}^{LDCmgm90}$ and $\Delta S_{11}^{LDCmgm90}$) and compare them with degree-1 coefficients of CSR and JPL RL06 GRACE Mascon solutions (Figure A1). To recover the corresponding total global effects of the original Mascon solution, we remove GAD products, reconsider the ICE-6G-D model and restore GAC products (total Mascon = original Mascon product—GAD + GIA + GAC). Our results, especially ΔS_{11}, have good agreement with other degree-1 coefficients.

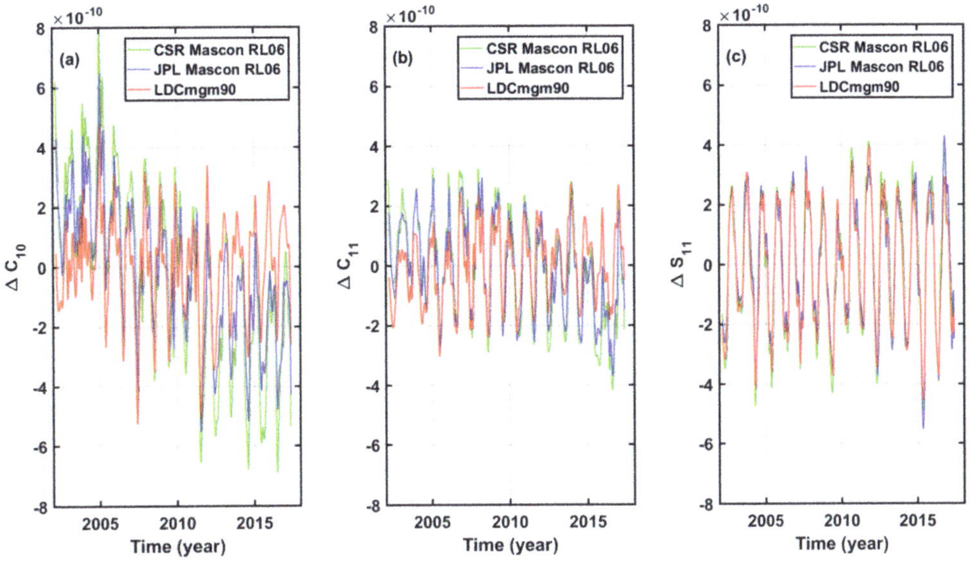

Figure A1. Time series of fully normalized degree-1 Stokes coefficients of CSR Mascon RL06 solutions (green line), JPL Mascon RL06 solution (blue line) and LDCmgm90 (red line) derived in this research according to Swenson et al. [49] spanning 04/2002~06/2017. (**a**) ΔC_{10}; (**b**) ΔC_{11}; (**c**) ΔS_{11}. Please notice that GAC and GIA are included in this figure.

The annual amplitudes and phases of LDCmgm90 degree-1 Stokes coefficients are compared with values from RL06 Mascon solutions from CSR and JPL in Table A1. Y-component annual fluctuation of LDCmgm90 total geocenter variation has good agreement with CSR and JPL Mascon solutions. However, phases of X- and Z-components of LDCmgm90 total geocenter variation are slightly larger than others. Our X-component trend also deviates from CSR and JPL Mascon solutions, while X-component and Z-component are similar to CSR and JPL Mascon solutions separately.

Table A1. Annual amplitudes, annual phase and linear trends of total geocenter variation of 04/2002~06/2017. CSR and JPL Mascon RL06 solutions are given as reference [1].

Model	X	σ_X	Y	σ_Y	Z	σ_Z
Annual Amplitude (unit: mm)						
CSR Mascon	2.06	0.02	3.06	0.02	2.84	0.02
JPL Mascon	1.82	0.02	2.88	0.02	1.91	0.02
LDCmgm90	1.52	0.01	2.93	0.01	1.91	0.02

Table A1. *Cont.*

Model	X	σ_X	Y	σ_Y	Z	σ_Z
Annual Phase (unit: day) [2]						
CSR Mascon	33.88	0.48	179.56	0.30	34.47	0.42
JPL Mascon	39.48	0.49	179.75	0.31	38.70	0.57
LDCmgm90	58.95	0.50	179.70	0.26	74.57	0.58
Linear Trend (unit: mm/year)						
CSR Mascon	−0.07	0.05	−0.46	0.05	1.01	0.06
JPL Mascon	−0.05	0.04	0.03	0.04	0.58	0.05
LDCmgm90	0.13	0.04	−0.37	0.04	0.62	0.06

[1] Values in this table are derived by least-square fitting, with σ as the formal errors of fitting. [2] Annual phases are defined by using sine function. Please notice that to describe the phase shift of the annual signal in its period, the unit of the annual phase is converted from radians to days by multiplying $365.25/2\pi$.

Appendix B. Sources of GPS Measurement Errors and Other Systematic Effects

Ray [1] listed many sources of measurement error and other systematic effects, such as draconitic errors, local site errors, temperature cycle errors and errors in IERS models. In this appendix, sources of annual errors are summarized briefly:

- Draconitic Year Errors

Draconitic year signal (period = 351.2 days) has strong spatial correlations in IGS position time series [3,4].

- Temperature Cycle

Thermal cycle of the Earth will cause expansions of antenna structure and surrounding bedrock [13].

- Errors in IERS Models

The recommended IERS Conventions (2010) models suffer weaknesses in various respects, including deficiencies and errors in some models as well as a lack of any standard model for some effects. Models with known errors include ocean tidal loading [16], diurnal and semidiurnal EOP tides that cause aliases into longer-period signals [53] though an alternative model has been put forward outside the formal conventions (https://ivscc.gsfc.nasa.gov/hfeop_wg/, accessed on 27 October 2021), possible lateral heterogeneity effects in the solid Earth tidal displacement [54] and the atmospheric pressure loading (e.g., S1, S2). Time variation of the low-degree geopotential terms has been recognized, but no standard model is recommended for all techniques. In fact, there have been no major updates to any IERS models in recent years apart from the return to a secular mean pole model in 2018 that replaced the ill-considered quadratic model recommended for several years. As a consequence, the techniques and analysis centers have adopted divergent approaches in some cases.

- Antenna-Related Effects

Local multipath errors and antenna calibration errors may add aliasing signals to observation records. Snow, ice cover and rain may also affect antennas by adding annual signals or changing sky visibility, especially in high latitudes [4]:

- Receiver Hardware Change [4]
- Troposphere Delay Mismodeling (see for example Martens et al. [41])
- Aliasing in Long-Term Time Series Stacking [4]
- Processing Differences among Analysis Centers (ACs)
- Clear differences between products for the same stations released by ACs sometimes exist [4].

References

1. Ray, J. *Precision and Accuracy of GNSS Positions*; Lecture Presented at School of Geodesy and Geomatics; Wuhan University: Wuhan, China, 4 January 2015.
2. Petit, G.; Luzum, B. *IERS Conventions (2010), IERS Technical Notes 36*; Verlag des Bundesamts für Kartographie und Geodäsie: Frankfurt, Germany, 2010; p. 179, ISBN 3-89888-989-6.
3. Ray, J.; Altamimi, Z.; Collilieux, X.; van Dam, T. Anomalous harmonics in the spectra of GPS position estimates. *GPS Solut.* **2008**, *12*, 55–64. [CrossRef]
4. Ray, J.; Rebischung, P.; Griffiths, J. IGS polar motion measurement accuracy. *Geod. Geodyn.* **2017**, *8*, 413–420. [CrossRef]
5. van Dam, T.; Wahr, J.; Milly, P.; Shmakin, A.; Blewitt, G.; Lavallee, D.; Larson, K.M. Crustal displacements due to continental water loading. *Geophys. Res. Lett.* **2001**, *28*, 651–654. [CrossRef]
6. Van Dam, T.; Collilieux, X.; Wuite, J.; Altamimi, Z.; Ray, J. Nontidal ocean loading: Amplitudes and potential effects in GPS height time series. *J. Geodyn.* **2012**, *86*, 1043–1057. [CrossRef]
7. Blewitt, G.; Lavallee, D.; Clarke, P.; Nurutdinov, K. A newglobal mode of Earth deformation: Seasonal cycle detected. *Science* **2001**, *294*, 2342–2345. [CrossRef]
8. Dong, D.; Fang, P.; Bock, Y.; Cheng, M.; Miyazaki, S. Anatomy of apparent seasonal variations from GPS-derived site position time series. *J. Geophys. Res.* **2002**, *107*, 2075. [CrossRef]
9. Kaniuth, K.; Vetter, S. Estimating atmospheric pressure loading regression coefficients from GPS observations. *GPS Solut.* **2006**, *10*, 126–134. [CrossRef]
10. Grapenthin, R.; Sigmundsson, F.; Geirsson, H.; Árnadóttir, T.; Pinel, V. Icelandic rhythmic: Annual modulation of land elevation and plate spreading by snow load. *Geophys. Res. Lett.* **2006**, *33*, L24305. [CrossRef]
11. Tregoning, P.; Watson, C. Atmospheric effects and spurious signals in GPS analyses. *J. Geophys. Res.* **2009**, *114*, B09403. [CrossRef]
12. King, M.A.; Watson, C.S.; Penna, N.T.; Clarke, P.J. Subdailysignals in GPS observations and their effect at semiannual and annual periods. *Geophys. Res. Lett.* **2008**, *35*, L03302. [CrossRef]
13. Yan, H.; Chen, W.; Zhu, Y.; Zhang, W.; Zhong, M. Contributions of thermal expansion of monuments and nearby bedrock to observed GPS height changes. *Geophys. Res. Lett.* **2009**, *36*, L13301. [CrossRef]
14. Jiang, Y.; Dixon, T.; Wdowinski, S. Accelerating uplift in the north Atlantic region as an indicator of ice loss. *Nat. Geosci.* **2010**, *3*, 404–407. [CrossRef]
15. Matsuo, K.; Heki, K. Time-variable ice loss in Asian high mountains from satellite gravimetry. *Earth Planet. Sci. Lett.* **2010**, *290*, 30–36. [CrossRef]
16. Yuan, L.G.; Chao, B.F. Analysis of tidal signals in surface displacement measured by a dense continuous GPS array. *Earth Planet. Sci. Lett.* **2012**, *355–356*, 255–261. [CrossRef]
17. van Dam, T.; Wahr, J. Displacement of the Earth's surface due to atmospheric loading: Effects on gravity and baseline measurements. *J. Geophys. Res.* **1987**, *92*, 1282–1286. [CrossRef]
18. Mangiarotti, S.; Cazenave, A. Annual vertical crustal motions predicted from surface mass redistribution and observed by space geodesy. *J. Geophys. Res.* **2001**, *106*, 4277–4291. [CrossRef]
19. Blewitt, G. Self-consistency in the reference frames, geocenter definition and surface loading of the solid earth. *J. Geophys. Res.* **2003**, *108*, 2103. [CrossRef]
20. Chen, W.; Ray, J.; Shen, W.B.; Huang, C.L. Polar motion excitations for an Earth model with frequency-dependent responses: 2 Numerical tests of the meteorological excitations. *J. Geophys. Res. Solid Earth* **2013**, *118*, 4995–5007. [CrossRef]
21. Chen, W.; Li, J.C.; Ray, J.; Cheng, M.K. Improved geophysical excitations constrained by polar motion observations and GRACE/SLR time-dependent gravity. *Geod. Geodyn.* **2017**, *6*, 11–22. [CrossRef]
22. Taplay, B.D.; Bettadpur, S.; Ries, J.C.; Thompson, P.F.; Watkins, M.M. GRACE measurements of mass variability in the Earth system. *Science* **2004**, *305*, 503–505. [CrossRef]
23. Davis, J.; Elósegui, P.; Mitrovica, J.; Tamisiea, M. Climate-driven deformation of the solid Earth from GRACE and GPS. *Geophys. Res. Lett.* **2004**, *31*, L24605. [CrossRef]
24. Yan, H.; Chen, W.; Yuan, L. Crustal vertical deformation response to different spatial scales of GRACE and GCMs surface loading. *Geophys. J. Int.* **2016**, *204*, 505–516. [CrossRef]
25. Fu, Y.; Freymueller, J.T. Seasonal and long-term vertical deformation in Nepal Himalaya constrained by GPS and GRACE measurements. *J. Geophys. Res.* **2012**, *117*, B03407. [CrossRef]
26. Chanard, K.; Fleitout, L.; Calais, E.; Rebischung, P.; Avouac, J.-P. Toward a global horizontal and vertical elastic load deformation model derived from GRACE and GNSS station position time series. *J. Geophys. Res. Solid Earth* **2018**, *123*, 3225–3237. [CrossRef]
27. Wahr, J.; Molenaar, M.; Bryan, F. Time variability of the earth's gravity field: Hydrological and oceanic effects and their possible detection using grace. *J. Geophys. Res. Solid Earth* **1998**, *103*, 30205–30229. [CrossRef]
28. Chen, J.L.; Wilson, C.R.; Famiglietti, J.S.; Rodell, M. Spatial sensitivity of GRACE time-variable gravity observations. *J. Geophys. Res.* **2005**, *110*, B08408.
29. Chen, W.; Luo, J.; Ray, J.; Yu, N.; Li, J.C. Multiple-data-based monthly geopotential model set LDCmgm90. *Sci. Data* **2019**, *6*, 228. [CrossRef] [PubMed]
30. Wu, X.; Heflin, M.B.; Ivins, E.R.; Fukumori, I. Seasonal and interannual global surface mass variations from multisatellite geodetic data. *J. Geophys. Res.* **2006**, *111*, B09401. [CrossRef]

31. Cheng, M.K.; Ries, J.C.; Tapley, B.D. Variations of the Earth's figure axis from satellite laser ranging and GRACE. *J. Geophys. Res.* **2011**, *116*, B01409. [CrossRef]
32. Cheng, M.K.; Tapley, B.D.; Ries, J.C. Deceleration in the Earth's oblateness. *J. Geophys. Res. Solid Earth* **2013**, *118*, 740–747. [CrossRef]
33. Bar-Sever, Y.E. A new model for GPS yaw attitude. *J. Geod.* **1996**, *70*, 714–723. [CrossRef]
34. Bar-Sever, Y.E.; Kuang, D. *New Empirically-Derived Solar Radiation Pressure Model for GPS Satellites*; Institute of Navigation: Portland, OR, USA, 2005; pp. 42–160. Available online: http://ipnpr.jpl.nasa.gov/progress_report/42-160/title.htm (accessed on 6 December 2020).
35. Bassiri, S.; Hajj, G.A. Higher-order ionospheric effects on the global positioning systems observables and means of modeling them. *Manuscr. Geodtica* **1993**, *18*, 280–289.
36. Bevis, M.; Businger, S.; Chiswell, S.; Herring, T.A.; Anthes, R.A.; Rocken, C.; Ware, R.H. GPS meteorology: Mapping zenith wet delays onto precipitable water. *J. Appl. Meteorol.* **1994**, *33*, 378–386. [CrossRef]
37. Blewitt, G. An automatic editing algorithm for GPS data. *Geophys. Res. Lett.* **1990**, *17*, 199–202. [CrossRef]
38. Kedar, S.; Hajj, G.; Wilson, B.; Heflin, M. The effect of the second order GPS ionospheric correction on receiver positions. *Geophys. Res. Lett.* **2003**, *30*, 1829. [CrossRef]
39. Kreemer, C.; Blewitt, G.; Klein, E. A geodetic plate motion and Global Strain Rate Model. *Geochem. Geophys. Geosystems* **2014**, *15*, 3849–3889. [CrossRef]
40. Sibois, A.; Selle, C.; Desai, S.; Sibthorpe, A.; Weiss, J. GSPM13: An updated empirical model for solar radiation pressure forces acting on GPS satellites. In Proceedings of the IGS Workshop 2014, Pasadena, CA, USA, 23–27 June 2014.
41. Martens, H.R.; Argus, D.F.; Norberg, C.; Blewitt, G.; Herring, T.A.; Moore, A.W.; Hammond, W.C.; Kreemer, C. Atmospheric pressure loading in GPS positions: Dependency on GPS processing methods and effect on assessment of seasonal deformation in the contiguous USA and Alaska. *J. Geod.* **2020**, *94*, 115. [CrossRef]
42. Petrov, L.; Boy, J.-P. Study of the atmospheric pressure loading signal in VLBI observations. *J. Geophys. Res.* **2004**, *109*, B03405.
43. Petrov, L. The International Mass loading Service. *arXiv* **2015**, arXiv:1503.00191.
44. Carrère, L.; Lyard, F. Modelling the barotropic response of the global ocean to atmospheric wind and pressure forcing—Comparisons with observations. *Gephys. Res. Lett.* **2003**, *30*, 1275–1278. [CrossRef]
45. Rodell, M.; Houser, R.; Jambor, U.; Gottschalck, J.; Mitchell, K.; Meng, C.-J.; Arsenault, K.; Cosgrove, B.; Radakovich, J.; Bosilovich, M.; et al. The Global Land Data Assimilation System. *Bull. Am. Meteorol. Soc.* **2004**, *85*, 381–394. [CrossRef]
46. Dill, R.; Dobslaw, H. Numerical simulations of global-scale high-resolution hydrological crustal deformations. *J. Geophys. Res. Solid Earth* **2013**, *118*, 5008–5017. [CrossRef]
47. Dobslaw, H.; Bergmann-Wolf, I.; Dill, R.; Poropat, L.; Thomas, M.; Dahle, C.; Esselborn, S.; König, R.; Flechtner, F. A new high-resolution model of non-tidal atmosphere and ocean mass variability for de-aliasing of satellite gravity observations: AOD1B RL06. *Geophys. J. Int.* **2017**, *211*, 263–269. [CrossRef]
48. Zhou, Y.; Yang, S.; Luo, J.; Ray, J.; Huang, Y.; Li, J. Global Glacial Isostatic Adjustment Constrained by GPS Measurements: Spherical Harmonic Analyses of Uplifts and Geopotential Variations. *Remote Sens.* **2020**, *12*, 1209. [CrossRef]
49. Swenson, S.C.; Chambers, D.P.; Wahr, J. Estimating geocenter variations from a combination of GRACE and ocean model output. *J. Geophys. Res. Solid Earth* **2008**, *113*, B08410. [CrossRef]
50. Han, D.; Wahr, J. The viscoelastic relaxation of a realistically stratified Earth, and a further analysis of post-glacial rebound. *Geophys. J. Int.* **1995**, *120*, 287–311. [CrossRef]
51. Farrell, W.E. Deformation of the Earth by Surface loads. *Rev. Geophys. Space Phys.* **1972**, *10*, 761–797. [CrossRef]
52. Chen, J.Y.; Pan, E.; Bevis, M. Accurate computation of the elastic load Love numbers to high spectral degree for a finely layered, transversely isotropic and self-gravitating Earth. *Geophys. J. Int.* **2018**, *212*, 827–838. [CrossRef]
53. Griffiths, J.; Ray, J.R. Sub-daily alias and draconitic errors in the IGS orbits. *GPS Solut.* **2013**, *17*, 413–422. [CrossRef]
54. Yuan, L.; Chao, B.F.; Ding, X.; Zhong, P. The tidal displacement field at Earth's surface determined using global GPS observations. *J. Geophys. Res. Solid Earth* **2013**, *118*, 2618–2632. [CrossRef]

remote sensing

Article

Displacement Characterization and Spatial-Temporal Evolution of the 2020 Aniangzhai Landslide in Danba County Using Time-Series InSAR and Multi-Temporal Optical Dataset

Jianming Kuang [1,2], Alex Hay-Man Ng [1,2,3,*] and Linlin Ge [2]

1 Department of Surveying Engineering, Guangdong University of Technology, Guangzhou 510006, China; jianming.kuang@student.unsw.edu.au
2 Geoscience Earth Observing System Group (GEOS), School of Civil and Environmental Engineering, University of New South Wales (UNSW), Sydney, NSW 2052, Australia; l.ge@unsw.edu.au
3 Key Laboratory for City Cluster Environmental Safety and Green Development of the Ministry of Education, Institute of Environmental and Ecological Engineering, Guangdong University of Technology, Guangzhou 510006, China
* Correspondence: hayman.ng@gdut.edu.cn

Abstract: On 17 June 2020, a large ancient landslide over the Aniangzhai (ANZ) slope, Danba County, Sichuan Province, China, was reactivated by a series of multiple phenomena, including debris flow triggered by heavy rainfall and flooding. In this study, Synthetic Aperture Radar (SAR) images acquired by the Sentinel-1A/B satellite and optical images captured by the PlanetScope satellites were jointly used to analyze and explore the deformation characteristics and the Spatial-Temporal evolution of the ANZ landslide before and after the multi-hazard chain. Several areas of pre-failure movements were found from the multi-temporal optical images analysis before the reactivation of the ANZ landslide. The large post-failure surface deformation over the ANZ slope was also retrieved by the optical pixel offset tracking (POT) technique. A major northwest movement with the maximum horizontal deformation of up to 14.4 m was found. A time-series InSAR technique was applied to analyze the descending and ascending Sentinel-1A/B datasets spanning from March 2018 to July 2020, showing that the maximum magnitudes of the Line of Sight (LoS) displacement velocities were −70 mm/year and 45 mm/year, respectively. The Spatial-Temporal evolution over the ANZ landslide was analyzed based on the time-series results. No obvious change in acceleration (precursory deformation) was detected before the multi-hazard chain, while clear accelerated deformation can be observed over the slope after the event. This suggested that heavy rainfall was the most significant triggering factor for the generation and reactivation of the ANZ landslide. Other preparatory factors, including the deformation behavior, the undercutting and erosion of the river and the outburst flood, the local terrain conditions, and earthquakes, might also have played an important role in the generation and reactivation of the landslide.

Keywords: Aniangzhai landslide; Danba County; multi-hazard chain; time-series InSAR; optical image analysis

Citation: Kuang, J.; Ng, A.H.-M.; Ge, L. Displacement Characterization and Spatial-Temporal Evolution of the 2020 Aniangzhai Landslide in Danba County Using Time-Series InSAR and Multi-Temporal Optical Dataset. *Remote Sens.* **2022**, *14*, 68. https://doi.org/10.3390/rs14010068

Academic Editor: Francesca Cigna

Received: 19 October 2021
Accepted: 21 December 2021
Published: 24 December 2021

1. Introduction

The movement of a wide range of ground elements, such as rock masses, soil, debris, or garbage, is referred to as a landslide [1]. Landslides are natural phenomena distributed all around the world; they pose a severe threat to lives and infrastructures and have caused enormous loss to our society in terms of safety and economy. A series of geological, hydro-meteorological, or even human-related factors, such as seismic and volcanic activity [2,3], climate change [4], heavy rainfall [5–7], subsurface and surface engineering work [8,9], deforestation [10], agricultural activities, and urbanization [11], have significantly contributed to the occurrence of landslides. The southwest part of China is prone to the occurrence of landslides because of the abundant precipitation and mountainous topography. Sichuan

Province is the one of the regions in the southwest part of China which frequently suffers from medium- and large-scale landslides. On 10 July 2013, days of heavy rainfall and floods triggered a landslide in Sanxi village, Dujiangyan city, Sichuan Province, which caused 44 fatalities, 117 missing people, and the destruction of 12 houses [12,13]. The catastrophic 2017 Maoxian landslide in Xinmo village, Sichuan Province, buried 64 houses, killed 10 people, and left 73 people missing [6,14]. Landslide mapping and monitoring over these landslide-prone regions is essential for understanding their failure mechanisms and their long- and short-term evolutions. The measurement of the surface displacement caused by landslides is the most intuitive indication for understanding the failure mechanisms and evolution patterns over a long and short period. Conventional field investigation or in situ measurements can indeed provide a high accuracy of measurements at centimetre to millimetre levels [15–17]. However, the several limitations of these techniques are that: (1) only measurements on a point-by-point basis can be provided; (2) there is a high demand for the labour force, the techniques are time-consuming, and there are expensive costs for the installation and maintenance of the survey network; and (3) plenty of landslides in remote and rural areas are inaccessible.

The Spaceborne Interferometric Synthetic Aperture Radar (InSAR) technique, as a radar remote-sensing technique, can measure surface deformation along the radar Line of Sight (LoS) direction at high accuracy with all-weather and all-day operational capability, wide spatial coverage, and high spatial resolution (up to 1 m) [18–22]. Multi-temporal InSAR, also called time-series InSAR, is an improved InSAR technique. This technique can significantly minimize the temporal and spatial decorrelation issues in the conventional InSAR technique. A variety of time-series InSAR approaches were developed by different researchers, for example the Permanent Scatterer InSAR (PSInSAR™) [23], the Small Baseline Subset Approach (SBAS) [24,25], and the SqueeSAR™ [26], which provide a chance for researchers to detect the subtle movement of slow-moving landslides and gain a retrospective view of the historic displacements in the temporal and spatial domain. Time-series InSAR has been widely and successfully applied in landslide investigation, including landslide detection [27–29], precursory deformation detection, and pre- and post-failure analysis [6,30,31], as well as landslide inventory, susceptibility, and hazard assessment [32].

For example, Intrieri et al. [30] have successfully applied the SqueeSAR™ to detect the precursors of failure before the 24 June 2017 Maoxian landslide in China; Zhou et al. [33] used the Stanford Method for Persistent Scatterer (StaMPS) small baselines subset method with Sentinel-1 images to detect the obvious spatiotemporal deformation of the Muyubao landslide in the Three Gorges Reservoir area in China; Ciampalini et al. [34] have demonstrated that the PSInSAR and SqueeSAR™ techniques, with the COSMO-SkyMed (CSK) images, can be used to further refine the Landslide susceptibility maps (LSM) of the Messina Province (Sicily, Italy); Dong et al. [35] developed a time-series InSAR method, known as coherent scatterer InSAR (CSI), which can jointly analyse the persistent scatterers (PS) and distributed scatterers (DS), and applied it to an investigation of the deformation characteristics of the Jiaju landslide in Danba County, China, using the ALOS PALSAR and ENVISAT ASAR datasets.

On the other hand, the optical remote-sensing technique can be used for visual interpretation of the Spatial-Temporal evolution of the landslide boundary and the detection of precursory features such as scarps and cracks over the slope surface. Besides, the optical pixel offset tracking (POT) technique can be applied to retrieve the horizontal displacement caused by slope movement with an accuracy of up to about 0.2 pixels of the optical images used [36–38]. The relatively large displacement (up to tens of metres) of landslide is not likely to be seen by the time-series InSAR technique due to high decorrelations. More importantly, the time-series InSAR technique is not sensitive to the movement in the north–south direction due to the near-polar orbits of SAR satellites [39]. Horizontal displacements measured by the optical POT technique are therefore an excellent supplement for the time-series InSAR technique in this regard. Hence, the time-series InSAR technique and the optical remote-sensing technique are complementary in their detection of Spatial-Temporal evolution and their displacement measurements of landslides at different stages.

On 17 June 2020, an intense, short-duration rainfall struck the region of the Aniangzhai and Guanzhou villages in Danba County, Sichuan Province, China, which triggered a serious multi-hazard chain along the Xiaojinchuan River and, most importantly, reactivated the ancient Aniangzhai (ANZ) landslide, posing a serious threat to 12 towns and 112 villages along the downstream of the Xiaojinchuan River, including over 20,000 people in 5000 families. Emergency field investigations were conducted shortly after the occurrence of the event, which provided valuable and reliable information for our study [40,41]. The multi-hazard chain was also interpreted and characterized with the use of seismic signal data [42]. However, these field survey results could not provide a deformation map with dense measurement points and only post-event information could be obtained. By using satellite remote-sensing data for monitoring the landslide, the characteristics of the historic deformation with dense measurement points could be reviewed, and unstable slopes could be identified in the early stage, which provided significant information for the early warning of landslides and the emergency response.

In this study, the time-series InSAR technique was applied to process two stacks of Sentinel-1A/B data (descending and ascending tracks) to map the surface displacement and explore the Spatial-Temporal evolution of the ANZ landslide. Besides that, a time-series optical images analysis, based on multi-temporal, high-resolution optical images acquired from the PlanetScope satellite, was also conducted to investigate the evolution of the landslide features over the ANZ slope. The large post-failure displacement over the slope was retrieved using the optical POT technique. The displacement characteristics, failure mechanism, and triggering factors of the ANZ landslide were further explored and discussed, based on the radar and optical remote-sensing results.

2. Study Area

As shown in Figure 1a, the ANZ landslide is located in the central-west section of the Sichuan Province, which is in the southwest part of China, with the location at 30.979° N, 102.024° E. The region lies at the transition zone between the Qinghai-Tibetan Plateau and the Sichuan Basin where the elevation difference is up to 7000 m (Figure 1b). At the same time, this region is located at the interaction zone of two large, active fault belts: the Xianshuihe-Xiaojiang fault in the west and the Longmenshan thrust fault belt in the east; this is a seismically active region with plenty of active faults around and numerous geohazards occurring. According to the United States Geological Survey (USGS), there have been 13 large historical earthquakes with a magnitude of over 6.0 on the Richter Scale recorded in this region, including the 1933 M_W 7.5 Diexi Earthquake on the Songpinggou fault [43], the 1973 M_S 7.6 Luhuo earthquake on the Xianshuihe fault [44], and the 2008 M_W 7.9 Wenchuan Earthquake on the Longmenshan fault [45], as marked by the yellow and blue stars in Figure 1b. The most recent event was recorded on 20 April 2013, with a distance of 110 km in the southeast and a magnitude of 6.6 (M_W).

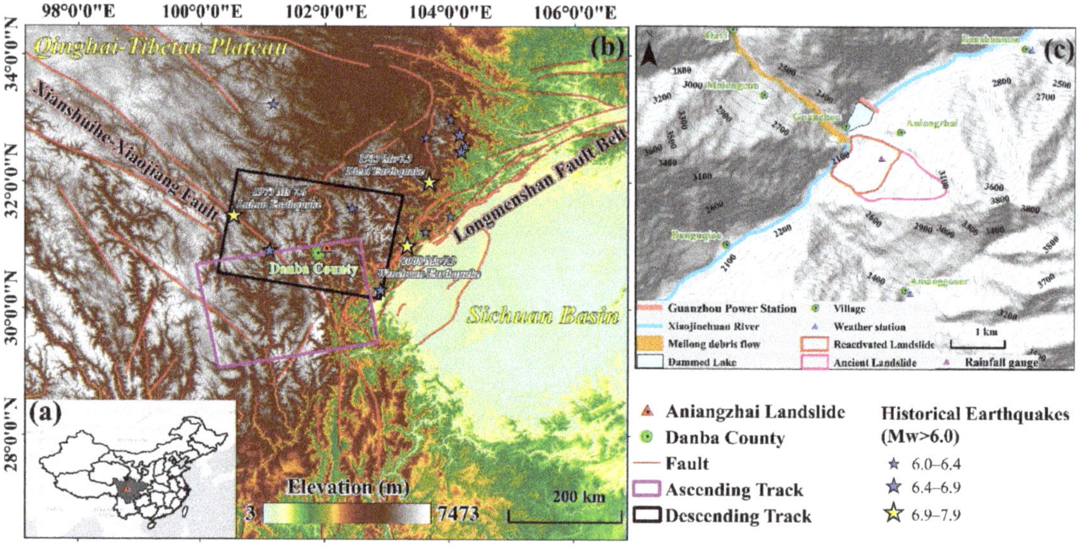

Figure 1. Geological setting of the study area: (**a**) overall location of the study area; (**b**) regional tectonic setting and topography map derived from the 1-arc-second Digital Elevation Model (DEM) of the Shuttle Radar Topography Mission (SRTM) [46]; (**c**) overview of the multi-hazard chain in the study area. Black lines indicate the contour lines.

Figure 1c depicts the detailed overview of the multi-hazard chain on 17 June 2020. The ANZ landslide is located on the left bank of the Xiaojinchuan River. The ANZ landslide is an ancient landslide that once blocked the Xiaojinchuan River, and it has experienced slow-creep deformation in recent years, according to local residents [40]. According to local records, at about 3:00 am on 17 June 2020, the intense rainfall at midnight had triggered the Meilong debris flow on the right bank of the Xiaojinchuan River. The hillslope failure with a high elevation brought the deposits of the Meilong debris flow rapidly rushing down to the Xiaojinchuan River. According to the field investigation [40], the riverbed was uplifted by approximately 8–12 m, which blocked the Xiaojinchuan River in the upstream and formed a large dammed lake just behind the Guanzhou power station. A new river channel was then formed over the deposits, and it flowed down to the toe of the ANZ landslide, which seriously washed and eroded the toe section. At about 11:00 p.m. local time on the same day, the dammed lake began to breach and caused a flood [41]. Under the continuous downcutting and erosion of the new river channel and outburst flood, the whole toe section of the ANZ slope collapsed on 18 June 2020. Then, the lower and middle sections of the ancient ANZ landslide were reactivated, and the affected area was named the reactivated ANZ landslide. That formed a steep face with a slope of about 60–70° and a height of 60 m at the toe of the ANZ landslide, according to the field investigation. In contrast, the slope in the middle and upper section was relatively gentle, ranging from 35° to 45°. The elevation difference of the ancient ANZ landslide is up to 1000 m, with about 3100 m at the top. According to the rainfall data recorded from the Danba Meteorological Bureau, the annual rainfall in the area of the ANZ landslide is about 532.7–823.3 mm, 80% of which is mainly concentrated between May and September [40]. The average annual temperature in this area is 14.2 °C. According to the filed investigation and published work, the lithology of the ANZ is mainly composed of the Lower Carboniferous Yaoji Formation (D1q) and the Quaternary sedimentary layers [41]. The bedrock is primarily composed of limestone, with a generally dipping attitude of $N50-60^{\circ}W/NE\angle 70-80^{\circ}$ and a visible thickness of 5–20 m. The avalanche deposits are widely spread over the surface of the slope, with the thickness ranging from 10 to 50 m. Fifty to eighty percent of this unit is rock fragments, which are mainly angular and sub-angular blocks from the detachment of the rock avalanche. The ANZ landslide is located on a northwest-facing slope, with an inclination of 50° at the top and 20° at the lower part.

3. Materials and Methods

3.1. Datasets

3.1.1. Optical Images

The PlanetScope satellite can acquire optical images with high resolution (3 m in pixel size) in near daily global coverage, which can be used as a useful tool for landslide detection and monitoring [47]. In this study, Planetscope images with four bands (blue, green, red, and near infrared) were collected from the Planet Labs Company under a research and education license. The level 3B surface reflectance product of PlanetScope was chosen as it has been orthorectified and pre-processed with geometric, radiometric, and atmospheric corrections [48]. A total of nine scenes of Planetscope images from 26 March 2020 to 27 July 2020 were selected with cloud-free (i.e., 0% cloud percentage) and minimal shadow condition over the ANZ slope. The temporal coverage spans both the pre-failure and post-failure stages of the multi-hazard chain. The detailed information of the PlanetScope images is shown in Table 1.

Table 1. Optical images acquired by the PlanetScope satellite for this study.

Stage	Imaging Date	Sun Elevation Angle (Degree)	Sun Azimuth Angle (Degree)	Instrument
Pre-failure	26 March 2020	55.1°	140.7°	PS2.SD
	20 April 2020	63.3°	131.9°	PS2.SD
	10 May 2020	64.7°	115.9°	PS2
	18 May 2020	63.1°	108.3°	PS2
	30 May 2020	66.7°	107.1°	PS2
	15 June 2020	66.8°	103.1°	PS2
	16 June 2020	66.8°	103.1°	PS2
Post-failure	24 June 2020	69.0°	105.5°	PS2
	27 July 2020	58.4°	104.2°	PSB.SD

3.1.2. Satellite SAR Datasets

In this study, both the descending and the ascending tracks of the C-band Sentinel-1A/B SAR images in the Terrain Observation by Progressive Scans (TOPS) mode were collected for time-series InSAR analysis. The TOPS SAR data can cover a swath width of 250 km at about 5 m by 20 m resolution in the range and azimuth directions, respectively [49]. The basic parameters of the SAR images are shown in Table 2. Fifty-eight images were included in the descending stack spanning from 27 March 2018 to 2 July 2020 and seventy-one images in the ascending stack for the period from 8 March 2018 to 7 July 2020. The spatial coverages of both tracks are shown in Figure 1b, with black and purple rectangles for the descending and ascending tracks, respectively.

Table 2. Basic parameters of SAR data.

Satellite	Orbit	Imaging Mode	Imaging Period	Heading Angle (°)	Incidence Angle (°)	Number of Images
Sentinel-1A/B	Descending	TOPS	27 March 2018–2 July 2020	169.6°	37.1°	58
	Ascending	TOPS	08 March 2018–7 July 2020	−9.8°	44.2°	71

3.1.3. Rainfall Data

The monthly rainfall spanning from 2018 to 2020 was calculated from the daily rainfall based on the CHIRPS (Climate Hazards Group InfraRed Precipitation with Station) data from the Google Earth Engine [50,51]. In addition, the daily rainfall data between 1 May 2020 and 17 June 2020, according to the Anianggouer station, were collected from the published paper [40], as shown in Figure 2a. The daily rainfall data recorded after the multi-hazard chain, from 22 June 2020 to 19 August 2020, were also collected from the locally installed rainfall gauge (Figure 2b) [41]. The locations of the Anianggouer station and locally installed rainfall gauge are shown in Figure 1c.

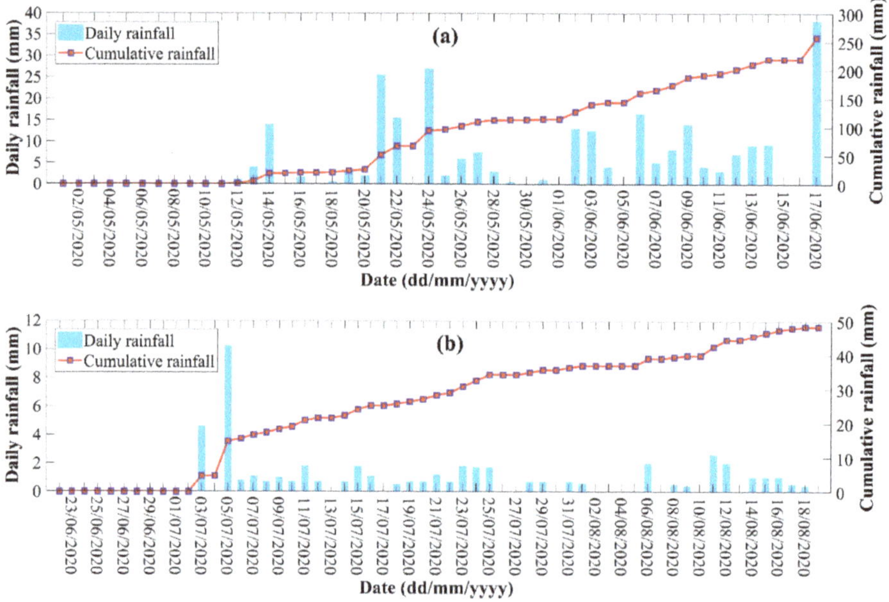

Figure 2. (**a**) Daily rainfall data between 1 May 2020 and 17 June 2020, before the multi-hazard chain, according to the Anianggouer weather station from the Danba Meteorological Bureau. The daily rainfall data were collected from Zhao et al., 2021 [40]. (**b**) Daily rainfall data between 22 June 2020 and 19 August 2020, after the multi-hazard chain, according to the installed rainfall gauge over the ANZ slope. These data were collected from Zhu et al., 2021 [41].

3.2. Optical Analysis Method

Three main procedures were conducted based on the collected optical images. Firstly, true color PlanetScope images were generated by compositing three bands (red, green, and blue) for the initial visual interpretation of the evolution of the surface features over the ANZ slope. Secondly, by using the vegetation index, the pre- and post-failure landslide features can be effectively mapped and detected, and the effect of the mountain shadows can be minimized. The Atmospherically Resistant Vegetation Index (*ARVI*) time series was derived and was then classified using unsupervised classification to investigate the Spatial-Temporal evolution of the landslide in the pre- and post-failure stages [52]. Finally, the optical pixel offset tracking (POT) technique was applied to obtain the post-failure displacement of the ANZ slope.

The interpretation of the landslide features based on true color images could often be influenced by the shadows in the images as the landslide generally occurred in the mountain region. Using the Normalized Difference Vegetation Index (NDVI) for the landslide interpretation could minimize the influence of the mountain shadows [53]. The *ARVI*, an improved index compared to the NDVI, uses blue light reflectance measurements to correct for the atmospheric scattering effects through a self-correction process. The *ARVI* calculation can be expressed as [54]:

$$\begin{aligned} \text{ARVI} &= (\rho_{nir} - \rho_{rb})/(\rho_{nir} + \rho_{rb}) \\ \rho_{rb} &= \rho_{red} - \gamma(\rho_{blue} - \rho_{red}) \end{aligned} \tag{1}$$

where ρ_{nir}, ρ_{red}, and ρ_{nir} are the near-infrared, red, and blue bands, respectively. The assumption that $\gamma = 1$ is often used unless the aerosol model is known a priori.

In this study, the *ARVI* was calculated based on Equation (1) using the collected PlanetScope images. These images were divided into two groups: the pre- and post-failure groups. The *ARVI* images in the pre-failure and post-failure groups were stacked to form the time-series *ARVI* with 6 and 2 layers, respectively, in a chronological order using the Layer Stack module in the ENVI software. Then, the K-means unsupervised classification method was applied to classify the pre-failure time-series *ARVI* and the post-failure time-series *ARVI* image. The number of classes was set to 5 with other the default parameters in the classification process.

COSI-Corr, a useful POT software for measuring sub-pixel ground surface deformation based on the image correlation method [55], was used to map the post-failure displacement over the ANZ slope. The PlanetScope image pair between 16 June 2020 (pre-failure) and 24 June 2020 (post-failure) was chosen as the input data, and the red band was used for the POT calculation. The combinations of the initial and final window sizes used in this study are 128 and 64, respectively, which can reduce the background noise and uncertainties [56]. Three layers of the POT results can be derived, including displacements in the east–west and north–south directions, and the signal-to-noise ratio (SNR). Positive values in the east–west and north–south displacement layers indicate eastward and northward movements, respectively. Displacements to the west and south are represented by negative values. The range of SNR values is from 0 to 1, and the higher value suggests the measurement with the higher quality. Measurements with an SNR value of less than 0.95 were masked in the east–west and north–south displacement datasets. To reduce the mismatch error for the image pair, a stable area was selected near the ANZ slope. The displacements derived from the stable area can be used to depress the background noise through removing the co-registration errors within the image pair used. In this case, the mean east–west and north–south displacement derived from stable area was used to correct the image mismatch and generate the final corrected east–west and north–south displacement. The standard deviations of the east–west and north–south displacements in the stable area were used justify the uncertainties of the results [56]. The overall horizontal displacement was then calculated based on the corrected east–west and north–south displacements.

3.3. Time-Series InSAR Analysis Method

The Persistent Scatterer (PS) and Small Baseline (SB) approaches were both exploited and jointly processed for the time-series analysis. The combination of PS and Coherent Scatterer (CS) points was applied to increase the density of the measurement points (MPs) in the rural and mountainous region. Three major processing steps are included: PS processing, SB processing, and combination processing of the PS and CS targets. The Sentinel-1 datasets were pre-processed using the TOPSAR stack processor module in the Interferometric synthetic aperture radar Scientific Computing Environment (ISCE) software to generate single-master and multiple-master differential interferograms stacks [57]. Time-series analysis was then conducted using the Stanford Method for Persistent Scatterers (StaMPS) software based on the differential interferograms generated [58].

As for the PS processing, the PS candidates were initially selected based on the Amplitude Dispersion Index (ADI) [23]. The pixels with ADI values over 0.4 were initially selected as the PS candidates. Then, the final PS targets are further selected based on the phase stability criterion [59].

For the case of the SB processing, the Spatial-Temporal distribution of the interferometric combination of the Small Baseline network is shown in Figure 3. The perpendicular baseline for over 90% of pairs was less than 100 m due to the precise orbit control of the Sentinel-1A/B satellites. To ensure low temporal decorrelation, the maximum temporal baseline was set to 120 days. The CS points were selected based on the amplitude difference dispersion ($D_{\Delta A}$), which is similar to D_A [58].

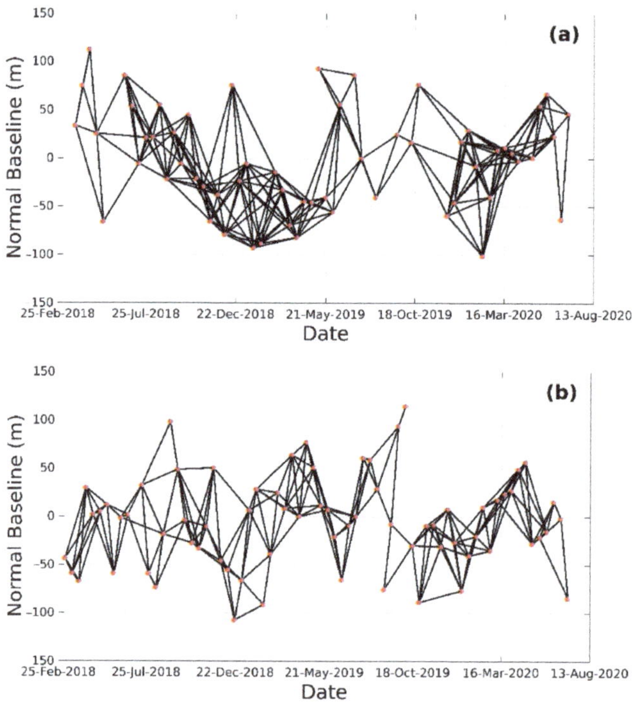

Figure 3. The interferometric combination of Small Baseline network for Sentinel-1A/B: (**a**) descending track and (**b**) ascending track. The red circles and black lines represent SAR image acquisitions and interferometric pairs, respectively.

For both the PS and the SB processing, the topographic effects were removed using the 1-arc-second SRTM DEM [46]. The overall wrapped differential phase φ_{Diff} of a pixel cell after flattening and removing the topographic phase can be expressed as follows:

$$\varphi_{Diff} = W\left\{\varphi_{Defo} + \varphi_{Atmos} + \varphi_{Orb} + \varphi_{DEM} + \varphi_N\right\} \quad (2)$$

where $W\{\}$ denotes the wrapping operator. φ_{Defo} represents the Line of Sight (LoS) deformation phase term. φ_{Atmos} indicates the atmospheric delay phase and φ_{Orb} is the phase caused by orbit error. φ_{DEM} represents the residual topographic phase caused by the DEM error and φ_N is the random noise phase, including thermal noise and co-registration error. The deformation phase can be estimated and separated from the other phase terms by iteratively filtering in the temporal–spatial domain. The detailed processing chain can be referred to in [58,60,61].

The displacement velocities measured from the times-series InSAR are along the radar Line-of-Sight (*LoS*) direction. Therefore, the displacement measured from the time-series InSAR is a composite of vertical, easting, and northing displacement components, which can be written as [62–66]:

$$\begin{bmatrix} cos(\theta) & -sin(\theta)cos(\alpha) & sin(\theta)sin(\alpha) \end{bmatrix} \begin{bmatrix} D_V \\ D_E \\ D_N \end{bmatrix} = D_{LoS} \quad (3)$$

where θ is the incidence angle and α is the heading angle of the satellite (positive clockwise from the north). D_V, D_E, and D_N are the vertical, eastward, and northward displacement, respectively. D_{LoS} is the displacement measured in the radar LoS direction. Ideally, measurements from at least three viewing geometries are needed to resolve the 3D deformation.

4. Results

4.1. Landslide Evolution and Deformation Mapped by Optical Datasets

4.1.1. Landslide Evolution Revealed by the Time-Series Optical Analysis

Visual interpretation based on multi-temporal optical images can provide useful information from identifying surface moving features, which could directly reflect the formation and evolution of landslides. As shown in Figure 4, true color images were synthesized with three bands: red (B3 at 590–670 nm), green (B2 at 500–590 nm), and blue (B1 at 455–515 nm). Before 17 June 2020, there were three flows on the pre-failure surfaces observed, with two small masses over the main slope and a long flow at the upper neck. After the multi-hazard chain occurred on 17 June 2020, the collapse area could be clearly observed at the toe of the landslide, and the Xiaojinchuan River extended much wider because of the flooding caused by the debris flow at the upstream.

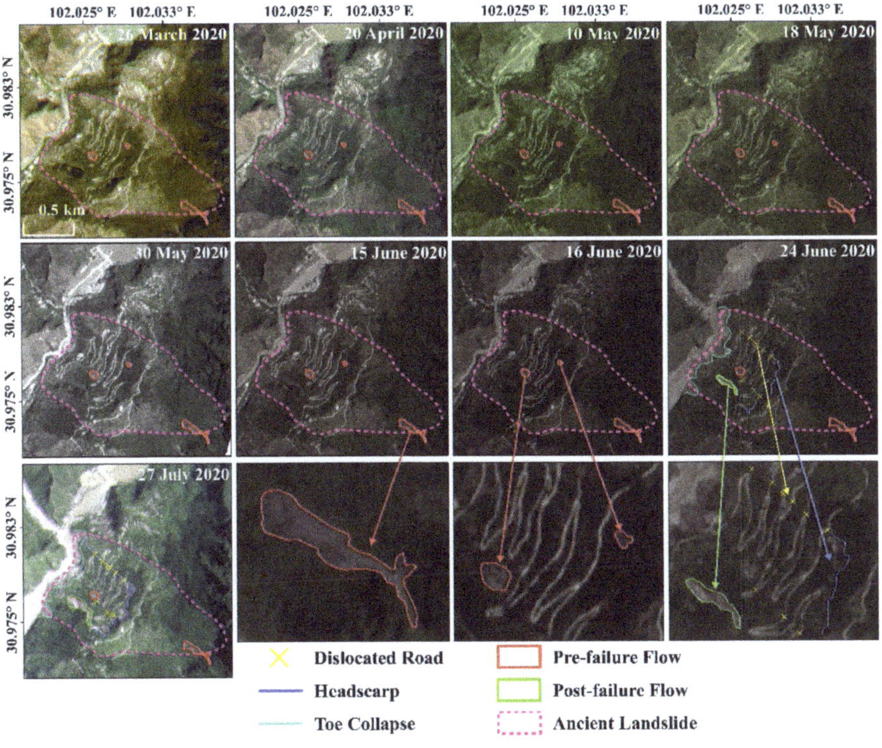

Figure 4. Time-series true color images derived from multi-temporal optical images from the PlanetScope satellite over the ANZ Landslide, with acquisition dates annotated in the upper-right corners.

A long stripping flow was detected close to the left boundary of the ancient landslide area. At the same time, multiple dislocations could be clearly observed over the roads at the middle of the slope, which was also consistent with the published field investigation [40]. Two obvious head scarps at the middle were detected on the post-failure image acquired on 24 June 2020, and it possibly penetrated down to the toe through the multiple road dislocations at the upstream section. Those two head scarps were connected from the image acquired on 27 July 2020. No obvious changes were seen at the pre-failure flowing mass on the upper neck.

Time-series *ARVI* images were generated based on high-resolution PlanetScope images acquired from 26 March 2020 to 27 July 2020, as shown in Figure 5. The pre- and post-failure flows over the slope can be obviously detected over the slope due to the bright background and reduced influence of the mountain shadows. It is noted that the head scarps were clearly observed to run through and down to the toe of the slope in the post-failure *ARVI* image. For further investigation of the evolution of the landslide features over the slope, the K-mean unsupervised classification based on the time-series *ARVI* images was conducted. Two groups were divided, corresponding to two different stages: the pre-failure stage (10 May 2020–16 June 2020) and the post-failure (24 June 2020–27 July 2020) stage, as shown in Figure 6. The *ARVI* image on 30 May 2020 was discarded in the K-mean unsupervised classification due to limited ground coverage. Five major classes were identified over the slope. Class 1, with a red colour, consists of ground features with the lowest *ARVI* value, including river and bare soil land. Class 2, with a pink colour, represents features with a medium-low *ARVI* value, including road, landslide scarps, and buildings. The landslide area includes flows, bare soil land, and landslide scarps. Therefore, the flows could be jointly covered by class 1 and 2 in this case. Class 4 and 5 represent ground features with a high *ARVI* value, including evergreen, tall, and dense vegetation over the slope. Class 3 is the transition zone among classes 1–2 and classes 4–5, possibly representing unstable vegetation cover over this zone.

Landslide scarps and flows can be distinguished from rivers, roads, and buildings as they are irregular features and located on the same slope. As shown in Figure 6a, the Xiaojinchuan River and roads, as linear features, are detected in class 1 and class 2, respectively. At the same time, three pre-failure flows were clearly detected during the pre-failure stage, with two at the centre slope and one at the upper slope. As for the post-failure stage on 24 June 2020 and 27 July 2020 in Figure 6b, the Xiaojinchuan River was obviously much wider than before due to the occurrence of flooding caused by the debris flow on the upstream. The post-failure flow next to the left boundary at the downstream section was observed, with an area up to 22,000 m^2. The pre-failure flow at the centre upstream was extended further when compared with that from the pre-failure stage, while no obvious change was observed for the pre-failure flow at the upper slope.

4.1.2. Post-Failure Displacement Detected from Optical Pixel Offset Tracking

Figure 7 shows the optical POT results between one pre-failure image on 16 June 2020 and one post-failure image on 24 June 2020. Positive values in the east–west and north–south measurements indicated that the slope moved towards the west and north directions, respectively. Westward and southward movements are represented by negative values. The standard deviation of the east–west and north–south displacements in the stable are 0.43 m and 0.49 m, respectively. These minor standard deviations indicated that the selected stable area is reliable and can be used to derive the corrected east–west and north–south displacements [56]. As shown in Figure 7a,b, post-failure movement over the reactivated landslide area can be clearly observed from both the west–east and north–south directions. The horizontal displacement is generated by combining these two measurements in different directions (Figure 7c). The ANZ slope mainly moved towards the northwest direction, with the maximum west and north displacement up to -13.7 m and 10.0 m, respectively. The horizontal displacement peaks at 14.4 m, and most of the large displacement ranging from 12.0 to 15.0 m, were found at the middle and upstream section of the reactivated landslide area. However, less than 4 m of horizontal displacements were observed in the upper section of the ancient landslide area during this period. The overall reactivated boundary of the ANZ landslide can be clearly observed from the ancient landslide area, showing an hourglass-shaped boundary from the top to the bottom.

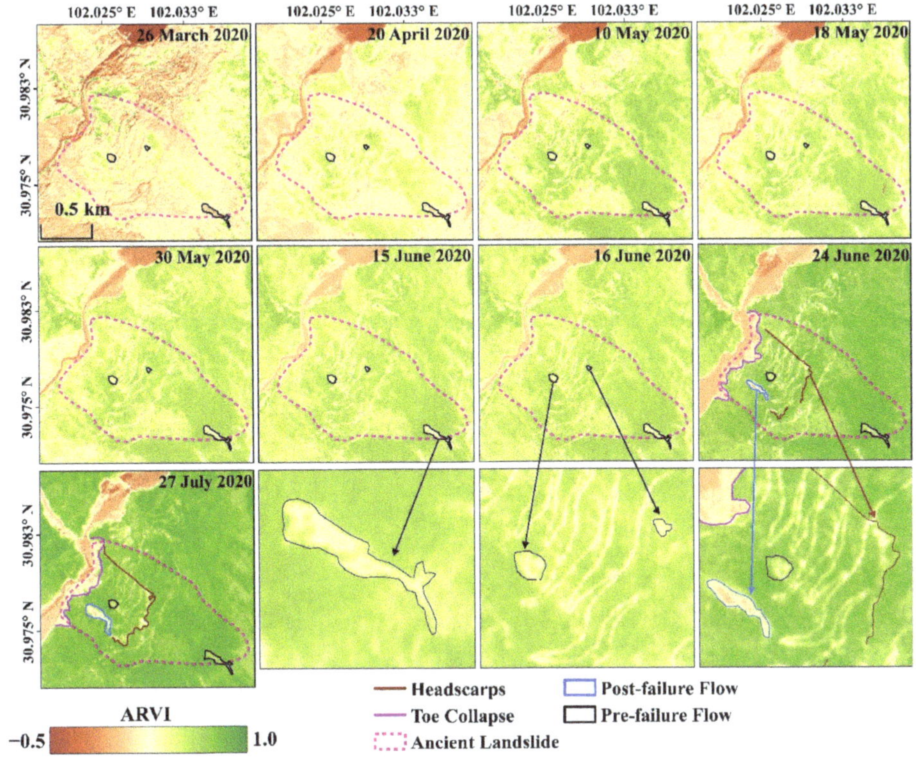

Figure 5. Time-series *ARVI* images derived from high-resolution optical images from the PlanetScope satellite over the ANZ landslide, with acquisition dates annotated in the upper-right corners.

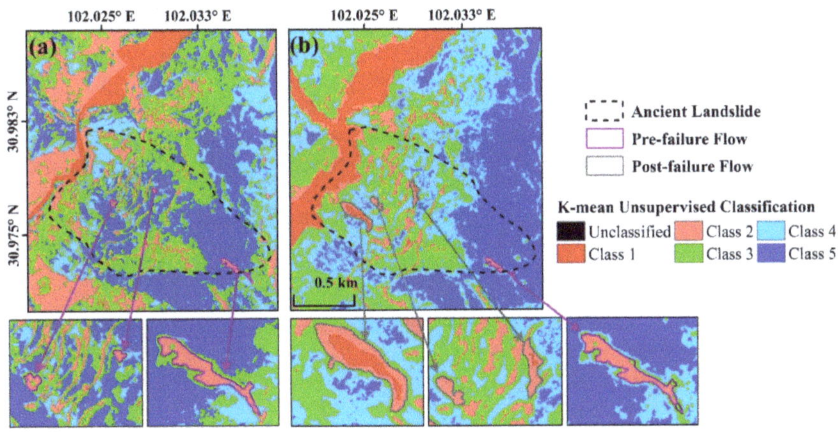

Figure 6. K-mean unsupervised classification based on the time-series *ARVI* images: (**a**) pre-failure stage (10 May 2020–16 June 2020); (**b**) post-failure stage (24 June 2020–27 July 2020).

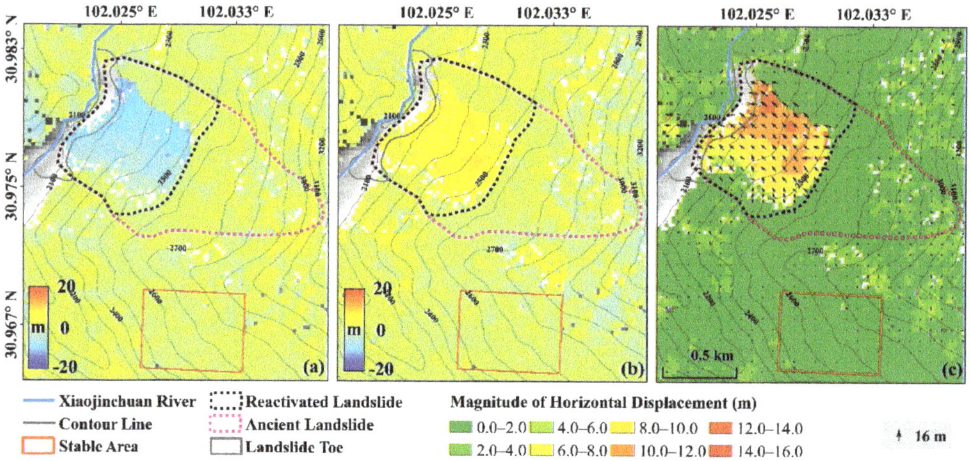

Figure 7. Post-failure deformation measured from image pair between 16 June 2020 and 24 June 24, acquired from the PlanetScope satellite: (**a**) west–east displacement; (**b**) north–south displacement; and (**c**) magnitude of horizontal displacement covered by the proportional black arrows indicating moving direction.

4.2. Deformation Velocity Measured by Time-Series InSAR Analysis

The LoS displacement annual-velocity map can be obtained from the time-series InSAR processing of the Sentinel-1A/B datasets from the descending and ascending orbits, as shown in Figure 8. The measurement points with a red colour (negative values) represent targets moving away from the satellite along the LoS direction, while those with a blue colour represent motion toward the satellite. Due to dense vegetation cover and steep topography over the region, only limited measurement points were obtained. According to the deformation characteristics and local topography, the ANZ landslide can be divided into four zones: the central deformation zone (I), the landslide toe collapse zone (II), the upstream deformation zone (III), and the landslide shoulder deformation zone (IV). Deformation caused by the slope movement can be clearly observed from the descending measurement at the central deformation zone (I), the landslide toe collapse zone (II), and the upstream deformation zone (III), mostly ranging from −20 to −50 mm/year, as shown in Figure 8a. The maximum magnitude of deformation velocity was up to −70 mm/year, which was found at landslide toe collapse zone (II) near the Xiaojinchuan River. On the other hand, the land surface was evolving less over the landslide shoulder deformation zone (IV), and the deformation velocity ranging from 15 to −30 mm/year was observed. As for the ascending measurement shown in Figure 8b, the overall detected deformation velocity is smaller than the descending measurement because of the lower sensitivity of the ascending SAR data for the northwest-facing slope [67]. The density of measurement points is also lower, relatively, than that from the descending orbit. The deformation rate of most measurement points was between 5 to 15 mm/year along the LoS direction. Similarly, the largest deformation was found at zone (II), with a rate of 45 mm/year along the LoS direction. In the upstream section of zone (II), where there was no measurement point detected from the descending SAR data, it was detected to be unstable, with most of the points ranging from 20–45 mm/year from the ascending data, which could probably be uplifted accumulation at the left bank of the Xiaojinchuan River. In addition, much smaller deformations were observed at zone IV, with the deformation rate ranging from −10 to 7 mm/year.

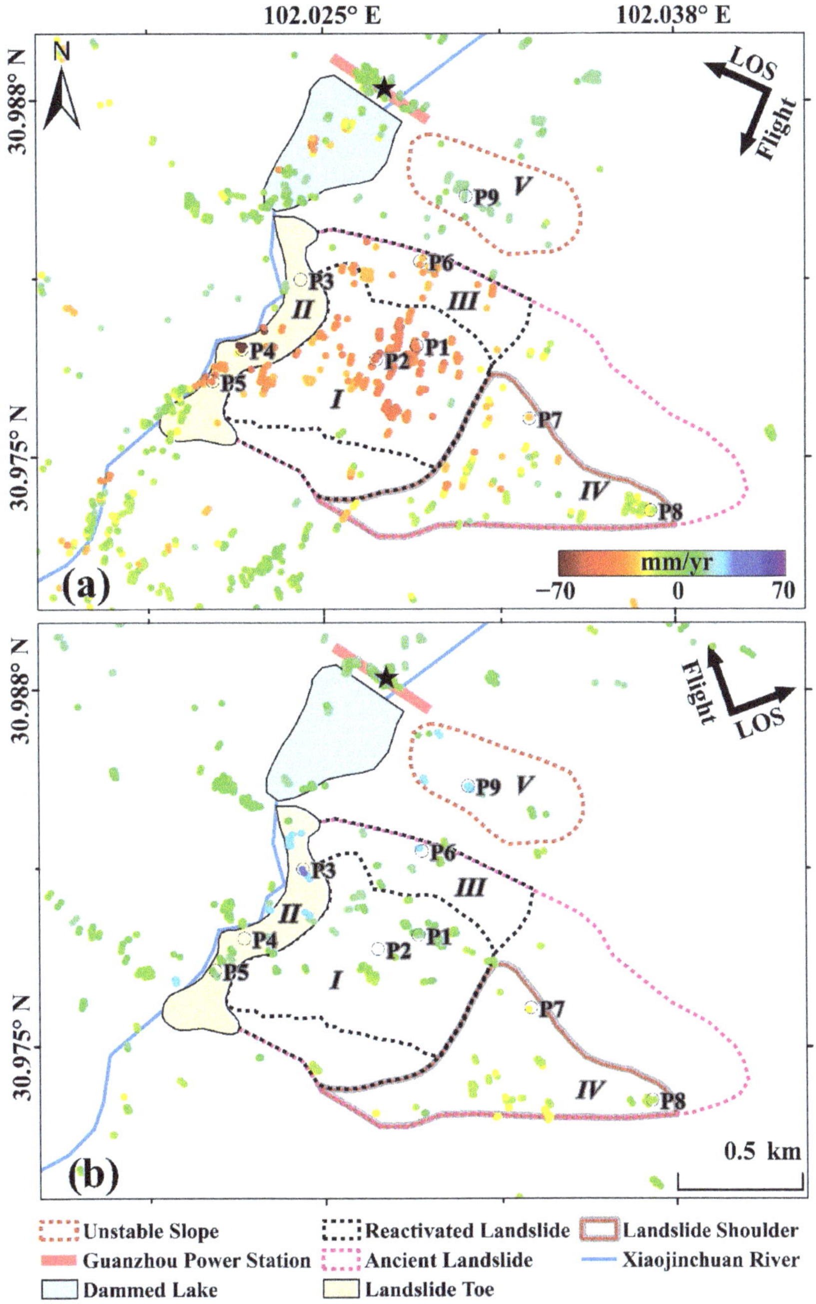

Figure 8. LoS deformation velocity detected from the Sentinel-1: (**a**) descending dataset and (**b**) ascending datasets. Black star represents the reference point, and the black circles represent the selected measurement points.

Beside the ANZ landslide, an unstable slope was detected 200 m away from the right side of the ANZ landslide, with its maximum deformation velocity up to 19 and

30 mm/year from the descending and ascending measurements, respectively. As both measurements showed positive movement over the detected unstable slope, such an area is likely to be an uplifted area. The location of this unstable slope is just above the dammed lake and next to the Guanzhou power station, which could form a secondary landslide and causes a serious threat to the station.

In many InSAR studies, the LoS displacement measured from time-series InSAR can be projected into the vertical direction by assuming that there was no displacement in horizontal direction [63,68], whereas in this case the horizontal displacement was obviously detected over the ANZ slope from the optical POT result (Figure 7) and could not be neglected. In order to conduct a cross-validation between the descending and ascending measurements, a stable area over the Guanzhou power station was selected. This power station is expected to be relatively stable (i.e., no deformation) during the InSAR acquisition period. Both the ascending and the descending measurements were first projected into the vertical direction by dividing the cosine of the local incidence angle, and the differences between the projected vertical displacements from both tracks were calculated. The standard deviation of the calculated differences is 5.5 mm/year. As shown in Figure 9, the differences were mainly distributed in the range of −5 mm/year to 5 mm/year, accounting for over 60% of the measurement points, indicating that the two measurements are highly correlated with each other and that the time-series InSAR results are reliable.

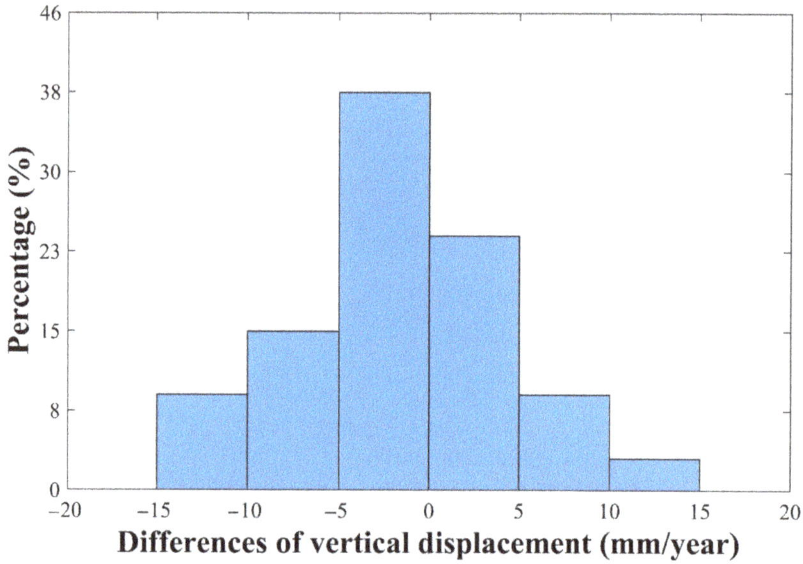

Figure 9. Cross validation based on the differences between the projected vertical displacements from the descending and ascending InSAR measurements.

5. Discussion

5.1. Temporal–Spatial Evolution of the ANZ Landslide

To further investigate the temporal evolution of the pre-failure slope stability and whether there were any precursory signals detected before the multi-hazard chain, a time-series analysis of selected measurement points was conducted (Figure 10). Several measurement points at four deformation zones of the ANZ landslide were selected, as marked by the black circles in Figure 8. In order to investigate the effect of local rainfall on the ANZ landslide, the monthly rainfall was calculated from the daily rainfall based on the CHIRPS (Climate Hazards Group InfraRed Precipitation with Station) data from the Google Earth Engine [50,51]. The mean of the measurement points in a specific spatial distance (with a radius of 30 m) was calculated to eliminate possible gross errors. Figure 10a shows the time-series deformation of selected points (P1 and P2) at zone (I). Due to different observation geometries of the descending and ascending data for the direction of the slope movement, the LoS time-series deformations at point P1 show opposite trends. That is consistent with the northwest movement detected from the optical POT result. No obvious acceleration in surface movement was detected from the descending and ascending measurements before the multi-hazard chain on 17 June 2020. Before the occurrence of the multi-hazard chain, only a 6 mm change in cumulative deformation was observed at point P1 between 1 June 2020 and 13 June 2020 from the ascending track. A sudden acceleration in surface movement can be observed at point P1 from the ascending track between 25 June 2020 and 7 July 2020, moving at approximately 14.5 mm along the LoS direction within the final acquisition (12 days). According to the local rainfall data collected from 22 June 2020 to 19 August 2020, a maximum daily rainfall of 10.2 mm was recorded on 5 July 2020 (Figure 2b). This sudden change in surface movement was responding to the intense daily rainfall recorded on 5 July 2020. An accelerated displacement was also observed from a field survey on 5 July [41], showing that our InSAR result is in good agreement with the field data. Conversely, a linear growth trend can be observed from the descending measurement at point P1 until 2 July 2020. A gentle acceleration can be observed at point P2 from the descending measurement between 8 June 2020 and 2 July 2020, and the cumulative deformation over the observation period is up to 110 mm.

At zone (II), the cumulative deformations at point P3 and P4 are 83 mm and −130 mm for the observation period, respectively, which are significantly larger than those at point P5 from both datasets, as shown in Figure 10b, and are probably caused by serious washout and erosion of the river and the outburst flood caused by the Meilong debris flow at the upstream and middle section of zone (II), where point P3 and P4 were located (Figure 8). The rainfall season started in June with the intense rainfall of 38.1 mm recorded before the multi-hazard chain on 17 June 2020. During this period, a clear acceleration in deformation can also be observed at point P4 from the descending track, falling about 17 mm between 8 June 2020 and 2 July 2020. A sharp drop in cumulative deformation was obtained from the ascending measurement at point P3 between 13 June 2020 and 25 June 2020 when the multi-hazard chain occurred, which could have been caused by the collapse of the landslide toe at the upstream section after the multi-hazard chain.

Figure 10c shows the time-series deformation at zone (III), the upstream section of the ANZ landslide. The cumulative deformations at point P6 are 46 mm and 71 mm for the descending and ascending measurements, respectively, which are both smaller than those at zones (I-II). Similarly, there is a sudden acceleration in displacement observed at point P6 from the ascending datasets from 1 June 2020, rising about 30 mm in about one month. This is consistent with the rainfall season in June and the intense daily rainfall recorded on 5 July 2020 (Figure 2). However, no obvious acceleration was detected from the descending datasets.

The landslide shoulder zone (IV) is located at the upper section of the ancient landslide area and close to the downstream section. Before June 2019, the cumulative displacements for all of the two selected points in both tracks were fluctuating within a range of −20 mm to 20 mm, which indicated that this section was in a relatively stable stage.

At point P7 (Figure 10d), both the descending and the ascending measurements captured a clear acceleration at the last three acquisitions, mainly after the multi-hazard chain and with the increasing monthly rainfall in June and July, moving 22 mm and 30 mm within one month, respectively. Point P8 is located at the pre-failure flow in the upper section detected from the optical image analysis (Figures 4 and 5). The cumulative deformation is smaller at point P8 in both datasets, mostly ranging from −30 to 30 mm. It is noted that a similar acceleration of deformation was also captured by the ascending measurements at point P8, rising about 25 mm in the last two acquisitions, which is consistent with the ascending measurement at point P7.

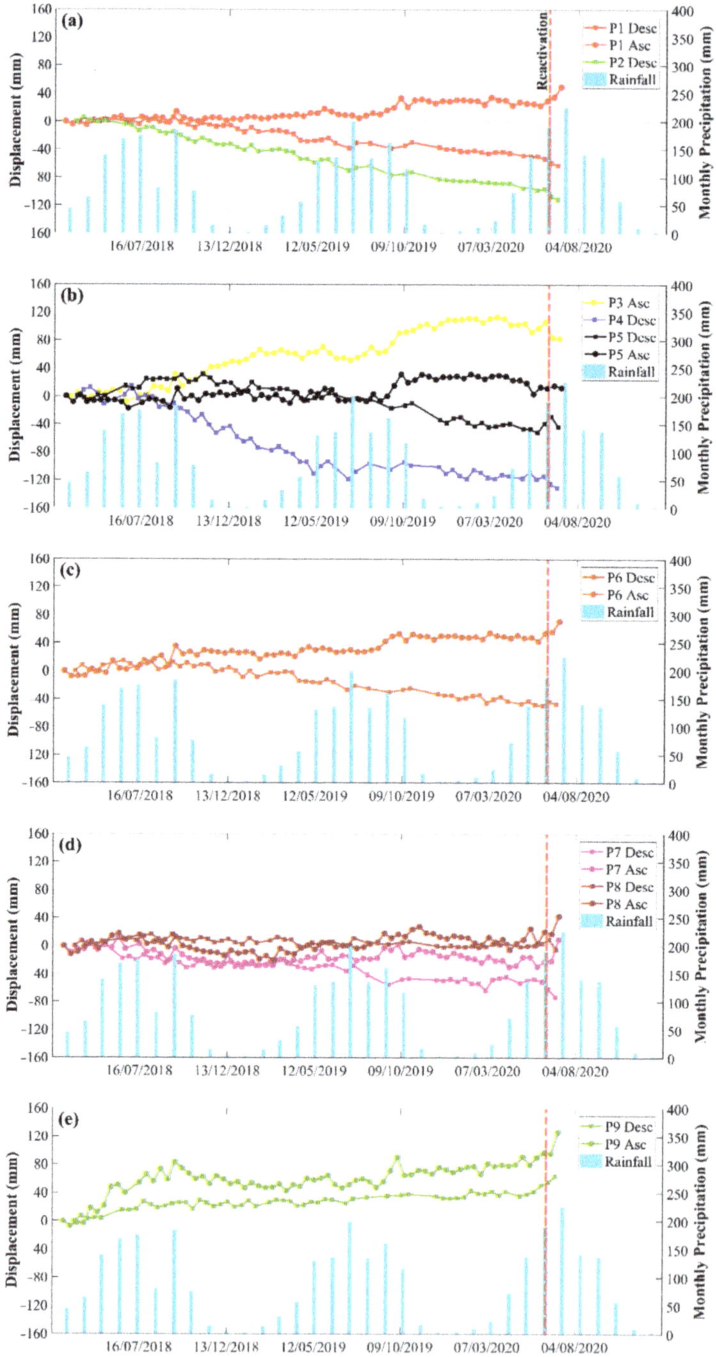

Figure 10. Pre-failure time-series plotting for selected measurement points from the different zones: (**a**) central zone (I); (**b**) landslide toe zone (II); (**c**) upstream zone (III); (**d**) landslide shoulder zone (IV); (**e**) unstable slope (V) zone in the pre-failure mean velocity map (Figure 8). Red dashed line indicates the reactivation of the ANZ landslide.

The time-series deformation over the detected unstable slope, located on the right side of the ANZ landslide, is shown in Figure 10e. The maximum cumulative deformations are 63 mm and 126 mm for the descending and ascending measurements, respectively. It was found that acceleration in deformation was detected from the descending measurement since the end of May in 2020, moving 22 mm within the final four acquisitions, which agreed with the increasing monthly rainfall in June. A dramatic change of cumulative deformation was also found at the final two acquisitions (25 June 2020 to 7 July 2020) from the ascending measurement, surging 30 mm within just 12 days, which is a response to the intense rainfall recorded on 5 July 2020, suggesting that this unstable slope may develop to failure after this event and that continuous monitoring should be further conducted over this site.

To further explore the deformation evolution in the spatial domain, the cumulative LoS deformation maps of the descending and ascending measurements are depicted in Figures 11 and 12, respectively. Sixteen acquisitions were selected and referred to the initial acquisition dates on 27 March 2018 and 8 March 2018 for the descending and ascending maps, respectively. From the joint analysis of the descending and ascending cumulative deformation maps, the central and toe sections of the ANZ slope were found to move initially in January 2019. It is noted from the descending map that the deformation detected at the central section gradually expanded to the surrounding areas with increasing cumulative deformation within the reactivated ANZ landslide area. Most of the points in the upper section were in a relatively stable state with much smaller cumulative deformations. Most importantly, no obvious accelerations were found in the cumulative deformation before the multi-hazard chain on 17 June 2020 from either of the measurements. However, clear changes in deformation can be observed over several sections of the ANZ slope from both measurements after the multi-hazard chain.

5.2. Deformation Mechanism and Triggering/Preparatory Factor Analysis

When comparing the Spatial-Temporal evolution between the different zones above, it is clearly seen that the movement of zones (I-II) at the central and toe sections of the ANZ landslide has occurred since September 2018, long before the obvious movement detected at zone (IV) in the upper section since June 2019. After the expansion of deformation at the central section and following the collapse of the toe section, the upper section gradually lost support at the bottom and presented movement. It is clearly seen that the deformation at the upper section was much smaller than that from the central and toe section, which indicated that the ANZ landslide was characteristic of the retrogressive failure mechanism [69]. After the occurrence of the multi-hazard chain on 17 June 2020, obvious accelerations of deformation from several sections of the ANZ landslide were clearly captured by the Spatial-Temporal analysis of the time-series InSAR result. It is evident that the reactivated ANZ landslide presented an acceleration between 17 June 2020 and the beginning of July 2020, which is highly consistent with the field survey [40,41]. At the same time, an unstable slope next to the ancient ANZ landslide was detected from the time-series InSAR analysis, and it also presented an accelerated deformation similar to that of the reactivated ANZ landslide after the multi-hazard chain. Therefore, continuous monitoring over this area is essential for further assessment of the post-failure stability and the detection of secondary landslides.

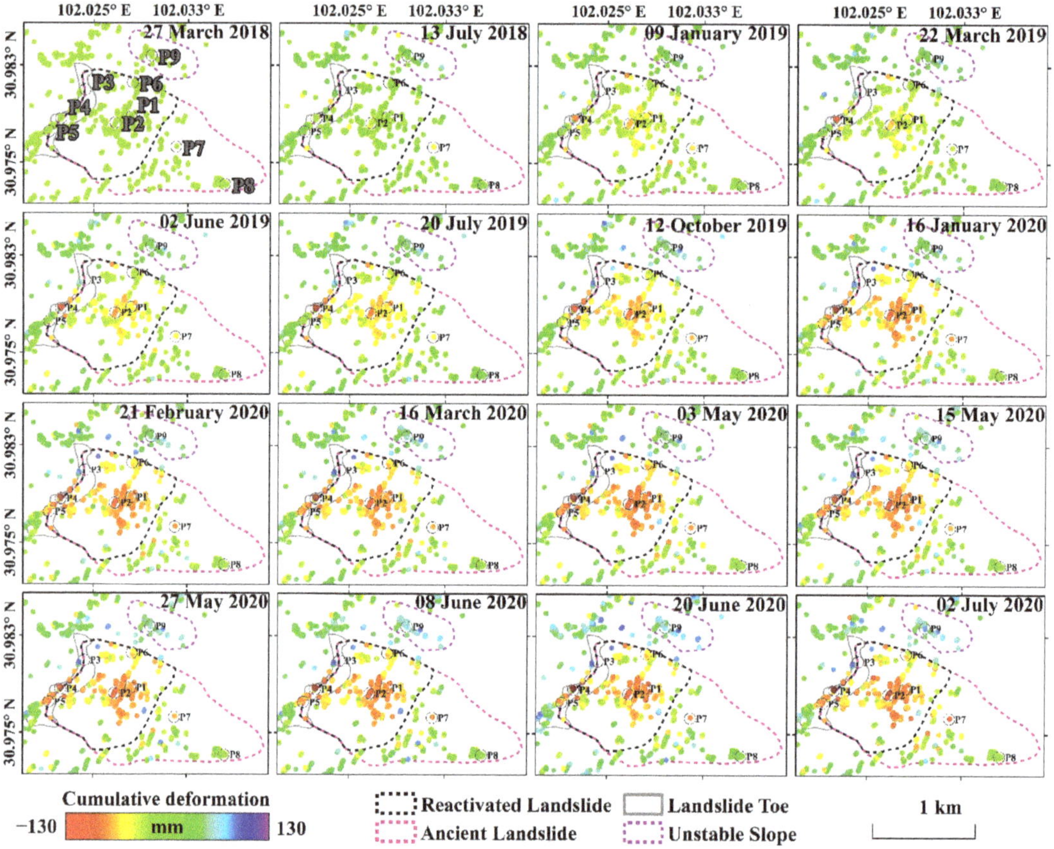

Figure 11. Descending cumulative LOS deformation of the measurement points located at the Aniangzhai slope for the 16 selected acquisition dates of Sentinel-1 descending datasets, referring to the first SAR image acquired on 27 March 2018.

It is noted that a recently published work [70] found a clear acceleration over the upper parts of the ANZ landslide starting in spring 2020 from the multi-temporal InSAR analysis with the descending Sentinel-1A dataset [70]. However, the InSAR results obtained in this study suggested that no obvious accelerations were detected before 17 June 2020, with both the descending and the ascending Sentinel-1A datasets, which is consistent with another published InSAR result with shorter temporal coverage (study period from 20 April 2018 to 20 May 2020) [41]. Such inconsistencies between the InSAR results could be caused by different datasets and the pre-processing and post-processing strategies. Most importantly, it is consistent that all three InSAR results confirmed the clear displacements over the ANZ slope before the event, ranging from −50 to −80 millimetres/year.

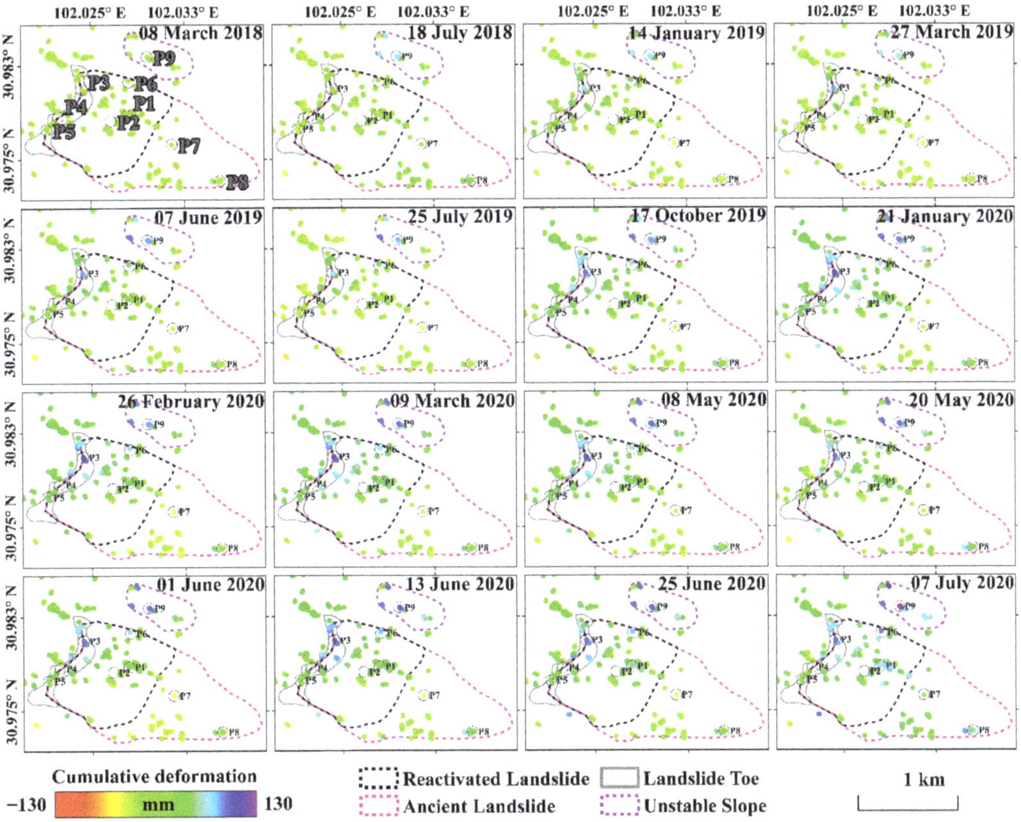

Figure 12. Ascending cumulative LOS deformation of the measurement points located at the Aniangzhai slope for the 16 selected acquisition dates of Sentinel-1 descending datasets, referring to the first SAR image acquired on 8 March 2018.

According to the rainfall data collected from the Anianggouer weather station, the cumulative rainfall from 1 May to 17 June 2020 was over 250 mm and the daily rainfall on 17 June 2020 was up to 38.1 mm, as shown in Figure 2a [40]. The Spatial-Temporal evolution of the ANZ landslide from the time-series InSAR analysis clearly revealed that there was no obvious acceleration (precursory signal) detected before the failure of the multi-hazard chain on 17 June 2020, which is consistent with the published results [41]. This also suggested that the intense rainfall from May to June 2020 was one of the most significant triggering factors for the ANZ landslide. In addition, several preparatory factors also contributed to the occurrence of the ANZ landslide. Firstly, the short, intense rainfall event on 17 June 2020 indeed triggered the failure of the Meilong debris flow, and then, the undercutting river and outburst flood from the dammed lake seriously damaged the stability of the toe section of the ANZ slope. The following collapse of the toe section further reactivated the ANZ landslide and formed the accelerations of deformation. Secondly, the ANZ landslide was located on a region with numerous active faults where numbers of strong earthquakes occurred. The frequent occurrence of earthquakes would significantly shake the ANZ slope and develop fractures and cracks over the slope, which could make the slope looser and slide more easily. At the same time, the dip angle of the landslide toe is up to 70°. It is prone to fail under the effect of gravity and long-term washout and erosion of the underlying river. The ANZ landslide body mainly consists of decimetre-scale angular and sub-angular blocks, which make it easier for the rainwater and groundwater to infiltrate into the surface layers of the slope. This would increase the weight of the soil in the surface layer and result in the reduction in the shear strength of the soil and gradually contribute to the generation of a sliding surface over the slope. Besides that, based on the high-resolution true colour images shown in Figure 4, multiple man-built roads laterally crossed the middle section of the ANZ slope. When intense rainfall occurs in a short duration, a large amount of rainwater will infiltrate the surface layer of the slope. These roads in the middle section could temporally block the flow of rainwater over the surface, which would probably have an influence on the movement of the superficial layers over the slope. It is also consistent with the larger magnitude of deformation in the middle section detected from the optical POT result in Figure 7. At the same time, from the true colour images, there are no roads built in the upper section. Therefore, the magnitude of deformation is much smaller from the optical POT result. The time-series InSAR result also confirmed that the cumulative deformation in the upper section is smaller than the lower section. Thus, it is believed that human activities might also have some influence on the more superficial movement of the slope. Above all, the ANZ landslide was likely triggered by the heavy rainfall, and several preparatory factors, including wash-out and erosion of the river and outburst flood, seismic activities, and local terrain conditions over the slope, also make significant contributions to this event.

6. Conclusions

In this study, the deformation characteristics and Spatial-Temporal evolution of the ANZ landslide were investigated in detail by the joint use of multi-temporal optical images and time-series InSAR analysis. The analysis of multi-temporal optical images clearly showed the spatial evolution of pre- and post-failure landslide features over the slope. Several pre-failure flows over the slope were observed. The post-failure displacement was retrieved with the optical POT technique. These optical POT results revealed that the maximum horizontal displacement was up to 14.4 m, and the ANZ slope mainly moved towards the northwest direction. The deformation velocity and time series of the reactivated ANZ landslide was also retrieved using time-series InSAR analysis with Sentinel-1A/B datasets from the descending and ascending tracks. The maximum magnitude of the LoS deformation velocity was up to −70 mm/year and 45 mm/year from the descending and ascending measurements, respectively. The large deforming areas were found at the landslide toe collapse zone. A large deforming area with a velocity of −20 to −50 mm/year was clearly observed from the descending measurement at the central zone over the ANZ slope. At the same time, an unstable slope was detected on the right side of the ANZ landslide. Most importantly, the time-series analysis of multiple sections over the reactivated ANZ landslide showed that no obvious accelerations of deformation (precursory deformation) were detected before the multi-hazard chain on 17 June 2020. However, after the occurrence of the multi-hazard chain, obvious accelerations can be observed over the ANZ slope, which suggested that the ANZ landslide entered the acceleration status after the reactivation. Through comparing the Spatial-Temporal evolution between different zones of the slope, it is evident that the ANZ landslide presented retrogressive failure mode. The Spatial-Temporal evolution and deformation behaviour derived from the time-series InSAR result suggested the occurrence of the ANZ landslide is not the result of a single triggering factor due to heavy rainfall, but a joint effect of several preparatory factors, including frequent seismic activities, serious wash-out, and erosion of the river and the outburst flood and local terrain conditions.

Satellite SAR interferometry demonstrated its effectiveness and usefulness for mapping and monitoring the surface deformation caused by landslides over a long period. At the same time, the optical remote-sensing technique can provide valuable supplementary information about landslide features and extremely large deformation-of-slope movements. On-going monitoring of the reactivated ANZ landslide is recommended as the ANZ slope presented a clear acceleration after the multi-hazard chain. Therefore, continuous monitoring of the slope is essential for the detection of further failure and for emergency response. Thus, further investigation of post-failure monitoring with the time-series InSAR technique and optical time-series analysis with the POT technique will be studied in our future research.

Author Contributions: Conceptualization, J.K. and A.H.-M.N.; methodology, J.K. and A.H.-M.N.; validation, J.K. and A.H.-M.N.; formal analysis, J.K. and A.H.-M.N.; investigation, J.K.; resources, J.K., A.H.-M.N. and L.G.; data curation, J.K.; writing—original draft preparation, J.K.; writing—review and editing, J.K., A.H.-M.N. and L.G.; supervision, A.H.-M.N. and L.G.; project administration, A.H.-M.N. All authors have read and agreed to the published version of the manuscript.

Funding: This research was funded by the Program for Guangdong Introducing Innovative and Entrepreneurial Teams (2019ZT08L213), Natural Science Foundation of Guangdong Province (grant number 2018A030310538 and 2021A1515011483).

Acknowledgments: The authors thank ESA for providing the Sentinel-1A/B data and the Planet Labs Company for providing the PlanetScope data under a research and education license.

Conflicts of Interest: The authors declare no conflict of interest.

References

1. Cruden, D.M. A simple definition of a landslide. *Bull. Eng. Geol. Environ.* **1991**, *43*, 27–29. [CrossRef]
2. Sato, H.; Harp, E. Interpretation of earthquake-induced landslides triggered by the 12 May 2008, M7. 9 Wenchuan earthquake in the Beichuan area, Sichuan Province, China using satellite imagery and Google Earth. *Landslides* **2009**, *6*, 153–159. [CrossRef]
3. Ren, Z.; Zhang, Z.; Yin, J. Erosion associated with seismically-induced landslides in the middle Longmen Shan region, eastern Tibetan plateau, China. *Remote Sens.* **2017**, *9*, 864. [CrossRef]
4. Patton, A.I.; Rathburn, S.L.; Capps, D.M. Landslide response to climate change in permafrost regions. *Geomorphology* **2019**, *340*, 116–128. [CrossRef]
5. Park, H.-J.; Jang, J.-Y.; Lee, J.-H. Physically based susceptibility assessment of rainfall-induced shallow landslides using a fuzzy point estimate method. *Remote Sens.* **2017**, *9*, 487. [CrossRef]
6. Dong, J.; Zhang, L.; Li, M.; Yu, Y.; Liao, M.; Gong, J.; Luo, H. Measuring precursory movements of the recent Xinmo landslide in Mao County, China with Sentinel-1 and ALOS-2 PALSAR-2 datasets. *Landslides* **2018**, *15*, 135–144. [CrossRef]
7. Fan, X.; Tang, J.; Tian, S.; Jiang, Y. Rainfall-induced rapid and long-runout catastrophic landslide on July 23, 2019 in Shuicheng, Guizhou, China. *Landslides* **2020**, *17*, 2161–2171. [CrossRef]
8. Yang, Z.; Li, Z.; Zhu, J.; Preusse, A.; Yi, H.; Hu, J.; Feng, G.; Papst, M. Retrieving 3-D large displacements of mining areas from a single amplitude pair of SAR using offset tracking. *Remote Sens.* **2017**, *9*, 338. [CrossRef]
9. Laimer, H.J. Anthropogenically induced landslides—A challenge for railway infrastructure in mountainous regions. *Eng. Geol.* **2017**, *222*, 92–101. [CrossRef]
10. García-Ruiz, J.M.; Beguería, S.; Arnáez, J.; Sanjuán, Y.; Lana-Renault, N.; Gómez-Villar, A.; Álvarez-Martínez, J.; Coba-Pérez, P. Deforestation induces shallow landsliding in the montane and subalpine belts of the Urbión Mountains, Iberian Range, Northern Spain. *Geomorphology* **2017**, *296*, 31–44. [CrossRef]
11. Ahmed, B.; Dewan, A. Application of bivariate and multivariate statistical techniques in landslide susceptibility modeling in Chittagong City Corporation, Bangladesh. *Remote Sens.* **2017**, *9*, 304. [CrossRef]
12. Jingtao, L.; Yuliang, C.; Jun, W.; Youming, W.; Bin, L.; Meng, W.; Lei, Y. Remote sensing investigation and formation mechanism on wulipo landslide of July 10, 2013 in Sanxi village, Dujiangyan, Sichuan province. *J. Eng. Geol.* **2014**, *22*, 1194–1203.
13. Chen, X.-Z.; Cui, Y.-F. The formation of the Wulipo landslide and the resulting debris flow in Dujiangyan City, China. *J. Mt. Sci.* **2017**, *14*, 1100–1112. [CrossRef]
14. Fan, X.; Xu, Q.; Scaringi, G.; Dai, L.; Li, W.; Dong, X.; Zhu, X.; Pei, X.; Dai, K.; Havenith, H.-B. Failure mechanism and kinematics of the deadly June 24th 2017 Xinmo landslide, Maoxian, Sichuan, China. *Landslides* **2017**, *14*, 2129–2146. [CrossRef]
15. Malet, J.-P.; Maquaire, O.; Calais, E. The use of Global Positioning System techniques for the continuous monitoring of landslides: Application to the Super-Sauze earthflow (Alpes-de-Haute-Provence, France). *Geomorphology* **2002**, *43*, 33–54. [CrossRef]
16. Jaboyedoff, M.; Ornstein, P.; Rouiller, J.-D. Design of a geodetic database and associated tools for monitoring rock-slope movements: The example of the top of Randa rockfall scar. *Nat. Hazards Earth Syst. Sci.* **2004**, *4*, 187–196. [CrossRef]
17. Squarzoni, C.; Delacourt, C.; Allemand, P. Differential single-frequency GPS monitoring of the La Valette landslide (French Alps). *Eng. Geol.* **2005**, *79*, 215–229. [CrossRef]
18. Bamler, R.; Hartl, P. Synthetic aperture radar interferometry. *Inverse Probl.* **1998**, *14*, R1–R54. [CrossRef]
19. Gabriel, A.K.; Goldstein, R.M.; Zebker, H.A. Mapping small elevation changes over large areas: Differential radar interferometry. *J. Geophys. Res. Solid Earth* **1989**, *94*, 9183–9191. [CrossRef]
20. Massonnet, D.; Feigl, K.L. Radar interferometry and its application to changes in the Earth's surface. *Rev. Geophys.* **1998**, *36*, 441–500. [CrossRef]

21. Rosen, P.A.; Hensley, S.; Joughin, I.R.; Li, F.; Madsen, S.N.; Rodriguez, E.; Goldstein, R.M. Synthetic aperture radar interferometry. *IOP Sci.* **1998**, *14*, R1. [CrossRef]
22. Simons, M.; Rosen, P. Interferometric synthetic aperture radar geodesy. In *Treatise on Geophysics—Geodesy*; Elsevier: Amsterdam, The Netherlands, 2007; pp. 391–446. ISBN 9780444527486.
23. Ferretti, A.; Prati, C.; Rocca, F. Permanent scatterers in SAR interferometry. *IEEE Trans. Geosci. Remote Sens.* **2001**, *39*, 8–20. [CrossRef]
24. Berardino, P.; Fornaro, G.; Lanari, R.; Sansosti, E. A new algorithm for surface deformation monitoring based on small baseline differential SAR interferograms. *IEEE Trans. Geosci. Remote Sens.* **2002**, *40*, 2375–2383. [CrossRef]
25. Mora, O.; Lanari, R.; Mallorqui, J.J.; Berardino, P.; Sansosti, E. A new algorithm for monitoring localized deformation phenomena based on small baseline differential SAR interferograms. In Proceedings of the IEEE International Geoscience and Remote Sensing Symposium, Toronto, ON, Canada, 24–28 June 2002; pp. 1237–1239.
26. Ferretti, A.; Fumagalli, A.; Novali, F.; Prati, C.; Rocca, F.; Rucci, A. A new algorithm for processing interferometric data-stacks: SqueeSAR. *IEEE Trans. Geosci. Remote Sens.* **2011**, *49*, 3460–3470. [CrossRef]
27. Zhang, J.; Zhu, W.; Cheng, Y.; Li, Z. Landslide Detection in the Linzhi–Ya'an Section along the Sichuan–Tibet Railway Based on InSAR and Hot Spot Analysis Methods. *Remote Sens.* **2021**, *13*, 3566. [CrossRef]
28. Dong, J.; Liao, M.; Xu, Q.; Zhang, L.; Tang, M.; Gong, J. Detection and displacement characterization of landslides using multi-temporal satellite SAR interferometry: A case study of Danba County in the Dadu River Basin. *Eng. Geol.* **2018**, *240*, 95–109. [CrossRef]
29. Shi, X.; Xu, Q.; Zhang, L.; Zhao, K.; Dong, J.; Jiang, H.; Liao, M. Surface displacements of the Heifangtai terrace in Northwest China measured by X and C-band InSAR observations. *Eng. Geol.* **2019**, *259*, 105181. [CrossRef]
30. Intrieri, E.; Raspini, F.; Fumagalli, A.; Lu, P.; Del Conte, S.; Farina, P.; Allievi, J.; Ferretti, A.; Casagli, N. The Maoxian landslide as seen from space: Detecting precursors of failure with Sentinel-1 data. *Landslides* **2018**, *15*, 123–133. [CrossRef]
31. Li, M.; Zhang, L.; Dong, J.; Tang, M.; Shi, X.; Liao, M.; Xu, Q. Characterization of pre-and post-failure displacements of the Huangnibazi landslide in Li County with multi-source satellite observations. *Eng. Geol.* **2019**, *257*, 105140. [CrossRef]
32. Scaioni, M.; Longoni, L.; Melillo, V.; Papini, M. Remote sensing for landslide investigations: An overview of recent achievements and perspectives. *Remote Sens.* **2014**, *6*, 9600–9652. [CrossRef]
33. Zhou, C.; Cao, Y.; Yin, K.; Wang, Y.; Shi, X.; Catani, F.; Ahmed, B. Landslide characterization applying Sentinel-1 images and InSAR technique: The Muyubao landslide in the three gorges reservoir area, China. *Remote Sens.* **2020**, *12*, 3385. [CrossRef]
34. Ciampalini, A.; Raspini, F.; Lagomarsino, D.; Catani, F.; Casagli, N. Landslide susceptibility map refinement using PSInSAR data. *Remote Sens. Environ.* **2016**, *184*, 302–315. [CrossRef]
35. Dong, J.; Zhang, L.; Tang, M.; Liao, M.; Xu, Q.; Gong, J.; Ao, M. Mapping landslide surface displacements with time series SAR interferometry by combining persistent and distributed scatterers: A case study of Jiaju landslide in Danba, China. *Remote Sens. Environ.* **2018**, *205*, 180–198. [CrossRef]
36. Casson, B.; Delacourt, C.; Allemand, P. Contribution of multi-temporal remote sensing images to characterize landslide slip surface—Application to the La Clapière landslide (France). *Nat. Hazards Earth Syst. Sci.* **2005**, *5*, 425–437. [CrossRef]
37. Delacourt, C.; Allemand, P.; Berthier, E.; Raucoules, D.; Casson, B.; Grandjean, P.; Pambrun, C.; Varel, E. Remote-sensing techniques for analysing landslide kinematics: A review. *Bull. Soc. Géol. Fr.* **2007**, *178*, 89–100. [CrossRef]
38. Lacroix, P.; Berthier, E.; Maquerhua, E.T. Earthquake-driven acceleration of slow-moving landslides in the Colca valley, Peru, detected from Pléiades images. *Remote Sens. Environ.* **2015**, *165*, 148–158. [CrossRef]
39. Wright, T.J.; Parsons, B.E.; Lu, Z. Toward mapping surface deformation in three dimensions using InSAR. *Geophys. Res. Lett.* **2004**, *31*, 1–5. [CrossRef]
40. Zhao, B.; Zhang, H.; Hongjian, L.; Li, W.; Su, L.; He, W.; Zeng, L.; Qin, H.; Dhital, M.R. Emergency response to the reactivated Aniangzhai landslide resulting from a rainstorm-triggered debris flow, Sichuan Province, China. *Landslides* **2021**, *18*, 1115–1130. [CrossRef]
41. Zhu, L.; He, S.; Qin, H.; He, W.; Zhang, H.; Zhang, Y.; Jian, J.; Li, J.; Su, P. Analyzing the multi-hazard chain induced by a debris flow in Xiaojinchuan River, Sichuan, China. *Eng. Geol.* **2021**, 106280. [CrossRef]
42. Yan, Y.; Cui, Y.; Liu, D.; Tang, H.; Li, Y.; Tian, X.; Zhang, L.; Hu, S. Seismic signal characteristics and interpretation of the 2020 "6.17" Danba landslide dam failure hazard chain process. *Landslides* **2021**, *18*, 2175–2192. [CrossRef]
43. Ren, J.; Xu, X.; Zhang, S.; Yeats, R.S.; Chen, J.; Zhu, A.; Liu, S. Surface rupture of the 1933 M 7.5 Diexi earthquake in eastern Tibet: Implications for seismogenic tectonics. *Geophys. J. Int.* **2018**, *212*, 1627–1644. [CrossRef]
44. Chen, X.; Zhou, Q.; Ran, H.; Dong, R. Earthquake-triggered landslides in southwest China. *Nat. Hazards Earth Syst. Sci.* **2012**, *12*, 351–363. [CrossRef]
45. Ge, L.; Zhang, K.; Ng, A.; Dong, Y.; Chang, H.-c.; Rizos, C. Preliminary results of satellite radar differential interferometry for the co-seismic deformation of the 12 May 2008 Ms8.0 Wenchuan earthquake. *Geogr. Inf. Sci.* **2008**, *14*, 12–19.
46. Farr, T.G.; Rosen, P.A.; Caro, E.; Crippen, R.; Duren, R.; Hensley, S.; Kobrick, M.; Paller, M.; Rodriguez, E.; Roth, L. The shuttle radar topography mission. *Rev. Geophys.* **2007**, *45*, 1–33. [CrossRef]
47. Tavakkoli Piralilou, S.; Shahabi, H.; Jarihani, B.; Ghorbanzadeh, O.; Blaschke, T.; Gholamnia, K.; Meena, S.R.; Aryal, J. Landslide detection using multi-scale image segmentation and different machine learning models in the higher himalayas. *Remote Sens.* **2019**, *11*, 2575. [CrossRef]

48. Marta, S. *Planet Imagery Product Specifications*; Planet Labs: San Francisco, CA, USA, 2018; p. 91.
49. Torres, R.; Snoeij, P.; Geudtner, D.; Bibby, D.; Davidson, M.; Attema, E.; Potin, P.; Rommen, B.; Floury, N.; Brown, M. GMES Sentinel-1 mission. *Remote Sens. Environ.* **2012**, *120*, 9–24. [CrossRef]
50. Funk, C.; Peterson, P.; Landsfeld, M.; Pedreros, D.; Verdin, J.; Shukla, S.; Husak, G.; Rowland, J.; Harrison, L.; Hoell, A. The climate hazards infrared precipitation with stations—A new environmental record for monitoring extremes. *Sci. Data* **2015**, *2*, 150066. [CrossRef]
51. Banerjee, A.; Chen, R.; Meadows, M.E.; Singh, R.; Mal, S.; Sengupta, D. An analysis of long-term rainfall trends and variability in the uttarakhand himalaya using google earth engine. *Remote Sens.* **2020**, *12*, 709. [CrossRef]
52. Yang, W.; Wang, Y.; Sun, S.; Wang, Y.; Ma, C. Using Sentinel-2 time series to detect slope movement before the Jinsha River landslide. *Landslides* **2019**, *16*, 1313–1324. [CrossRef]
53. Fiorucci, F.; Ardizzone, F.; Mondini, A.C.; Viero, A.; Guzzetti, F. Visual interpretation of stereoscopic NDVI satellite images to map rainfall-induced landslides. *Landslides* **2019**, *16*, 165–174. [CrossRef]
54. Tanre, D.; Holben, B.N.; Kaufman, Y.J. Atmospheric correction algorithm for NOAA-AVHRR products: Theory and application. *IEEE Trans. Geosci. Remote Sens.* **1992**, *30*, 231–248. [CrossRef]
55. Leprince, S.; Ayoub, F.; Klinger, Y.; Avouac, J.-P. Co-registration of optically sensed images and correlation (COSI-Corr): An operational methodology for ground deformation measurements. In Proceedings of the 2007 IEEE International Geoscience and Remote Sensing Symposium, Barcelona, Spain, 23–28 July 2007; pp. 1943–1946.
56. Yang, W.; Wang, Y.; Wang, Y.; Ma, C.; Ma, Y. Retrospective deformation of the Baige landslide using optical remote sensing images. *Landslides* **2020**, *17*, 659–668. [CrossRef]
57. Fattahi, H.; Agram, P.; Simons, M. A network-based enhanced spectral diversity approach for TOPS time-series analysis. *IEEE Trans. Geosci. Remote Sens.* **2016**, *55*, 777–786. [CrossRef]
58. Hooper, A. A multi-temporal InSAR method incorporating both persistent scatterer and small baseline approaches. *Geophys. Res. Lett.* **2008**, *35*. [CrossRef]
59. Hooper, A.; Segall, P.; Zebker, H. Persistent scatterer interferometric synthetic aperture radar for crustal deformation analysis, with application to Volcán Alcedo, Galápagos. *J. Geophys. Res. Solid Earth* **2007**, *112*, 1–21. [CrossRef]
60. Hooper, A.; Zebker, H.A. Phase unwrapping in three dimensions with application to InSAR time series. *JOSA A* **2007**, *24*, 2737–2747. [CrossRef] [PubMed]
61. Bekaert, D.; Walters, R.; Wright, T.; Hooper, A.; Parker, D. Statistical comparison of InSAR tropospheric correction techniques. *Remote Sens. Environ.* **2015**, *170*, 40–47. [CrossRef]
62. Fialko, Y.; Simons, M.; Agnew, D. The complete (3-D) surface displacement field in the epicentral area of the 1999 Mw7. 1 Hector Mine earthquake, California, from space geodetic observations. *Geophys. Res. Lett.* **2001**, *28*, 3063–3066. [CrossRef]
63. Ng, A.H.-M.; Ge, L.; Li, X.; Abidin, H.Z.; Andreas, H.; Zhang, K. Mapping land subsidence in Jakarta, Indonesia using persistent scatterer interferometry (PSI) technique with ALOS PALSAR. *Int. J. Appl. Earth Obs. Geoinf.* **2012**, *18*, 232–242. [CrossRef]
64. Bayramov, E.; Buchroithner, M.; Kada, M.; Zhuniskenov, Y. Quantitative Assessment of Vertical and Horizontal Deformations Derived by 3D and 2D Decompositions of InSAR Line-of-Sight Measurements to Supplement Industry Surveillance Programs in the Tengiz Oilfield (Kazakhstan). *Remote Sens.* **2021**, *13*, 2579. [CrossRef]
65. Xiong, L.; Xu, C.; Liu, Y.; Wen, Y.; Fang, J. 3D displacement field of Wenchuan Earthquake based on iterative least squares for virtual observation and GPS/InSAR observations. *Remote Sens.* **2020**, *12*, 977. [CrossRef]
66. Fuhrmann, T.; Garthwaite, M.C. Resolving three-dimensional surface motion with InSAR: Constraints from multi-geometry data fusion. *Remote Sens.* **2019**, *11*, 241. [CrossRef]
67. Colesanti, C.; Wasowski, J. Investigating landslides with space-borne Synthetic Aperture Radar (SAR) interferometry. *Eng. Geol.* **2006**, *88*, 173–199. [CrossRef]
68. Du, Z.; Ge, L.; Ng, A.H.-M.; Zhu, Q.; Yang, X.; Li, L. Correlating the subsidence pattern and land use in Bandung, Indonesia with both Sentinel-1/2 and ALOS-2 satellite images. *Int. J. Appl. Earth Obs. Geoinf.* **2018**, *67*, 54–68. [CrossRef]
69. Marko, K.; Tiit, H.; Peeter, T.; Volli, K. Analysis of a retrogressive landslide in glaciolacustrine varved clay. *Eng. Geol.* **2010**, *116*, 109–116. [CrossRef]
70. Xia, Z.; Motagh, M.; Li, T.; Roessner, S. The June 2020 Aniangzhai landslide in Sichuan Province, Southwest China: Slope instability analysis from radar and optical satellite remote sensing data. *Landslides* **2021**, 1–17. [CrossRef]

remote sensing

MDPI

Article

Construction of "Space-Sky-Ground" Integrated Collaborative Monitoring Framework for Surface Deformation in Mining Area

Yueguan Yan [1], Ming Li [1,*], Linda Dai [2], Junting Guo [3], Huayang Dai [1] and Wei Tang [1]

[1] College of Geoscience and Surveying Engineering, China University of Mining & Technology (Beijing), Beijing 100083, China; yanyueguan@cumtb.edu.cn (Y.Y.); dhy@cumtb.edu.cn (H.D.); weitang@cumtb.edu.cn (W.T.)
[2] School of Management, China University of Mining & Technology (Beijing), Beijing 100083, China; BQT2000503020@student.cumtb.edu.cn
[3] State Key Laboratory of Water Resource Protection and Utilization in Coal Mining, Beijing 102209, China; junting.guo.a@chnenergy.com.cn
* Correspondence: bqt2000204053@student.cumtb.edu.cn; Tel.: +86-131-2666-2613

Abstract: Ground deformation measurements in mining areas play a key role in revealing the surface subsidence law, retrieving the subsidence parameters, warning of geological disasters and restoring the surface ecology. With the development of science and technology, there have emerged a great number of monitoring techniques and buildings of diverse protection levels. The diversity of monitoring techniques and the multiplicity of monitoring objects have brought challenges for surface deformation monitoring in the coal industry. Based on the existing deformation monitoring techniques, this paper established a framework of "space-sky-ground" collaborative monitoring system in mining area. We also constructed an AHP-TOPSIS (Analytic Hierarchy Process method-Technique for Order Preference by Similarity to an Ideal Solution) preference model of "space-sky-ground" collaborative monitoring of surface deformation in mining area, and carried out engineering application. Our study shows that the framework of the "space-sky-ground" collaborative monitoring system for surface subsidence in mining areas established in this paper, combined with the AHP-TOPSIS monitoring preference model, which can fully combine the advantages of each monitoring technique, overcome the limitations of a single monitoring technique, comprehensively obtain the surface subsidence data and work out the surface deformation subsidence pattern. This information provides a data and technical support for surface environment management.

Keywords: "space-sky-ground" collaborative monitoring framework; mining subsidence; preference model; AHP-TOPSIS

Citation: Yan, Y.; Li, M.; Dai, L.; Guo, J.; Dai, H.; Tang, W. Construction of "Space-Sky-Ground" Integrated Collaborative Monitoring Framework for Surface Deformation in Mining Area. *Remote Sens.* **2022**, *14*, 840. https://doi.org/10.3390/rs14040840

Academic Editors: Alex Hay-Man Ng, Linlin Ge, Hsing-Chung Chang, Zheyuan Du and Francesca Cigna

Received: 25 December 2021
Accepted: 7 February 2022
Published: 10 February 2022

1. Introduction

Despite the development of science and technology, coal-oil-gas still plays a key role in government development strategies, which cannot be replaced within a short-time period. What is more, due to its abundant reserves and relatively low cost, coal occupies a crucial position in energy consumption. Large-scale exploitation of coal resources, however, has resulted in a series of negative consequences, including land subsidence and collapse, damage of ground infrastructure and geological hazards—which pose a threat to the sustainable development of local environment and ecology. Government departments and relevant industry associations have established a great number of laws and regulations to ensure the sustainability of coal mining [1]. However, coal mining still affects the normal operation of local buildings and damages ecological balance. Therefore, it is necessary to set up a framework of "space-sky-ground" collaborative monitoring system to obtain accurate surface subsidence to fully understand the surface subsidence pattern in mining areas, which is important for ecological restoration and mining subsidence management.

Vertical land subsidence and horizontal movement are the two basic movement quantities of surface deformation in mining areas [2,3]. Conventional measurement includes two conventional instruments, levelling and total station, which have been developed from traditional optical instruments to current electronic automatic measurement instruments, thanks to the development of production and continuous progress of science and technology. In the monitoring of surface subsidence in mining area, the arrangement of surface observation station is divided into profile line and mesh. The profile line observation station can obtain the surface movement pattern of the main section, but the surface deformation pattern of the whole basin cannot be recognized; the mesh observation station can obtain the surface deformation information of the whole mining subsidence pattern, but the arrangement process is limited by the topographic factors. The workload is large, time-consuming and labor-intensive, and the deformation law of the discrete points obtained cannot completely present the surface of the mining area. Moreover, the traditional measurement cannot collect data under all-time conditions [4]. Therefore, the application of new technologies to surface deformation monitoring in mining areas has become a new development trend.

Emerging technologies have great potential for monitoring surface deformation and can significantly optimize data-collecting abilities, making it possible to obtain a full range of multi-temporal subsidence patterns in subsidence areas. The Global Navigation Satellite System (GNSS) can rendezvous the location of unknown points on the ground (user receivers) using known spatial positions of more than three satellites [5], and with the availability and selectivity of a sufficient number of satellites, GNSS can be applied to the monitoring of ground subsidence. Parul R. Patel analyzed the correlation between ground subsidence, oil and gas production and pressure depletion [6]. Researchers also have applied GNSS technology to aquifer compaction ground subsidence due to groundwater pumping, and settlement monitoring in metal mining [7–9]; Interferometric Synthetic Aperture Radar (InSAR) is a new technique for ground deformation monitoring with accuracy at the centimeter or even millimeter level [7].Three-dimensional laser scanning technology (TLS) is a convenient and accurate way to obtain the distance of all targets by acquiring the real three-dimensional coordinates of the targets, and it shows advantages such as fast speed, short cycle time and no need for contact in the monitoring of surface subsidence areas in mining areas. It can also obtain the three-dimensional point cloud data of the whole area, which is able to reflect more comprehensive deformation information of the subsidence area [10], which has been applied to smart grid 3D model reconstruction, geological disaster mapping, monitoring and modeling [11–13]; Unmanned Aerial Vehicle (UAV) photogrammetry system, which generates various data sets containing point clouds, 4D products, etc. [14–16], has been widely used as a fast and flexible data acquisition system for remote sensing of mine sites, aerial physical prospecting and topographic surveys [17–20]. The mentioned modern observation techniques have been widely used in surface deformation monitoring. Many scholars have contributed to the comprehensive understanding of surface deformation by improving observation accuracy through data acquisition and data processing methods [21–28]. Meanwhile, many studies simultaneously use one or two traditional measurement techniques and modern techniques to study and analyze surface subsidence, but few studies combine the advantages of various techniques for analysis [29–32]. Since different techniques have different advantages, it is often difficult to obtain the comprehensive surface subsidence pattern in mining areas by using a single deformation monitoring techniques.

The diversity of emerging monitoring technologies and the diversity of monitoring targets bring challenges for surface subsidence monitoring. How can different monitoring technologies be combined with the factors affecting them to better serve surface subsidence monitoring in mining areas? The AHP (analytic hierarchy process method) has important advantages in terms of weighting procedures for pairwise comparisons and has been widely used in areas such as decision analysis of supply chain networks and mining methods [33–35]. TOPSIS (Technique for Order Preference by Similarity to an Ideal Solution) selects the optimal solution by ranking it with the similarity to the ideal solution [36,37]. AHP-TOPSIS is a widely accepted multi-attribute decision making technique that considers the ideal and anti-ideal solutions, and the calculation process is easy to program [38,39]. The TOPSIS method is able to combine multiple indicator factors of a solution, but for multi-factor analysis and weight calculation it is difficult and complex, the AHP method overcomes this disadvantage and ensures the global optimality of the algorithm. AHP-TOPSIS has been widely used in other fields [40–43]. Therefore, this paper introduces the AHP-TOPSIS model into the preferred model of surface subsidence monitoring techniques in mining areas.

Based on the characteristics of surface deformation in the mine area and the advantages of existing monitoring techniques, this paper established the framework of the "space-sky-ground" collaborative monitoring system for surface subsidence, and con-structed an AHP-TOPSIS surface monitoring method preferential model under three constraints: natural factors, target constraints and site needs. The established "space-sky-ground" collaborative monitoring framework and the preferred model were applied to the surface subsidence monitoring of the 401 working face of Shendong coal mine, and we obtained the results of a single technology, and the surface subsidence basin and surface discontinuous deformation information of the mine area under the "space-sky-ground" collaborative monitoring framework. We also discussed the advantages of integrating multiple monitoring technologies under the "space-sky-ground" collaborative monitoring framework.

2. Methodology

2.1. Surface Deformation in Mining Area

Surface deformation process: Before the mining of coal resources, the underground rock was in a relatively balanced state. Underground mining of ore bodies destroys the original stress balance state of surrounding rock, causes stress redistribution, and results in a new balance. In this process, rock movement and destruction are also called strata movement. Underground mining is the driving force of rock movement, which is the internal cause of surface subsidence, and surface subsidence is the external performance of rock movement. Mining subsidence from mined-out area to the surface of surrounding rock, in general, is the process of attenuation of the degree of damage and expansion of the range of movement, which has a great impact on the local environment, resources, land and human activities. After the coal was mined out, the overlying strata collapsed and filled the goaf, as shown in Figure 1.

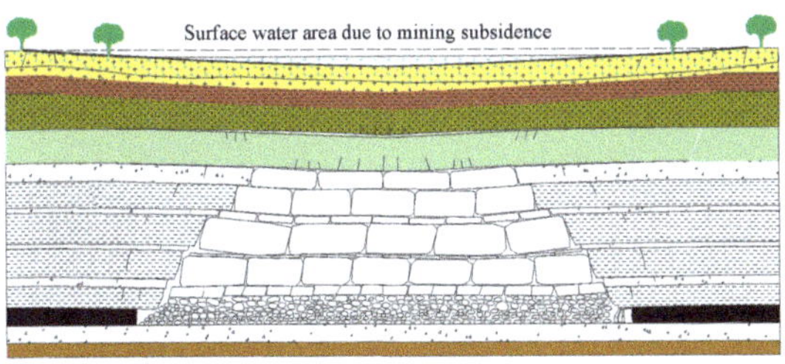

Figure 1. Mining damage transfer process.

Surface deformation is an external form if damage caused by coal mining, and it can be categorized into continuous and discontinuous deformation:

(1) Continuous deformation: Subsidence is a general description of movement and deformation. According to the different ways, directions and properties of movement and deformation, the subsidence conditions of rock strata and surface are described as subsidence, inclination, curvature, horizontal movement and horizontal deformation. The above-mentioned five basic movement and deformation variables are usually used. Damage of continuous deformation to the surface is shown in Figure 2a–c. The

trajectory of the surface point depends on the relationship between the surface point and the relative position of the working face in time-space. In general, the moving vector of each point on the surface above the bending zone points to the center of the moving basin from both its starting and ending relative positions.

(2) Discontinuous deformation: Discontinuous deformation refers to the large surface displacement over limited surface area with steps or discontinuities in the surface profile, discontinuous deformation within underground coal mining rock formations contains: (a) crown hole; (b) chimney caving; (c) plug subsidence; (d) solution cavities; (e) block caving; (f) progressive hanging wall caving [44]. In China, it is generally considered that the main manifestations of surface discontinuous deformation are mining cracks, step cracks and collapse pits [45,46]. As a typical manifestation of discontinuous deformation, surface mining cracks are prevalent and worthy of study. The discussion in this paper focuses on surface mining cracks in discontinuous deformation. Discontinuous deformation is generated with certain conditions; the surface mining fracture characteristics are closely related to the location of generation; the fracture zone is generated periodically with the surface advance in the direction of working face advance; the mining cracks are developed in the form of fracture zones; the topsoil properties have a significant influence on the discontinuous development characteristics. The damage of discontinuous deformation on the surface is shown in Figure 2d,e.

Figure 2. Surface subsidence and damages in mining areas: (**a**) Surface water area; (**b**) Road cracks; (**c**) Housing cracks; (**d**) Surface cracks; (**e**) Collapse pits.

2.2. "Space-Sky-Ground" Collaborative Monitoring Framework of Mining Subsidence

Based on the characteristics of surface deformation in the mine area and existing monitoring techniques (as shown in Table 1), the "space-sky-ground" collaborative monitoring framework was established. The "space-sky-ground" collaborative monitoring framework of surface subsidence in mining area is composed of three parts: space monitoring components, sky monitoring components and ground monitoring components. It combines, coordinates and complements various technologies such as observation network control, traditional point line observation, regional interference measurement, UAV measurement of surface, TLS measurement of surface shallow mining cracks, and realizes the function of comprehensive and three-dimensional monitoring of surface movement, deformation and shallow cracks. Through the efficient cooperation of space monitoring components, sky monitoring components and ground monitoring components, the spatial-temporal dynamic evolution process of surface deformation and ground cracks development in the mining process is effectively monitored, and the deformation and failure law of surface subsidence in mining area is obtained, as is shown in Figure 3. The space monitoring component, the sky monitoring component and the ground monitoring component are described in detail as follows:

Table 1. Common ground deformation monitoring techniques.

Monitoring Method	Advantages	Accuracy
GNSS and CORS	Simple operation; Small workload [47–49]	5 mm
D-InSAR	Low cost; Long period; Global surface deformation [26,30,50]	1–3 cm
SBAS-InSAR [51]	Enhancement of SAR data usage; Low cost	<1 cm
TLS [52–55]	Long range; High accuracy	1 cm
UAV [56,57]	High speed and efficiency; Flexibility	10–15 cm
Levelling and Total station	High monitoring accuracy; Accurate reflection of the law of settlement	1 mm
GNSS and RTK	High speed and efficiency [58]	2 cm
Steel rule and GNSS	High accuracy; Law of accurate reflection [59]	0.1 mm

The abbreviation in Table 1: CORS (Continuously Operating Reference Stations), D-InSAR (Differential Interferometry Synthetic Aperture Radar), SBAS-InSAR (Small Baseline Subset Interferometry Synthetic Aperture Radar), RTK (Real Time Kinematic).

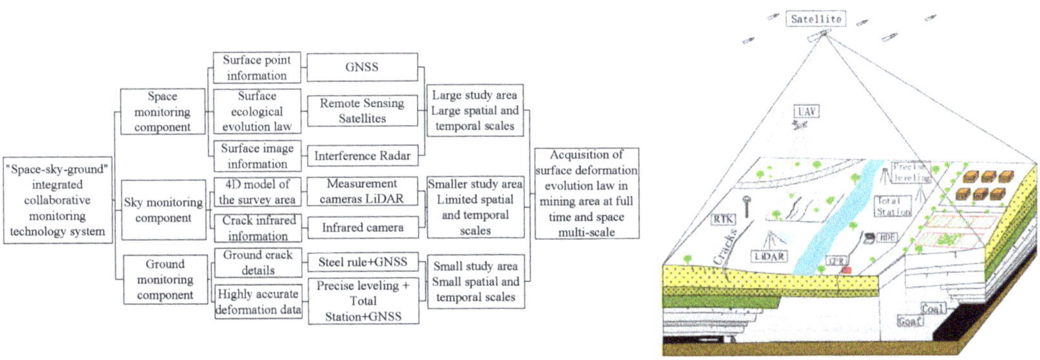

(a) (b)

Figure 3. "Space-sky-ground" collaborative monitoring framework for surface subsidence in mining area: (**a**) the content of "Space-sky-ground" collaborative monitoring framework; (**b**) the diagram of "Space-sky-ground" monitoring framework.

Space monitoring components can obtain large-scale, high-precision and high-spatial-resolution surface deformation. Compared with traditional observation technology, InSAR has incomparable advantages for mine subsidence monitoring in a large area and complex terrain conditions. The basic principle of InSAR is to obtain two radar images of the same area by using two pairs of antennas for simultaneous observation or two near-parallel observations, and to obtain high-precision and high-resolution ground elevation or surface deformation information by obtaining the phase difference between two echo signals corresponding to the same target and combining with orbital data. InSAR technology is not only used in topographic mapping, but also widely used in land subsidence, change detection, seismic activity, digital elevation modeling and many other fields. Surface deformation monitoring is important for the safety of mine sites, since the maintenance of the stability of mine topography is the basis for conducting normal production and life. With the development of radar remote sensing technology, InSAR technology is widely used in the subsidence deformation monitoring of mines, especially the time series InSAR analysis methods (PS-InSAR, SBAS, etc.) can invert the historical evolution of ground subsidence based on time series SAR data with the accuracy of millimeter level. GNSS provides a unified datum for multi-source monitoring data and performs geolocation information alignment up and down the well.

Sky monitoring components use UAVs with measurement cameras to navigate and photograph within a specified range to record the extent of surface crack development and spatial spreading characteristics of the subsidence area. According to the different needs of the observer, the measurement cameras are classified into the generally visible camera, 3D camera, infrared camera and multispectral camera. Compared with the traditional manual measurement, the UAV with a measurement camera can quickly and efficiently obtain the information of ground crack and accurately save the GPS information, which has its unique advantages in places where humans cannot reach.

Ground monitoring components are based on the precise levelling and total station, steel ruler and GNSS precision measurement, and use the ground conventional measurement technology (levelling + wire), steel ruler and vernier caliper to measure the surface subsidence area and crack distribution with high precision.

Duration of surface deformation monitoring: the duration of surface subsidence monitoring is closely related to the geological mining conditions, the duration of the working face mining and the quality of the monitoring technology. In addition, the following steps are assisted: (1) after determining the monitoring object (mining workings), the characteristics of surface deformation under similar conditions should be determined by consulting the relevant literature, and the probability integral method should be used to make dynamic forecast of surface deformation when available; (2) at the early stage of mining, precision leveling measurement should be selected to monitor the location of the opencut to determine the time when the surface displacement caused by underground coal mining is reflected to the surface; (3) Integrate the results of (1) and (2) analysis, consider the advantages and disadvantages of each monitoring method, and determine the time of surface monitoring using different monitoring techniques.

2.3. *Preference Model*

2.3.1. Infactors Determination and Analysis

We determined the indicators analysis by AHP. AHP is a flexible quantitative method, which has become one of the commonly used methods in risk assessment [60–64]. Based on the relevant literature and engineering requirements to determine the influencing factors affecting the surface subsidence monitoring program [14,65–67], a total of three primary indicators and 16 secondary indicators are included, as shown in Figure 4 (natural factors and target constraints are determined in advance):

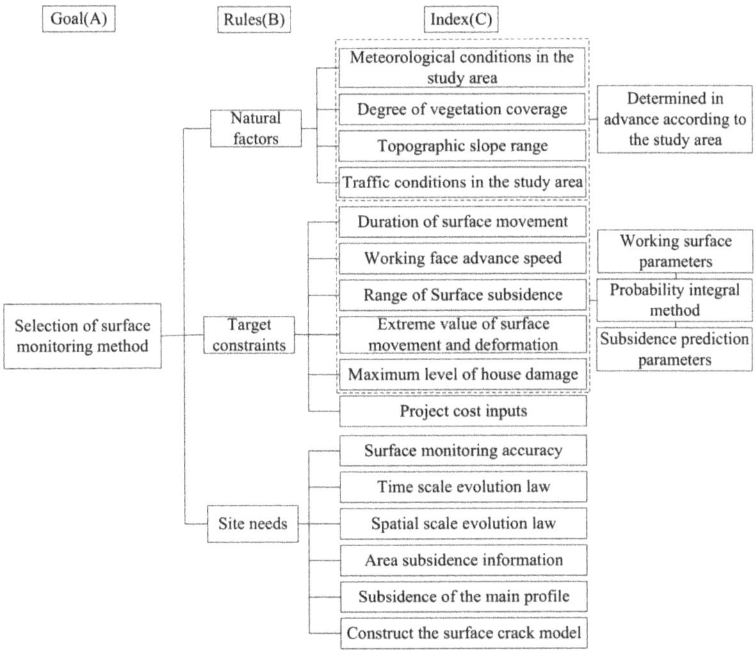

Figure 4. Influencing factors for the preference of surface subsidence monitoring methods selection.

Based on the criterion and judgment principle of two-by-two comparison, fuzzy mathematical theory was applied to derive the criteria for comparing the importance of indicators. The factors of the previous level are used as the benchmark for the factors of the next level, and the judgment matrix is D after the two-by-two comparison.

$$D = \begin{bmatrix} X_{11} & X_{12} & \cdots & X_{1n} \\ X_{21} & X_{22} & \cdots & X_{2n} \\ \vdots & \vdots & \ddots & \vdots \\ X_{m1} & X_{m2} & \cdots & X_{mn} \end{bmatrix} = \begin{bmatrix} \frac{X_1}{X_1} & \frac{X_1}{X_2} & \cdots & \frac{X_1}{X_n} \\ \frac{X_2}{X_1} & \frac{X_2}{X_2} & \cdots & \frac{X_2}{X_n} \\ \vdots & \vdots & \ddots & \vdots \\ \frac{X_n}{X_1} & \frac{X_n}{X_2} & \cdots & \frac{X_n}{X_n} \end{bmatrix} \quad (1)$$

In Equation (1), D is the analytic hierarchy process judgment matrix, $X_{ij}(i = j = 1, 2, \ldots, n)$ is the significance of the comparison between two adjacent indicators.

Normalization and consistency testing are performed:

$$\overline{W}_i = \sqrt[n]{\prod_{j=1}^{n} X_{ij}}, W_i = \overline{W}_i / \sum_{j=1}^{n} \overline{W}_j \quad (2)$$

$$\lambda_{max} = \sum_{i=1}^{n} \frac{(DW)_i}{nW_i} \quad (3)$$

$$R_C = \frac{I_C}{I_R} = \frac{\lambda_{max} - n}{(n-1) I_R} \quad (4)$$

where n is the order of the judgment matrix, $n = 1, 2, \ldots, n$, W_i is the normalized judgment matrix, λ_{max} is the maximum eigenvalue of the judgment matrix, I_C is the consistency test index, I_R is the random consistency indicator, R_C is the consistency ratio of the judgment matrix.

The maximum eigenvalues and eigenvector are derived by normalizing the indicators affecting the surface monitoring program, as shown in Table 2. The results are tested for consistency, and the results are shown in Table 3.

Table 2. Maximum eigenvalue and eigenvector of preferred index.

Judgment Matrix	Maximum Eigenvalue	Eigenvector
A (Preferred monitoring method)	3.0536	0.1571, 0.2493, 0.5936
B1 (Natural factors)	4.1023	0.1306, 0.5492, 0.2302, 0.0900
B2 (Targets constraints)	6.5616	0.0603, 0.0440, 0.1162, 0.2355, 0.1829, 0.3611
B3 (Site needs)	6.4613	0.3466, 0.1784, 0.0821, 0.1528, 0.1607, 0.0676

Table 3. Consistency check calculation results.

Index	A	B1	B2	B3
I_C	0.0268	0.0341	0.1123	0.0923
I_R	0.0462	0.0379	0.0906	0.0744

Based on the calculation results of AHP, the optimal weight of each preferred index is analyzed, and the results are shown in Figure 5. Figure 5 shows that among the natural factors, vegetation coverage has the greatest weight on the indexes. Among the target constraints, the project cost inputs have a great influence on the selection of monitoring methods. Among the site needs, the surface monitoring accuracy is the most important factor, and the accuracy factor is essential for understanding the mining displacement mechanism and mining-related geohazard prevention and control. The results of the analysis in Figure 5 are generally consistent with reality and prove the feasibility of using AHP to determine the weights of the surface monitoring program.

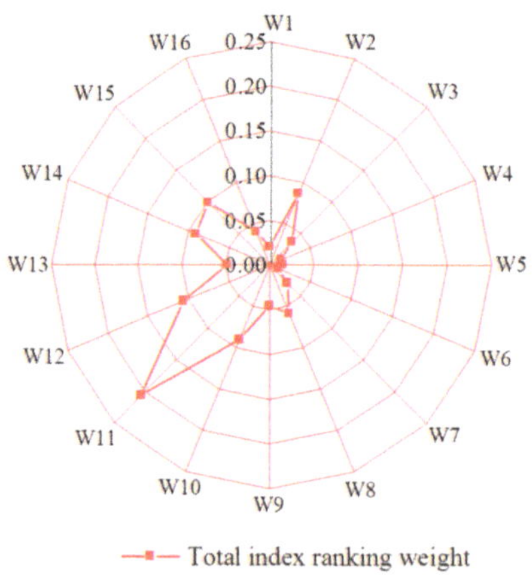

Figure 5. Total index ranking weight.

2.3.2. Comprehensive Preferred Model Based on AHP-TOPSIS

The TOPSIS method can comprehensively consider a variety of index factors of the combined monitoring program, but it is difficult and complicated for multi-factor analysis and weight calculation, while the AHP method overcomes this drawback and ensures the global optimality of the algorithm. The model is applied to the surface monitoring program selection in mining areas, which can make the selection system more scientific, comprehensive and accurate, and the evaluation results are more in line with the reality. The steps of the combined surface subsidence monitoring program optimization using AHP-TOPSIS are as follows:

Step 1: Calculate the weighted normalized decision matrix.

Setting different monitoring programs sets $A = \{A_1, A_2, \ldots, A_m\}$ and influencing factors sets is $X = \{X_1, X_2, \ldots, X_n\}$. Based on the "space-sky-ground" collaborative monitoring framework to determine different monitoring programs and impact factors index matching degree, the matching degree is 1–5 levels. We have established the matching matrix between monitoring programs and impact factors, and calculated the weighted standardized decision matrix:

$$A = \begin{bmatrix} X_{11} & X_{12} & \ldots & X_{1n} \\ X_{21} & X_{22} & \ldots & X_{2n} \\ \vdots & \vdots & \ddots & \vdots \\ X_{m1} & X_{m2} & \ldots & X_{mm} \end{bmatrix} \tag{5}$$

$$C = \left(c_{ij}\right)_{m \times n} = \begin{bmatrix} w_1 b_{11} & w_2 b_{12} & \ldots & w_n b_{1n} \\ w_1 b_{21} & w_2 b_{22} & \ldots & w_n b_{2n} \\ \vdots & \vdots & \ddots & \ddots \\ w_1 b_{m1} & w_2 b_{m2} & \ldots & w_n b_{mn} \end{bmatrix} \tag{6}$$

where C is the weighted standardized decision matrix, W_i is the total ranking weight of indicators, b_{ij} is the standardized decision matrix.

Step 2: Determine the closeness of different monitoring programs.

Calculating the distance of different monitoring programs from the positive and negative ideal solutions and determining the closeness of the monitoring program to the positive ideal solution:

$$d_i^{+} = \sqrt{\sum_{i=1}^{n} (c_{ij} - c_j^{+})^2}, d_i^{-} = \sqrt{\sum_{i=1}^{n} (c_{ij} - c_j^{-})^2} \tag{7}$$

$$E_i^{+} = \frac{d_i^{-}}{d_i^{+} + d_i^{-}}, 0 \leq E_i^{+} \leq 1 \tag{8}$$

In Equations (7) and (8), c_j^{+} and c_j^{-} are the corresponding elements of C^{+} and C^{-} respectively, The closer E_i^{+} is to 1, the nearer the monitoring program is to the ideal monitoring program.

Step 3: Establishing a preference evaluation matrix.

Combined with TOPSIS method, we determine the closeness of each monitoring program and AHP method to determine the eigenvector, the integrated monitoring method evaluation result vector F is derived, and the expression is as follows. The integrated preference evaluation matrix is established.

$$F = WE \tag{9}$$

where E is the evaluation matrix formed by proximity analysis, W is the weight of rule layer calculated by AHP method.

Step 4: Preferred monitoring program.

The optimal monitoring program is selected by combining the result vector F calculated in Step 3.

3. Case Study

3.1. Study Area

Shangwan minefield is located in the southeast of Dongsheng minefield, eastern part of Ordos Plateau, and most of its administrative jurisdiction belongs to Ulanmulun town of

Igingholo Banner. The elevation is high in the northwest part and low in the southeast part which shows the shape of a slope. On the east side of 401, the fully mechanized mining face of Shangwan minefield is 402, the fully mechanized mining face which has not been circled yet; on the south side is minefield boundary, adjacent to Wenjiata minefield of Ulan Group; on the west side is 412, the fully mechanized mining face of planning and design, in which the digging has not started yet; on the north side is the minefield boundary, adjacent to QIlianta minefield. The fully mechanized mining face is arranged along with the inclination, with a strike length of 299.2 m, an advance length of 5254.8 m, and a design mining height of 8.8 m. The ground elevation of the fully mechanized face is 1188–1300 m, and the floor of the coal seam floor is 1043–1066 m. The 401 working face layout is shown in Figure 6:

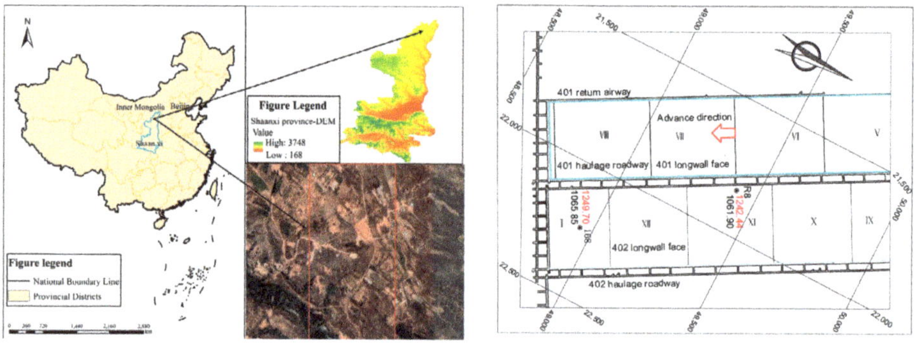

(a) (b)

Figure 6. Location of Shangwan Mine and 401 working face layouts: (**a**) Location of Shangwan Mine; (**b**) 401 working face layouts.

3.2. Preference of Surface Subsidence Monitoring Program

Based on the "space-sky-ground" collaborative monitoring framework established in Section 2.2, five monitoring techniques were initially identified based on the objective factors of the project: GNSS and CORS, InSAR, TLS, UAV, Steel ruler and GNSS, and now according to the actual project we need to select four combination monitoring techniques to monitor the surface subsidence of 401 working face in Shendong mine area. There are five combination programs in Table 4.

Table 4. Monitoring method combination comparison table.

Program	GNSS and CORS	InSAR	TLS	UAV	Steel Rule and GNSS
I	√	√	√	√	
II	√	√	√		√
III	√	√		√	√
IV	√		√	√	√
V		√	√	√	√

In the framework of "space-sky-ground" collaborative monitoring, based on the information on the study area, probability integral method and the prediction parameters of the adjacent working face, the reference values of the first 10 of the 16 evaluation indicators are determined. Then, the initial evaluation matrix is constructed for natural factor indicators, target constraints indicators and site needs indicators respectively, and the matrix is weighed and standardized to calculate the closeness of each scheme. According to TOPSIS, the optimal evaluation matrix for the monitoring scheme of 401 working face in Shendong mine is determined.

$$E = \begin{bmatrix} 0.1835 & 0.3542 & 0.2751 & 0.0000 & 0.1835 \\ 0.2043 & 0.1498 & 0.1548 & 0.2375 & 0.2536 \\ 0.0863 & 0.2357 & 0.1501 & 0.2432 & 0.2848 \end{bmatrix} \quad (10)$$

The weight of the preferred model criterion layer (layer B) for the selection of surface monitoring program in mining areas determined using the AHP method is shown in the following equation.

$$W = \begin{bmatrix} 0.1571 & 0.2493 & 0.5936 \end{bmatrix} \quad (11)$$

So:

$$F = WE = \begin{bmatrix} 0.1302 & 0.2329 & 0.1709 & 0.2036 & 0.2614 \end{bmatrix} \quad (12)$$

In summary, the comprehensive superiority of each program can be obtained, and the order of superiority of the surface subsidence monitoring program at the 401 working face of Shendong mine is: Program V > Program II > Program IV > Program III > Program I. The monitoring scheme is shown in Figure 7.

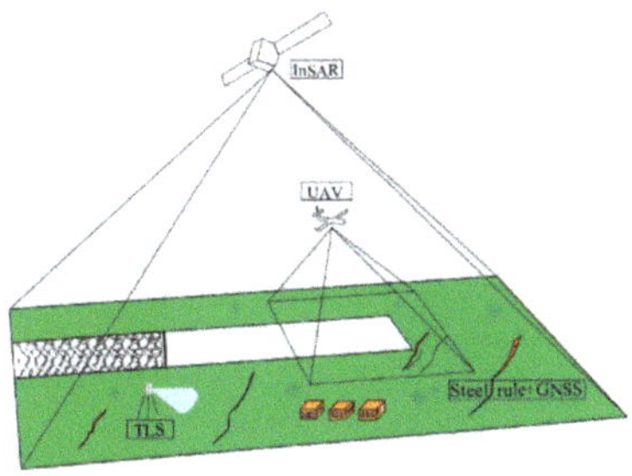

Figure 7. Diagram of the optimized solution for "space-sky-ground" collaborative monitoring at working face 401.

Meanwhile, based on the "space-sky-ground" collaborative monitoring framework, the surface deformation monitoring period of 401 working face was determined. The 401 working face mining is a shallow buried thick coal seam mining type, this type of surface subsidence has the characteristics of early initiation of subsidence, fast subsidence speed and short time to reach the maximum subsidence value; combined with the results of precision levelling at the location of the opencut, the advantages and disadvantages of monitoring technology and the actual project needs, the 401 working face surface subsidence monitoring time was determined. The period studied in this experiment is the position of the working face from the opencut advance to near 1000 meters.

3.3. *Surface Subsidence Monitoring*

3.3.1. InSAR Subsidence Monitoring

In this experiment, we used 30 SAR images from the Sentinel-1 satellite for the time interval 14 October 2017 to 21 October 2018, with images covering the entire Shendong mining area. The Sentinel-1 satellite orbits in a near-polar sun-synchronous orbit at an altitude of approximately 700 km, with a revisit period of 12 days. The Sentinel-1 satellite employs stringent orbit control techniques to provide very high-quality interferometric SAR image data. The satellite has multiple imaging modes, allowing for different polarizations such as monitorization and dual polarization. Details of the SAR images from the experiment are shown in Table 5. The DEM used in the experiment is the 30m resolution SRTM DEM data provided by NASA, which is used to remove the reference terrain phase, and the precision orbital POD data, which is released 21 days after the SAR imaging, to attenuate the orbital phase error.

Table 5. Experimental use of SAR image data details.

Number	Date	Path-Frame	Orbital Direction	Polarization Mode	Model	Incidence Angle
1	14 October 2017	11-126	Ascending	VV	IW	40-41
2	26 October 2017	11-126	Ascending	VV	IW	40-41
3	7 November 2017	11-126	Ascending	VV	IW	40-41
4	19 November 2017	11-126	Ascending	VV	IW	40-41
5	1 December 2017	11-126	Ascending	VV	IW	40-41
6	13 December 2017	11-126	Ascending	VV	IW	40-41
7	25 December 2017	11-126	Ascending	VV	IW	40-41
8	6 January 2018	11-126	Ascending	VV	IW	40-41
9	30 January 2018	11-126	Ascending	VV	IW	40-41
10	11 February 2018	11-126	Ascending	VV	IW	40-41
11	23 February 2018	11-126	Ascending	VV	IW	40-41
12	7 March 2018	11-126	Ascending	VV	IW	40-41
13	19 March 2018	11-126	Ascending	VV	IW	40-41
14	31 March 2018	11-126	Ascending	VV	IW	40-41
15	12 April 2018	11-126	Ascending	VV	IW	40-41
16	24 April 2018	11-126	Ascending	VV	IW	40-41
17	6 May 2018	11-126	Ascending	VV	IW	40-41
18	18 May 2018	11-126	Ascending	VV	IW	40-41
19	30 May 2018	11-126	Ascending	VV	IW	40-41
20	11 June 2018	11-126	Ascending	VV	IW	40-41
21	23 June 2018	11-126	Ascending	VV	IW	40-41
22	5 July 2018	11-126	Ascending	VV	IW	40-41
23	29 July 2018	11-126	Ascending	VV	IW	40-41
24	10 August 2018	11-126	Ascending	VV	IW	40-41
25	22 August 2018	11-126	Ascending	VV	IW	40-41
26	3 September 2018	11-126	Ascending	VV	IW	40-41
27	15 September 2018	11-126	Ascending	VV	IW	40-41
28	27 September 2018	11-126	Ascending	VV	IW	40-41
29	9 October 2018	11-126	Ascending	VV	IW	40-41
30	21 October 2018	11-126	Ascending	VV	IW	40-41

3.3.2. TLS Subsidence Monitoring

The TLS was implemented three times, the first period was from 28 May 2018 to 1 June 2018, the second period was from 3 July 2018 to 6 July 2018, and the third period was from 11 September 2018 to 15 September 2018. Based on the actual terrain conditions, a suitable racking station location and scanning resolution will be selected. After the scanning station location were determined, the actual scanning was prioritized according to the distance from the workings, with the areas closer to the workings being significantly affected by mining and prioritized for scanning. As the terrain varies from one period to another, the scanning stations will be adjusted according to changes in the terrain. The station layout is shown in Figure 8a, and the point cloud data were classified, denoised, resampled, spliced and treated with other internal processing. To study the impact of mining on the deformation of the area close to the working face, the scanning stations and the location of the strike and tendency observation lines were combined to conduct comparative analysis of settlement changes along the strike and tendency observation lines respectively, and the relationship between the strike observation line K, tendency observation line L and the location of the point cloud model is shown in Figure 8b.

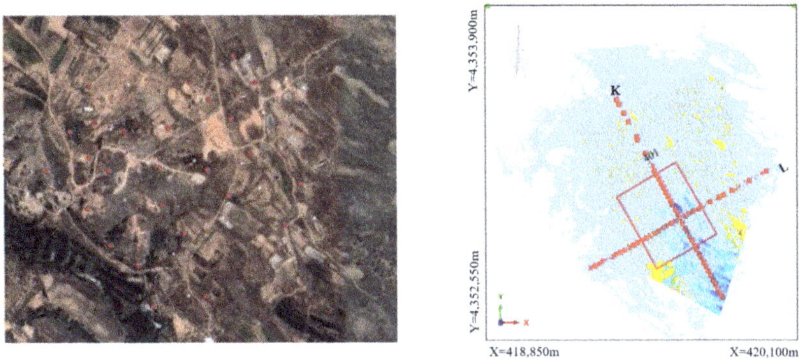

(a) (b)

Figure 8. TLS deployment and regional location map: (**a**) TLS distribution; (**b**) Selection area with the point cloud model, working surface relationship map.

3.3.3. UAV Subsidence Monitoring

The M600Pro professional DJI UAV with Tau2-640R HD infrared camera, Z3 visible camera and tilt 3S 3D camera were used to achieve infrared observation of surface cracks and modeling of surface 3D morphology. The UAV and field flight path are shown in Figure 9.

(a) (b)

Figure 9. Photos and flight path of UAV on site: (**a**) UAV photo; (**b**) Flight path.

The crack observation of visible and infrared imaging of the whole study area was carried out according to the designed flight path. Based on the observation using the aerial photography of the UAV, the Trimble® Geo 7x handheld was used for the uninterrupted measurement and positioning of the surface cracks in combination with the manual measurement data of the surface cracks.

3.3.4. Steel Ruler and GNSS Subsidence Monitoring

The steel ruler and GNSS were used to measure the surface cracks dynamically and record the changes in crack opening and drop, the crack width measurement method is shown in Figure 10. The cracks are divided into central dynamic cracks and boundary cracks for zoned measurements to track the dynamic change of "creation-expansion- closure" of the cracks in different areas of the ground. Daily records are kept of the opening and step-off of the same crack at the same location, while information on the mining advance rate and mine pressure is collected to analyze the relationship between the coupling of high-intensity mining overburden transport and cracks.

(**a**) (**b**)

Figure 10. Surface crack drop and width measurement:(**a**) Measurement of ground crack tension; (**b**) Ground crack step drop measurement.

4. Results

4.1. Surface Subsidence Monitoring from Single Technique

4.1.1. Surface Deformation Monitoring Results of Mining Area Based on InSAR Technology

The SAR images were analyzed to obtain deformation rates and cumulative deformation maps, as shown in Figure 11. In Figure 11a, there are 15 areas of ground subsidence due to coal mining in the area (black dashed ovals in the figure), with a maximum ground subsidence rate of −326 mm/yr. At the same time, there is surface uplift in five areas (red dashed ovals in the figure), with a maximum uplift rate of 125 mm/yr. In Figure 11b, the maximum cumulative subsidence detected from InSAR reached 365 mm and the maximum cumulative surface uplift reached 137 mm. As can also be seen in Figure 11, the large gradient of deformation in the middle of the mine area resulted in interferometric decoherence and gaps in some areas where deformation information was not available.

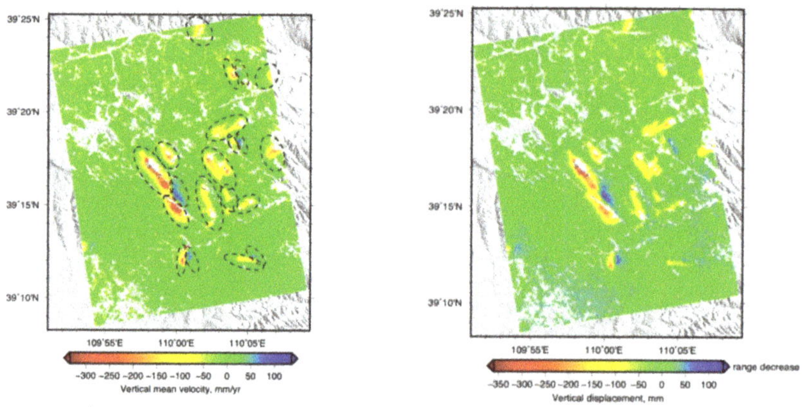

(**a**) (**b**)

Figure 11. InSAR monitoring results in Shendong Mining Area: (**a**) Rate of vertical surface deformation; (**b**) Accumulated surface deformation.

4.1.2. TLS Deformation Results

The point cloud data of Phase I and Phase III of the K and L lines were collated, and the amount of subsidence at each point on the two profile lines was calculated, as shown in Figure 12a. The point cloud data settlement comparison analysis is based on Riscan PRO software and color separation display is gained according to different settlement amounts, the results are shown in Figure 12b.

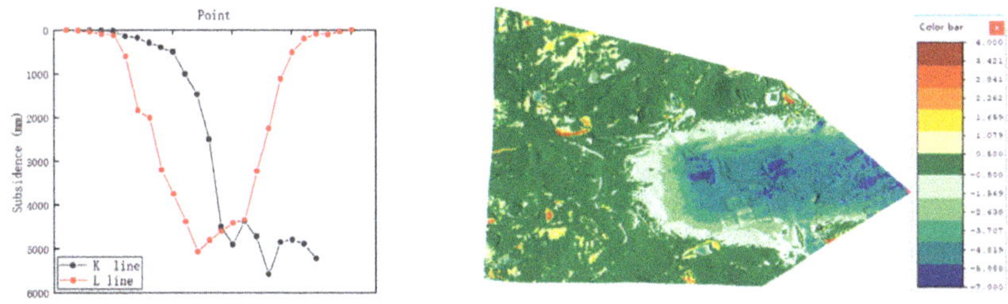

(**a**) (**b**)

Figure 12. Schematic diagram of surface subsidence at the mine area: (**a**) K line and L line; (**b**) Surface area subsidence.

Figure 12 shows that the subsidence in most areas of the region is between 2–4 m, with a small amount of subsidence greater than 5 m, and there are also a small number of elevation anomalies—the area with an elevation rise of 3–4 m in the figure. The reasons for this phenomenon are: (1) Because this area is at the edge of the scanning, error of triangulation network construction is caused by point cloud voids. As this area is at the edge of the scanning range, the point cloud data is less, and the modeling of the triangulation network will lift the surface of the feature and produce errors. (2) The error of feature change is caused by human activities, such as the demolition of houses and the accumulation of construction debris in the vicinity. Meanwhile, we can find that TLS monitors surface subsidence in small areas with high accuracy, and the monitoring area is relatively small compared with InSAR. If large surface deformation needs to be acquired, TLS technology will be limited.

4.1.3. UAV Deformation Results

(1) Surface cracks

Monitoring data was selected for 9 May and 2 June 2018, when the workings were advanced to 155 m and 310 m from the opencut. The specific surface development was plotted on the corresponding location of the working face to study the correlation law between mining on the working face and surface crack development [68], as shown in Table 6.

Table 6. Variation of the development regularity of surface cracks with the advanced distance of working face.

Advance the Position	155 m	310 m
Crack development law	The working face advances slowly, the old roof is damaged by mining to a relatively large extent, and the overlying rock and surface cracks are fully developed, resulting in the development of cracks ahead of the working face.	All three marginal cracks beyond the working face are developed beyond the mining boundary, and the new ground cracks continue to develop 38 m forward.
Schematic diagram	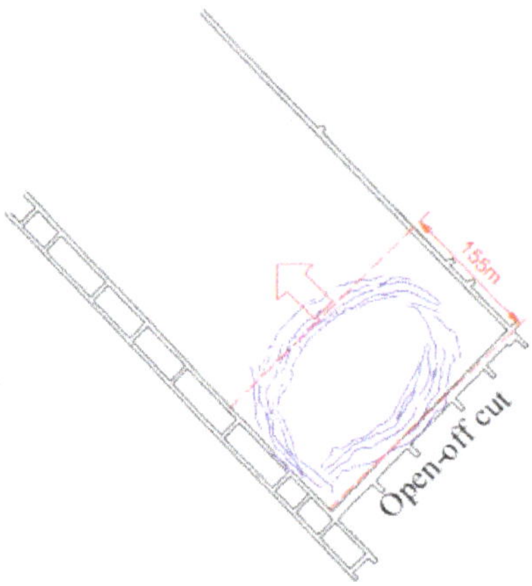	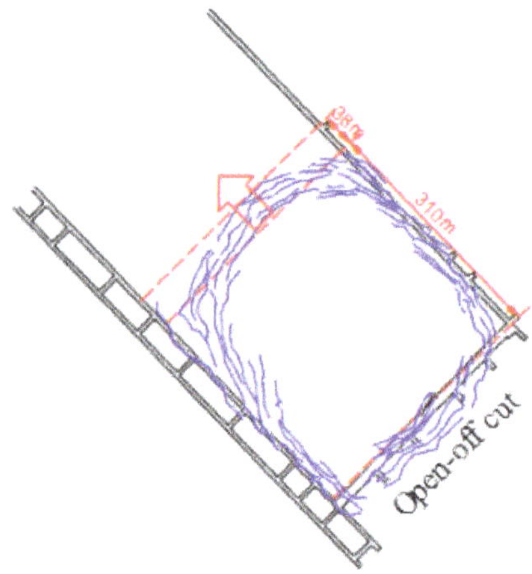

(2) Construction of 3d surface model

Based on the visible photos taken by the on-site UAV and the 3D model results produced by ContextCapture Center software, the survey area 0.05 m resolution DOM (Digital Orthophoto Map), 1m resolution DSM (Digital Surface Model) and 1m resolution DEM (Digital Elevation Model) are generated, as shown in Figure 13.

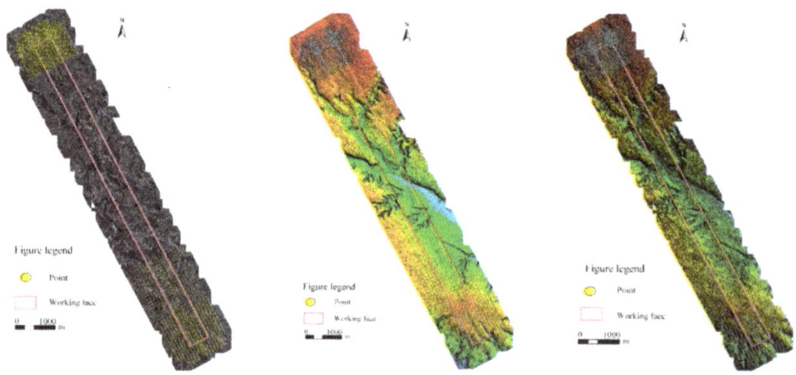

(a) (b) (c)

Figure 13. 3D model results: (**a**) DOM; (**b**) DEM; (**c**) DSM.

4.1.4. Deformation from Steel Ruler and GNSS

Based on the site situation of 401 working face of Shangwan mine in Shendong mining area, field data collection, collation and analysis of surface cracks caused by mining influence were carried out from 19 April 2018 to 3 August 2018, a total of 106 days. A total of 1878 (groups) cracks were collected. The crack development from 1 June 2018 to 7 June 2018 was selected to analyze the expansion pattern of surface cracks, as shown in Figure 14.

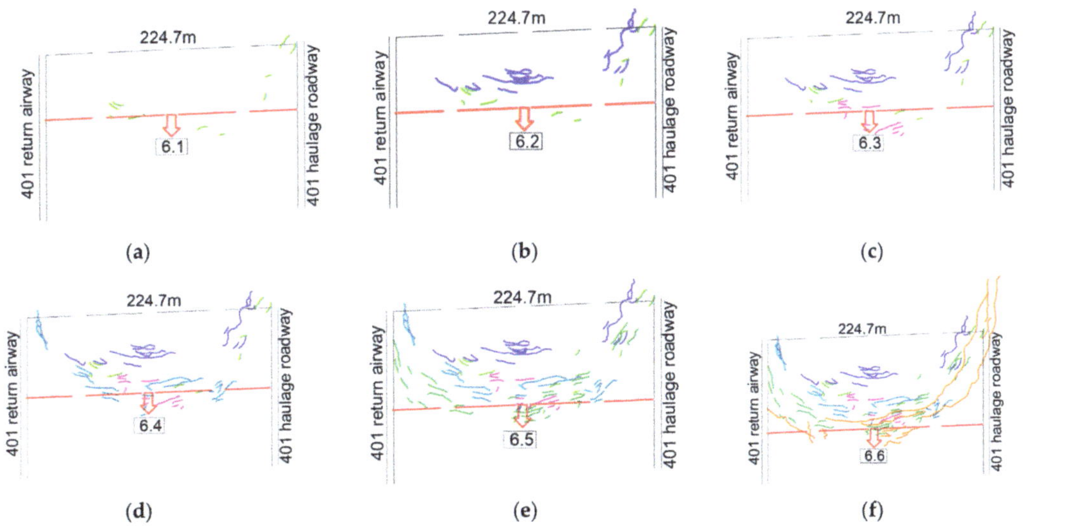

Figure 14. The extension process of surface cracks: (**a**): surface cracks directly above the working face overtake the development of the advancing position of the working face; (**b**) during the advancing process of the working face, surface cracks expand along the advancing direction of the working face on the one hand, and on the other hand gradually expand gradually from directly above the working face to both sides of the mining area until near the boundary of the mining area; (**c**) surface cracks develop suddenly directly in front of the working face, indicating that the development of cracks with suddenness; (**d**) surface cracks develop in the direction of the side of the mining area and the two roadways, but the cracks hardly develop forward; (**e**) surface cracks continue to develop at the boundary of the mining area; (**f**) the generation of cracks at the boundary of the mining area usually delays the advancing position of the working face, and large cracks are formed at the surface.

Figure 14 shows that with the advancement of the working face, the expansion pattern of surface cracks in the middle of the mining area is as follows: firstly, they protrude a distance in front of the working face and produce multiple tensile cracks. After that, with the advancement of the working face, the surface cracks develop in the direction of the mining area and the haulage roadway, then the surface deformation increases, and the original cracks will be connected to form large cracks. With the advancement of the working face, the expansion of surface cracks repeats the previous pattern, and the overall form of surface cracks in the mining area shows the arc-shaped distribution. We can find that the steel ruler and GNSS crack observation has significant advantages for the study of short-term and marginal fracture development patterns at the surface.

4.2. Deformation Results from the "Space-Sky-Ground" Collaborative Monitoring Framework

There are some limitations in relying on a single monitoring technique to obtain surface subsidence information, such as: (1) the limitation of InSAR loss of coherence cannot obtain the deformation information of the large deformation; (2) the difficulty of TLS moving stations to obtain the full range of surface subsidence basins; (3) the UAV equipped with thermal infrared camera cannot obtain the moving basin edge cracks, etc. Therefore, this section fused InSAR and TLS monitor data of surface subsidence basin. The feasibility of combining InSAR technology to identify large cracks of surface deformation is discussed by combining the cracks of easy access to InSAR data. The final shape of surface cracks near 1000 m of the location of opencut is obtained by combining the advantages of UAV, steel ruler and GNSS technology crack monitoring techniques.

4.2.1. Results of Continuous Surface Deformation

Based on the "space-sky-ground" monitoring framework, the input TLS and InSAR data are fused into a whole according to the correlation, forming an integrated dataset containing "space-sky-ground" multi-source measurement information; the coarse fused dataset is used as a benchmark, starting from the low-resolution and wide-range InSAR remote sensing image data. The coarse fused dataset is used as the benchmark, starting from the low-resolution and wide-range InSAR remote sensing data, and gradually correcting the accuracy with the higher resolution TLS data. The corrected data contains the characteristic information of each data and makes good use of the correlation between different data to achieve the effect of "space-sky-ground" integration and multisource data linkage, and finally obtains the surface subsidence cloud map of the whole mine area, as shown in Figure 15. The acquisition of the overall overview of the surface subsidence within the mine area provides basic data for the protection of the ecological restoration of the mine area.

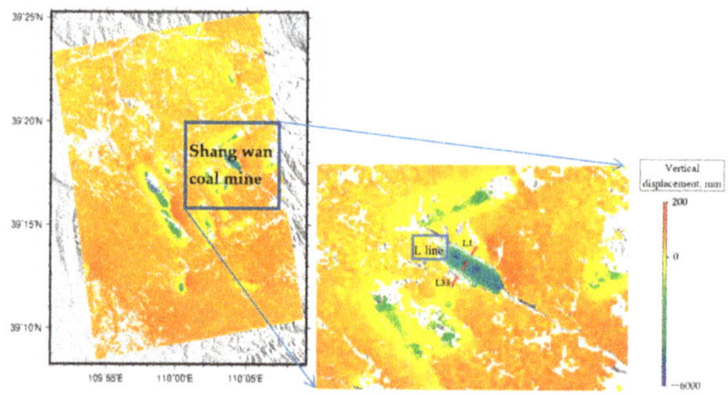

Figure 15. Surface subsidence map of working face based on InSAR and TLS data fusion.

Figure 15 shows that: the average subsidence rate in the mining area of 401 working face is about 300 mm/a, which is higher than the subsidence in other locations in the mine area, and it is known from the regional measurement results that the surface subsidence of 401 working face is 2–4 m, and the subsidence in a small amount of area is more than 5 m. To further analyze the data accuracy of the fusion monitoring results, the maximum subsidence of the strike line and the strike line are compared with the tendency and the control points obtained from the level measurement in the same period, respectively. The results are shown in Table 7, and the L line of the profile was taken to compare the fused subsidence data with the levelling data, and the graphs are plotted in Figure 16.

Table 7. Comparison of maximum subsidence values monitored by level data and fused data.

Maximum Subsidence Value	Point Cloud Data (InSAR and TLS)	Levelling Data
K line	5579 mm (K36)	5562 mm (K36)
L line	5067 mm (L22)	4942 mm (L22)

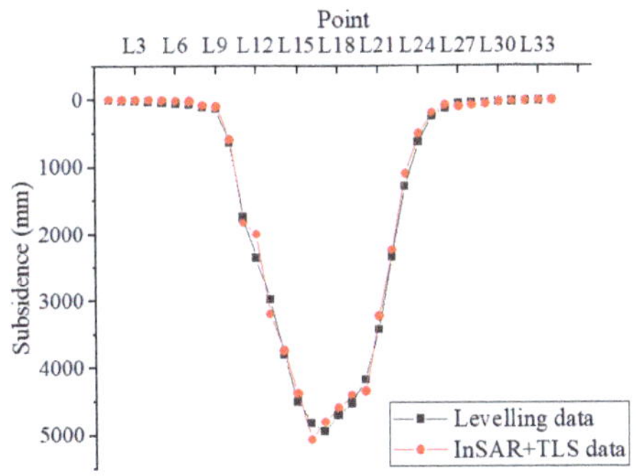

Figure 16. Comparison of levelling monitoring results and data fusion monitoring results.

Figure 16 and Table 7 compare the monitoring results of the two profiles of the fused data with the levelling data. The subsidence trends of the two fused profiles are consistent with the levelling subsidence trends, and the relative error of the difference between the maximum subsidence values is 2.5%. Therefore, it is feasible to obtain the surface subsidence basin in a large surface area through data fusion using two monitoring techniques, InSAR and TLS, in the framework of "space-sky-ground" cooperative monitoring.

4.2.2. Discontinuous Deformation Identification and Distribution

(1) Surface discontinuous deformation identification

To investigate the characteristic relationship between surface mining cracks and InSAR deformation rate at the working face, the deformation gradient and surface mining cracks are analyzed. Among them, the surface deformation gradient is elaborated from three time periods, namely, early recovery period, active period and downhole stoppage. The surface deformation gradient can be expressed as Equation (13).

$$i = \tan^{-1}\frac{\Delta W}{\Delta U} \tag{13}$$

In Equation (13), ΔW is the vertical settlement of radar pixel points, ΔU is the horizontal movement of the corresponding point, The value range of i is 0–180.

Because SBAS-InSAR obtained the results with 20 m × 20 m resolution after multi-view processing, the deformation results were considered only from the distribution location of cracks, and the characteristics such as width of cracks and depth of subsurface development were ignored. In addition, to clarify the correlation between surface cracks and InSAR analysis results, only the InSAR deformation monitoring results of surface cracks with a width of 50 mm or more are considered.

Figure 17 shows that the surface cracks were damaged due to highly intensive underground mining activities, the overall shape of the cracks is approximately "circular" in distribution, and the "diameter" of the circle is larger than the width of the working face, but smaller than the influence range of the mining area. In other words, the fracture range is generally larger than the width of the working face, but it occurs within the surface deformation range, which indicates that the surface deformation affects the distribution of cracks.

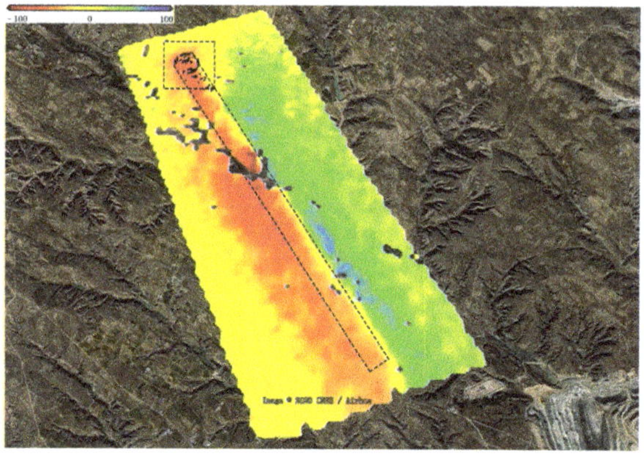

Figure 17. Surface deformation rate (mm/y) and cracks superimposed map.

To further analyze the influence relationship between surface deformation rate and cracks, the schematic surface deformation gradient diagrams of different mining stages were selected in conjunction with the working face advancement time (from March 2018 to September 2019): among them, the schematic distribution of surface deformation gradient at the early stage of recovery (10 August 2018), the active period of recovery (18 February 2019), and the stopping stage (10 September 2019) are shown in Figure 18a–c respectively. Figure 18 shows that the area with larger surface deformation gradient gradually expands with the advancement of the mining face, and the gradient development direction is more consistent with the mining direction. The shape of the surface deformation gradient as a whole is "long", and the surface cracks in the area of the open cut are mainly distributed in and around the location of the larger vertical deformation gradient and are ahead of the range of the large surface deformation gradient.

(**a**) (**b**) (**c**)

Figure 18. Gradient map of surface deformation: (**a**) Early period (10 August 2018): Cracks at the location of the opencut are distributed in an approximate circle, with a diameter greater than the width of the working face; (**b**) Active period (18 February 2019): the crack gradient distribution map of the surface was developing along the direction of workface advance as the workface continues to advance; (**c**) Downhole stoppage (10 September 2019): the surface deformation gradient map has a general "long strip" distribution.

In general, detailed information on cracks is measured using steel rulers, and GNSS measures information on the location of cracks. All information on cracks being recorded in detail requires a lot of resources, and the repair of large cracks deserves more attention. The use of gradients to identify cracks proposed in this section provides a new way of thinking about the identification of large surface cracks and provides reference for the next step of surface crack filling.

(2) Surface discontinuous deformation distribution

With the combination of the surface UAV thermal infrared crack measurement data and the measurement of key area with the assistance of GNSS and Steel ruler, the distribution of surface cracks in the working face was mapped and the width of the cracks was categorized by grade and indicated by different colors. The final mapping result is shown in Figure 19, which provides a reference for fully understanding the distribution of surface cracks with the mining situation of the working face.

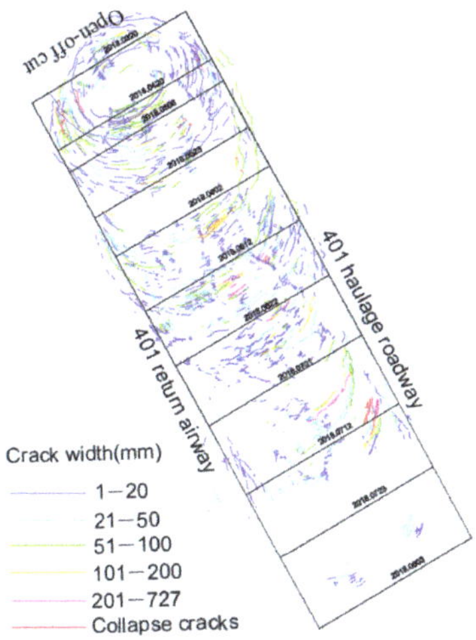

Figure 19. Distribution map of cooperative monitoring cracks at 401 working face of Shangwan Mine in Shendong Mining Area.

The mining of shallow buried thick coal seams in western mines (working face 401 in this paper) is characterized by rapid advancement and large surface deformation. The development of surface fractures is a dynamic process, and it is important to obtain the true shape of surface cracks distribution in a timely and accurate manner. There are shortcomings in the monitoring results of single crack monitoring technology: the traditional field survey method of ground cracks has a high accuracy but is slow, long and costly, and it is a great challenge to comprehensively understand the development laws of cracks with the advancement of the working surface in a short period of time by using the traditional ground crack survey method; the use of UAV equipped with thermal infrared and visible light cameras to monitor cracks is limited by the edge recognition algorithm, small cracks and shallow buried cracks are not identified at the edge of the ground. Therefore, by combining traditional ground crack survey methods and UAV fracture monitoring technology, the final distribution law of surface cracks can be accurately obtained by applying UAV technology to monitor fractures in the center of the subsidence basin and at the inflection point, and applying traditional ground fracture survey methods to cracks at the edge of the subsidence basin, in response to the characteristics of shallow buried thick coal seam mining in western mining areas.

5. Discussion

Currently, surface subsidence monitoring in mining areas is based on only one or two technical techniques, and it is difficult to obtain the surface area deformation law in the subsidence affected area. In this paper, based on the characteristics of many kinds of surface subsidence monitoring objects and various monitoring techniques, we have proposed the "space-sky-ground" collaborative monitoring framework for surface subsidence monitoring in mining areas. And we constructed an AHP-TOPSIS preference model of "space-sky-ground" integrated collaborative monitoring of surface deformation in mining area and carried out engineering applications. The following are the issues that deserve further discussion in this paper.

(1) Based on the characteristics of surface deformation in mining areas and existing surface deformation monitoring technologies, this paper established the "space-sky-ground" collaborative monitoring framework. This framework is an important guideline for the preference model of "space-sky-ground" collaborative monitoring and the integration of various technologies. The determination of the weighting of different programs in the monitoring program selection model needs to be combined with the "space-sky-ground" collaborative monitoring framework, so as to better utilize the advantages of each monitoring technology, for example, in Section 3.2 of this paper, the weighting process of the "spatial scale evolution law" is based on the "space-sky-ground" collaborative monitoring framework, and program 1 (GNSS and CORS, InSAR, TLS, UAV) is less weighted than the other schemes because of the lack of small-scale measurement techniques; the integration of multiple technologies needs to fully consider the time scale and spatial scale of each monitoring method in the "space-sky-ground" cooperative monitoring framework, so as to obtain the best data fusion results and comprehensively obtain the surface subsidence pattern.

The "space-sky-ground" collaborative monitoring framework includes space monitoring components, sky monitoring components and ground monitoring components, and the monitoring tools include existing monitoring technologies for surface movement and deformation. In addition, the monitoring framework is not limited to surface deformation monitoring in the mine area, but can be expanded to cover the entire mine area, which requires additional monitoring tools such as underground mineral pressure and underground overburden, as shown in Figure 20.

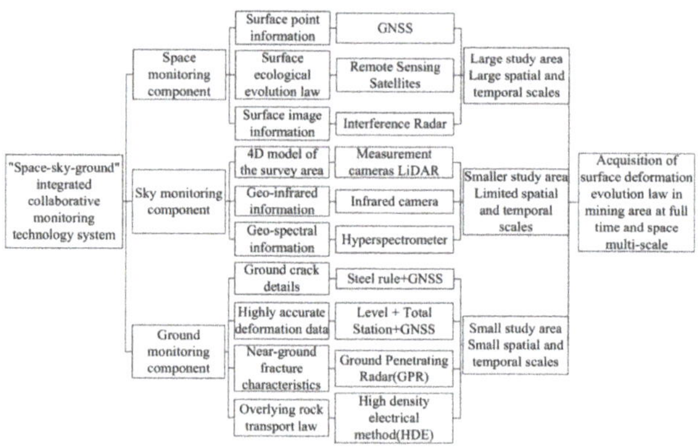

Figure 20. "Space-sky-ground" collaborative monitoring framework for in mining area.

(2) The AHP-TOPSIS model has many applications in other areas of program selection. In this paper, the AHP-TOPSIS model is chosen as the preferred model for the "space-sky-ground" collaborative monitoring framework. The purpose of this paper is to apply the AHP-TOPSIS model in the "space-sky-ground" collaborative monitoring framework to prefer a monitoring program. The preferred model directly determines the results of "space-sky-ground" collaborative monitoring framework and should be taken seriously. As the field of artificial intelligence continues to receive attention, artificial intelligence algorithms such as neural networks and support vector machines will significantly improve the rationality and accuracy of monitoring method selection.

(3) InSAR technology has the advantages of easy data acquisition and large monitoring period span, which can monitor long-period surface subsidence and be used to compare the difference of deformation patterns with the adjacent unmined areas,

but InSAR is limited by decoherence and cannot monitor continuous large surface deformation in a short period of time (Figure 21a). TLS monitoring has the advantage of acquiring surface subsidence information with high accuracy, and the fusion of the two surface subsidence monitoring results can accurately obtain the large-scale surface subsidence laws (Figure 21b). Combining the fused monitoring results of InSAR and TLS provides a new way of thinking for the later analysis of the surface subsidence laws over a large area of the working face. The results provide a foundation for global planning to protect the ecological environment of the mining area and provide data for overburden damage control and ecological restoration in the area, which cannot be achieved by using a single TLS or InSAR monitoring technique.

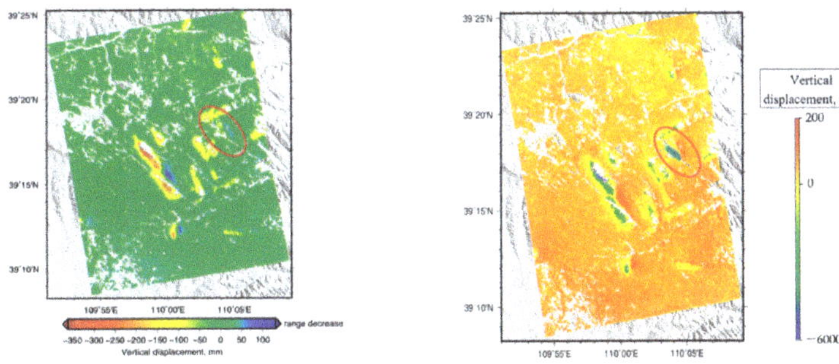

(a) (b)

Figure 21. Surface subsidence monitoring results at the mine area:(**a**) Surface subsidence monitoring from single technique (InSAR); (**b**) Surface subsidence monitoring from "space-sky-ground" integrated collaborative monitoring framework.

(4) Combining InSAR monitoring technology with the advantages of easy data acquisition and large observation time range, InSAR is used to calculate the surface deformation gradient, analyzing the location of surface cracks generated in the mine area. This technique provides a new idea for other working faces to use InSAR technology to identify the location of surface cracks in subsidence areas and provides a reference for the filling of surface cracks and the restoration of surface ecology. Figure 22 shows the process of manual treatment of the cracks identified by InSAR.

(a) (b)

Figure 22. Crack manual management process: (**a**) Filled cracks; (**b**) Ecological restoration.

(5) By combining traditional ground crack survey methods and UAV crack monitoring technology, the final distribution law of surface cracks can be accurately obtained by applying UAV technology to monitor cracks in the center of the subsidence basin and at the inflection point and applying traditional ground fracture survey methods to cracks at the edge of the subsidence basin, in response to the characteristics of shallow buried thick coal seam mining in western mining areas. The "space-sky-ground"

collaborative monitoring framework gives full play to the advantages of the emerging technology, making it possible to comprehensively understand the development law of cracks and laying the foundation for the analysis and management of cracks in large areas of coal mine subsidence areas.

6. Conclusions

This paper establishes the "space-sky-ground" collaborative monitoring framework for surface subsidence in mining area and sets up a surface monitoring method preference model based on AHP-TOPSIS. The engineering application results show that based on the "space-sky-ground" collaborative monitoring framework proposed in this paper, the preferred monitoring program suitable for this mine area can be selected by using the surface monitoring program preference model. The practical application was carried out at 401 working face of Shendong mine area, and the distribution pattern of discontinuous deformation and large-scale surface deformation results in this mine area were obtained, the discontinuous large deformation was identified. The obtained results provide data to support the crack management in this mine area. The research work of this paper can provide reference for the acquisition of basic data of surface ecological restoration in similar mines. The results of this paper are only an experiment of the "space-sky-ground" collaborative monitoring framework, the framework can be appropriately adjusted according to the needs of the actual application and analysis. The "space-sky-ground" collaborative monitoring framework established increases the possibility of a comprehensive understanding of surface subsidence laws in mining areas. With the accumulation of field experience and the continuous efforts of researchers, the performance of "space-sky-ground" collaborative monitoring framework will be better and this method will be used more frequently in the monitoring of surface subsidence in mining area. The significant innovations and conclusions of this paper are as follows.

(1) Based on the characteristics of surface deformation in mining areas and existing monitoring technologies, this paper established the "space-sky-ground" collaboration monitoring framework in mining areas. The monitoring framework consists of three parts: space monitoring component, sky monitoring component and ground monitoring component, moreover, it basically includes the existing surface monitoring technologies, and new monitoring technologies can be added to the framework in the future according to their own characteristics. The framework of this paper is an important guide to the analysis of the influencing factors of the "space-sky-ground" collaborative monitoring preference program, the establishment of the preference model and the implementation steps of the preference model. At the same time, the proposed framework presents new thinking to recognize the multi-temporal and multi-scale surface subsidence patterns in mining areas.

(2) Determine the influencing factors affecting the surface preference monitoring program, a total of 3 primary indicators and 16 secondary indicators, the primary indicators include: natural factors, target constraints and site needs. The importance of the 16 secondary indicators was analyzed using AHP. Meanwhile, combining the advantages of AHP-TOPSIS preference model, the AHP-TOPSIS preference model was constructed based the "space-sky-ground" collaborative monitoring framework, and four steps of monitoring scheme preference based on AHP-TOPSIS model were proposed. The practical application was carried out in the surface monitoring program preference of 401 working face in Shendong mine.

(3) Based on the preferred monitoring program under the "space-sky-ground" collaborative monitoring framework at the 401 working face of the Shendong mine, the information on the distribution of discontinuous deformation and the results of continuous deformation of large surface area in the whole area of the mine were obtained, and the discontinuous large deformation was identified. The results obtained provide a data support for the crack management in the mine area, and provide an example for the further extension of the "space-sky-ground" collaborative monitoring results

proposed in this paper. The "space-sky-ground" collaborative monitoring framework established in this paper helps scholars to understand the surface subsidence pattern in mining areas at multiple scales and provide a technical and data support for surface ecological restoration.

Author Contributions: Conceptualization, Y.Y. and H.D.; methodology, Y.Y.; software, W.T.; validation, Y.Y.; formal analysis L.D.; investigation, J.G.; resources, Y.Y. and W.T.; writing—original draft preparation, M.L.; writing—review and editing, L.D. and H.D.; supervision, H.D.; funding acquisition, Y.Y. and H.D. All authors have read and agreed to the published version of the manuscript.

Funding: This research was funded by the Open Fund of State Key Laboratory of Water Resource Protection and Utilization in Coal Mining (Grants nos. WPUKFJJ2019-17 and GJNY-20-113-20), the Fundamental Research Funds for the Central Universities (2021YQDC09), National Natural Science Foundation of China (grant numbers 51574242 and 51404272).

Institutional Review Board Statement: Informed consent was obtained from all subjects involved in the study.

Informed Consent Statement: Informed consent was obtained from all subjects involved in the study.

Data Availability Statement: The data used to support the findings of this study are available from the corresponding author upon request.

Acknowledgments: The authors thank ASI for providing the COSMO-SkyMed data and ESA for providing the Sentinel-1 data.

Conflicts of Interest: The authors declare no conflict of interest.

References

1. Chen, Y.; Tao, Q.; Liu, G.; Wang, L.; Wang, F.; Wang, K. Detailed mining subsidence monitoring combined with InSAR and probability integral method. *Chin. J. Geophys.* **2021**, *64*, 3554–3566. [CrossRef]
2. Dai, H.; Guo, J.; Yi, S.; Wang, G.; Liu, A.; Kong, B.; Zou, B. The mechanism of strata and surface movements induced by extra-thick steeply inclined coal seam applied horizontal slice mining. *J. China Coal Soc.* **2013**, *38*, 1109–1115. [CrossRef]
3. Dai, H. Mining subsidence variables and their time-space relationship description. *J. China Coal Soc.* **2018**, *43*, 450–459. [CrossRef]
4. Alam, M.S.; Kumar, D.; Chatterjee, R.S. Improving the capability of integrated DInSAR and PSI approach for better detection, monitoring, and analysis of land surface deformation in underground mining environment. *Geocarto Int.* **2021**, *2021*, 1864028. [CrossRef]
5. Minami, M.; Morikawa, H.; Aoyama, T.; Mizumachi, M. An adaptive multipath estimation/elimination technique for GPS signals reception. *Electron. Commun. Jpn.* **2003**, *86*, 74–82. [CrossRef]
6. Patel, P.R.; Kulkarni, M.N. Preliminary Results of GPS Studies for Monitoring Land Subsidence Over the Shallow Gas Reservoir in India. *Surv. Rev.* **2013**, *40*, 356–365. [CrossRef]
7. Zhao, H.; Ma, F.; Zhang, Y.; Gue, J. Monitoring and Assessment of Ming Subsidence in a Metal Mine in China. *Environ. Eng. Manag. J.* **2014**, *13*, 3015–3024. [CrossRef]
8. Hung, W.; Hwang, C.; Chang, C.; Yen, J.; Liu, C.; Yang, W. Monitoring severe aquifer-system compaction and land subsidence in Taiwan using multiple sensors: Yunlin, the southern Choushui River Alluvial Fan. *Environ. Earth Sci.* **2009**, *59*, 1535–1548. [CrossRef]
9. Kim, D.; Lee, H.; Okeowo, M.A.; Basnayake, S.; Jayasinghe, S. Cost-effective monitoring of land subsidence in developing countries using semipermanent GPS stations: A test study over Houston, Texas. *J. Appl. Remote Sens.* **2017**, *11*, 026033. [CrossRef]
10. Ao, J.; Wu, K.; Wang, Y.; Li, L. Subsidence Monitoring Using Lidar and Morton Code Indexing. *J. Surv. Eng.* **2016**, *142*, 06015002. [CrossRef]
11. Chen, S.; Wang, C.; Dai, H.; Zhang, H.; Pan, F.; Xi, X.; Yan, Y.; Wang, P.; Yang, X.; Zhu, X.; et al. Power Pylon Reconstruction Based on Abstract Template Structures Using Airborne LiDAR Data. *Remote Sens.* **2019**, *11*, 1579. [CrossRef]
12. Jones, L.; Hobbs, P. The Application of Terrestrial LiDAR for Geohazard Mapping, Monitoring and Modelling in the British Geological Survey. *Remote Sens.* **2021**, *13*, 395. [CrossRef]
13. Tiwari, A.; Narayan, A.B.; Dwivedi, R.; Dikshit, O.; Nagarajan, B. Monitoring of landslide activity at the Sirobagarh landslide, Uttarakhand, India, using LiDAR, SAR interferometry and geodetic surveys. *Geocarto Int.* **2018**, *35*, 535–558. [CrossRef]
14. Dawei, Z.; Lizhuang, Q.; Demin, Z.; Baohui, Z.; Lianglin, G. Unmanned Aerial Vehicle (UAV) Photogrammetry Technology for Dynamic Mining Subsidence Monitoring and Parameter Inversion: A Case Study in China. *IEEE Access* **2020**, *8*, 16372–16386. [CrossRef]
15. Ignjatović Stupar, D.; Rošer, J.; Vulić, M. Investigation of Unmanned Aerial Vehicles-Based Photogrammetry for Large Mine Subsidence Monitoring. *Minerals* **2020**, *10*, 196. [CrossRef]

16. Mateos, R.M.; Azañón, J.M.; Roldán, F.J.; Notti, D.; Pérez-Peña, V.; Galve, J.P.; Pérez-García, J.L.; Colomo, C.M.; Gómez-López, J.M.; Montserrat, O.; et al. The combined use of PSInSAR and UAV photogrammetry techniques for the analysis of the kinematics of a coastal landslide affecting an urban area (SE Spain). *Landslides* **2016**, *14*, 743–754. [CrossRef]
17. Park, S.; Choi, Y. Applications of Unmanned Aerial Vehicles in Mining from Exploration to Reclamation: A Review. *Minerals* **2020**, *10*, 663. [CrossRef]
18. Rauhala, A.; Tuomela, A.; Davids, C.; Rossi, P. UAV Remote Sensing Surveillance of a Mine Tailings Impoundment in Sub-Arctic Conditions. *Remote Sens.* **2017**, *9*, 1318. [CrossRef]
19. Vlastelica, G.; Matoš, B.; Pikelj, K.; Lužar-Oberiter, B.; Kordić, B. Integration of Terrestrial Laser Scanning and UAS Photogrammetry in Geological Studies: Examples from Croatia. *Period. Polytech. Civ. Eng.* **2019**, *63*, 989–1003. [CrossRef]
20. Yao, H.; Qin, R.; Chen, X. Unmanned Aerial Vehicle for Remote Sensing Applications—A Review. *Remote Sens.* **2019**, *11*, 1443. [CrossRef]
21. Chen, J.; Zhou, Y.; Chen, G.; Hao, M. Decades of Ground Deformation in the Weihe Graben, Shaanxi Province, China, in Response to Various Land Processes, Observed by Radar Interferometry and Levelling. *Remote Sens.* **2021**, *13*, 2374. [CrossRef]
22. Chen, Y.; Tong, Y.; Tan, K. Coal mining deformation monitoring using SBAS-InSAR and offset tracking: A case study of Yu County China. *IEEE J. Sel. Top. Appl. Earth Obs. Remote Sens.* **2020**, *13*, 10. [CrossRef]
23. Diao, X.; Wu, K.; Zhou, D.; Li, L. Integrating the probability integral method for subsidence prediction and differential synthetic aperture radar interferometry for monitoring mining subsidence in Fengfeng, China. *J. Appl. Remote Sens.* **2016**, *10*, 016028. [CrossRef]
24. Furst, S.L.; Doucet, S.; Vernant, P.; Champollion, C.; Carme, J.L. Monitoring surface deformation of deep salt mining in Vauvert (France), combining InSAR and leveling data for multi-source inversion. *Solid Earth* **2021**, *12*, 15–34. [CrossRef]
25. He, Q.; Zhang, Y.; Wu, H.; Luo, Y. Mining Subsidence Monitoring with Modified Time-Series SAR Interferometry Method Based on the Multi-Level Processing Strategy. *IEEE Access* **2021**, *9*, 106039–106048. [CrossRef]
26. Liu, X.; Huang, J. An Improved Multi-Platform Stacked D-InSAR Method for Monitoring the Three-Dimensional Deformation of the Mining Area. *IEEE Access* **2021**, *9*, 66878–66890. [CrossRef]
27. Ng, A.H.-M.; Ge, L.; Li, X.-j. Assessments of land subsidence in the Gippsland Basin of Australia using ALOS PALSAR data. *Remote Sens. Environ.* **2015**, *159*, 15. [CrossRef]
28. Yang, Z.; Li, Z.; Zhu, J.; Yi, H.; Hu, J.; Feng, G. Deriving Dynamic Subsidence of Coal Mining Areas Using InSAR and Logistic Model. *Remote Sens.* **2017**, *9*, 125. [CrossRef]
29. Dwornik, M.; Bała, J.; Franczyk, A. Application of a New Semi-Automatic Algorithm for the Detection of Subsidence Areas in SAR Images on the Example of the Upper Silesian Coal Basin. *Energies* **2021**, *14*, 3051. [CrossRef]
30. Fan, H.D.; Cheng, D.; Deng, K.Z.; Chen, B.Q.; Zhu, C.G. Subsidence monitoring using D-InSAR and probability integral prediction modelling in deep mining areas. *Surv. Rev.* **2015**, *47*, 438–445. [CrossRef]
31. Jiang, C.; Wang, L.; Yu, X. Retrieving 3D Large Gradient Deformation Induced to Mining Subsidence Based on Fusion of Boltzmann Prediction Model and Single-Track InSAR Earth Observation Technology. *IEEE Access* **2021**, *9*, 87156–87172. [CrossRef]
32. Li, T.; Zhang, H.; Fan, H.; Zheng, C.; Liu, J. Position Inversion of Goafs in Deep Coal Seams Based on DS-InSAR Data and the Probability Integral Methods. *Remote Sens.* **2021**, *13*, 2898. [CrossRef]
33. Balt, K.; Goosen, R.L. MSAHP: An approach to mining method selection. *J. South. Afr. Inst. Min. Metall.* **2020**, *120*, 451–460. [CrossRef]
34. Reich, J.; Kinra, A.; Kotzab, H.; Brusset, X. Strategic global supply chain network design how decision analysis combining MILP and AHP on a Pareto front can improve decision making. *Int. J. Prod. Res.* **2021**, *59*, 1557–1572. [CrossRef]
35. Satty, T.L. How to make a decision: The analytic hierarchy process. *Interfaces* **1994**, *24*, 19–43. [CrossRef]
36. Chen, C. Extensions of the TOPSIS for group decision-making under fuzzy environment. *Fuzzy Set. Syst.* **2000**, *114*, 1–9. [CrossRef]
37. Gul, M.; Celik, E.; Aydin, N.; Taskin Gumus, A.; Guneri, A.F. A state of the art literature review of VIKOR and its fuzzy extensions on applications. *Appl. Soft Comput.* **2016**, *46*, 60–89. [CrossRef]
38. Onut, S.; Soner, S. Transshipment site selection using the AHP and TOPSIS approaches under fuzzy environment. *Waste Manag.* **2007**, *28*, 1552–1559. [CrossRef]
39. Pazand, K.; Hezarkhani, A. Porphyry Cu potential area selection using the combine AHP-TOPSIS methods: A case study in Siahrud area (NW, Iran). *Earth Sci. Inform.* **2015**, *8*, 207–220. [CrossRef]
40. Hanine, M.; Boutkhoum, O.; Tikniouine, A.; Agouti, T. An Application of OLAP/GIS-Fuzzy AHP-TOPSIS Methodology for Decision Making: Location Selection for Landfill of Industrial Wastes as a Case Study. *KSCE J. Civ. Eng.* **2016**, *21*, 2074–2084. [CrossRef]
41. Mishra, P.K.; Nadda, R.; Kumar, R.; Rana, A.; Sethi, M.; Ekileski, A. Optimization of multiple arcs protrusion obstacle parameters using AHP-TOPSIS approach in an impingement jet solar air passage. *Heat Mass Transf.* **2018**, *54*, 3797–3808. [CrossRef]
42. Yu, J.; Wang, L.; Gong, X. Study on the Status Evaluation of Urban Road Intersections Traffic Congestion Base on AHP-TOPSIS Modal. *Procedia Soc. Behav. Sci.* **2013**, *96*, 609–616. [CrossRef]
43. Zhuang, Z.-Y.; Lin, C.-C.; Chen, C.-Y.; Su, C.-R. Rank-Based Comparative Research Flow Benchmarking the Effectiveness of AHP–GTMA on Aiding Decisions of Shredder Selection by Reference to AHP–TOPSIS. *Appl. Sci.* **2018**, *8*, 1974. [CrossRef]
44. Brady, B.H.G.; Brown, E.T. *Rock Mechanics for Underground Mining*; George Allen &Unwin: London, UK, 1985.

45. Yin, D.; Chen, S.; Li, B.; Guo, W. Bed separation backfill to reduce surface cracking due to mining under thick and hard conglomerate: A case study. *R. Soc. Open Sci.* **2019**, *6*, 190880. [CrossRef] [PubMed]
46. Zuo, J.; Yu, M.; Li, C.; Sun, Y.; Hu, S.; Li, Z. Analysis of Surface Cracking and Fracture Behavior of a Single Thick Main Roof Based on Similar Model Experiments in Western Coal Mine, China. *Nat. Resour. Res.* **2021**, *30*, 657–680. [CrossRef]
47. Wang, G.; Wu, Q.; Li, P.; Cui, X.; Gong, Y.; Zhang, J.; Tang, W. Mining Subsidence Prediction Parameter Inversion by Combining GNSS and DInSAR Deformation Measurements. *IEEE Access* **2021**, *9*, 89043–89054. [CrossRef]
48. Lian, X.; Li, Z.; Yuan, H.; Hu, H.; Cai, Y.; Liu, X. Determination of the Stability of High-Steep Slopes by Global Navigation Satellite System (GNSS) Real-Time Monitoring in Long Wall Mining. *Appl. Sci.* **2020**, *10*, 1952. [CrossRef]
49. Tao, T.; Liu, J.; Qu, X.; Gao, F. Real-time monitoring rapid ground subsidence using GNSS and Vondrak filter. *Acta Geophys.* **2018**, *67*, 133–140. [CrossRef]
50. Yao, J.; Yao, X.; Wu, Z.; Liu, X.; Diraco, G. Research on Surface Deformation of Ordos Coal Mining Area by Integrating Multitemporal D-InSAR and Offset Tracking Technology. *J. Sens.* **2021**, *2021*, 660922. [CrossRef]
51. Pawluszek-Filipiak, K.; Borkowski, A. Integration of DInSAR and SBAS Techniques to Determine Mining-Related Deformations Using Sentinel-1 Data: The Case Study of Rydułtowy Mine in Poland. *Remote Sens.* **2020**, *12*, 242. [CrossRef]
52. Anders, K.; Marx, S.; Boike, J.; Herfort, B.; Wilcox, E.J.; Langer, M.; Marsh, P.; Höfle, B. Multitemporal terrestrial laser scanning point clouds for thaw subsidence observation at Arctic permafrost monitoring sites. *Earth Surf. Processes Landf.* **2020**, *45*, 1589–1600. [CrossRef]
53. Barbarella, M.; Fiani, M.; Lugli, A. Landslide monitoring using multitemporal terrestrial laser scanning for ground displacement analysis. *Geomat. Nat. Hazards Risk* **2013**, *6*, 398–418. [CrossRef]
54. Lian, X.; Hu, H. Terrestrial laser scanning monitoring and spatial analysis of ground disaster in Gaoyang coal mine in Shanxi, China: A technical note. *Environ. Earth Sci.* **2017**, *76*, 287. [CrossRef]
55. Matwij, W.; Gruszczyński, W.; Puniach, E.; Ćwiąkała, P. Determination of underground mining-induced displacement field using multi-temporal TLS point cloud registration. *Measurement* **2021**, *180*, 109482. [CrossRef]
56. Puniach, E.; Gruszczyński, W.; Ćwiąkała, P.; Matwij, W. Application of UAV-based orthomosaics for determination of horizontal displacement caused by underground mining. *ISPRS J. Photogramm. Remote Sens.* **2021**, *174*, 282–303. [CrossRef]
57. Zhou, T.; Hasheminasab, S.M.; Habib, A. Tightly-coupled camera/LiDAR integration for point cloud generation from GNSS/INS-assisted UAV mapping systems. *ISPRS J. Photogramm. Remote Sens.* **2021**, *180*, 336–356. [CrossRef]
58. Jing-xiang, G.; Hong, H. Advanced GNSS technology of mining deformation monitoring. *Procedia Earth Planet. Sci.* **2009**, *1*, 1081–1088. [CrossRef]
59. Dai, H.; Luo, J.; Guo, J.; Yan, Y.; Zhang, W.; Zhu, Y. In site surveying and study on development laws of surface cracks by high—intensity mining in Shangwan Mine. *Coal Sci. Technol.* **2020**, *48*, 124–129. [CrossRef]
60. Aminbakhsh, S.; Gunduz, M.; Sonmez, R. Safety risk assessment using analytic hierarchy process (AHP) during planning and budgeting of construction projects. *J. Saf. Res.* **2013**, *46*, 99–105. [CrossRef]
61. Du, Y.; Chen, W.; Cui, K.; Zhang, K. Study on Damage Assessment of Earthen Sites of the Ming Great Wall in Qinghai Province Based on Fuzzy-AHP and AHP-TOPSIS. *Int. J. Archit. Herit.* **2019**, *14*, 903–916. [CrossRef]
62. Ercanoglu, M.; Kasmer, O.; Temiz, N. Adaptation and comparison of expert opinion to analytical hierarchy process for landslide susceptibility mapping. *Bull. Eng. Geol. Environ.* **2008**, *67*, 565–578. [CrossRef]
63. Tuzkaya, G.; Onut, S.; Tuzkaya, U.R.; Gulsun, B. An analytic network process approach for locating undesirable facilities: An example from Istanbul, Turkey. *J. Environ. Manag.* **2008**, *88*, 970–983. [CrossRef] [PubMed]
64. Vahidnia, M.H.; Alesheikh, A.A.; Alimohammadi, A. Hospital site selection using fuzzy AHP and its derivatives. *J. Environ. Manag.* **2009**, *90*, 3048–3056. [CrossRef]
65. Chen, Y.; Yu, S.; Tao, Q.; Liu, G.; Wang, L.; Wang, F. Accuracy Verification and Correction of D-InSAR and SBAS-InSAR in Monitoring Mining Surface Subsidence. *Remote Sens.* **2021**, *13*, 4365. [CrossRef]
66. Gu, Y.; Zhou, D.; Zhang, D.; Wu, K.; Zhou, B. Study on subsidence monitoring technology using terrestrial 3D laser scanning without a target in a mining area: An example of Wangjiata coal mine, China. *Bull. Eng. Geol. Environ.* **2020**, *79*, 3575–3583. [CrossRef]
67. Ishwar, S.G.; Kumar, D. Application of DInSAR in mine surface subsidence monitoring and prediction. *Curr. Sci.* **2017**, *112*, 46–51. [CrossRef]
68. Zhang, C.; Zhao, Y.; He, X.; Guo, J.; Yan, Y. Space-sky-surface integrated monitoring system for overburden migration regularity in shallow-buried high-intensity mining. *Bull. Eng. Geol. Environ.* **2020**, *80*, 1403–1417. [CrossRef]

remote sensing

Article

The Current Crustal Vertical Deformation Features of the Sichuan–Yunnan Region Constrained by Fusing the Leveling Data with the GNSS Data

Yong Zhang [1,2], Caijun Xu [1,*], Zhijiang Zheng [3], Hongbao Liang [3] and Shuang Zhu [3]

1 School of Geodesy and Geomatics, Wuhan University, Wuhan 430079, China; zhangy@cea.gov.cn
2 China Earthquake Administration, Beijing 100036, China
3 The First Monitoring and Application Center, China Earthquake Administration, Tianjin 300180, China; jorry@fmac.ac.cn (Z.Z.); hongbao_cea@fmac.ac.cn (H.L.); shzhu1026@fmac.ac.cn (S.Z.)
* Correspondence: cjxu@sgg.whu.edu.cn; Tel.: +86-27-68778805

Abstract: This study uses the least squares collocation method to fuse the leveling vertical deformation velocity in the Sichuan–Yunnan region with the GNSS observations of this region from 320 stations in the China Crustal Movement Observation Network (CMONOC) and the China Continental Tectonic Environment Monitoring Network (CMTEMN) from 1999 to 2017. Such fusion is to improve the accuracy of the vertical deformation rates in large spatial scales. The fused vertical deformation results show that: (1) the fused deformation field has a uniform spatial distribution, and shows detailed change characteristics of key regions; (2) the current vertical crustal motion in this region is featured by the contemporaneous occurrence of crustal compression, shortening and uplift and basin extensional subsidence; (3) most areas in this region experience uplifts, as the lateral push of the Qinghai–Tibet Plateau was blocked by the Sichuan Basin. The areas on the northwest side of the Longmenshan fault and the Lijiang-Xiaojinhe fault are dominated by uplifts, with the velocity of 1.5 mm/a–5.5 mm/a, and the region on the southeast side has slight uplifts, with the velocity of 1.0 mm/a–1.5 mm/a; (4) many areas have high gradient vertical deformation, especially the region close to the Wenshan fault and on the two sides of the Yarlung Zangbo fault that has the value of 3.0–4.0 $\times$ 10^{-8}/a, deserving further attention to be paid to the long-term earthquake hazards.

Keywords: GNSS; leveling; Sichuan–Yunnan region; vertical deformation

Citation: Zhang, Y.; Xu, C.; Zheng, Z.; Liang, H.; Zhu, S. The Current Crustal Vertical Deformation Features of the Sichuan–Yunnan Region Constrained by Fusing the Leveling Data with the GNSS Data. *Remote Sens.* **2022**, *14*, 1139. https://doi.org/10.3390/rs14051139

Academic Editor: Alex Hay-Man Ng

Received: 23 January 2022
Accepted: 23 February 2022
Published: 25 February 2022

1. Introduction

The Sichuan–Yunnan region (SYR) is located on the southeastern margin of the Qinghai–Tibet Plateau (QTP) in southwestern China, and is the edge of the area bearing the interaction between the Eurasian plate and the Indian plate. The northward subduction of the Indian plate underneath the Eurasian continent resulted in the northeastward extrusion of the QTP and pushed the plateau materials flow to the east, which, however, was blocked by the hard South China block. As the main channel of the lateral extrusion in the middle QTP, the SYR experienced particularly strong crustal motions, so this region has great topographic fluctuation and frequent seismic activity. There are many large-scale faults with great slip rates [1–7]. The material extrusion from the middle QTP together with the soft lower crust cause the strong compression of the Longmenshan area in the southeast of the Bayankala block, forming an active compression fault system consisting of the Longmenshan fault and the Minjiang fault, etc. Inside the SYR, the faults are characterized by the extrusion caused by strike-slip. From these faults, a high-speed left-lateral strike-slip fault system has developed, consisting of the Ganzi–Yushu fault, the Xianshuihe fault, the Anning River fault, and the Zemuhe fault, Daliangshan fault, Xiaojiang fault, Litang fault, Lijiang-Xiaojinhe fault, and East Kunlun fault (eastern segment). A corresponding right-lateral strike-slip fault system has also developed, consisting of the Zhongdian fault, the

Red River fault, the Qujiang fault zone, the Nujiang fault zone, and the Jinsha River fault. The left-lateral strike-slip of the SYR together with the right-lateral strike-slip of the other blocks contribute to the conjugate strike-slip of the southern and western Yunnan areas. The faults experiencing left-lateral strike-slip are the Wanding fault, the Nanting River fault, the Daluo fault, the Jinghong fault, the Dien Bien Phu fault, and that experiencing the right-lateral strike-slip is the Longling-Lancang fault. Since the last century, 122 earthquakes of magnitude 6.0–6.9 and 23 earthquakes of magnitude 7.0–7.9 have occurred in the region, and the largest recorded earthquakes in the SYR are the 1833 Songming M 8.0 earthquake and the 2008 Wenchuan Mw8.0 earthquake.

Accommodating the strongest crustal deformation and seismic activity in the QTP, the SYR has been a focus for studying the present-day crustal deformation model and tectonic evolution of the QTP. GPS observations show a large-scale southeastward lateral movement on the southeastern margin of the QTP [8,9]. The "rigid block extrusion mode" [10,11] and the "lower crustal flow mode" [12–15] were developed to constrain the internal deformation and lateral extrusion of the QTP, respectively. The "lower crustal flow model" believes that the middle and lower crusts of the QTP are weak, where the crust thickening and shortening on the eastern margin of the plateau are most likely to occur [14,16,17]. This model has been widely accepted as it can well explain the crustal deformation characteristics observed on the surface. Furthermore, seismologic observations [18–24] and magnetotelluric studies [25–29] have reported low-velocity and high-conductivity/low-resistivity layers in the middle and lower crusts, which confirm the existence of the middle and lower crust flow.

Although the "lower crustal flow model" can explain the deformation characteristics of the eastern margin of the QTP well, whether the lower crustal flow in the eastern QTP has entered the SYR and how it distributes are still debated in the geosciences field. Some scholars have suggested that the lower crust of the eastern margin of the QTP is unlikely to have large-scale ductile flows due to the constraints of faults and tectonic boundaries. There may be flows along two arc-shaped channels, which may have passed through the Jinsha River–Red River fault to southern Yunnan [29]. Some scholars believed that the lower crustal flow from the eastern QTP was blocked by the Lijiang-Xiaojinhe fault and stopped at the northern Yunnan [30]. Studying the crustal movements and the deep driving mechanism in the SYR helps to understand the dynamic process of the southeastern boundary of the QTP, and also helps to estimate the future seismic risks of the key structure in China.

The vertical crustal deformation is important to the study of regional subsidence, crustal evolution and fault motion. The research on seismic activity trends based on vertical deformation provides essential information for estimating the regional strong earthquake trends and detecting dangerous areas. The SYR is a seismically active zone, where several strong earthquakes occurred in the past two decades, including the 2008 Wenchuan Mw 8.0 earthquake, 2013 Lushan Mw 7.0 earthquake and 2017 Jiuzhaigou Mw 7.0 earthquake. These earthquakes brought great damages. Therefore, it is of great significance to analyze the vertical movement characteristics of the major structural faults in the region, and to further explore their correlation with strong earthquakes. The National Leveling Network, Earthquake Leveling Monitoring Network, China Crustal Movement Observation Network (CMONOC) and China Continental Tectonic Environment Monitoring Network (CMTEMN) form a relatively complete deformation observation network for the SYR. The rich archive leveling and GNSS observations can be used for studying the crustal vertical deformation characteristics.

The traditional precision leveling and GNSS were separately used to study the vertical crustal deformation [31–33], but their advantages cannot be fully explored. As shown in Figure 1, the leveling route in SYR was generally arranged along the main structural belt, resulting in some observation gaps in the middle part of the secondary blocks. The GNSS stations were uniformly distributed, covering the observation gaps in the leveling route. The two observations are spatially complementary, so combining them can solve the problem of lacking velocity reference when using only leveling observations, and slow down the error transferring and accumulation. It also can constrain the deformed surface, so as to obtain a regional vertical motion field with higher spatial resolution and reasonability. Here, we use the least squares collocation method to obtain the vertical subsidence rate of the SYR, combining high-precision leveling data [31,34] with the observation of 320 GNSS stations in the CMONOC and the CMTEMN from 1999 to 2017. On this basis, we analyze the vertical motion characteristics and dynamic mechanism of this region.

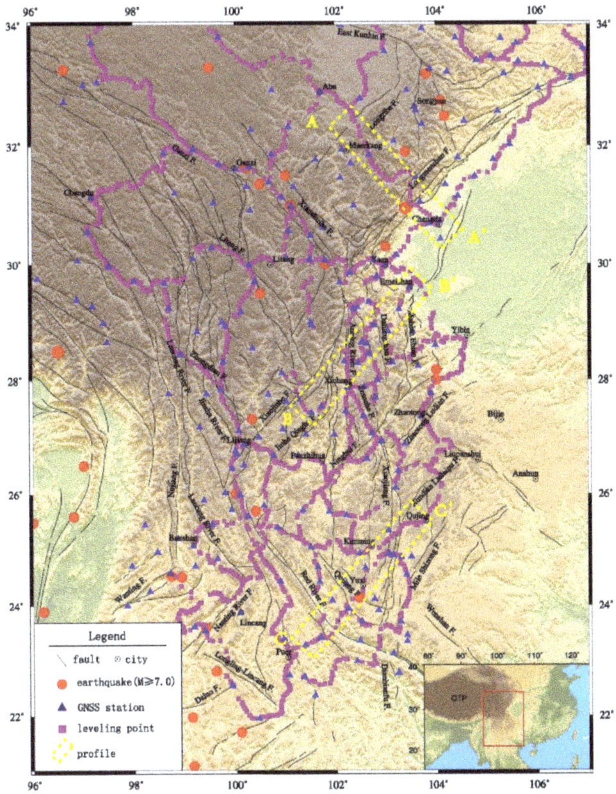

Figure 1. The tectonic setting of the Sichuan–Yunnan region (SYR). Triangles and squares denote GNSS stations and leveling points, respectively. Red dots denote the M $\geq$ 7.0 earthquakes occurred in this region since 1900. Profiles AA′, BB′ and CC′ will be discussed in Section 3.1. Inset shows the location of the study area in China.

2. Data and Processing

2.1. Leveling Data

The leveling data are from Hao et al. [31], containing the leveling vertical velocities of 3386 measuring points from 1977 to 2011. In addition, the velocities of 98 measuring points from 1990 to 2015 provided by Su et al. [34] are used as supplement. Therefore, the leveling data of 3484 measuring points in total during 1970s–2010s are used for this study. All leveling routes have the data accumulation of more than 2 periods to ensure the reliability of the results. A leveling network of several thousand kilometers usually takes many years to complete the first observation. To solve the time inconsistency, we use the linear dynamic adjustment method to obtain the average vertical deformation rates in a long period. For the rank deficiency of the normal equation, we adopt the rate of a small number of GNSS observation stations as the a priori constraint. The GNSS vertical velocity values (vertical motion velocity under ITRF2005) are from Wang [26]. Hao et al. [31] provided the data of eight GNSS sites, including CHDU, XIAG, KUNM, XNIN, JB08, JB09, JB27, JB35 and JB40, and Su et al. [34] provide the data of 4 GNSS sites, which are XIAG, KUNM, JB40 and SCJL. Theoretically, after the constrained adjustment, the benchmarks for the two datasets are the same. On the basis of Hao's results, this study uses the flux isostasy theory [35–37] to obtain the leveling results under the unified benchmark of the two datasets. The velocity uncertainty is estimated by the dynamic leveling adjustment. On this basis, we remove the gross errors and obtain the vertical motion vectors from the remaining 3278 measuring points (Figure 2a).

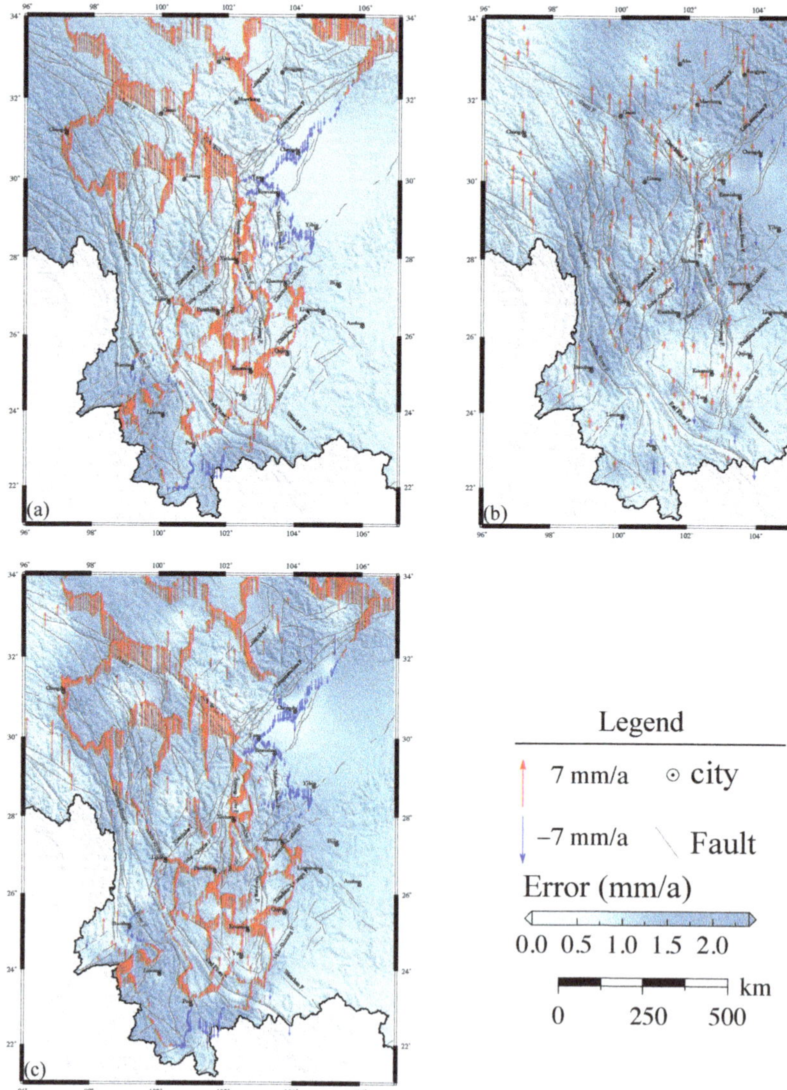

Figure 2. The vertical motion vectors of the SYR obtained (**a**) from the leveling data, (**b**) from the GNSS data, and (**c**) by fusing the leveling and GNSS results.

2.2. GNSS Data

This study collects the observations of 320 GNSS stations in the CMONOC and CMTEMN for the SYR from 1999 to 2017 and adopts unified models to process the global International GNSS Service (IGS) station data and the GNSS data by using the GAMIT /GLOBK10.61 software [38,39]. The steps are as follows: ① simultaneously solve the satellite orbits, earth orientation parameters, station coordinates, tropospheric delay and horizontal gradient parameters. Use loosely constrained algorithms for the station positions (5 cm for IGS core stations and 10 cm for other stations); ② fix the ambiguity; ③ set the satellite masking angle as 10°; ④ set the sampling interval as 30 s, and reweight the observations according to the post-test phase residuals [40]; ⑤ calculate the earth tide correction, ocean tide correction and pole tide correction. The ocean tide load model adopts FES2004 [41]; ⑥ adopt GMF as the tropospheric mapping function [42]. The atmospheric tide load correction and non-tidal atmospheric load correction are not considered; ⑦ adopt the absolute antenna phase center correction model; ⑧ use GLOBK for benchmark transformation to obtain the station coordinate time series with respect to ITRF2014 [43]. During the transformation, 7 parameters, including the benchmark transformation estimates, translation, rotation and scale factor, are used to reduce the datum distortion caused by the unmodeled non-tidal atmospheric pressure load; ⑨ select 77 IGS stations distributed evenly around the world as the frame sites to achieve the self-consistency of the results from the multiphase observations. Finally, the epochs of the stations with the error bigger than 5 mm north, 5 mm east and 20 mm up are removed to obtain the precise coordinate sequence of the GNSS station.

The coseismic deformation and post-seismic deformation of large earthquakes are useful for understanding the long-term tectonic deformation. From 94 continuous GNSS stations in the SYR, we obtain the tectonic velocity by the following principles: for the observations from the stations in the near field of some large earthquakes, such as the 2008 Wenchuan M8.0 earthquake, only the pre-earthquake data are used to avoid the influence of post-earthquake deformation; a small number of stations with no pre-earthquake observations or only with the first-phase observation data are discarded; the data from the stations in the far field are corrected by the estimated coseismic displacement. We analyze the GNSS deformation time series to obtain reliable vertical velocity, during which only the coseismic deformation and unexplained step effects are considered. The vertical velocity is obtained by Formula (1).

$$y(t_i) = a + bt_i + c\sin(2\pi t_i) + d\cos(2\pi t_i) + e\sin(4\pi t_i) + f\cos(4\pi t_i) + \sum_{j=1}^{n_g} g_j h(t_i - t_{gj}) \quad (1)$$

where a is the initial position, b is the rate of linear motion, c and d are the annual term coefficients, e and f are the semi-annual term coefficients, g is the magnitude of jth step caused by the replacement of equipment or the coseismic displacement and h is the Heaviside step function.

In order to ensure the reliability of the vertical velocity from the GNSS station, we use time series analysis to evaluate the observations in terms of observation time span, number of observation periods and station stability, etc. The selected GNSS stations should satisfy the following requirements: the weighted root mean square (WRMS) of the coordinate is less than 3.3 mm horizontally and 10 mm vertically; the observation time is longer than 7 years; the station has at least 4 observation periods. Finally, 226 GNSS stations with low observation noise and little interference from annual motion are selected. The linear motion rate is estimated without considering the annual motion. Figures 3 and 4 shows the time series and fitting results of the coordinates of some campaign and continuous stations, respectively. The residuals after fitting are small, with the horizontal WRMS smaller than 3 mm and the vertical WRMS smaller than 8.0 mm. These results demonstrate that the above processing scheme can obtain reliable horizontal and vertical movement trends of the stations.

In addition, in order to increase the spatial resolution of GNSS stations, we adopt the vertical velocity of 441 GNSS stations from 1998 to 2013 from Liang et al. [44]. Following the regional flux isostasy theory, we superimpose the calculated data on Liang's velocity field to obtain the GNSS vertical velocity in the SYR from 1990s to 2010s. Then, the gross error is removed to obtain the GNSS vertical motion vector (Figure 2b).

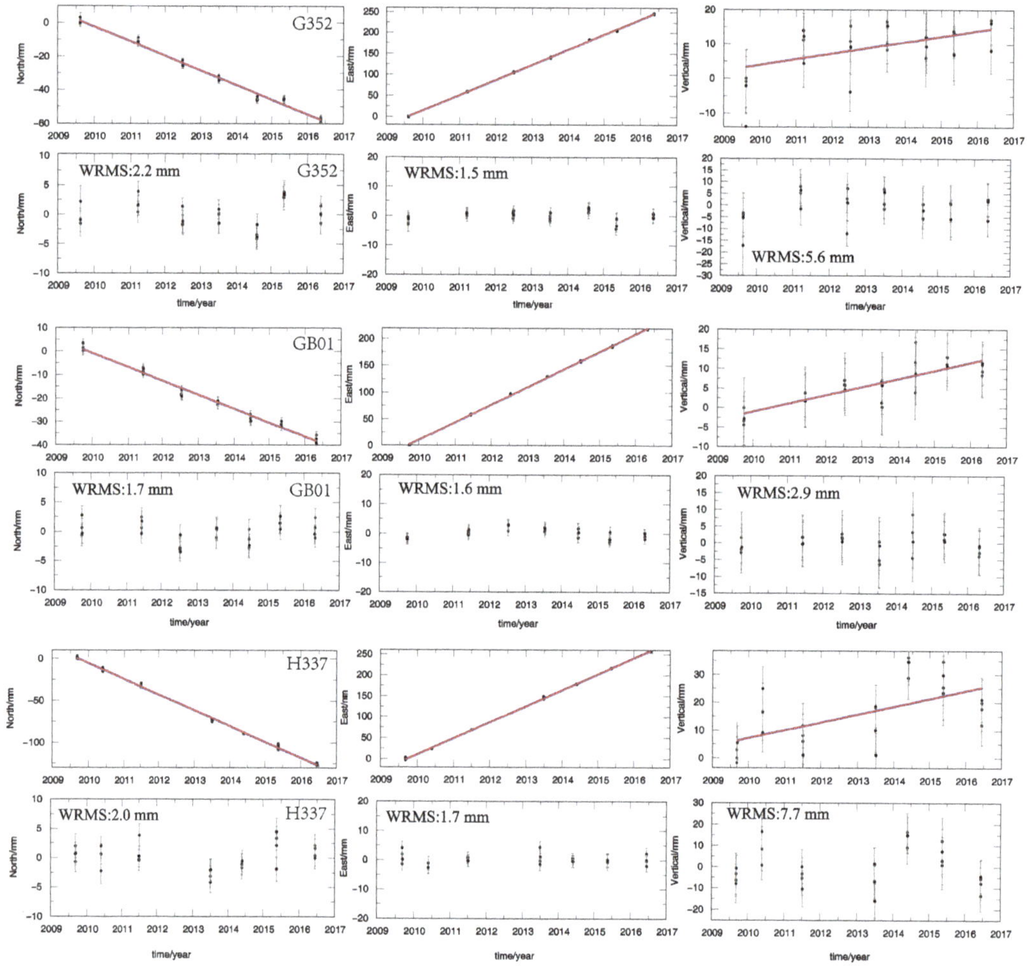

Figure 3. Time series and errors of the north (**left**), east (**middle**) and vertical (**right**) components of three GNSS campaign sites. The first two rows are for station G352, the third and fourth rows are for station GB01, and the fifth and sixth rows are for station H337. The red lines indicate the estimated velocities. The WRMS of every component are shown at the left up corner of error panels.

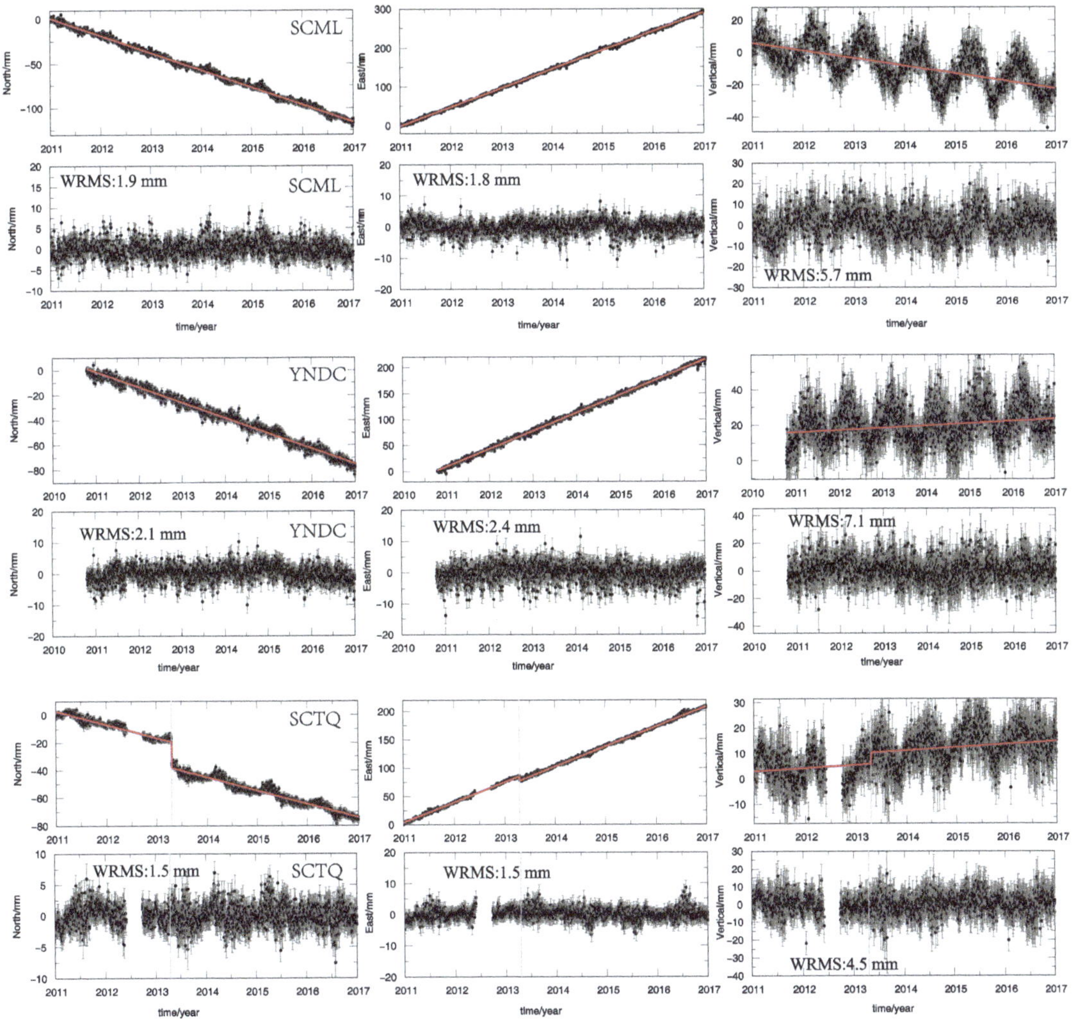

Figure 4. Time series and errors of three GNSS continuous sites: SCML (first two rows), YNDC (third and fourth rows), SCTQ (fifth and sixth rows). Others are the same as Figure 3.

2.3. Fusion of the Leveling Results and the GNSS Results

The 1970s–2010s leveling vertical velocity obtained by the linear dynamic adjustment method and the 1990s–2010s GNSS vertical velocity obtained using the time series analysis model are the annual average velocity of the long-time vertical motion and can reflect the background characteristics of the current vertical crustal movement in the SYR. In order to increase the spatial resolution of the deformation velocity by exploring the advantage of the above two results, we develop the least squares collocation method from the method of Vestol [45], which is based on multiquadric function [46–52].

The error equations for the two types of data are as follows

$$\left.\begin{aligned} V_i &= \sum_{i=1}^{m} C_i Q_i(x,y) + S_i - l_i \\ V_j &= \sum_{i=1}^{m} C_j Q_j(x,y) + S_j - l_j \end{aligned}\right\} \Rightarrow \left.\begin{aligned} V_1 &= A_1 X + B_1 S - L_1 \\ V_2 &= A_2 X + B_2 S - L_2 \end{aligned}\right\} \Rightarrow V = AX + BS - L \quad (2)$$

where $Q_i(x,y) = ((x-x_i)^2 + (y-y_i)^2)^{\frac{1.5}{2}} + 1$. V_i is the error equation for the leveling results. It is obtained by Helmert iterative adjustment, and the corresponding weight matrix is P_1; V_j is the error equation for the GNSS results, and the weight matrix is P_2. l is the observation, C is the kernel function coefficient, $Q(x,y)$ is the kernel function, S is the random signal, A is the design matrix, X is the unknown parameter, B is the coefficient matrix and L is the observation Vector, $V = AX + BS - L$ is the overall error equation, with the weight matrix p.

From Formula (2), we obtain

$$\begin{cases} \hat{X} = (A^T C_{XX}^{-1} A)^{-1} A^T C_{XX}^{-1} L \\ \hat{S} = C_{SS} B^T C_{XX}^{-1} (I - A\hat{X}) \end{cases} \tag{3}$$

where $C_{XX} = C_{\Delta\Delta} + BC_{SS}B^T$, with $C_{\Delta\Delta}$ and C_{SS} denoting the covariance matrix of the observed value and of the random signal, respectively.

The mean square error of the unit weight is

$$\begin{bmatrix} \hat{\sigma}_{\Delta}^{2(i)} \\ \hat{\sigma}_{S}^{2(i)} \end{bmatrix} = \begin{bmatrix} n - 2tr(N^{-1}N_{\Delta}) + tr(N^{-1}N_{\Delta})^2 & tr(N^{-1}N_{\Delta}N^{-1}N_S) \\ tr(N^{-1}N_{\Delta}N^{-1}N_S) & n - 2tr(N^{-1}N_S) + tr(N^{-1}N_S)^2 \end{bmatrix} \begin{bmatrix} V^T P_{\Delta} V \\ \hat{S}^T P_S \hat{S} \end{bmatrix} \tag{4}$$

where

$$N = \begin{bmatrix} A^T P_{\Delta}^{(i)} A & A^T P_{\Delta}^{(i)} B \\ B^T P_{\Delta}^{(i)} A & B^T P_{\Delta}^{(i)} B + P_S^{(i)} \end{bmatrix} N_{\Delta} = \begin{bmatrix} A^T P_{\Delta}^{(i)} A & A^T P_{\Delta}^{(i)} B \\ B^T P_{\Delta}^{(i)} A & B^T P_{\Delta}^{(i)} B \end{bmatrix} N_S = \begin{bmatrix} \mathbf{0} & \mathbf{0} \\ \mathbf{0} & P_S^{(i)} \end{bmatrix}$$

The adaptive factor is

$$\alpha^{(i)} = \hat{\sigma}_S^{2(i)} / \hat{\sigma}_{\Delta}^{2(i)} \tag{5}$$

The two types of observations are divided into two groups, and we apply iterative calculation to each group using the adaptive least squares collocation method [53,54]. The specific process is as follows: ① solve the least squares collocation configuration by formulas (2) and (3); ② solve the posterior unit weight error $\hat{\sigma}_{\Delta}^{(i)}, \hat{\sigma}_{S}^{(i)}$ and adaptive factor $\alpha^{(i)}$ by formulas (4) and (5); ③ determine whether $|\alpha^{(i)} - 1| < \varepsilon$ works, where ε is a small constant, which is 0.001 in this paper. If the conditions are satisfied, the iteration stops; if the conditions are not satisfied, let $P_{\Delta}^{(i+1)} = \alpha^{(i)} P_{\Delta}^{(i)}$, modify the weight of the observed value and repeat steps ①–③.

3. Results and Discussion

3.1. The Current Vertical Crustal Deformation in the SYR

The leveling vertical velocity ranges from −2.3 to 6.5 mm/a, and the error ranges from ±0.5 to ±2.1 mm/a, with an average error of ±1.2 mm/a (Figure 2a). The GNSS vertical velocity ranges from −3.6 mm/a to 4.9 mm/a, and the error ranges from ±0.1 mm/a to ±2.4 mm/a, with an average error of ±0.9 mm/a (Figure 2b). As Figure 2 shows, the two results are complementary in space. After fusion, the leveling observation increases the spatial density of the GNSS observation across the faults, and the GNSS observation makes the spatial distribution of the leveling observation more even. The vertical deformation velocity derived from these two datasets in the eastern margin of the QTP and its surrounding areas are consistent. Generally, the deformation in the SYR is dominated by uplifts. The uplift rates are large in eastern Tibet and the northern part of the rhombus block. Some regions have subsidence, including the Sichuan Basin and its periphery, the Jinghong–Pu'er–Jiangcheng area in southern Yunnan, and the Yunlong–Baoshan–Changning area.

In this paper, the least squares collocation method based on the multiquadric function is adopted to fuse the results of the two datasets with the adaptive Helmert iterative weighting. During the fusion, the two results constrain each other to ensure the comparability. The velocity after fusion ranges from −3.1 mm/a to 6.3 mm/a (Figure 2c). We generate the contour and gradient maps from the fused vertical motion rates by the least squares collocation method (Figure 5). On the whole, most areas in the SYR show vertical deformation patterns dominated by uplifts. Specifically, the eastern Tibet and the northern part of the rhombus block in the SYR have rapid uplift, but the Sichuan Basin and its periphery, the Jinghong–Pu'er–Jiangcheng area in southern Yunnan and the Yunlong–Baoshan–Changning area show subsidence.

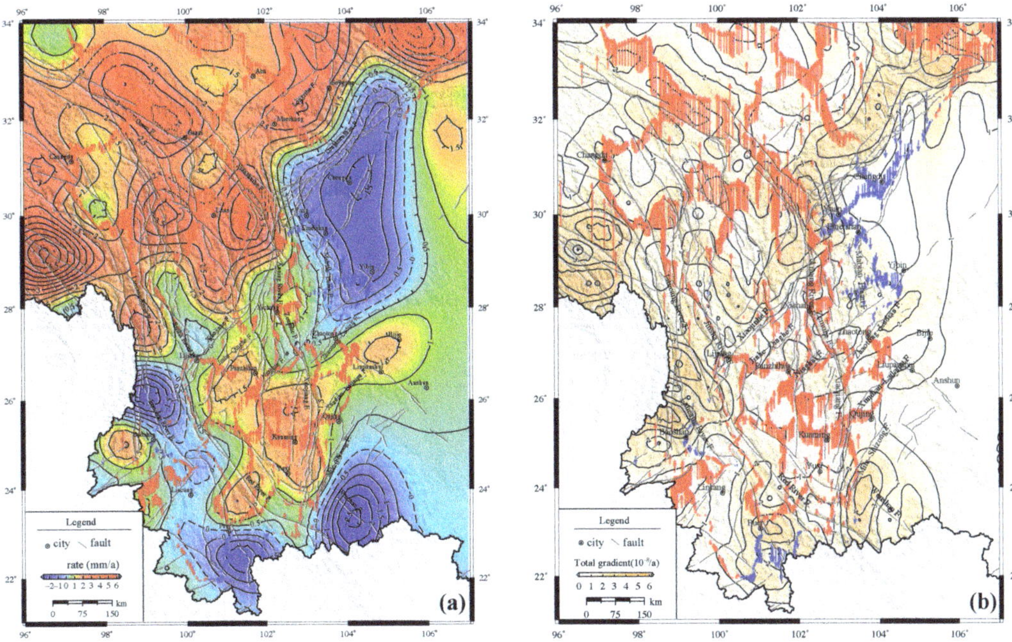

Figure 5. (**a**) Contour and (**b**) gradient map of the vertical motion velocity in the eastern margin of the Qinghai–Tibet Plateau and its surrounding areas. Solid black lines indicate the contours of vertical motion velocities in (**a**) and gradients in (**b**). The blue and red arrows are the vertical velocity after fusing the GNSS and the leveling data, respectively.

We combine the reported horizontal crustal movement features in the SYR [55] with our vertical deformation velocities to analyze the crustal movement features in the SYR: the vertical uplift is very obvious in the areas with strong horizontal crustal compression; the subsidence appears in the areas with extensive horizontal tension; the horizontal and vertical crustal movements are spatially coincided. These features reflect that the SYR was pushed and rotated by the QTP and blocked and squeezed by the South China block. In such an environment, the vertical crustal movement presents the contemporaneous occurrence of crustal compression, shortening and uplift and basin extensional subsidence. The detailed characteristics of the vertical deformation are as follows:

(1) The vertical velocity of the eastern Longmenshan fault is −2 mm/a~−1 mm/a, and that of the subblocks from the western Longmenshan fault to the southeastern Longriba fault is 1 mm/a~3 mm/a, showing rapid uplifts. The results are consistent with those of Hao [31] and Du et al. [56]. The two sides of the Longmenshan fault show very obvious vertical motion. The northwestern part, in which the 2008 Wenchuan M8.0 earthquake occurred, is also in a high gradient area of vertical velocity ($2.0 \sim 3.0 \times 10^{-8}$/a).

(2) The average deformation velocity is 2 mm/a in the areas to the north of the eastern section of the East Kunlun fault, from the eastern section of the Ganzi–Yushu

fault to the Xianshuihe fault and the Kangding–Yajiang–Litang–Batang area. The maximum velocity, 4 mm/a, is observed at the eastern section of the Ganzi–Yushu fault, but it rapidly drops to about 2 mm/a at the junction with the Xianshuihe fault, and then gradually rises to 3 mm/a~4 mm/a along the Xianshuihe fault, showing a segmentation feature. The results of the Gongga Mountain area, a part of the southeastern section of the Xianshuihe fault, are consistent with those of Wang [26] (1.8 mm/a~3.6 mm/a) and Wang et al. [57] (about 4 mm/a), but are smaller than those of Su et al. [34] (5 mm/a~6 mm/a) and Wang et al. [58] (7.4 mm/a). The overall uplift also confirms that during the plastic flow, the material under the QTP crust was compressed near the boundary due to the blockage of the hard Sichuan Basin and exerted great upward fluid pressure on the elastic upper crust [26,59].

(3) The uplift rate is 1.5 mm/a~2.0 mm/a for the Anning River fault, the Zemuhe fault and the area surrounded by the Daliangshan fault, smaller than the 2.5 mm/a~3.0 mm/a of Wang et al. [57].

(4) The interior area of the rhombus block is dominated by uplift, with a rate ranging from 0.5 mm/a to 2.0 mm/a. The rates vary greatly between the north and south parts. The velocity of the block boundary is smaller than the internal part, which is 0.5 mm/a~1.5 mm/a. The vertical deformation velocity on the west side is smaller than that on the east side. The whole rhombus block presents an E-shaped (with the opening to the north) uplift.

(5) The whole Western Sichuan Plateau has rapid uplifts. The uplift velocity gradually increases from the west (1 mm/a) to the east (4 mm/a), and that of the north block is 3 mm/a.

(6) The Longmenshan fault and the junction between the Chengdu Plain and the western mountains, namely the Jiangyou–Beichuan–Dujiangyan–Mabian–Zhaotong area, have a subsidence velocity of 2 mm/a; the Longling-Lancang River fault and the area near the Daluo fault has a subsidence velocity of 3 mm/a; the average subsidence rate in the east of the Lancang River fault and the Jinghong–Pu'er–Jiangcheng area is 1.5 mm/a; on the west side of the northern section of the Red River fault, namely the area surrounded by Dali–Yunlong–Tengchong–Lincang, the subsidence rate is smaller than 1 mm/a; in the area on the east of the Jinsha River fault zone, that is, from Xiangcheng to Zhongdian, the subsidence rate is smaller than 1 mm/a.

To further explore the vertical deformation features of some important sections in the SYR, we extract three profiles, AA′, BB′ and CC′, in the eastern Bayankala block, the eastern part of the Sichuan–Yunnan border and the southern Yunnan area, respectively (Figure 1). Vertical motions along these three profiles are shown in Figure 6. Profile AA′ starts from the Aba area of Sichuan and trends NW–SE to Chengdu, with a total length of about 300 km, crossing the Longriba and Longmenshan faults. Profile BB′ starts from Yanyuan and trends NE–SW to Yuexi, with a total length of about 370 km, crossing the Anning River and the Mabian-Ebian fault. Profile CC′ starts from Mojiang and trends NE–SW to Qujing, with a total length of about 380 km, crossing the Red River and Xiaojiang faults. The vertical deformation characteristics reflected by the profiles are as follows:

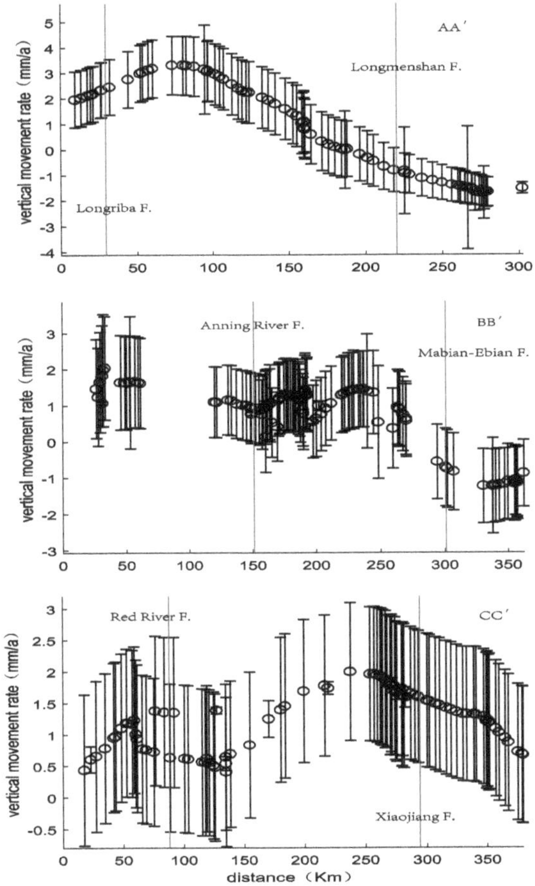

Figure 6. Three profiles, AA′, BB′ and CC′, showing the vertical motions in the SYR. Profile locations are shown in Figure 1.

(1) Profile AA′ shows that the two sides of the Longmenshan fault have opposite vertical motions, with the east side subsiding (−2 mm/a~−1 mm/a) and the west side uplifting (1 mm/a~3 mm/a). The two sides of the Longriba fault also have opposite vertical motions, but the rate difference is smaller, as the uplifting rate on the west side is 2 mm/a~2.5 mm/a, and the subsiding rate on the east side is 2.5 mm/a~3 mm/a. In general, the area on the west of the Longmenshan fault has rapid uplift, especially the area near the Longriba fault with a rate of 2 mm/a~3 mm/a. The area with the maximum uplift rate area is located 100 km east of the Longriba fault. This suggests the contemporaneous occurrence of crustal compression, shortening and uplift and basin extensional subsidence, as the Bayankala block's eastward motion was blocked by the South China block. The rapid material and energy accumulation are favorable for the occurrence of strong earthquakes.

(2) Profile BB′ shows that the vertical velocity of the east side of the Mabian-Ebian fault is −2 mm/a~−1 mm/a, and that of the west side is 0.5 mm/a~1.5 mm/a. The two sides show significant vertical motion difference. On the two sides of the Anning River fault, the vertical movement difference is not significant, both showing uplifts with a rate of about 1 mm/a. This suggests that the series of thrust faults on the border of Sichuan

and Yunnan experienced compression and rising caused by the eastward movement of the northwestern Sichuan block. There may be risks of regional strong earthquakes.

(3) Profile CC′ shows that the vertical movement rate of the east side of the Xiaojiang fault is 0.5 mm/a~1.5 mm/a, and that of the west side is 0.5 mm/a~2.0 mm/a, which are not significantly different. The vertical movement rate on the south side of the Red River fault zone is 0.5 mm/a~1.5 mm/a, and the vertical difference between the two sides is not significant.

3.2. Kinetic Mechanism

The SYR is located on the southeastern margin of the QTP, where the lateral extrusion of the QTP caused by the northward push of the Indian plate was blocked by the stable South China block, resulting in the plateau eastern margin continuously uplifting [31,44]. Such a movement pattern has seen no great changes since the Pliocene [17,60]. The collision between the Indian plate and the QTP plate not only caused the crust vertical movement, but also controlled the fault motion mode and the horizontal movement of each block. Most of the material from the QTP moved southward. The rheological strength of the middle and lower crust in the SYR is weaker than that of other crusts. The crust thickness in the region is 20~30 km less than that of the Sichuan Basin and South China block, and the elevation difference is about 3000 m, causing lateral pressure difference. Under this pressure difference, the weak rheological layer in the middle and lower crust flowed, which enhanced the clockwise rotation of the lithosphere in the southeastern QTP around the eastern Himalayan tectonic junction. Such rotation caused the deformation of the upper brittle crust and the movement of the faults, which finally showed as a huge left-lateral strike-slip arc fault structure [61].

The material from the eastern margin of the QTP flowed to three directions. Some moved northeastward and interacted strongly with the Ordos Basin. Some moved to western Sichuan and strongly coupled with the Longmenshan fault, and the other moved to Yunnan after a sharp turn. In the Longmenshan area north to the Xianshuihe fault, the material accumulated in the middle and lower crust after being blocked by the hard crust of the Sichuan Basin, resulting in sharp crust thickening and strong surface uplifts, forming the towering Longmenshan Mountains [14,16,62–64]. In the SYR south to the Xianshuihe fault, the crust was relatively soft and the material flow channels in the middle and lower crusts were open. Affected by the drag of this pipeline flow, the SYR had a clockwise rotation, and the crust gradually thickened on a scale of thousands of kilometers. A large-scale of ground surface had uplifted, forming the broad southeastern margin of the QTP, featured by similar geomorphology and blurred boundary [14,16,17,63,64].

3.3. Distribution of Lower Crustal Flows

Deep geophysical exploration revealed that the middle and lower crusts on the eastern margin of the QTP have seismic low-velocity [21,23,28], high Poisson's ratio [19,27] and high conductivity/low resistivity [25,27,28]. These conditions favor partial melting, especially if there are salt-bearing fluids. Partial melting reduces the strength of the middle and lower crusts and promotes laminar flow [61].

In this study, the finite difference method was adopted to calculate the effective elastic thickness of the lithosphere in the SYR from the EIGEN6C4 Bouguer gravity anomaly and SIO V15.1 topographic data (Figure 7). The effective elastic thickness of the lithosphere reflects the relation between the mechanical strength of the lithosphere [65,66] in the geological time scale (>1 Ma) and the composition of crust–mantle matter, thermal state, and the crust–mantle coupling extent [67,68]. Combined with the reported results in seismology and magnetotelluric fields, we infer that the extruded material from the QTP was blocked by the Sichuan Basin, and accumulated in the Litang–Daocheng–Lijiang area, which showed as prominent surface uplifts and large-scale internal crustal deposits with low velocity and high conductivity/low resistivity. A small part of the material may have moved northward through the Kangding–Daofu area in the Xianshuihe fault, but most of the material moved southward. After being blocked by the southern Yunnan block, a part of the material flowed to Tengchong in the southwest and the rest flowed southeast to the Panzhihua–Dongchuan area.

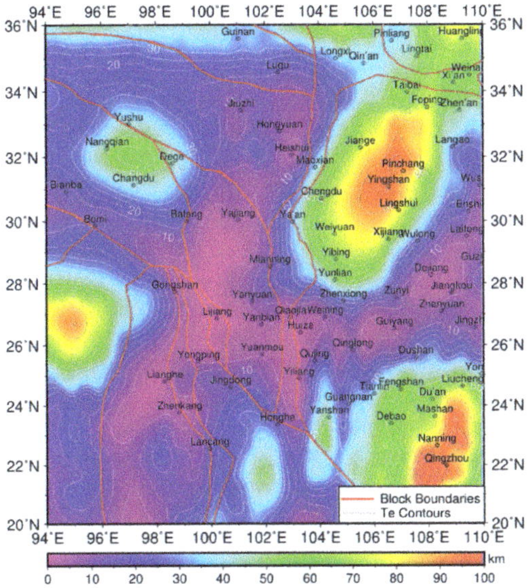

Figure 7. The effective elastic thickness of the lithosphere in the SYR, calculated from the EIGEN6C4 Bouguer gravity anomaly and SIO V15.1 topographic data.

The vertical crustal movement features in the SYR (Figure 2) suggest that the southeastern margin of the QTP was blocked by the Sichuan Basin, so most of the material flowed southward along the arc-shaped channel into southern Yunnan, some of which was blocked by the Lijiang-Xiaojinhe fault. The rest continued to move southward, but was blocked by the southern Yunnan block. Some crossed the Red River fault and flowed to Pu'er in the southwest, and the rest crossed the Xiaojiang fault and flowed to Liupanshui in the southeast. In the SYR, due to the blockage of the Lijiang-Xiaojinhe fault, the material crossed the Jinsha River–Red River fault and the Xiaojiang fault in the southwest and southeast directions, respectively.

4. Conclusions

Fusing the precise leveling data of the 1970s–2010s and the GNSS data of the 1990s–2010s by the least squares collocation method, we obtain the background field of the vertical crustal movements in the SYR, which reflects the features of the current crustal vertical movements in this region. These features coincided with the neotectonic movement background that the plateau and mountains uplifted and basins subsided. We come to the following conclusions:

(1) Calculating the vertical deformation field of the regional crust by fusing leveling data and GNSS data can use the advantages of the two data. The obtained deformation has more uniform spatial distribution, and shows more detailed change characteristics of the key regions, than the results obtained from either data. Furthermore, it takes into account the vertical deformation characteristics of the whole region and some important areas and sections, to obtain more comprehensive and detailed vertical deformation.

(2) Under the combined contribution of the pushing and rotation of the QTP and the blocking and compression of the South China block, the current vertical crustal move-

ment in the SYR presents the contemporaneous occurrence of crustal compression, shortening and uplift and basin extensional subsidence.

(3) In the study area, the current vertical crustal movements have two different trends on the two sides of the boundary formed by the Longmenshan fault and the Lijiang-Xiaojinhe fault. The region on the northwest side is dominated by rapid uplift, and the uplift rate gradually increases from 1.5 mm/a on the west side of the Western Sichuan Plateau to 5.5 mm/a on the east side. The material from the QTP crust flowed southeastward, and was blocked by the South China block in the east. It exerted great upward fluid pressure on the elastic upper crust, and led to the uplift of the whole region. However, the west side of this region bore smaller blockages and formed smaller upward fluid pressure than that on the east side. The region on the southeast side of the boundary, namely the central Yunnan and South China blocks, had slight uplifts, with a rate of 1.0 mm/a~1.5 mm/a. The surrounding areas had dominating subsidence, especially the Longling-Lancang fault and the Daluo fault, with the subsidence rate of 0.5 mm/a~2.0 mm/a. The reason is that the material under the QTP crust was divided into two branches in this area, and flowed in the southwest and southeast directions separately. This bifurcation released the upward fluid pressure and caused the subsidence of the elastic upper crust.

(4) This region has many areas with high gradient vertical deformation, including that on the northwest side of the Longmenshan fault (2.0~3.0 $\times$ 10^{-8}/a), and the intersection of the Xianshuihe, Longmenshan and Anning River faults (2.0~3.0 $\times$ 10^{-8}/a), the area near the Wenshan fault (3.0~4.0 $\times$ 10^{-8}/a), the area near Jinggu fault (2.0~4.0 $\times$ 10^{-8}/a), west side of the Jinsha River fault (2.0~5.0 $\times$ 10^{-8}/a) and both sides of the Yarlung Zangbo fault (3.0~4.0 $\times$ 10^{-8}/a).

(5) The vertical movements in the SYR suggest that the material from the QTP was blocked by the Sichuan Basin and caused the uplift of the southeastern margin of the QTP. The material that flowed to the south was blocked by the southern Yunnan block. One branch passed through the Red River fault and flowed to Pu'er in the southwest. The other branch passed through the Xiaojiang fault and flowed to the Liupanshui fault in the southeast. The surface deformation confirms the "crustal flow model", but whether these low-velocity areas are connected to each other and where they are from remain unclear. These issues may be solved by simultaneous detection and comprehensive constraints using methods such as comprehensive geophysical profile detection and dense array detection.

Author Contributions: Conceptualization, Y.Z. and C.X.; methodology, Y.Z. and Z.Z.; software, Z.Z., H.L. and S.Z.; validation, Y.Z. and Z.Z.; formal analysis, Y.Z.; writing—original draft preparation, Y.Z. and C.X.; writing—review and editing, Y.Z., C.X. and Z.Z. All authors have read and agreed to the published version of the manuscript.

Funding: This work is supported by the National Natural Science Foundation of China under Grants No. 41721003 and No. 42074007.

Institutional Review Board Statement: Not applicable.

Informed Consent Statement: Not applicable.

Data Availability Statement: All data used in this manuscript can be archived at the Figshare website (https://doi.org/10.6084/m9.figshare.18871442.v1, last accessed on 1 February 2022) for free public access.

Acknowledgments: We appreciate the GPS observations from the Crustal Movement Observation Network of China and the China Continental Tectonic Environment Monitoring Network. We thank Ming Hao at the Second Monitoring Center, China Earthquake Administration, for providing the leveling data. All figures were constructed using Generic Mapping Tools (GMT) version 4.5.1 [69].

Conflicts of Interest: The authors declare no conflict of interest.

References

1. Li, P. *Xianshuihe-Xiaojiang Fault Zone*; Earthquake Press: Beijing, China, 1993. (In Chinese)
2. Lacassin, R.; Replumaz, A.; Leloup, P. Hairpin river loops and slip-sense inversion on southeast Asian strike-slip faults. *Geology* **1998**, *26*, 703–706. [CrossRef]
3. Xu, X.; Wen, X.; Zheng, R.; Ma, W.; Song, F. The latest tectonic changes of active blocks in the Sichuan-Yunnan region and their dynamic sources. *Sci. China (Ser. D)* **2003**, *33*, 151–162.
4. Xu, X.; Wen, X.; Yu, G.; Zheng, R.; Luo, H.; Zheng, B. Average slip rate, seismic rupture segmentation and recurrence characteristics of the Litang fault zone in western Sichuan. *Sci. China (Ser. D)* **2005**, *48*, 1183–1196.
5. Xu, X.; Zhang, P.; Wen, X.; Qing, Z.; Chen, G.; Zhu, A. Basic characteristics of active structures in western Sichuan and its adjacent areas and strong earthquake recurrence model. *Earthq. Geol.* **2005**, *27*, 446–461. (In Chinese)
6. Wen, X.; Xu, X.; Zheng, R.; Xie, Y.; Wan, C. The average slip rate of the Ganzi-Yushu fault and the rupture of modern earthquakes. *Chin. Sci. (Ser. D)* **2003**, *33*, 276–288.
7. Lei, J.; Zhao, D. Structural heterogeneity of the Longmenshan fault zone and the mechanism of the 2008 Wenchuan earthquake (Ms 8.0). *Geochem. Geophys. Geosyst.* **2009**, *10*, 1–17. [CrossRef]
8. Wang, Q.; Zhang, P.-Z.; Freymueller, J.T.; Bilham, R.; Larson, K.M.; Lai, X.; You, X.; Niu, Z.; Wu, J.; Li, Y.; et al. Present-Day Crustal Deformation in China Constrained by Global Positioning System Measurements. *Science* **2001**, *294*, 574–577. [CrossRef]
9. Gan, W.; Zhang, P.; Shen, Z.-K.; Niu, Z.; Wang, M.; Wan, Y.; Zhou, D.; Cheng, J. Present-day crustal motion within the Tibetan Plateau inferred from GPS measurements. *J. Geophys. Res. Earth Surf.* **2007**, *112*, D17301. [CrossRef]
10. Tapponnier, P.; Molnar, P. Active faulting and tectonics in China. *J. Geophys. Res. Earth Surf.* **1977**, *82*, 2905–2930. [CrossRef]
11. Avouac, J.-P.; Tapponnier, P. Kinematic model of active deformation in central Asia. *Geophys. Res. Lett.* **1993**, *20*, 895–898. [CrossRef]
12. Royden, L. Coupling and decoupling of crust and mantle in convergent orogens: Implications for strain partitioning in the crust. *J. Geophys. Res. Earth Surf.* **1996**, *101*, 17679–17705. [CrossRef]
13. England, P.; Molnar, P. Active Deformation of Asia: From Kinematics to Dynamics. *Science* **1997**, *278*, 647–650. [CrossRef]
14. Clark, M.; Royden, L. Topographic ooze: Building the eastern margin of Tibet by lower crustal flow. *Geology* **2000**, *28*, 703–706. [CrossRef]
15. Klemperer, S. Crustal flow in Tibet: Geophysical evidence for the physical state of Tibetan lithosphere, and inferred patterns of active flow. *Geol. Soc. Lond. Spec. Publ.* **2006**, *268*, 39–70. [CrossRef]
16. Clark, M.; Bush, J.; Royden, L. Dynamic topography produced by lower crustal flow against rheological strength hetero-geneities bordering the Tibetan Plateau. *Geophys. J. Int.* **2005**, *162*, 575–590. [CrossRef]
17. Schoenbohm, L.; Burchfiel, B.C.; Liangzhong, C. Propagation of surface uplift, lower crustal flow, and Cenozoic tectonics of the southeast margin of the Tibetan Plateau. *Geology* **2006**, *34*, 813–816. [CrossRef]
18. Wang, C.; Wu, J.; Lou, H.; Zhou, M.; Bai, Z. Crustal P-wave velocity structure in western Sichuan-eastern Tibet. *Sci. China Ser. D Earth Sci.* **2003**, *33*, 181–189. (In Chinese)
19. Wang, C.; Lou, H.; Lv, Z.; Wu, J.; Chang, L.; Dai, S.; You, H.; Tang, F.; Zhu, L.; Silver, P. S-wave velocity structure of the crust and upper mantle in the eastern Qinghai-Tibet Plateau: The deep environment of the lower crustal flow. *Sci. China Ser. D Earth Sci.* **2008**, *38*, 22–32.
20. Wang, C.; Yang, W.; Wu, J.; Ding, Z. Research on the lithospheric structure and earthquakes in the north-south tectonic belt. *Chin. J. Geophys.* **2015**, *58*, 3867–3901. (In Chinese)
21. Yao, H.; Van Der Hilst, R.D.; Montagner, J.-P. Heterogeneity and anisotropy of the lithosphere of SE Tibet from surface wave array tomography. *J. Geophys. Res. Earth Surf.* **2010**, *115*, B12307. [CrossRef]
22. Bao, X.; Sun, X.; Xu, M.; Eaton, D.W.; Song, X.; Wang, L.; Ding, Z.; Mi, N.; Li, H.; Yu, D.; et al. Two crustal low-velocity channels beneath SE Tibet revealed by joint inversion of Rayleigh wave dispersion and receiver functions. *Earth Planet. Sci. Lett.* **2015**, *415*, 16–24. [CrossRef]
23. Liu, Q.; Van, D.; Li, Y.; Yao, H.; Chen, J.; Guo, B. Eastward expansion of the Tibetan Plateau by crustal flow and strain par-titioning across faults. *Nat. Geosci.* **2014**, *7*, 361–365. [CrossRef]
24. Wang, S.; Xu, X.; Hu, J. A review of research on crustal structure and dynamical models on the southeastern margin of the Qinghai-Tibet Plateau. *Chin. J. Geophys.* **2015**, *58*, 4235–4253. (In Chinese)
25. Sun, J.; Jin, G.; Bai, D.; Wang, L. The detection of the electrical structure of the crust and upper mantle on the eastern margin of the Qinghai-Tibet Plateau and its tectonic significance. *Sci. China Ser. D Earth Sci.* **2003**, *33*, 173–180. (In Chinese)
26. Wang, M. Refinement Analysis of GPS Observation Results and Research on Present-Day Crustal Deformation Field in Mainland China. Ph.D. Thesis, Institute of Geology, China Earthquake Administration, Beijing, China, 2009. (In Chinese).
27. Xu, L.; Rondenay, S.; van der Hilst, R.D. Structure of the crust beneath the southeastern Tibetan Plateau from teleseismic receiver functions. *Phys. Earth Planet. Inter.* **2007**, *165*, 176–193. [CrossRef]
28. Zhao, G.; Chen, X.; Wang, L.; Wang, J.; Tang, J.; Wan, Z.; Zhang, J.; Zhan, Y.; Xiao, Q. Evidence from electromagnetic de-tection of the crustal "tubular flow" layer on the eastern margin of the Qinghai-Tibet Plateau. *Sci. Bull.* **2008**, *53*, 345–350. [CrossRef]
29. Bai, D.; Unsworth, M.; Meju, M.A.; Ma, X.; Teng, J.; Kong, X.; Sun, Y.; Sun, J.; Wang, L.; Jiang, C.; et al. Crustal deformation of the eastern Tibetan plateau revealed by magnetotelluric imaging. *Nat. Geosci.* **2010**, *3*, 358–362. [CrossRef]
30. Chen, Y.; Zhang, Z.; Sun, C.; Badal, J. Crustal anisotropy from Moho converted Ps wave splitting analysis and geodynamic implications beneath the eastern margin of Tibet and surrounding regions. *Gondwana Res.* **2013**, *24*, 946–957. [CrossRef]

31. Hao, M. Studying the present vertical movement of the crust in the eastern margin of the Qinghai-Tibet Plateau using precise leveling data. In Proceedings of the 28th Annual Meeting of Chinese Geophysical Society, Beijing, China, 17–20 October 2012. (In Chinese).
32. Zhao, B.; Nie, Z.; Huang, Y.; Wang, W.; Zhang, C.; Tan, K.; Du, R. Present-day vertical motion in North China revealed by large-scale GPS. *Geod. Geodyn.* **2014**, *34*, 35–39. (In Chinese)
33. Qin, S.; Wang, W.; Song, S. Comparative study on vertical deformation based on GPS and leveling data. *Geod. Geodyn.* **2018**, *9*, 11–16. (In Chinese) [CrossRef]
34. Su, G.; Tian, X.; Sun, Q.; Deng, D.; Chang, L.; Wang, S. Research on vertical deformation and main fault activity in Si-chuan-Yunnan region. *Geod. Geodyn.* **2017**, *37*, 686–691. (In Chinese)
35. Yang, G. Application of field theory method in data processing of vertical crustal deformation. *Crustal Deform. Earthq.* **1991**, *11*, 61–67. (In Chinese)
36. Zhang, Q.; Fan, Y. Equilibrium Theory and Analysis Model of Vertical Crustal Movement. *J. Surv. Map-Ping* **2001**, *30*, 233–237. (In Chinese)
37. Zhang, Q.; Fan, Y.; Zhao, C. Analysis model of crustal vertical deformation field based on flux equilibrium datum. *J. Wuhan Univ. (Inf. Sci. Ed.)* **2004**, *29*, 231–234. (In Chinese)
38. Herring, T.; King, R.; Floyd, M.; McClusky, S. *GAMIT Reference Manual GPS Analysis at MIT(Release 10.6)*; Department of Earth, Atmosphere, and Planetary Sciences Massachusetts Institute Technology: Cambridge, UK, 2015.
39. Herring, T.; King, R.; McClusky, S. *GLOBK Reference Manual. Global Kalman filter VLBI and GPS Analysis Program (Release 10.6)*; Department of Earth, Atmosphere, and Planetary Sciences Massachusetts Institute Technology: Cambridge, UK, 2015.
40. Tregoning, P.; Watson, C. Atmospheric effects and spurious signals in GPS analyses. *J. Geophys. Res. Earth Surf.* **2009**, *114*. [CrossRef]
41. Lyard, F.; Lefevre, F.; Letellier, T.; Francis, O. Modelling the global ocean tides: Modern insights from FES2004. *Ocean Dyn.* **2006**, *56*, 394–415. [CrossRef]
42. Boehm, J.; Niell, A.; Tregoning, P.; Schuh, H. Global Mapping Function (GMF): A new empirical mapping function based on numerical weather model data. *Geophys. Res. Lett.* **2006**, *33*, L07304. [CrossRef]
43. Altamimi, Z.; Rebischung, P.; Métivier, L.; Collilieux, X. ITRF2014: A new release of the International Terrestrial Reference Frame modeling nonlinear station motions. *J. Geophys. Res. Solid Earth* **2016**, *121*, 6109–6131. [CrossRef]
44. Liang, S.; Gan, W.; Shen, C.; Xiao, G.; Liu, J.; Chen, W.; Ding, X.; Zhou, D. Three-dimensional velocity field of present-day crustal motion of the Tibetan Plateau derived from GPS measurements. *J. Geophys. Res. Solid Earth* **2013**, *118*, 5722–5732. [CrossRef]
45. Vestol, O. Determination of Postglacial Land Uplift in Fennoscandia from Leveling, Tide-gauges and Continuous GPS Sta-tions using Least Squares Collocation. *J. Geod.* **2006**, *80*, 248–258. [CrossRef]
46. Hardy, R. Multi quadratic equations of topography and other irregular surfaces. *J. Geophys.* **1971**, *76*, 1905–1915. [CrossRef]
47. Hardy, R. *Geodetic Applications of Multiquadric Equations*; ERI. Project 10705; American (NTIS PB 255296); Iowa State University: Ames, IA, USA, 1976.
48. Hardy, R. *The Application of Multi-Quadric Equations and Point Mass Anomaly Models to Crustal Movement Studies*; NOAA Technical Report NOS 76, NGS 11; NOAA: Washington, DC, USA, 1978.
49. Yang, G.; Huang, L. Preliminary Numerical Study on Several Properties of Multifaceted Functions in Rate Surface Fitting Method. *Crustal Deform. Earthq.* **1990**, *10*, 70–82. (In Chinese)
50. Huang, L.; Tao, B.; Zhao, C. The application of polyhedral function fitting in the study of vertical crustal motion. *J. Surv. Mapp.* **1993**, *22*, 25–32. (In Chinese)
51. Tao, B.; Du, F. Study on the vertical movement of the present crust in the Xianshuihe area by the multi-faceted function method. *J. Wuhan Univ. Surv. Mapp.* **1991**, *10*, 12–21. (In Chinese)
52. Tao, B.; Yao, Y. Parameter estimation based on multifaceted kernel function configuration model. *J. Wuhan Univ. (Inf. Sci. Ed.)* **2003**, *28*, 547–550. (In Chinese)
53. Yang, Y.; Zeng, A.; Zhang, J. Adaptive collocation with application in height system transformation. *J. Geod.* **2009**, *83*, 403–410. [CrossRef]
54. Guo, C.; Nie, J.; Wang, B.; Jiang, G. Fitting of GPS level and gravitational geoid using adaptive least squares configuration. *J. Wuhan Univ. (Inf. Sci. Ed.)* **2013**, *38*, 44–47. (In Chinese)
55. Wang, M.; Shen, Z.-K. Present-Day Crustal Deformation of Continental China Derived From GPS and Its Tectonic Implications. *J. Geophys. Res. Solid Earth* **2020**, *125*, e2019JB018774. [CrossRef]
56. Du, F.; Wen, X.; Zhang, P.; Wang, Q. Interseismic deformation across the Longmenshan fault zone before the 2008 Wen-chuan M8.0 earthquake. *Chin. J. Geophys.* **2009**, *52*, 2729–2738. (In Chinese)
57. Wang, S.; Jiang, F.; Hao, M.; Zhu, L. Study on the present three-dimensional crustal movement characteristics of the eastern margin of the Qinghai-Tibet Plateau. *Chin. J. Geophys.* **2013**, *56*, 3334–3345. (In Chinese)
58. Wang, Q.; Cui, D.; Wang, W.; Zhang, S.; Liu, W.; Shi, Q. Research on present-day vertical crustal movement in western Sichuan. *Sci. China Ser. D* **2008**, *038*, 598–610.
59. Burchfiel, B.; Royden, L.; Van Der Hilst, R.; Hager, B.; Chen, Z.; King, R.; Li, C.; Lu, J.; Yao, H.; Kirby, E. A geological and geophysical context for the Wenchuan earthquake of 12 May 2008, Sichuan, People's Republic of China. *GSA Today* **2008**, *18*, 4. [CrossRef]

60. Wang, E.; Burchfiel, B.C. Late Cenozoic to Holocene deformation in southwestern Sichuan and adjacent Yunnan, China, and its role in formation of the southeastern part of the Tibetan Plateau. *GSA Bull.* **2000**, *112*, 413–423. [CrossRef]
61. Zhang, P. Present tectonic deformation, strain distribution and deep dynamic process in the eastern margin of the Qing-hai-Tibet Plateau and western Sichuan. *Sci. China Earth Sci.* **2008**, *38*, 1041–1056.
62. Royden, L.H.; Burchfiel, B.C.; King, R.W.; Wang, E.; Chen, Z.; Shen, F.; Liu, Y. Surface Deformation and Lower Crustal Flow in Eastern Tibet. *Science* **1997**, *276*, 788–790. [CrossRef]
63. Clark, M.K.; Schoenbohm, L.; Royden, L.H.; Whipple, K.X.; Burchfiel, B.C.; Zhang, X.; Tang, W.; Wang, E.; Chen, L. Surface uplift, tectonics, and erosion of eastern Tibet from large-scale drainage patterns. *Tectonics* **2004**, *23*, 1–21. [CrossRef]
64. Kirby, E.; Reiners, P.; Kro, M.; Whipple, K.; Hodges, K.; Farley, K.; Tang, W.; Chen, Z. Late Cenozoic evolution of the eastern margin of the Tibetan Plateau: Inferences from 40Ar/39Ar and (U-Th)/He thermochronology. *Tectonics* **2002**, *21*, 1–20. [CrossRef]
65. Forsyth, D.W. Subsurface loading and estimates of the flexural rigidity of continental lithosphere. *J. Geophys. Res. Earth Surf.* **1985**, *90*, 12623–12632. [CrossRef]
66. Jiao, S.; Jin, Z. Study on effective elastic thickness of continental lithosphere and its dynamic significance. *Geol. Sci. Technol. Inf.* **1996**, *15*, 8–12. (In Chinese)
67. Burov, E.B.; Diament, M. The effective elastic thickness (Te) of continental lithosphere: What does it really mean? *J. Geophys. Res. Earth Surf.* **1995**, *100*, 3905–3927. [CrossRef]
68. An, M.; Shi, Y. Study on the thickness distribution of the lithosphere in mainland China. *Geosci. Front.* **2006**, *13*, 23–30. (In Chinese)
69. Wessel, P.; Smith, W.H.F.; Scharroo, R.; Luis, J.; Wobbe, F. Generic Mapping Tools: Improved Version Released. *EOS Trans. Am. Geophys. Union* **2013**, *94*, 409–410. [CrossRef]

remote sensing

MDPI

Article

Measuring Land Surface Deformation over Soft Clay Area Based on an FIPR SAR Interferometry Algorithm—A Case Study of Beijing Capital International Airport (China)

Xuemin Xing [1,2,3,4], Lingjie Zhu [1,3,5], Bin Liu [1,3,6,*], Wei Peng [1,3], Rui Zhang [1] and Xiaojun Ma [1,3]

1 School of Traffic and Transportation Engineering, Changsha University of Science and Technology, Changsha 410114, China
2 Hunan Key Laboratory of Remote Sensing of Ecological Environment in Dongting Lake Area, Changsha 410114, China
3 Laboratory of Radar Remote Sensing Applications, Changsha University of Science and Technology, Changsha 410014, China
4 Hunan International Scientific and Technological Innovation Cooperation Base of Advanced Construction and Maintenance Technology of Highway, Changsha University of Science & Technology, Changsha 410114, China
5 School of Geosciences and Info-physics, Central South University, Changsha 410083, China
6 Engineering Laboratory of Spatial Information Technology of Highway Geological Disaster Early Warning in Hunan Province, Changsha University of Science & Technology, Changsha 410114, China
* Correspondence: binliu@csust.edu.cn

check for updates

Citation: Xing, X.; Zhu, L.; Liu, B.; Peng, W.; Zhang, R.; Ma, X. Measuring Land Surface Deformation over Soft Clay Area Based on an FIPR SAR Interferometry Algorithm—A Case Study of Beijing Capital International Airport (China). *Remote Sens.* **2022**, *14*, 4253. https://doi.org/10.3390/rs14174253

Academic Editors: Alex Hay-Man Ng, Linlin Ge, Hsing-Chung Chang and Zheyuan Du

Received: 13 July 2022
Accepted: 24 August 2022
Published: 29 August 2022

Abstract: Long-term settlement monitoring of infrastructure built in soft clay areas is of great importance. When using InSAR technology for soft clay settlement monitoring, deformation modeling is a key process. In most InSAR deformation modeling, each component of the total deformation is expressed directly with a fixed functional model in phase functions and assumed to occupy an equal weight. This causes equal weight assumption uncertainty and ignores the actual certain contribution of each phase component related to certain deformational factors. Moreover, the commonly used mathematical empirical models in traditional InSAR are not suitable for describing the nonlinear characteristics of the temporal settlement evolution for soft clay. To address these limitations, we propose an SAR interferometry algorithm, namely, FIPR (FastICA Poisson-curve reciprocal accumulation method), which separates the original InSAR signal based on FastICA to extract each deformation component, and then the models can each extract deformation components and estimate the unknown parameters based on a reciprocal accumulation method. Each independent component and the obtained deformation parameters are used to generate the final deformation time series. Both simulated and real data experiments were designed. The simulated experimental results indicated that the sICA (spatial independent component analysis) separated results were much closer to the original signals than those of the tICA (temporal independent component analysis), with their RMSE lower than 2 mm, and the sICA is thus more highly recommended. Beijing Capital International Airport in China was selected as the study area in the real data experiment. Using 24 high-resolution TerraSAR-X radar satellite images from January 2012 to February 2015, the time-series deformation was obtained, with the maximum cumulative subsidence of 126 mm. The modeling accuracy for the proposed model was estimated as ±2.6 mm, with an improvement of 36.6% compared to the EWA-LM (linear model with equal weight accumulation) algorithm and 16.1% compared to the EWA-PC (Poisson curve with equal weight accumulation) algorithm. The RMSE with external leveling measurements was estimated as ±1.0 mm, with 69.7% improvement compared to EWA-LM and 50% to EWA-PC. This indicated that FIPR can reduce the uncertainty of artificial assumptions in deformation modeling and improve the accuracy of deformation analysis for highways in soft clay areas, providing a reference for road maintenance and management.

Keywords: InSAR; independent component analysis; soft clay; highway; deformation monitoring

1. Introduction

Long-term settlement monitoring of soft clay in urban areas is of great significance in preventing urban safety accidents and ensuring the quality of urban engineering construction [1–4]. Interferometric Synthetic Aperture Radar (InSAR) has been widely applied to ground surface movement monitoring with its advantages of large spatial coverage, high imaging resolution, and non-intrusive surveying. MT-InSAR (Multi-temporal Interferometric Synthetic Aperture Radar) technology, represented by persistent scatterer InSAR (PSI) [5], small baseline subsets InSAR (SBAS-InSAR) [6], and other advanced InSAR techniques, with its advantages of large spatial coverage, non-contact observation, and high accuracy for earth observation, has been proven to possess high capability in large urban infrastructure monitoring [7–13].

In InSAR data processing, deformation modeling is a crucial step. Most InSAR deformation models used for highway deformation monitoring are single mathematical empirical models (such as linear models [6], seasonal models [14], polynomial models [15], etc.) or their combinations, which ignore the specific deformation mechanisms of soft clays and are not able to describe the real temporal displacement law of the soft soil. This may affect the accuracy of the final derived deformation time series and counteract the subsequent highway deformation prediction and interpretation during the post-construction period. To correct these deficiencies, Zhu et al. proposed a temporal InSAR deformation model based on Poisson's curve and successfully implemented it in the soft clay highway deformation estimation, which showed an accuracy improvement of 67% compared to the traditional InSAR model [16]. However, in the study, each deformation component, such as the Poisson model-related component and the linear component, was directly superimposed on the established InSAR phase functions under the assumption of equal weights. Actually, the information of the differential interference phase was considerably complex, so the simple equal weight composition for all the deformation components under the equal weight assumption could not reflect the actual characteristics of deformation composition and introduced uncertainties and errors. If each signal in the InSAR phase can be accurately separated, accurate deformation-related components can be extracted and then modeled according to the displacement characteristics, the significance of the model will be enhanced, and the accuracy of the final acquired deformation will be greatly improved.

Independent component analysis (ICA) is a commonly used blind signal separation (BSS) technique employed to decompose mixed random signals into linear combinations of independent components. It has been widely used in image extraction [17], seismic wave decomposition [18], and measurement data processing [19–24]. Scholars have applied ICA to the separation of the InSAR phase and deformation signals to assist in the analysis of geophysical phenomena. Bellator successfully used ICA to extract the digital elevation models from SAR data [25]; Ebmeier et al. applied ICA in InSAR volcano monitoring [26] and proved that ICA was capable of analyzing and separating volcanic signals accurately, even under atmospheric noise interference situations; Gaddes et al. applied ICA to the 2015 Wolf volcanic eruption and developed an algorithm based on ICA technique to identify the volcanic eruptions [27]; Peng et al. utilized ICA in the separation of the mixed InSAR time-series signals and proved that ICA could not only reveal the existence of two different spatio-temporal deformation characteristics in the Willcox Basin but also filter the residuals in InSAR observations [28]. The above studies fully validated the feasibility and reliability of the idea that introducing ICA for InSAR phase separation can assist the subsequent deformation extraction and analysis.

To compensate for the uncertainty of artificial assumptions with equal weight modeling in traditional InSAR data processing and to improve the accuracy of deformation estimation, we propose an ICA-based deformation estimation method for soft clay regions—FIPR (FastICA Poisson-curve reciprocal accumulation method). First, we use the FastICA algorithm to separate the original InSAR signals to extract and analyze the deformation signals in InSAR data. Then, we use the Poisson curve to model the extracted deformation components and the periodic function to model the environment-related components, and, we then introduce the reciprocal accumulation algorithm to estimate the model parameters. Finally, we obtain the total time-series deformation results. The FIPR SAR Interferometry Algorithm proposed here can achieve high-accuracy in deformation monitoring and future deformation prediction. The proposed method can assist in revealing the information of deformational magnitude, boundary of occurred scope and area, suggesting potential risk areas and suspected hidden danger points, identifying the occurred and potential geological disaster areas to a certain extent, and providing guidance for the related government and constructions to carry out in-situ emphasis of monitoring areas.

2. Methodology

2.1. InSAR Signal Separation Based on FastICA

ICA is a data-driven method based on blind source separation. Its basic principle is to separate the multi-channel observation signals into statistically independent components according to a certain optimization criterion method [29,30]. Compared with the model-driven method, it can effectively detect signal features without any prior conditions, such as a functional model and the weight of each independent variable [31]. Ebmeier et al. proved that in the interferogram formed by the complex conjugation of two SAR radar images, each pixel can be regarded as a linear combination of specific points and noise signals in the time series of multiple potential deformation sources [26]. If these noise signals and deformation sources are statistically independent, spatially and temporally, the restoration of the original signal can be regarded as the problem of blind source separation. The relationship between the InSAR time-series phase signals and the independent components can be expressed as:

$$X = A \cdot S \tag{1}$$

where X is the obtained unwrapped temporal InSAR phase signals; A is the mixing matrix, where the rows represent the spatial responses of the independent components; S is the independent component matrix, where each row represents an independent component.

The original InSAR data are a set of image sequences that should be squeezed into a two-dimensional matrix $X{\cdot}(M \times N_{pixel})$ in spatial ICA (sICA) or $X{\cdot}(N_{pixel} \times M)$ in temporal ICA (tICA) processing. The different arrangement of X determines that ICA has different detection methods in the response domain of InSAR data. sICA considers that multiple regions related to deformation signals, noise, and other interference factors are independent of each other. It considers all N pixels at M time points as mixed observation vectors for independent component analysis and separates the images of M independent components. Among them, the number of mixed signals is the same as the number of interferograms (M), and each signal is sampled on N pixels. tICA considers that the time series related to deformation signals, noise, and other interference factors is independent. The mixed time series formed by N pixels sampled in M time points is decomposed as an observation vector to separate the time series of n independent components. Since the number of time-series InSAR observational data in the spatial dimension is much larger than that of the temporal dimension, and the signals of ground deformations are strongly independent of the noise signals, scholars generally use sICA to analyze InSAR data [26,28,32].

With the unwrapped time-series InSAR phases generated via the SBAS technique as the input original signals, the FastICA algorithm based on negative entropy is utilized here to decompose the time-series phases into statistically independent sets of components [25–27]. FastICA is a widely used fast optimization iterative ICA algorithm with good stable capability [33–36]. The InSAR phase time series for N unwrapped interferograms on the ith image can be represented as a vector x_i:

$$x_i = \left(\varphi_i^1, \varphi_i^2, \ldots, \varphi_i^N\right) \tag{2}$$

X_i consists of M vectors of interferograms, which can be expressed as:

$$X_M = (x_1, x_2, \ldots, x_M) = (\varphi_1, \varphi_2, \ldots, \varphi_M) \tag{3}$$

The unwrapped phase φ_i can be regarded as a linear combination of statistically independent components, which can be expressed as:

$$\varphi_i = \varphi_{i,defo} + \varphi_{i,topo} + \varphi_{i,orb} + \varphi_{i,atm} + \varphi_{i,noise} \tag{4}$$

where $\varphi_{i,defo}$ denotes the deformation-related component, which consists of soft clay settlement-related components φ_{defo_sof} and environmental-related components φ_{defo_per}:

$$\varphi_{defo} = \varphi_{defo\_sof} + \varphi_{defo\_per} \tag{5}$$

Considering (3) and (4), (1) can be written as:

$$X_{M\times N} = A_{M\times i} \cdot S_{i\times N} = A_{M\times i} \times \begin{pmatrix} \varphi^1_{i,defo\_sof}, \varphi^1_{i,def\_per}, \varphi^1_{i,topo}, \varphi^1_{i,orb}, \varphi^1_{i,atm}, \varphi^1_{i,noise} \\ \vdots \\ \varphi^N_{i,defo\_sof}, \varphi^N_{i,def\_per}, \varphi^N_{i,topo}, \varphi^N_{i,orb}, \varphi^N_{i,atm}, \varphi^N_{i,noise} \end{pmatrix} \tag{6}$$

For InSAR signals, deformation phases and noise-related signals can be considered statistically spatio-temporally independent. The InSAR observed signal can be decomposed into the combination of each component using the immobile points iterative algorithm in FastICA according to the characteristics of each component. Each row of A represents a spatial response of each independent component. The extracted interested deformation component $\varphi_{i,defo}$ can be used for subsequent physical modeling and deformation interpretation.

2.2. Time-Series Modeling of the Deformation Component

As discussed above, after φ_{defo_sof} and φ_{defo_per} are extracted through FastICA, the soft clay physical settlement-related component D_{sof} can be modeled based on the mechanism of soft soil. According to [16], the Poisson function is utilized here, which can be written as:

$$D_{sof} = \frac{\lambda}{4\pi}\varphi_{defo\_sof} = \frac{D_0}{1 + ae^{-bt}} \tag{7}$$

where λ is the radar wavelength, $D_{sof}(t)$ is the accumulated subsidence at moment t along the vertical direction; D_0 is the maximum settlement value of the surface; a and b are the shape parameters of the Poisson curve. The unknown parameters here are D_0, a, b [16,35]. The periodical function is used here to describe the environmental-related deformation component, which can be written as [14]:

$$D_{per} = a_1 t + a_2 \sin\frac{2\pi}{T}t + a_3 \cos\frac{2\pi}{T}t \tag{8}$$

where a_1, a_2, a_3 denote the seasonal coefficients. Considering (7) and (8), (5) can be rewritten as:

$$D = \frac{D_0}{1 + ae^{-bt}} + a_1 t + a_2 \sin\frac{2\pi}{T}t + a_3 \cos\frac{2\pi}{T}t. \tag{9}$$

Supposing N interferometric pairs are generated, the unknown parameters here are the Poisson curve parameters (D_0, a, b) and periodical function parameters (a_1, a_2, a_3). When the number of equations is higher than (or equal to) 6, the unknown parameters over all the high coherent points can be estimated.

2.3. Model Parameters Estimation Based on Reciprocal Accumulation Method

Solving the unknown parameters (D_0, a, b, a_1, a_2, a_3) in Equation (9) is a nonlinear parameter estimation problem, and the exponential term will aggravate the solving difficulties. Here, the Reciprocal Accumulation Method (RAM) is introduced to substitute for the GA algorithm [37]. The basic thought for RAM is transforming the Poisson curve modeled deformation time series based on reciprocal accumulation, and then solving the parameters based on their mathematical relationships, which can avoid the disadvantages of the time-consuming iterative process, high dependence on the initial searching range of unknown parameters, and poor accuracy of final solutions for GA algorithm. When using RAM, it is first necessary to divide the number of input time-series observations into three equal durations. Consequently, each period of the deformation sequence contains $r = \frac{n}{3}$ terms. Suppose $y_1, y_2, y_3, \ldots, y_n$ represent the generated deformation sequences, each divided duration can be denoted as:

$$\begin{aligned} s_1 &= t_1, t_2, \ldots, t_r, \\ s_2 &= t_{r+1}, t_{r+2}, \ldots, t_{2r}, \\ s_3 &= t_{2r+1}, t_{2r+2}, \ldots, t_{3r}. \end{aligned} \tag{10}$$

At this point, the reciprocal form of the Poisson function (7) can be written as:

$$\frac{1}{y_t} = \frac{1}{D_0} + \frac{ae^{-bt}}{D_0}. \tag{11}$$

The reciprocal accumulations for the three divided durations can be expressed as:

$$S_1 = \sum_{t_1}^{t_r} \frac{1}{y_t} = \frac{r}{D_0} + \frac{a}{D_0} \sum_{t_1}^{t_r} e^{-bt} = \frac{r}{D_0} + \frac{ae^{-b}\left(1 - e^{-rb}\right)}{D_0\left(1 - e^{-b}\right)}, \tag{12}$$

$$S_2 = \sum_{t_{r+1}}^{t_{2r}} \frac{1}{y_t} = \frac{r}{D_0} + \frac{ae^{-(r+1)b}\left(1 - e^{-rb}\right)}{D_0\left(1 - e^{-b}\right)}, \tag{13}$$

$$S_3 = \sum_{t=2r+1}^{t_{3r}} \frac{1}{y_t} = \frac{r}{D_0} + \frac{ae^{-(2r+1)b}\left(1 - e^{-rb}\right)}{D_0\left(1 - e^{-b}\right)}. \tag{14}$$

The parameter D_0, a and b can be estimated based on S_1 *to* S_3 by the following equations:

$$D_0 = \frac{r}{S_1 - \frac{(S_1 - S_2)^2}{(S_1 - S_2) - (S_2 - S_3)}}, \tag{15}$$

$$a = \frac{(S_1 - S_2)^2 D_0 (1 - e^{-b})}{((S_1 - S_2) - (S_2 - S_3))e^{-b}(1 - e^{-rb})}, \tag{16}$$

$$b = \frac{\ln \frac{S_1 - S_2}{S_2 - S_3}}{r}. \tag{17}$$

It should be noted that two requirements must be satisfied for RAM: (1) The number of datasets or periods in the time series should be a multiple of 3, so the total number can be divided into three durations. (2) Independent variable temporal intervals should be equal and continuous. Since the RAM is based on the algebraic relation between the model parameters, no initial values for the parameters and searching iterations are needed. Consequently, it can significantly improve the calculating efficiency and parameter estimation accuracy. After the Poisson function parameters are estimated, the remaining unknown periodic parameters (a_1, a_2, a_3) can be estimated by the least square algorithm.

2.4. Flow Chart of FIPR Algorithm and Processing Steps

Figure 1 shows the flow chart of the FIPR algorithm. The key steps are as follows:

(1) Unwrapped phase time series of differential interferometry are generated using the SBAS technique;

(2) Spatial ICA phase separation based on FastICA: generating the spatial independent component and temporal response of each component;

(3) Deformation modeling and parameter estimation: Deformation modeling for the extracted soft clay-related components with Poisson function and the environmental-related components with periodical functions; model parameters estimation using the reciprocal accumulation method, and generating the time series of each defor-mation component based on the estimated parameters;

(4) Deformation generation using the GARN algorithm and traditional equal-weighted accumulation modeling;

(5) Comparison and analysis of the total time-series deformation generated by the FIPR algorithm and the traditional equal-weighted accumulation modeling algorithm.

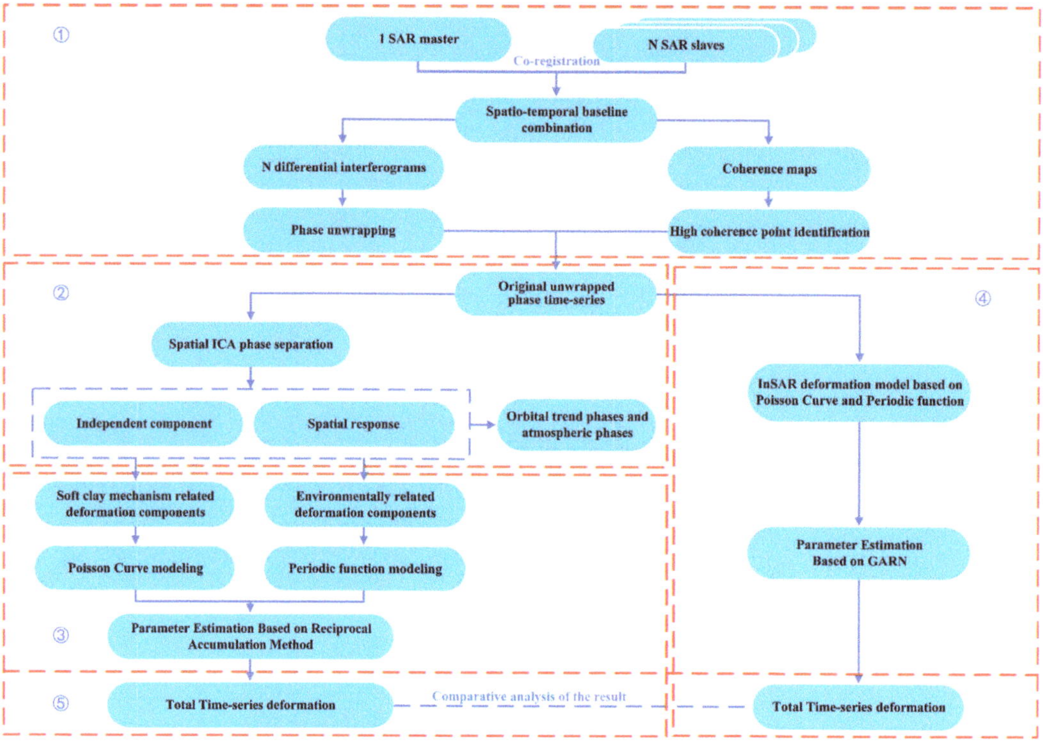

Figure 1. Flow chart of the FIPR algorithm.

3. Simulated Experiment

To verify the effect of ICA on the separation of deformation phases and guide the real data experiments, a simulation experiment was designed and executed here. Two groups (six images for each group) of temporal deformation fields were simulated based on the Poisson function represented as (7) and the periodical function represented as (8) with a region of 100 × 100 pixel area. The linear orbital error components were modeled with a linear trend mathematical function $z = jx + ky + c$, where j and k are random numbers following a normal distribution. The satellite parameters used here were set according to the images acquired in the real data experiment (TerraSAR X-band, Stripmap descending). After superimposing the three simulated original signals, two independent groups of experiments were conducted with temporal ICA (tICA) and spatial ICA (sICA) processing. Figure 2 shows partial results using sICA and tICA separation. It is clearly seen from the figure that the sICA separation fits better with the original signals compared to tICA.

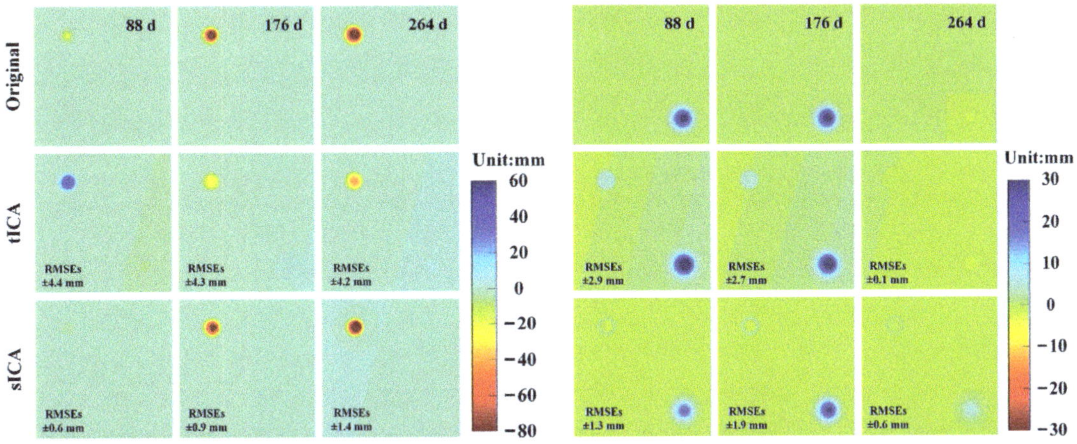

Figure 2. Comparison of the separated results of tICA and sICA at each temporal date (**left**: Poisson-related components; **right**: periodical deformation components).

To quantitatively evaluate the accuracy of the two groups of results, the RMSE (root mean square error) between the extracted deformation components and the simulated real values was calculated for each temporal acquisition. Figure 3 shows the sICA-separated results including the spatial responses (top) and time series (middle) for the two related deformation components, and the comparison between the RMSE of tICA (the length of each blue histogram) and sICA (the length of each top red histogram) generated results. The spatial response was normalized to −1 to 1. The overall RMSE of the Poisson-related components obtained by sICA was estimated as less than 2 mm, which was significantly lower than tICA, which had an RMSE higher than 6 mm. The maximum improvement for sICA was 85.8% at day 120. Similarly, for the periodical deformation components, the effect of sICA was still greatly improved, with an overall RMSE of 3 mm and a 22.3% improvement compared with that of tICA. This indicated that for the Poisson deformation signals, the sICA-separated results were much closer to the original signals than that of tICA. Since the high coherence points in the acquired InSAR dataset were spatially densely distributed but temporally sparsely discontinuous, sICA performed better for the signal separation than tICA [27]. Consequently, the spatial ICA algorithm is more recommended for our subsequent real data experiments.

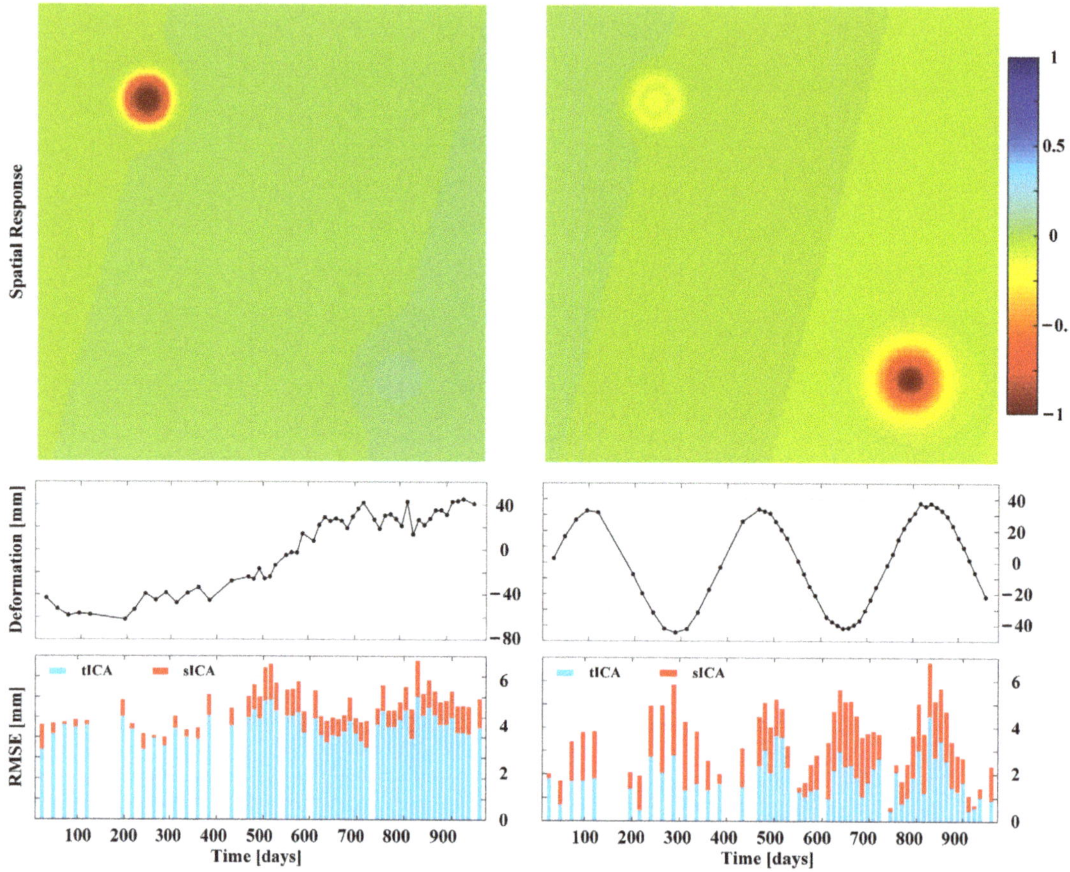

Figure 3. ICA separated results (**top-left**: spatial response of Poisson-related components, and **top-right**: spatial response of periodic deformation components; **middle-left**: time series of Poisson-related components and **middle-right**: time series of periodic deformation components; **bottom-left**: comparison of tICA and sICA-separated results of Poisson-related components at each temporal point; **bottom-right**: comparison of tICA and sICA separation results of periodic deformation components at each temporal point (the length of each blue histogram represents the RMSE value of tICA, whereas the top red one represents that of sICA).

4. Real Data Experiments

4.1. Study Area and Data Processing

The experimental area selected in this paper was around the Beijing Capital Airport. The soil in this area is mainly soft clay. The capital airport is the largest aviation port in China, and from 1978 to 2018, the annual passenger throughput of the Beijing Capital International Airport expanded from 1.03 to 110 million, ranking 1st in Asia and 2nd in the world. Figure 4 shows the geographic location of the study area. The capital airport is located in a flat area (within the elevation range of 22.0–35.4 m) between the alluvial fan of the Wenyu River and the Chaobai River, with a complex geological environment and partially crossing poor geological areas, such as lotus ponds and fish ponds. According to our in situ investigation and consultation of the reference geological materials, it was learned that the geological components of the roadbed were mostly newly deposited clayey and sandy soils [38–40]. Under the combined influence of natural and human activities, the equilibrium of the airport surface was highly susceptible to subsidence and disruption, which requires necessary long-term stability monitoring. A total of 24 TerraSAR-X-band Stripmap ascending images (track number 157), from 22 January 2012 to 6 February 2015, were collected and utilized. All the SAR interferometry processing was conducted via GAMMA software, and 23 high-quality unwrapped interferometric pairs were finally selected and generated for the subsequent processing [41].

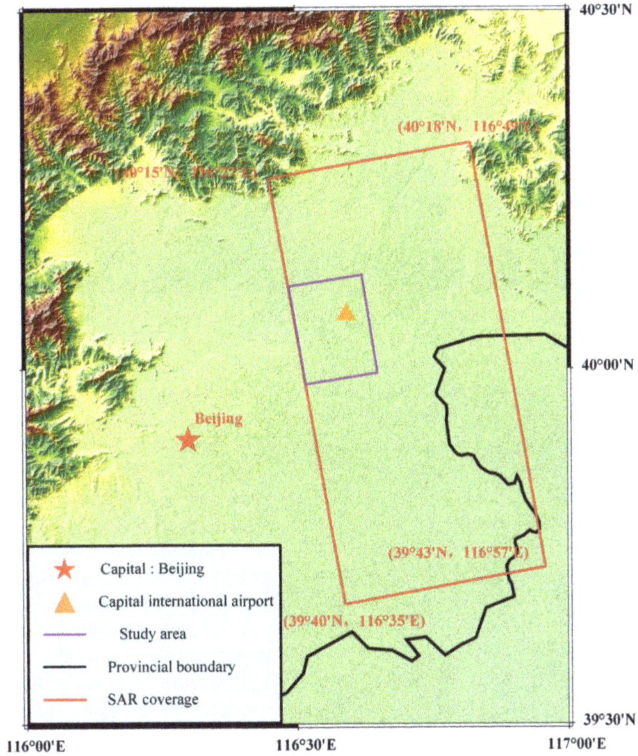

Figure 4. The geographic location of the study area.

4.2. *InSAR Phase-Independent Component Analysis*

As discussed in Section 3, the sICA algorithm was utilized in the real data experiments to conduct the independent phase component separation. For the specific implementation, to determine the appropriate number of independent components for our experiments, the contribution rates for each group of experiments under the specific number of components were calculated in advance. It was shown that for the five independent components, the cumulative contribution rate could reach 88%, indicating that when the number was greater than five, the remaining contribution of components occupied less than 12%, which could be ignored. The contribution of the first five independent components was defined according to the ratio value defined by Liu et al. [23], which were estimated as 8.56, 5.14, 8.25, 7.08, and 6.70, orderly. The smaller the ratio value, the greater the contribution of the corresponding independent component. Consequently, the number of independent components was set as five in our experiment. Figure 5 shows the obtained spatial response (top) and temporal response (bottom) of the five independent components (ICs) via the FastICA algorithm (the spatial response was normalized from −1 to 1). From Figure 5 we can see that for IC1 and IC3 there are no obvious spatial correlation characteristics in the spatial response image (top-left) and no significant variation pattern in the time-series response curve, inferring that the IC1 and IC3 signal may be related to the residual orbital error and partly topography-related tropospheric delay effects. For IC2 and IC5, the overall spatial response of most points in regions A and B showed a similar linear displacement trend, which was suggested to be related to the long-term deformation displacement, as their time-series responses showed an "S" shaped trend, reflecting a high dependence on time. Areas A and B were two typical subsiding areas, which will be further analyzed.

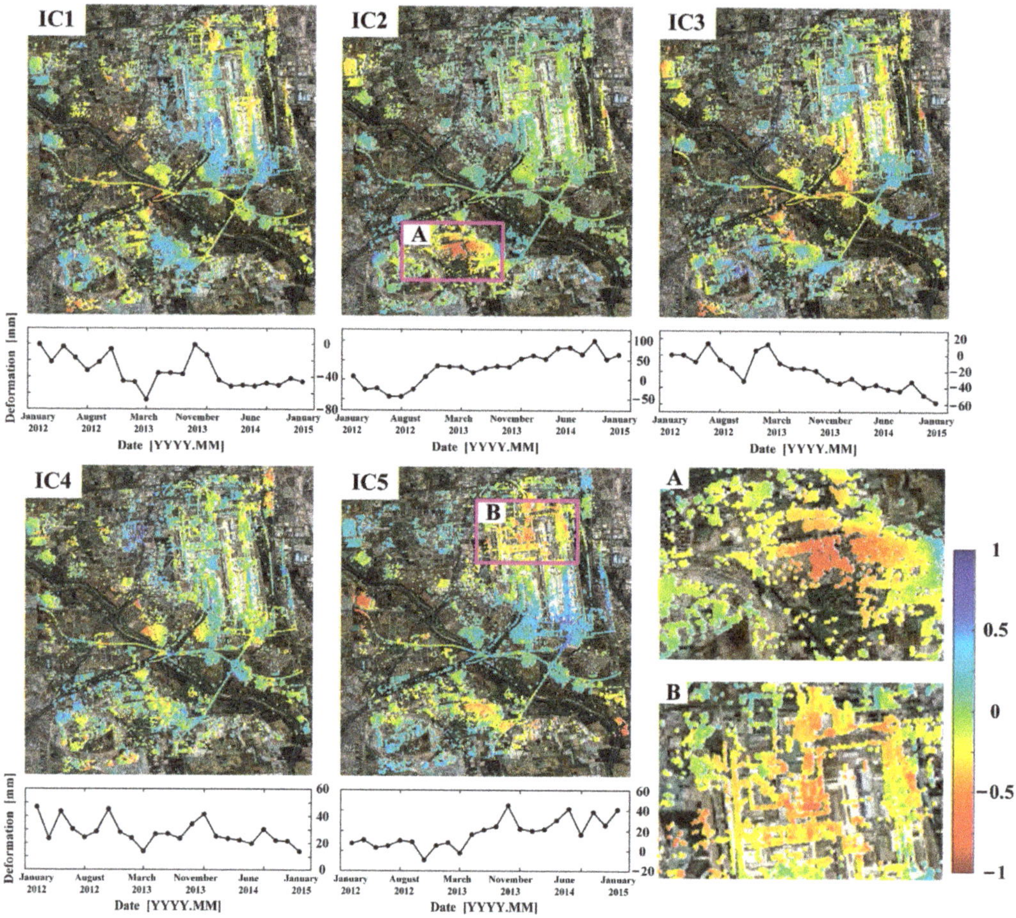

Figure 5. The spatial–temporal response of each independent component (**top**: spatial response of each independent component; **bottom**: time series of each independent component; Areas A and B were two typical subsiding areas.).

In Figure 5 we can also see that the IC4 signal showed an obvious periodic variation (as shown in the bottom-left of the figure); therefore, it is suggested as the environment-related deformation component. To reveal the correlation between the periodic deformation component and the precipitation, we conducted a correlation analysis between the extracted periodic deformation component and the precipitation in the area, which is shown in Figure 6. In the figure we can see that an obvious correlation with a 5-month lag effect was revealed. The average correlation coefficient between the periodic deformation component and precipitation was estimated as 0.536 and can be improved to 0.761 after the correction of the five-month lag.

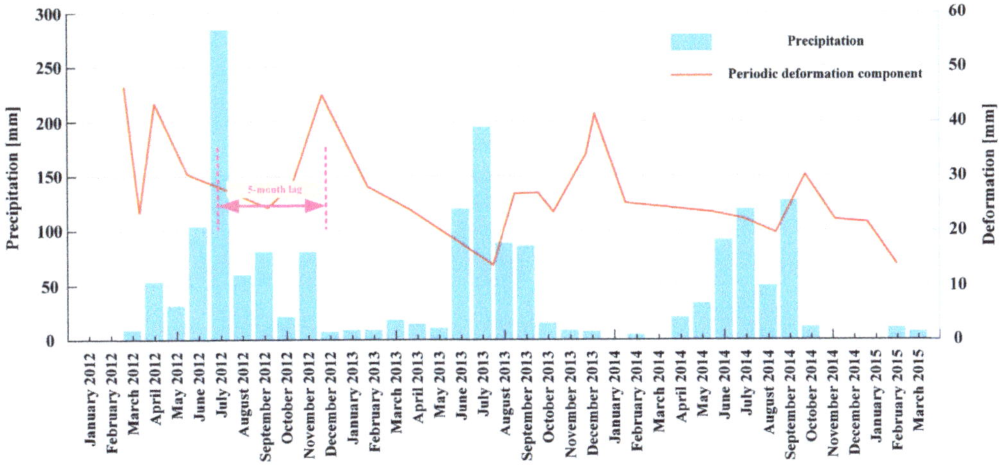

Figure 6. Correlation analysis between the periodic deformation components and precipitation.

4.3. Analysis of the Generated Total Time-Series Deformation via FIPR

Figure 7 shows the generated total time-series deformation results from 22 January 2012 to 6 February 2015 obtained via the FIPR algorithm. From the spatial color distribution, we can see that two obvious subsidence funnels were detected: the Jinzhan area (area A) and the northwest of the capital airport (area B), which are also revealed in Figure 5 (the enlarged image of A and B at the bottom-right of Figure 5). The area of Jinzhan mainly presented a color pattern of red-orange to crimson, and the overall average settlement over the area accumulated to 45 mm, with a maximum settlement of 126 mm on 6 February 2015. In contrast, the settlement in the northwestern part of the airport was significantly weaker, mainly orange-yellow, with overall average subsidence of higher than 40 mm and a maximum subsidence of 62 mm. The Jinzhan area suffered the most obvious ground subsidence in the Beijing plain, where the subsidence was suggested as mainly related to serious groundwater exploitation over the area. The development of ground settlement was easily induced by excessive groundwater exploitation, where the change in groundwater level accelerated the surface settlement [42]. In contrast, the degree of groundwater exploitation in the northwestern area of the capital airport was relatively weak, inducing a relatively stable phenomenon of surface deformation compared with the Jinzhan area. Judging from the temporal change in the color pattern over the area, the overall area stayed stable from 22 January to 20 September 2012, while from 20 September to 18 September 2013, the deformation increased under a gradually increased subsiding velocity; from 18 September 2013 to 6 February 2015, the deformation rate generally followed a trend from gradually decreasing to stable, which is consistent with the nonlinear subsiding law of soft clay consolidation disciplines.

4.4. Accuracy Analysis

To verify the accuracy of our proposed FIPR algorithm, first, we conducted a correlation analysis between the FIPR-generated time-series deformation results and the elevation of groundwater level at each monitoring well for groundwater pumping. The correlation and consistency between the deformation and the groundwater level depths can reflect the accuracy of the generated subsidence. Figure 8 shows the results of the average subsidence rate obtained via the FIPR algorithm, which was used as a background image to show the locations of six feature points near the monitoring wells (see P1–P6 in Figure 8 for the location of the feature points), where the feature points P1–P3 are located in the subsidence funnel of Jinzhan district (area A), and P4–P6 are located in the subsidence funnel of the northwest of the capital airport (area B). The pressurized groundwater measurements for the six monitoring wells within the subsidence funnels are drawn in Figures 9 and 10 (the purple dashed lines) and compared with the deformation results generated via traditional EWA-LM (linear model with equal weight accumulation) and EWA-PC (Poisson curve with equal weight accumulation). It is clear from the subfigures in Figures 9 and 10 that the blue line shows the best consistency with the purple dashed line (which denotes the underground water level) compared with the red and the green lines. This indicates that the FIPR-generated results showed the best accuracy. To quantitatively represent the correlation between the obtained deformation results at each feature point and the groundwater level, we calculated the exact correlation coefficients, which are shown in Table 1. The results indicated that the FIPR-generated deformations showed the best correlation coefficient, estimated as 0.946, whereas the maximum coefficient of EWA-LM and EWA-PC were 0.793 and 0.907, respectively. The reason for this phenomenon is suggested to be related to the artificial uncertainty in modeling induced by artificial assumptions for different physical deformation components via EWA-LM and EWA-PC. The uncertainty can be removed by the ICA technique and can be referenced in revising the weights of different deformation components in EWA-LM and EWA-PC.

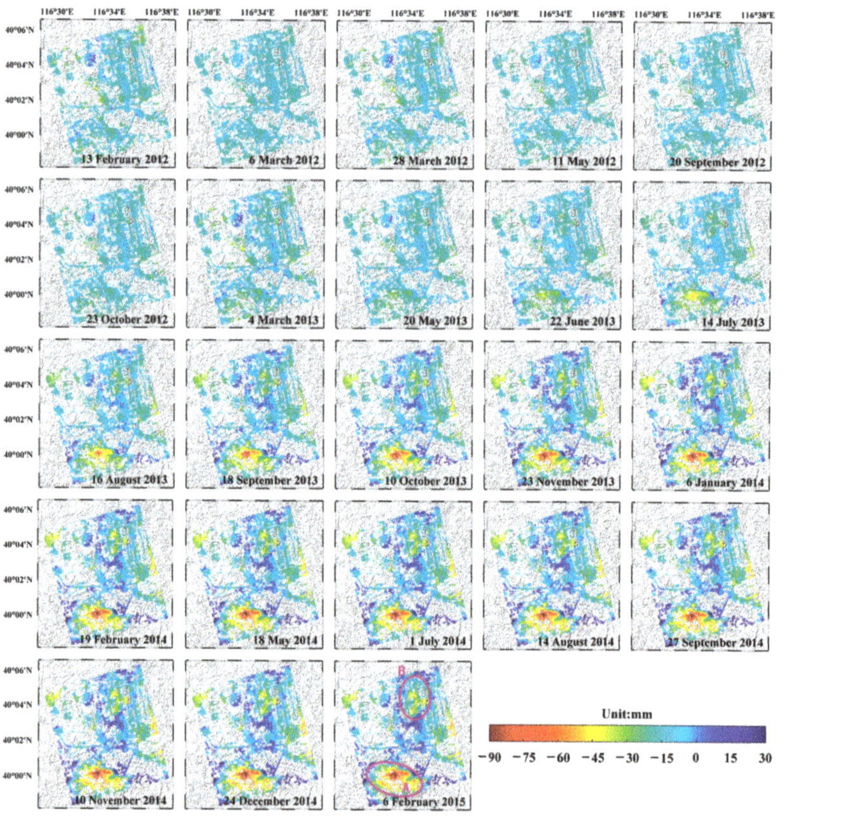

Figure 7. Time-series deformation results over the study area (with reference to 22 January 2012; Areas A and B were two typical subsiding areas.).

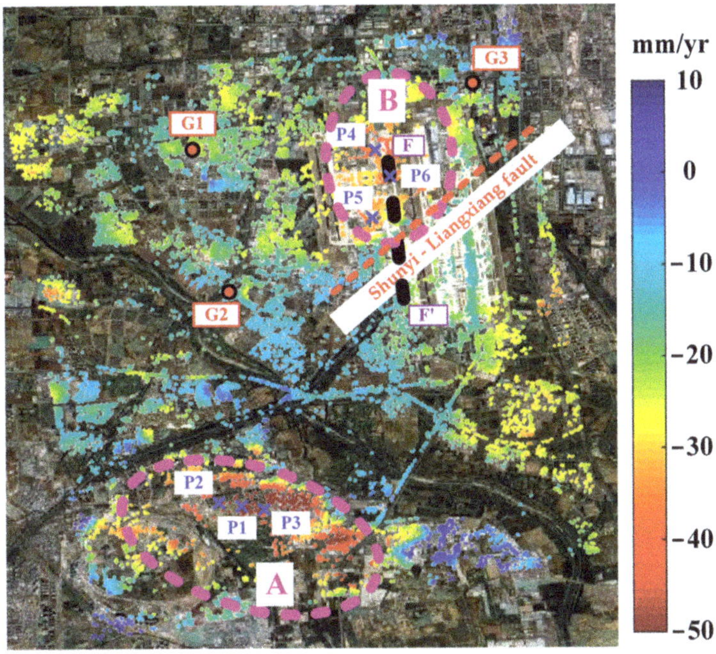

Figure 8. Locations of the ground-monitoring wells and ground benchmarks with the mean velocity map as background (Areas A and B were two typical subsiding areas; F-F′ is Shunyi-Liangxiang fault; G1–G3 is the position of three benchmarks; P1–P6 represent six feature points.).

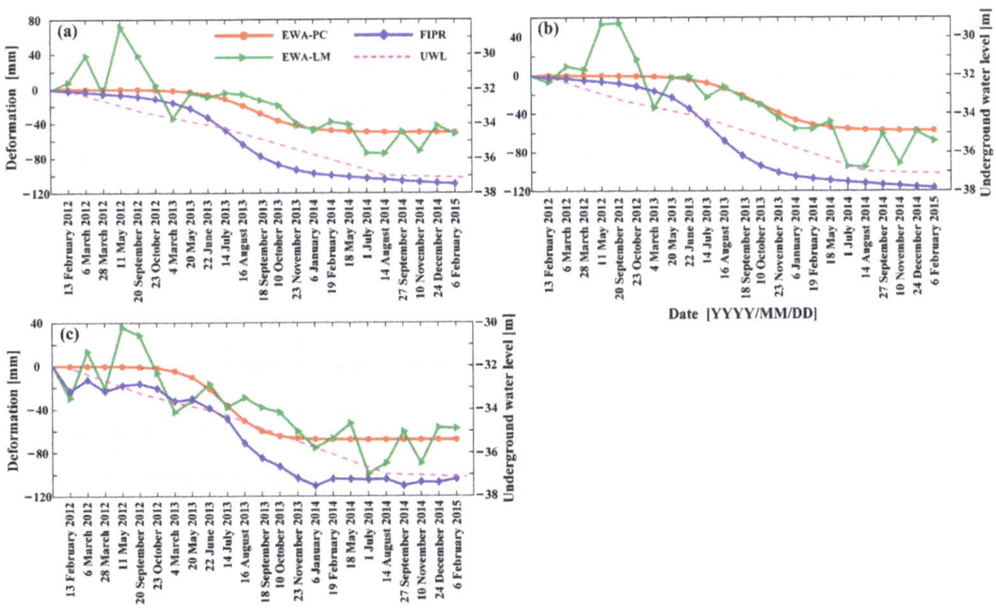

Figure 9. Correlation analysis between the time-series deformation results and the elevations of groundwater level (EWA-LM: linear model with equal weight accumulation; EWA-PC: Poisson curve with equal weight accumulation; UWL: elevation of groundwater level) at (**a**): P1; (**b**): P2; (**c**): P3.

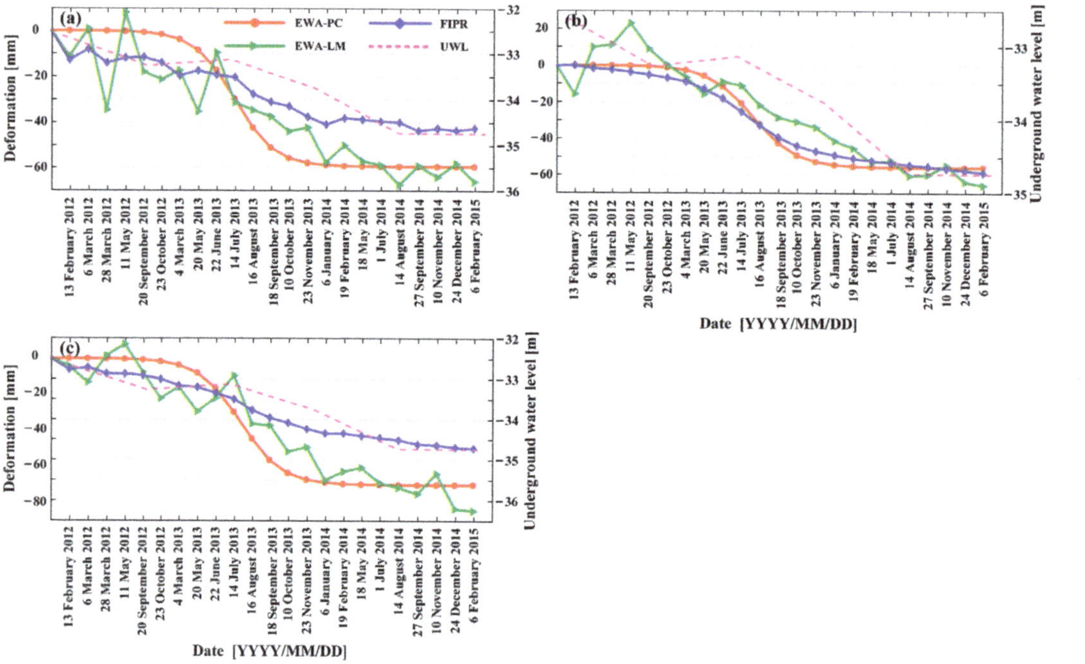

Figure 10. Correlation analysis between the time-series deformation results and the elevations of groundwater level (EWA-LM: linear model with equal weight accumulation; EWA-PC: Poisson curve with equal weight accumulation; UWL: elevation of groundwater level) (**a**): P4; (**b**): P5; (**c**): P6.

Table 1. Correlation coefficients between the deformation obtained by the three methods and the groundwater level.

Points	EWA-LM	EWA-PC	FIPR
P1	0.775	0.810	0.898
P2	0.793	0.835	0.895
P3	0.784	0.758	0.871
P4	0.779	0.818	0.932
P5	0.777	0.907	0.937
P6	0.786	0.802	0.946

To verify the modeling accuracy of FIPR, the magnitude of the high-pass (HP) deformation component was utilized here. According to [43], the HP deformation component can characterize the modeling accuracy of the deformation models. The smaller the HP deformation component, the higher the modeling accuracy. The HP deformation of each interferogram obtained through the FIPR was compared with those of EWA-PC and EWA-LM. Figure 11 shows a comparison of the HP deformation over all the high coherence points for each interferogram. As shown in Figure 11, for all the interferograms, the root mean square (RMS) of the HP deformation generated by the FIPR was estimated as ±2.6 mm, less than that of the EWA-LM, indicating that the FIPR was more suitable for the time-series deformation in the test area. The RMS of the HP deformation was estimated as ±4.1 mm for EWA-LM and ±3.1 mm for EWA-PC. The accuracy of FIPR was an improvement of 36.6% compared to that of EWA-LM and 16.1% compared to that of EWA-PC.

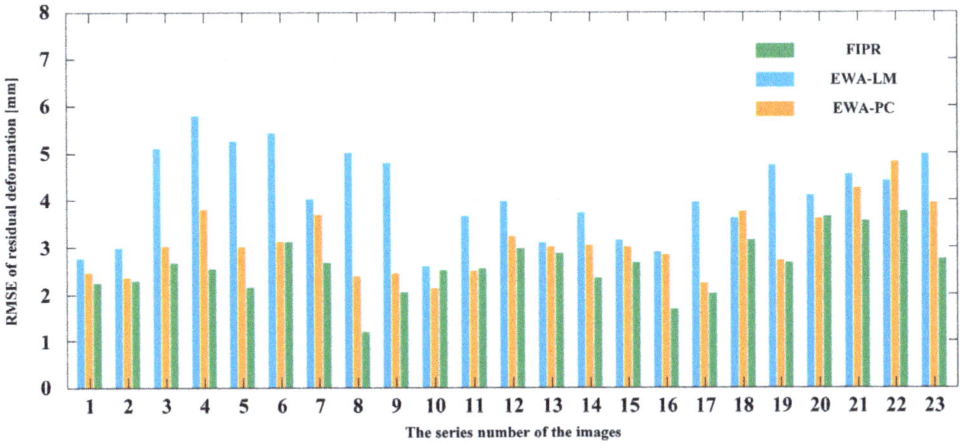

Figure 11. RMS of HP deformation comparison of 23 SAR images (EWA-LM: linear model with equal weight accumulation; EWA-PC: Poisson curve with equal weight accumulation).

In addition, we also collected the ground leveling measurements at three benchmarks in the test area to evaluate the external accuracy of our proposed algorithm. The locations of the three benchmarks are shown in Figure 8 (G1, G2, and G3), which are mainly distributed around the Capital International Airport. The temporal span of the leveling measurements was from September 2012 to September 2013. To conduct an accurate comparison, we transferred the generated line of sight (LOS) deformation into vertical displacement according to the equation $S_{LOS} = S_V \cos\theta$ and extracted the six dates of the measurements, which spanned periods consistent with the obtained SAR acquisitions. The comparison results are shown in Figure 12. As seen in the figure, the FIPR showed the best consistency with the leveling measurements. The bottom-left corner of Figure 12 shows the quantitative comparison results of the RMSE on the benchmarks. According to our calculation, the RMSE of FIPR was estimated as ±1.0 mm; compared with that of EWA-PC, the accuracy was improved by 50%, while the RMSE of EWA-PC was ±3.3 mm, with an improvement of 69.7%.

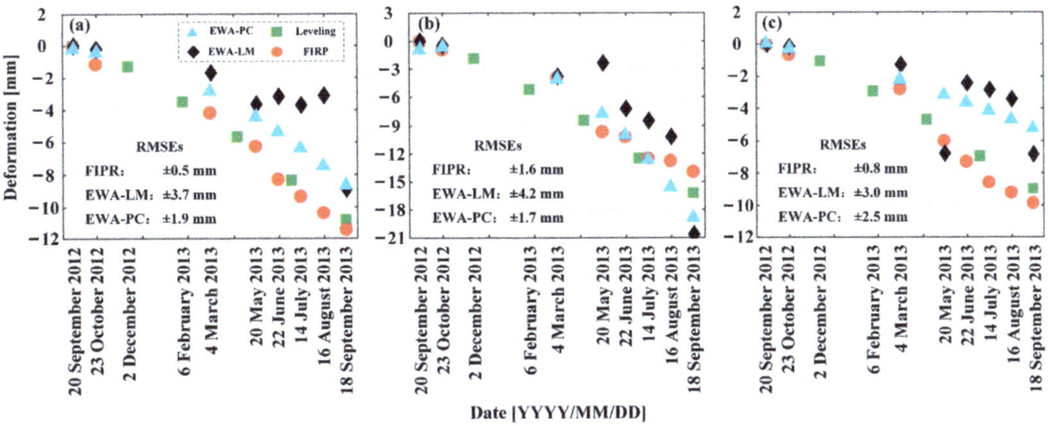

Figure 12. Time-series deformation results compared with leveling measurements on benchmarks. (**a**) G1 in Figure 8. (**b**) G2. (**c**) G3.

5. Discussions

5.1. Potential Reasons for the Derived Deformation

According to previous studies [44,45], the extent and rate of ground subsidence in the Beijing plain area were directly related to groundwater overexploitation. The pore elastic response of the soft clay layer to groundwater extraction acts as an elastic recoverable settlement or an inelastic, permanent compression-induced surface deformation. When the aquifer system exhibits continuous compression associated with a declining groundwater table, recoverable permanent subsidence of the soft clay layer occurs. As shown in Figure 8, the Jinzhan area (area A in Figure 5) is the most obvious subsiding area in the Beijing plain. Groundwater in the Jinzhan area is mainly endowed with the loose pore medium of the quaternary system, with fine particles of quaternary sediments and a large thickness of compressible layers. The quaternary system contains plentiful layers of pore aquifers, including both sand and gravel layers with coarser particles and good water-richness, and the clay layers with finer particles and higher compressibility, which were formed under the alluvial and flooding effects of the Yongding River and Chaobai River. The deposition environment is considerably complex. Affected by tectonic activities and the intensity of groundwater exploitation, the ground subsidence in the Jinzhan area maintains a serious level all year and is highly related to the variations in groundwater levels. Using the collected Beijing Water Resources Bulletin data from 2012 to 2015 provided by the Beijing Municipal Water Bureau [46], we obtained the overall depth of the groundwater level in the Capital International Airport area and compared it with the extracted time-series subsidence at the feature points (P1–P6 in Figure 8) for a detailed analysis. As shown in Figure 10, the time-series variations of groundwater level and the FIPR-derived subsidence of the feature points showed good consistency. When the groundwater level decreased from groundwater overexploitation, the voids between soil particles were compressed under an external constant load; thus, the deformation at this stage (before 2014) was characterized by a fast sinking rate; from July 2014, the natural compression of the soil layer reached its limit and the pore ratio decreased to a constant lowest value, causing the groundwater level to be stable. Consequently, the deformation was characterized by a decreasing subsiding velocity and tended to be stable afterward.

For other areas around the capital airport, the ground deformation was also mainly related to the underground responses of the soft clay with groundwater exploitation. As shown in Figure 10, we can see that from 2012 to 2013, the buried depth of the groundwater level in the capital airport area decreased significantly with reference to the beginning of the year, followed by a slight increase. The ground surface of the capital airport area remained stable during this stage; in contrast, in 2013, the serious over-extraction of groundwater led to a 1.5 m drop in the groundwater level, which induced serious surface subsidence of a maximum of 60 mm around the capital airport. At the end of 2014, water resources were sufficient in Beijing during the "South-to-North Water Diversion" project, which mitigated water use and groundwater exploitation in Beijing, and accordingly, the groundwater level tended to slowly stabilize. Consequently, the ground surface settlement maintained a stable status at this time.

In addition to the abovementioned effects of anthropogenic groundwater extraction, another geological-related reason was suggested here. From Figure 8 we can see that there is a clear linear boundary in the northwestern part of the capital airport (marked as a red dotted line), which is a potential correlation between the surface subsidence and the underground geological faults. According to [45], the Shunyi-Liangxiang fault was drawn with a red dashed line in Figure 8, located in the airport area. The Shunyi-Liangxiang fault crosses the entire airport and divides it into two parts with obviously different subsidence rates. To reveal the subsidence characteristics for the two different areas of the airport quantitively, a cross-section of the airport runway (marked as F-F′) was extracted and shown in Figure 13. From the figure, we can see that the Northwest part of the airport showed relatively serious subsidence with a maximum rate of 50 mm/yr, whereas the Eastern part had a relatively uniform and stable rate of lower than 10 mm/yr. The clearly different subsiding rates on both sides of the fault indicated that it was still active, and its location may impose a significant impact on the spatial distribution of ground subsidence. Therefore, close and continuous long-term subsidence monitoring of the airport runway in this area will provide an important reference for safety operations.

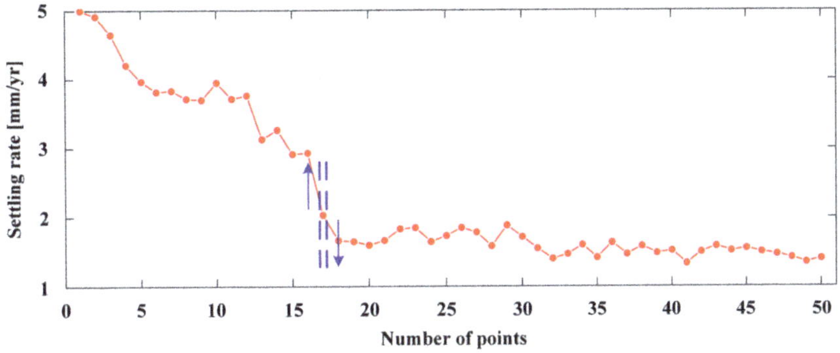

Figure 13. Cross-section analysis along F-F′ in Figure 8.

5.2. Applicability Analysis for ICA

(1) FastICA was used to separate the original InSAR signals to extract the exact component of each physical deformation signal based on each independent component, which can help to accurately determine the exact weight of each deformation signal related to different physical factors. The exact weight (or certain contribution of each component) can be utilized to guide subsequent deformation modeling, which can greatly reduce the uncertainty of artificial equal weight modeling assumptions. The results showed that the accuracy was significantly improved compared to the traditional InSAR equal weight modeling;

(2) The number of independent components (ICs) played an important role in ICA-based time-series InSAR analysis. Both too high and too low numbers of ICs will limit the accuracy of signal extraction. The number of ICs was mostly obtained through empirical and experimental calculations. The ideal number of ICs in this study was set as five, which was determined by experiments with different samples of numbers of ICs;

(3) In this paper, spatial ICA was more reliable than temporal ICA. The main reason was that the spatial independence of the deformation signal was better than the temporal independence in the study area, as revealed by the spatial–temporal expanding characteristics of the two subsiding funnels shown in Figure 7. As shown, during the monitoring period, the area of subsidence in areas A and B gradually expanded from 1.9 and 2.3 km^2 to 12.3 and 9.6 km^2, respectively, with clear independent subsiding boundaries; in contrast, according to our temporal correlation analysis, the correlation coefficient between the Poisson-related deformation component and the environment-related component was 0.43, indicating that the two signals were not independent (still showing a high correlation and not perfectly separated temporally). Therefore, spatial ICA separation was suggested here to produce a better time-series deformation signal related to different physical causes.

6. Conclusions

A time-series InSAR deformation estimation algorithm, namely, FIPR (FastICA Poisson-curve reciprocal accumulation method), was proposed in this paper. Each deformation component was extracted through the FIPR from the original InSAR unwrapped phase signals based on FastICA, and each extracted component was modeled based on its physical characteristics. The Poisson curve was utilized to model the soft soil-related physical deformation component, and the reciprocal accumulation method (RAM) was introduced to estimate the model parameters to generate the total time-series deformation.

Both simulated and real data experiments were designed and conducted. The simulated experimental results indicated that the sICA-separated signals were much closer to the original than those of tICA, with the RMSE lower than 2 mm, and accordingly were more highly recommended. In the real data experiment, the area around Beijing Capital International Airport in China was selected as the study area. With the use of 24 high-resolution TerraSAR-X radar satellite images, the deformation sequences from January 2012 to February 2015 were obtained with a maximum cumulative subsidence of 126 mm. The spatial–temporal evolution characteristics of the deformation were analyzed. Two obvious subsidence funnels were detected, where the Jinzhan region suffered the most obvious subsidence, with an accumulated settlement of up to 126 mm. The reason for the subsidence was suggested as related to serious overexploitation of the groundwater.

To evaluate the accuracy of the proposed algorithm, the RMS of HP deformations and the RMSE from the external leveling measurements were estimated. The accuracy of the proposed model was estimated as $\pm$2.6 mm, with an improvement of 36.6% compared to that of the EWA-LM (linear model with equal weight accumulation) algorithm and 16.1% compared to the EWA-PC (Poisson-curve with equal weight accumulation) algorithm. The RMSE from external leveling measurements was estimated as $\pm$1.0 mm, with 69.7% improvement compared to EWA-LM and 50% to EWA-PC. It can be inferred that RAM can avoid the disadvantages of traditional GA and improve the accuracy and efficiency of calculation. FIPR can reduce the uncertainty of artificial assumptions in deformation modeling and improve the accuracy of deformation analysis for soft clay areas, also providing a reference for road maintenance and traffic safety management.

7. Patents

There are patents resulting from the work reported in this manuscript. We have applied for a Chinese patent (202110368285.4) entitled "A monitoring method of soft clay deformation using time-series InSAR technology."

Author Contributions: Conceptualization, X.X. and L.Z.; methodology, X.X. and B.L.; software, X.M. and B.L.; validation, W.P. and L.Z.; formal analysis, X.X. and L.Z.; investigation, X.X.; resources, B.L. and X.M.; data curation, X.M.; writing—original draft preparation, L.Z. and X.M.; writing—review and editing, X.X., W.P. and R.Z.; visualization, W.P.; supervision, X.X.; project administration, X.X.; funding acquisition, X.X.,W.P. and R.Z. All authors have read and agreed to the published version of the manuscript.

Funding: This work was supported by the National Natural Science Foundation of China (No. 42074033, 41701536, 61701047, 41674040, 41904003), the Natural Science Foundation of Hunan Province (No. 2022JJ30589, 2017JJ3322, 2019JJ50639), Hunan International Scientific and Technological Innovation Cooperation Base of Advanced Construction and Maintenance Technology of Highway (Changsha University of Science and Technology) (kfj190805).

Acknowledgments: The TerraSAR dataset used in this study was provided by the DLR (Deutsches Zentrum für Luftund Raumfahrt, No. MTH3393, MTH3728).

Conflicts of Interest: The authors declare no conflict of interest.

References

1. Lin, H.; Ma, P.F.; Wang, W.X. Urban Infrastructure Health Monitoring with Spaceborne Multi-temporal Synthetic Aperture Radar Interferometry. *Acta Geo. Cartogr. Sin.* **2017**, *46*, 1421–1433.
2. Pan, W.Q. Monitoring and analysis of ground settlement during pipe roof construction of pipe-jacking groups in soft soil areas. *Chin. J. Geotech. Eng.* **2019**, *41*, 201–204.
3. Xing, X.M.; Chang, H.C.; Chen, L.F.; Zhang, J.H.; Yuan, Z.H.; Shi, Z.N. Radar Interferometry Time Series to Investigate Deformation of Soft Clay Subgrade Settlement—A Case Study of Lungui Highway, China. *Remote Sens.* **2019**, *11*, 429. [CrossRef]
4. Yu, J.; Sun, M.; He, S.; Huang, X.; Wu, X.; Liu, L. Accumulative Deformation Characteristics and Microstructure of Saturated Soft Clay under Cross-River Subway Loading. *Materials* **2021**, *14*, 537. [CrossRef] [PubMed]

5. Ferretti, A.; Prati, C.; Rocca, F. Permanent scatterers in SAR interferometry. *IEEE Trans. Geosci. Remote Sens.* **2001**, *39*, 8–20. [CrossRef]
6. Berardino, P.; Fornaro, G.; Lanari, R.; Sansosti, E. A new algorithm for surface deformation monitoring based on small baseline differential SAR interferograms. *IEEE Trans. Geosci. Remote Sens.* **2002**, *40*, 2375–2383. [CrossRef]
7. Luo, Q.; Zhou, G.; Perissin, D. Monitoring of Subsidence along Jingjin Inter-City Railway with High-Resolution TerraSAR-X MT-InSAR Analysis. *Remote Sens.* **2017**, *9*, 717. [CrossRef]
8. Qin, X.Q.; Yang, M.S.; Wang, H.M.; Yang, T.L.; Lin, J.X.; Liao, M.S. Application of High-resolution PS-InSAR in Deformation Characteristics Probe of Urban Rail Transit. *Acta Geo. Cartogr. Sin.* **2016**, *45*, 713–721.
9. Fan, H.D.; Gao, X.X.; Yang, J.K.; Deng, K.Z.; Yu, Y. Monitoring mining subsidence using a combination of phase-stacking and offset-tracking methods. *Remote Sens.* **2015**, *7*, 9166–9183. [CrossRef]
10. Qin, X.Q.; Liao, M.S.; Zhang, L.; Yang, M.S. Structural health and stability assessment of high-speed railways via thermal dilation mapping with time-series InSAR analysis. *IEEE J. Sel. Top. Appl. Earth Obs. Remote Sens.* **2017**, *10*, 2999–3010. [CrossRef]
11. Qin, X.Q.; Zhang, L.; Yang, M.S.; Luo, H.; Liao, M.S.; Ding, X.L. Mapping surface deformation and thermal dilation of arch bridges by structure-driven multi-temporal DInSAR analysis. *Remote Sens. Environ.* **2018**, *216*, 71–90. [CrossRef]
12. Chen, B.B.; Gong, H.L.; Chen, Y.; Lei, K.C.; Zhou, C.F.; Si, Y.; Li, X.J.; Pan, Y.; Gao, M.M. Investigating land subsidence and its causes along Beijing high-speed railway using multi-platform InSAR and a maximum entropy model. *Int. J. Appl Earth Obs. Geoinf.* **2021**, *96*, 102284. [CrossRef]
13. Wang, R.; Yang, T.L.; Yang, M.S.; Liao, M.S.; Lin, J.X. A safety analysis of elevated highways in Shanghai linked to dynamic load using long-term time-series of InSAR stacks. *Remote Sens Lett.* **2019**, *10*, 1133–1142. [CrossRef]
14. Li, S.S.; Li, Z.W.; Hu, J.; Sun, Q.; Yu, X.Y. Investigation of the Seasonal oscillation of the permafrost over Qinghai-Tibet Plateau with SBAS-InSAR algorithm. *Chin. J. Geophys.* **2013**, *56*, 1476–1486.
15. Zhang, Y.H.; Wu, H.A.; Sun, G.T. Deformation Model of Time Series Interferometric SAR Techniques. *Acta Geo. Cartogr. Sin.* **2012**, *41*, 864–869.
16. Zhu, L.J.; Xing, X.M.; Zhu, Y.K.; Peng, W.; Yuan, Z.H.; Xia, Q. An advanced time-series InSAR approach based on poisson curve for soft clay highway deformation monitoring. *IEEE J. Sel. Top. Appl. Earth Obs. Remote Sens.* **2021**, *14*, 7682–7698. [CrossRef]
17. Hoyer, P.O.; Hyvärinen, A. Independent component analysis applied to feature extraction from colour and stereo images. *Netw. Comput. Neural Syst.* **2000**, *11*, 191. [CrossRef]
18. De, L.E.; De, M.S.; Falanga, M.; Palo, M.; Notes, A. Decomposition of high-frequency seismic wavefield of the Strombolian-like explosions at Erebus volcano by independent component analysis. *Geophys. J. Int.* **2009**, *177*, 1399–1406.
19. Besic, N.; Vasile, G.; Chanussot, J.; Stankovic, S. Polarimetric incoherent target decomposition by means of independent component analysis. *IEEE Trans. Geosci. Remote Sens.* **2014**, *53*, 1236–1247. [CrossRef]
20. Forootan, E.; Kusche, J. Separation of global time-variable gravity signals into maximally independent components. *J. Geod.* **2012**, *86*, 477–497. [CrossRef]
21. Dai, W.J.; Huang, D.W.; Cai, C.S. Multipath mitigation via component analysis methods for GPS dynamic deformation monitoring. *GPS Solut.* **2014**, *18*, 417–428. [CrossRef]
22. Wen, H.J.; Huang, Z.W.; Wang, Y.L.; Liu, H.L.; Zhu, G.B. *Application of Independent Component Analysis in GRACE-Derived Water Storage Changes Interpretation: A Case Study of the Tibetan Plateau and Its Surrounding Areas*; Springer: Cham, Switzerland, 2015; pp. 179–188.
23. Liu, B.; Dai, W.J.; Liu, N. Extracting seasonal deformations of the Nepal Himalaya region from vertical GPS position time series using independent component analysis. *Adv. Space Res.* **2017**, *60*, 2910–2917. [CrossRef]
24. Luo, F.X.; Dai, W.J.; Tang, C.P.; Huang, D.W.; Wu, X.X. EMD-ICA with Reference Signal Method and Its Application in GPS Multipath. *Acta Geo. Cartogr. Sin.* **2021**, *41*, 366–371.
25. Ballatore, P. Extracting digital elevation models from SAR data through independent component analysis. *Int. J. Remote Sens.* **2011**, *32*, 3807–3817. [CrossRef]
26. Ebmeier, S.K.; Biggs, J.; Mather, T.A.; Amelung, F. *Applicability of InSAR to Tropical Volcanoes: Insights from Central America. Geological Society*; Special Publications: London, UK, 2013; Volume 380, pp. 15–37.
27. Gaddes, M.E.; Hooper, A.; Bagnardi, M.; Inman, H.; Albino, F. Blind signal separation methods for InSAR: The potential to automatically detect and monitor signals of volcanic deformation. *J. Geophys. Res. Solid Earth.* **2018**, *123*, 10226–10251. [CrossRef]
28. Peng, M.M.; Lu, Z.; Zhao, C.Y.; Motagh, M.; Lin, B.; Conway, B.D.; Chen, H.Y. Mapping land subsidence and aquifer system properties of the Willcox Basin, Arizona, from InSAR observations and independent component analysis. *Remote Sens. Environ.* **2022**, *271*, 112894. [CrossRef]
29. Jutten, C.; Herault, J. Blind separation of sources, Part I: An adaptive algorithm based on neuromimetic architecture. *Signal Process.* **1991**, *24*, 1–10. [CrossRef]
30. Zhu, K.; Zhang, X.; Sun, Q.; Wang, H.; Hu, J. Characterizing Spatiotemporal Patterns of Land Deformation in the Santa Ana Basin, Los Angeles, from InSAR Time Series and Independent Component Analysis. *Remote Sens.* **2022**, *14*, 2624. [CrossRef]
31. Stone, J.V. Independent component analysis: An introduction. *Trends Cogn. Sci.* **2002**, *6*, 59–64. [CrossRef]

32. Maubant, L.; Pathier, E.; Daout, S.; Radiguet, M.; Doin, M.-P.; Kazachkina, E.; Kostoglodov, V.; Cotte, N.; Walpersdorf, A. Independent Component Analysis and Parametric Approach for Source Separation in InSAR Time Series at Regional Scale: Application to the 2017–2018 Slow Slip Event in Guerrero (Mexico). *J. Geophys. Res. Solid Earth.* **2020**, *125*, e2019JB018187. [CrossRef]
33. Hyvarinen, A. Fast and robust fixed-point algorithms for independent component analysis. *IEEE Trans. Neural Netw.* **1999**, *10*, 626–634. [CrossRef] [PubMed]
34. Hyvärinen, A.; Oja, E. Independent component analysis: Algorithms and applications. *Neural Netw.* **2000**, *13*, 411–430. [CrossRef]
35. Ma, X.J.; Liu, B.; Dai, W.J.; Kuang, C.L.; Xing, X.M. Potential Contributors to Common Mode Error in Array GPS Displacement Fields in Taiwan Island. *Remote Sens.* **2021**, *13*, 4221. [CrossRef]
36. Ge, R.Y.; Wang, Y.B.; Zhang, J.P.; Yao, L.; Zhang, H.; Long, Z.Y. Improved FastICA algorithm in fMRI data analysis using the sparsity property of the sources. *J. Neurosci. Meth.* **2016**, *263*, 103–114. [CrossRef] [PubMed]
37. Zai, J.M.; Mei, G.X. Forecast method of settlement during the complete process of construction and operation. *Rock. Soil Mech.* **2000**, *21*, 322–325.
38. Chen, B.B.; Gong, H.L.; Lei, K.C.; Li, J.W.; Zhou, C.F.; Gao, M.L.; Guan, H.L.; Lv, W. Land subsidence lagging quantification in the main exploration aquifer layers in Beijing plain, China. *Int. J. Appl. Earth Obs. Geoinf.* **2019**, *75*, 54–67. [CrossRef]
39. Chen, M.; Tomás, R.; Li, Z.H.; Motagh, M.; Li, T.; Hu, L.Y.; Gong, H.Y.; Li, X.J.; Yu, J.; Gong, X.L. Imaging land subsidence induced by groundwater extraction in Beijing (China) using satellite radar interferometry. *Remote Sens.* **2016**, *8*, 468. [CrossRef]
40. Gao, M.L.; Gong, H.L.; Chen, B.B.; Zhou, C.F.; Chen, W.F.; Liang, Y.E.; Shi, M.; Si, Y. InSAR time-series investigation of long-term ground displacement at Beijing Capital International Airport, China. *Tectonophysics* **2016**, *691*, 271–281. [CrossRef]
41. Costantini, M.; Rosen, P.A. A Generalized Phase Unwrapping Approach for Sparse Data. In Proceedings of the International Geoscience and Remote Sensing Symposium, Hamburg, Germany, 28 June–2 July 1999; pp. 267–269.
42. Chen, B.B.; Gong, H.L.; Li, X.J.; Lei, K.C.; Zhu, L.; Gao, M.L.; Zhou, C.F. Characterization and causes of land subsidence in Beijing, China. *Int. J. Remote Sens.* **2017**, *38*, 808–826. [CrossRef]
43. Zhao, R.; Li, Z.W.; Feng, G.C.; Wang, Q.J.; Hu, J. Monitoring surface deformation over permafrost with an improved SBAS-InSAR algorithm: With emphasis on climatic factors modeling. *Remote Sens. Environ.* **2016**, *184*, 276–287. [CrossRef]
44. Zhou, C.F.; Gong, H.L.; Chen, B.B.; Zhu, F.; Duan, G.Y.; Gao, M.L.; Lu, W. Land subsidence under different land use in the eastern Beijing plain, China 2005–2013 revealed by InSAR timeseries analysis. *GISci. Remote Sens.* **2016**, *53*, 671–688. [CrossRef]
45. Gao, M.L.; Gong, H.L.; Li, X.J.; Chen, B.B.; Zhou, C.F.; Shi, M.; Lin, G.; Chen, Z.; Ni, Z.Y.; Duan, G.Y. Land subsidence and ground fissures in Beijing capital international airport (bcia): Evidence from quasi-ps insar analysis. *Remote Sens.* **2019**, *11*, 1466. [CrossRef]
46. Beijing Water Authority. Available online: http://swj.beijing.gov.cn (accessed on 11 December 2021).

remote sensing

MDPI

Article

Quantitative Evaluation of Environmental Loading Products and Thermal Expansion Effect for Correcting GNSS Vertical Coordinate Time Series in Taiwan

Bin Liu [1,2], Xiaojun Ma [1,2], Xuemin Xing [1,2,*], Jianbo Tan [1,2], Wei Peng [1,2] and Liqun Zhang [1,2]

1 Engineering Laboratory of Spatial Information Technology of Highway Geological Disaster Early Warning in Hunan Province, Changsha University of Science & Technology, Changsha 410114, China
2 School of Traffic and Transportation Engineering, Changsha University of Science & Technology, Changsha 410114, China
* Correspondence: xuemin.xing@csust.edu.cn

Abstract: We explored the driving factors of nonlinear signals in vertical coordinate sequences of stations in a Taiwan global navigation satellite system (GNSS) network, including atmospheric loading (ATML), hydrological loading (HYDL), and non-tidal ocean loading (NTOL) effects. At the same time, we used the finite element analysis software MIDAS to quantify the vertical displacements of different types of monuments due to the thermal expansion effect, including deep drilled braced (DDB) and short drilled braced (SDB). By quantitatively comparing the correction results of GNSS time series with different single mass loading models, we found that there was little difference in the correction of different environmental loading products. We compared different combinations of each loading product to correct the GNSS time series, and finally selected the best combination suitable for Taiwan GNSS network, that is, ATML (GFZ_ECMWF IB) + HYDL (IMLS_MERRA2) + NTOL (IMLS_MPIOM06). We found that the spatial and temporal models of ATML and NTOL are very similar, with non-tidal atmospheric loading and non-tidal ocean loading working together, a pattern that may be related to tropical cyclones. Both models also showed good correction effect on GNSS stations in the western plain of Taiwan, but with limited correction effect in the eastern part of Taiwan. This may be due to the influence of the subtropical monsoon climate in Taiwan and the barrier of the central mountain range, resulting in obvious differences between eastern and western Taiwan. The hydrological loading was found to act in the opposite way to the thermal expansion effect in the temporal domain, indicating that some displacements in hydrological loading may cancel out displacements caused by the thermal expansion effect. This aspect of displacement is not included in the hydrological loading model but should be considered when accurately estimating the temporal and spatial variation of water storage capacity in Taiwan using GNSS observed displacements.

Keywords: GNSS; environmental loading; thermal expansion effect; monument types

Citation: Liu, B.; Ma, X.; Xing, X.; Tan, J.; Peng, W.; Zhang, L. Quantitative Evaluation of Environmental Loading Products and Thermal Expansion Effect for Correcting GNSS Vertical Coordinate Time Series in Taiwan. *Remote Sens.* **2022**, *14*, 4480. https://doi.org/10.3390/rs14184480

Academic Editors: Alex Hay-Man Ng, Linlin Ge, Hsing-Chung Chang and Zheyuan Du

Received: 8 August 2022
Accepted: 5 September 2022
Published: 8 September 2022

1. Introduction

Moving masses of surface fluid layers interact with the solid Earth, resulting in periodic weak deformation and surface displacements [1]. Permanent reference stations on the ground can capture these non-tectonic deformation signals by long-term, continuous, and high-precision monitoring. Previous studies have identified three main classes of nonlinear changes causing the nonlinear variation in Global Navigation Satellite System (GNSS) derived coordinate time series. The first is error caused by imperfect GNSS data processing models or other systematic errors, such as high-order ionospheric delay errors, multipath effects [2–5], etc. The second is due to real nonlinear movement of the reference stations, including periodic displacements of the reference stations caused by geophysical effects, such as environmental loading (atmospheric loading, hydrological loading, or non-tidal ocean loading) and thermal expansion effects of bedrock and the monument [6–11], exhibited as seasonal periodic oscillation in the GNSS coordinate time series. The third is the periodic changes caused by observation noise. By studying the periodic deformation of environmental loads, the annual periodic amplitude of non-tectonic deformation of GNSS can be effectively reduced for improving the velocity estimation accuracy.

There has been increased attention paid to quantitative analysis of the contributions of different factors to GNSS coordinate time series, especially for the environmental loading effect. However, the correction effect of environmental loading deformations for GNSS coordinate time series varies greatly in different regions [11–14]. Moreover, for the quantitative calculation of loading deformation and displacements, the use of different Earth models, reference frames, mass loading data models, and coastline data from different institutions also give different calculation results. Li et al. [13] studied the individual contributions of five different continental water storage loading models, six different non-tidal atmospheric loading models, and five different non-tidal ocean loading models to the 220 GNSS coordinate time series in the Crustal Movement Observation Network of China (CMONOC), considering the joint contributions of different permutations and combinations of load products to GNSS time series. The results have a RMS reduction in the vertical component of 20%. Wu et al. [14] compared the contributions of environmental loading sequences from the German Research Center for Geosciences (GFZ) and the School and Observatory of Earth Sciences (EOST) to GNSS vertical time series. Compared with the RMS reduction value for individual stations, the combined model using sequences from both GFZ and EOST was better than use of GFZ alone to correct coordinate time series. Results can also be influenced by the independent climatic and topographic characteristics of different regions. Jiang et al. [11] considered the differences of hydrological data, grid interpolation, and terrain correction in the optimum model data (OMD) model for reduced scattering at 74% of global stations. One suggestion from the above results is that the loading sequences from different organizations should be adapted to local conditions, and the correction of GNSS stations coordinate time series should be carefully selected for different regions.

In addition to environmental loading, the thermoelastic deformation of the antenna monument of GNSS reference station and its bedrock driven by periodic temperature changes also contribute to nonlinear signals in time series data [7,15]. On the seasonal scale, the annual amplitude caused by the thermal expansion effect of the monument can exceed 5 mm for GNSS reference stations where the monuments are located at high latitudes [10]. Dong et al. [7] found that the annual vertical deformation of the surface caused by a change of surface temperature is less than 0.56 mm based on the semi-infinite space model. Yan et al. [16] calculated the influence of temperature change on the underground bedrock and cement pier of GNSS station in China, and showed that the maximum annual amplitude of vertical displacements caused by thermal expansion effect can reach 2.8 mm. Based on the three-dimensional full space thermoelastic deformation model, Tan et al. [17] calculated the influence of temperature change on the three-dimensional annual amplitude for China. The results showed that temperature change was an important factor causing continental surface deformation in addition to mass loading. Jia et al. [18] quantitatively analyzed the difference of vertical deformation between GNSS and Gravity Recovery and Climate Experiment (GRACE), and analyzed the influence of thermal expansion effect on vertical displacements of GNSS and estimation of vertical loading deformation of surface by GRACE. The results showed more consistent vertical deformation of GNSS and GRACE for the Chinese mainland after correcting for the thermal expansion effect in GNSS time series data. Boccara et al. [19] compared the differences between numerical models of land surface temperature from different sources at the global scale, and found that the annual amplitude of land surface temperature data provided by the European Centre for Medium-Range Weather Forecasts (ECMWF) was closest to the measured temperature data. Vertical displacements due to thermal expansion effect include the monument (TEM) on the ground and the bedrock (TEB) underground, but not any displacements caused by different specifications and materials of the monument. The GNSS network in Taiwan Province, China is operated by multiple organizations, so Rao et al. [20] classified more than 90% of the base types of stations and studied the performance of GNSS coordinate time series for the different bases. They found that the coordinate time series of metal rod monument has significant annual periodic variation, and the months of annual amplitude uplift and subsidence are similar to temperature variation.

Taiwan Province is located in an active tectonic area. With continuous collision between Luzon arc and Chinese mainland margin, there are many thrust faults and folds. Since 1990, there have been many destructive earthquakes with magnitude greater than 6. To obtain a more reliable velocity field, there is a significant need for more accurate estimation of nonlinear signals such as environmental loads and geophysical factors in GNSS coordinate sequence data. Kumar et al. [21] studied the sources of GNSS common-mode errors (CME) in Taiwan Province, and showed that 90% of the CME variations in Taiwan are related to non-seasonal ATML displacements. Ma et al. [22] used independent component analysis (ICA) to separate the signals related to atmospheric and hydrological loading from GNSS residual coordinate time series in Taiwan Province. Lai et al. [23] used data from GNSS stations in Taiwan to study inversion terrestrial water storage (TWS). The results showed that GNSS shows large annual vertical displacements in the plain area of southwest Taiwan and the Huadong Valley in the east of Taiwan. Using GNSS, GRACE, and hydrological (precipitation, GLDAS and LSDM assimilation models, and in situ groundwater level) datasets, Hsu et al. [24] systematically studied the spatial and temporal changes of water storage capacity in Taiwan. The results showed basically consistent spatial patterns of annual water storage capacity estimated by GNSS, GLDAS, and precipitation data, indicating that there is significant seasonal hydrological loading fluctuation in Taiwan.

Based on the findings of these previous studies, we analyzed the influence of atmospheric loading (ATML), hydrological loading (HYDL), and non-tidal ocean loading (NTOL) on Taiwan GNSS time series data. We considered the influence of temperature change on different bases of the stations, and accurately calculated the change in vertical displacements due to temperature changes of GNSS stations. We studied 84 GNSS sites in Taiwan Province from 1 January 2008 to 31 December 2012. These data were corrected using six ATML, five HYDL, and three NTOL products provided by International Mass Loading Service (IMLS), GFZ, and EOST. Displacements caused by thermal expansion effect (TEE) were characterized. Finally, the RMS reduction value was used to evaluate the influence of non-tectonic signals on GNSS sequences. By comparing the mass loading models provided by different institutions, we determined a group of combined mass loading models suitable for correcting GNSS time series in Taiwan Province. At the same time, we quantified the vertical displacements of DDB and SDB monument stations affected by temperature changes. These would be helpful to study the driving factors in nonlinear signals in GNSS vertical coordinate time series in Taiwan Province.

2. Data and Methodology

2.1. GNSS Data

The Taiwan continuous GNSS Network was established by the Institute of Earth Sciences (IESAS) in 1990. Currently, it consists of 539 stations that belong to different institutions and are operated by the Central Weather Bureau (CWB), IESAS, the Central Geological Survey (CGS), and the Ministry of the Interior (MOI). We utilized the classification of stations base forms devised by Rao et al. [20] and classified the stations into six types (Figure 1): deep drilled braced (DDB) monument (38), short drilled braced (SDB) monument (217), concrete pillar (30), roof (188), metal rod (14), and other forms (52).

After the above classification of station bases, we selected 84 DDB and SDB stations with high data quality for follow-up analysis. DDB and SDB stations were selected because these were constructed according to fixed specifications of the Central Weather Bureau and the Southern California Integrated GPS Network (SCIGN) (Figure 2a,b). DDB and SDB stations differ in that the DDB base contains five 12 m 316 stainless steel pipe supports that are buried 10 m deep underground. SDB contains only four stainless steel pipe supports, buried 1 to 2 m underground. Due to limited funds, regional restrictions, and construction difficulties in high mountain areas, GNSS stations in Taiwan have mostly adopted the SDB pedestal. The advantage of this pedestal is that these stations are mainly distributed in suburbs and mountainous areas far away from cities, and this kind of pedestal is relatively stable and is less noisy. Although roof pedestals account for 35% of GNSS observation network in Taiwan, these stations are mostly located in cities and areas with many buildings. To calculate the thermal expansion effect, many factors must be considered, so roof stations were not considered in this study.

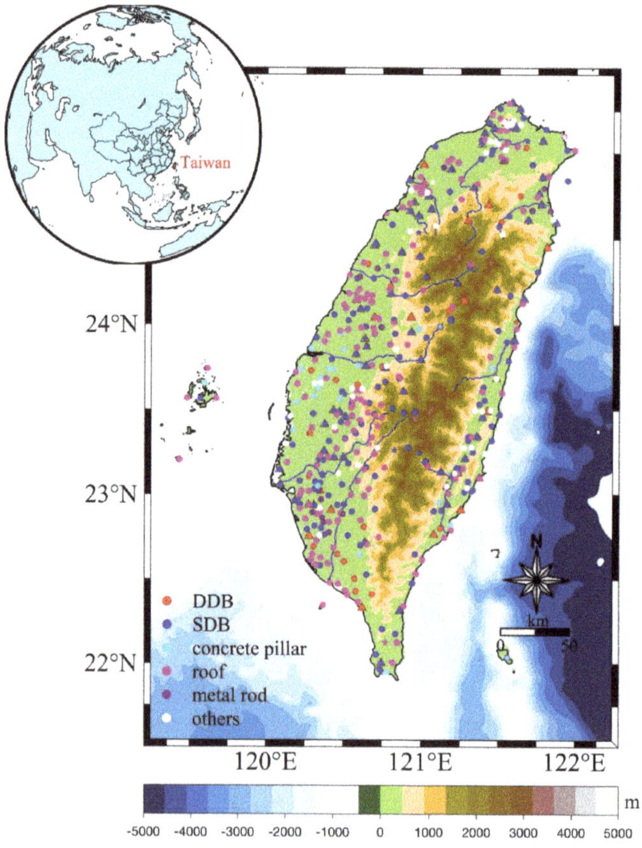

Figure 1. Distribution map of GNSS stations monument types in Taiwan Province. The circle represents the stations that are still in operation, the pentagram represents the stations that have been deactivated, and the triangle represents the stations used in this paper. Different colors represent different monument types: red are the deep drilled braced (DDB) monument stations, blue are the short drilled braced (SDB) monument stations, cyan are the concrete pillar monument stations, magenta are the roof stations, and purple are the metal rod type stations.

The GNSS data from 2008 to 2012 is processed using the GAMIT/GLOBK software packages [25], and a composite daily solution in the ITRF 2008 reference framework [26] is generated for subsequent time series analysis. IERS 2003 model and FES 2004 models were used for data processing to correct for polar tide and sea-ocean tidal loading [27,28]. Separate documents for each month provided by different institutions. After cubic spline interpolation, we obtained the predicted time series of loading products at each GNSS station. Our goal was to study the influence of environmental loading and thermal expansion on GNSS time series. After eliminating outliers, we used the following standard procedure to fit the time series data [29–31], and remove the trend term and offset terms of the coordinate time series.

$$y(t_i) = a + bt_i + c\sin(2\pi t_i) + d\cos(2\pi t_i) + e\sin(4\pi t_i) + f\cos(4\pi t_i) + \sum_{j=1}^{n_g} g_i H(t_i - T_{gi}) + \varepsilon_i \quad (1)$$

where $t_i (i = 1, 2, \cdots, N)$ is the daily solution epoch units of years, a and b are the intercept and linear rate, c and d are annual term coefficients, e and f are semi-annual coefficients, $H(t_i - T_{gi})$ is a Heaviside step function, ε_i is the model residual.

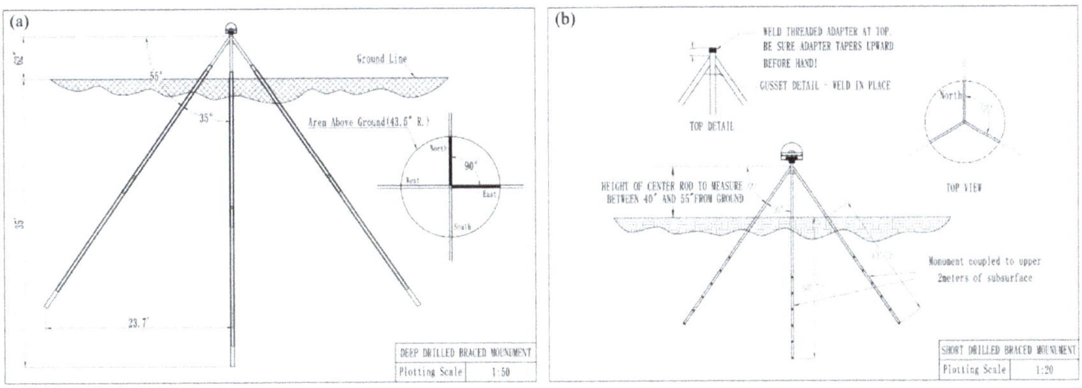

Figure 2. Design drawing of instrument monument. (**a**) It is a deep drilled braced monument station. The DDB monument is composed of five pillars, each of which extends 35 feet underground and 62 inches on the ground. (**b**) It is a short drilled braced monument station. The SDB monument is composed of four pillars, each of which extends 60–72 inches underground and 40–55 inches on the ground. Refer to UNAVCO website for more details (http://www.unavco.org/ (accessed on 7 August 2022)).

2.2. Surface Mass Loading Effects

Independently derived global models of major seasonal mass loading processes were used to assess the specific contribution of these load sources to the vertical data of Taiwan Province's GNSS observation network from 2008 to 2012. In this study, we focused on the seasonal movements in the vertical direction. We used the reanalysis numerical weather model MERRA2 and the operational semi-frozen numerical weather model GEOSF-PIT provided by the International Mass Loading Service (IMLS, http://massloading.net/#Download (accessed on 7 August 2022)) [32–35] to calculate the atmospheric mass loading displacements and the hydrological loading displacements, respectively. Non-tidal ocean loading displacements were calculated using the MPIOM06 model provided by the Max Planck Institute for Meteorology and GFZ. The load products provided by IMLS are built on a regular $2' \times 2'$ global grid, and pressure loading data corresponding to any point on Earth are available online on demand. The School and Observatory of Earth Sciences (EOST, http://loading.u-strasbg.fr/ (accessed on 7 August 2022)) [36] provides global displacement maps with a resolution of 0.5 degrees, with predicted displacements provided in ascii and netcdf data formats in the center of figure (CF) and center of mass (CM) reference frames. The ATML were loaded by modified inverted barometer model (ATMIB), induced oceanic loading (ATMMO), and hydrological loading including GLDAS/Noah models, with ERA interim 6-hour sampling by adjusting monthly model and deducting air tides (ERAin). Non-tidal ocean loading includes estimates of the circulation and climate of the ocean (ECCO1) and ECCO2 (follow-on ECCO, phase II). The German Research

Center for Geosciences (GFZ, http://rz-vm115.gfz-potsdam.de:8080/repository (accessed on 7 August 2022)) [37] provides non-tidal atmospheric loading products for atmospheric surface pressure enforced by ECMWF, hydrological loading products for calculations using the Land Surface Discharge Model (LSDM), and non-tidal ocean loading products for the EMPIOM ocean circulation model for atmospheric data operated by ECMWF. GFZ can also provide the calculation of loading products at GNSS sites. The above environmental loading time series were derived in the CF reference frame, as detailed in Table 1.

Table 1. Environmental loading product provided by EOST, GFZ, and IMLS.

Institution	Type	Model	Spatiotemporal Resolution
EOST	ATML	ECMWF(IB)	$0.5° \times 0.5°/3$ h
	ATML	ECMWF	$0.5° \times 0.5°/3$ h
	ATML	ERA interim	$0.5° \times 0.5°/6$ h
	HYDL	ERA interim	$0.5° \times 0.5°/6$ h
	HYDL	GLDAS/Noah	$0.5° \times 0.5°/3$ h
	NTOL	ECCO2	$0.5° \times 0.5°/24$ h
GFZ	ATML	ECMWF	$0.5° \times 0.5°/3$ h
	HYDL	LSDM	$0.5° \times 0.5°/24$ h
	NTOL	EMPIOM	$1° \times 1°/3$ h
IMLS	ATML	GEOSFPIT	$2' \times 2'/3$ h
	ATML	MERRA2	$2' \times 2'/6$ h
	HYDL	GEOSFPIT	$2' \times 2'/3$ h
	HYDL	MERRA2	$2' \times 2'/3$ h
	HYDL	MPIOM06	$2' \times 2'/3$ h

2.3. Temperature and Precipitation Data

The temperature data we used were from the reanalyzed datasets provided by the European Center for Medium-Range Weather Forecasts (ECMWF). ECMWF data provide a land surface temperature grid with the highest resolution of $0.125^{\circ} \times 0.125^{\circ}$, and time resolution of four points per day, measuring 0, 600, 1200, and 1800 h (UTC) (https://apps.ecmwf.int/datasets/interim-full-daily (accessed on 7 August 2022)). In this experiment, we selected $0.5^{\circ} \times 0.5^{\circ}$ resolution temperature grid data at 2 m on the ground at 600 (UTC), sampled every day from 1 January 2008 to 31 December 2012.

The spatial resolution was $0.1° \times 0.1°$, and the rainfall data sampled month by month were provided by the global precision measurement mission (GPM, https://disc.gsfc.nasa.gov/datasets?keywords=GPM&page=1 (accessed on 7 August 2022)).

2.4. Methodology

2.4.1. Calculation of Thermal Expansion Displacement

The effect of thermal expansion caused by temperature change on the vertical displacements of GNSS stations was divided into aboveground and underground contributions [10]. Generally, for the length change caused by expansion with heat and contraction with cold of the monument with GNSS antenna above the ground surface, the calculation equation is as follows [10,38]:

$$\Delta L = \alpha L \Delta T \tag{2}$$

where ΔL is the length change of the monument, the linear thermal expansion coefficient $\alpha = 1.2 \times 10^{-5}/°\text{C}$, ΔT is the temperature change, as the difference between the average value in the relative observation period, and L is the height of the monument.

DDB and SDB are composed of five and four stainless steel pipes, respectively. Compared with a single cement column, the connection between the bottom of SDB and DDB base rod and the ground is simulated by fixed support, with the top of the rod is rigidly connected. The whole structure is a high-order statically indeterminate structure, and the additional internal force generated by temperature change can lead to different displacements of the top of the rod from that of a straight rod. Therefore, it is not appropriate to consider the length of cement pier as a separate parameter in Equation (2), but the displacement change caused by expansion with heat and contraction with cold of stainless steel pipe remains proportional to the temperature difference. We used the finite element analysis software MIDAS to model two different types of pedestal models and added a temperature field. We calculated a total vertical displacement of the two models for a temperature difference of +10 °C, and calculated the daily vertical displacements of DDB and SDB pedestals according to Equation (3).

$$\Delta l = \frac{\Delta T}{10\ ^\circ\mathrm{C}} \times d_{SDB/DDB} \tag{3}$$

where $d_{SDB/DDB}$ is the vertical displacement calculated for DDB or SDB models in MIDAS software.

The influence of temperature change on bedrock through heat conduction must also be considered, as well as the resulting change in vertical displacement of GNSS station as follows [7]:

$$\Delta h = \frac{1+\delta}{1-\delta}\alpha A_T\sqrt{\frac{\kappa}{\omega}}\cos\left(\omega t - \varphi_T - \frac{\pi}{4}\right) \tag{4}$$

where $\delta \approx 0.266$, the thermal diffusion coefficients $\kappa = 1\ \mathrm{mm}^2/\mathrm{s}$, A_T, ω, and φ_T represent the amplitude, angular frequency, and initial phase of annual term of temperature change, respectively. The total displacements caused by thermal expansion effect can be obtained by adding Equations (3) and (4).

2.4.2. Evaluation Metrics

To make a quantitative comparison between the GNSS time series and a mass loading displacement, we defined a misfit function for the two datasets [39]:

$$D = \frac{1}{n}\sqrt{\sum_{i=1}^{n}\frac{\left(c_{i2}{}^2 + d_{i2}{}^2\right)\left[\left(c_{i1} - c_{i2}\right)^2 + \left(d_{i1} - d_{i2}\right)^2\right]}{\max\left(c_{i2}{}^2 + d_{i2}{}^2\right)}} \tag{5}$$

where c_{i1}, c_{i2}, d_{i1}, and d_{i2} are the amplitudes of the harmonic components of the two compared datasets defined in Equation (1), and n is equal to 84. A lower value of D indicates a better consistency of the periodic components between the two datasets.

We used RMS reduction value to evaluate the correction effect of products with different environmental loading. The RMS reduction value can be expressed as follows [40]:

$$RMS_{reduction} = \frac{RMS_{GNSS} - RMS_{(GNSS-loading)}}{RMS_{GNSS}} \tag{6}$$

where RMS_{GNSS} and $RMS_{(GNSS-loading)}$ are the RMS values of the vertical GNSS time series before and after ATML, HYDL, NTOL, and TEE correction. A larger $RMS_{reduction}$ value represents a better correction effect.

3. Results

3.1. Quantitative Assessment of Environmental Loading Effects

3.1.1. ATML Effects

As shown in Figure 3a, taking the ATML provided by IMLS_GEOSFPIT as an example, the annual vertical movements caused by atmospheric loading are in the range of 1–1.6 mm, with the peak uplift time around July. The amplitudes and phases distribution of GNSS time series are quite messy, and, except, for a few GNSS stations whose annual vertical movements reach 11 mm, most stations are in the range of 3–6 mm. The phases are bounded by a central mountain range, with a peak uplift time in western Taiwan from May to June and that in eastern Taiwan from February to April. Thus, atmospheric loading has a positive effect on the vertical time series of GNSS stations in western Taiwan, but a negative effect on the stations in eastern Taiwan.

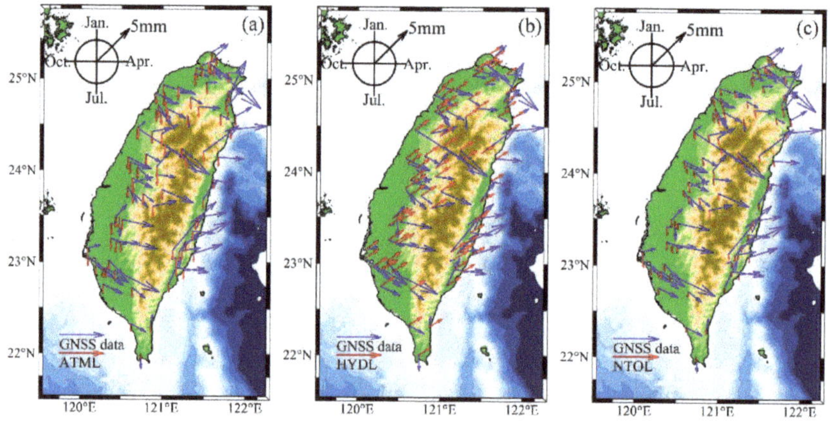

Figure 3. Environmental loading models and GNSS time series amplitudes and phases distribution. The arrow in the figure points to the month with the largest surface displacements uplift. The month rotates clockwise, and the length of the arrow represents the amplitude. According to Equation (1), the amplitude $A = \sqrt{c^2 + d^2}$, annual phase $\varphi = \tan^{-1}(d/c)$. (**a**) ATML (IMLS_GEOSFPIT). (**b**) HYDL (IMLS_MERRA2). (**c**) NTOL (GFZ_EMPIOM).

We used Equation (5) to quantify the temporal and spatial correlation between six ATML products and vertical GNSS time series. The results are shown in Table 2.

Table 2. Consistency between CWS products and GNSS vertical time series.

ATML Model	D	RMS_Reduction		
		Max (%)	Min (%)	Mean Forward (%)
ECMWF IB (EOST)	0.2085	6.29	−4.77	2.47
ECMWF	0.2073	8.16	−5.95	3.03
ERA interim	0.2091	6.25	−4.74	2.47
ECMWF IB (GFZ)	0.2081	5.94	−3.54	2.39
GEOSFPIT	0.2080	5.84	−4.29	2.23
MERRA2	0.2083	6.16	−4.25	2.45

The results of the RMS value reduction of the vertical GNSS stations coordinate time series are shown in Figure 4 and Table 2, where upward red arrows indicate positive correction and the downward blue arrows indicate negative correction. For visualization purposes, we used the adjustable tension continuous curvature splines [41] to interpolate the RMS reductions for each station displayed as a color profile. The RMS values of the vertical GNSS time series corrected by EOST_ECMWF IB, EOST_ECMWF, EOST_ERA interim, GFZ_ECMWF IB, IMLS_GEOSFPIT, and IMLS_MERRA2 were reduced by −4.77–6.29%, −5.95–8.16%, −4.74–6.25%, −3.54–5.94%, −4.29–5.84%, and −4.25–6.16%, respectively, and the mean RMS reduction values of forward correction results were 2.47%, 3.03%, 2.47%, 2.39%, 2.23%, and 2.45%, respectively. According to the correlation between the six ATML products and GNSS data and the reduction of RMS value after correction, ATML (EOST_ECMWF) products have the best positive effect for the correction of seasonal changes.

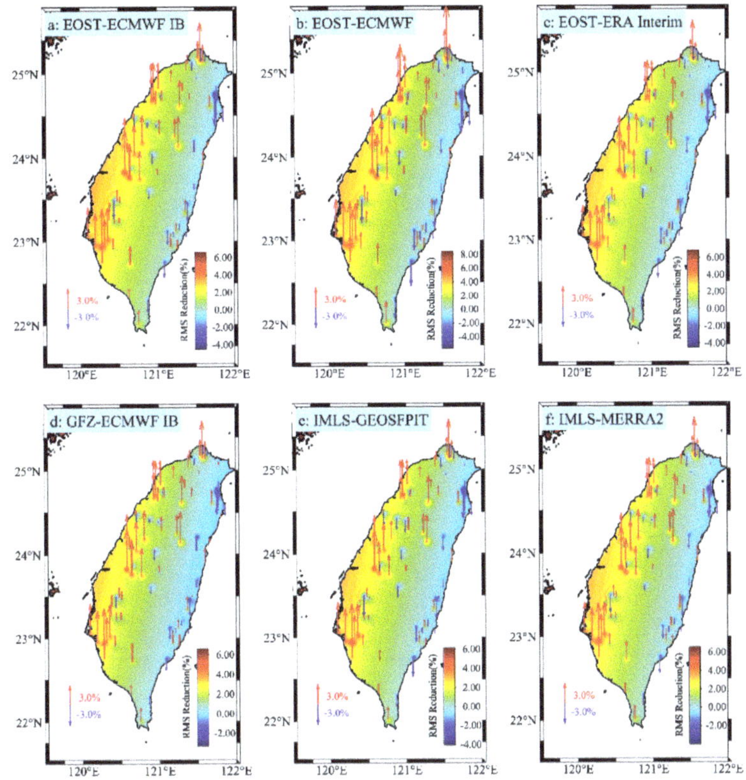

Figure 4. The RMS reduction value after GNSS time series correction of six ATML products. The color contours show the interpolated of RMS reduction, and the red (up) arrow indicates the positive response of ATML to GNSS, and the blue (down) arrow indicates the negative response. (**a–f**) respectively represent EOST_ECMWF IB, EOST_ECMWF, EOST_ERA interim, GFZ_ECMWF IB, IMLS_GEOSFPIT, and IMLS_MERRA2.

3.1.2. HYDL Effects

As shown in Figure 3b, for HYDL provided by IMLS_MERRA2 as an example, we can see annual vertical movements caused by hydrological loading of 1.6–2.5 mm for the whole station area, with phases around February. Compared with the western part of Taiwan, the eastern part of Taiwan is closer to the hydrological loading phases. Hydrological loading plays a positive role for most stations in Taiwan Province, so hydrological loading has a significant impact on the change of GNSS time series. We used Equation (5) to quantify the temporal and spatial correlation between five HYDL products and vertical GNSS time series, and the results are shown in Table 3.

Table 3. Consistency between HYDL products and GNSS vertical time series.

HYDL Model	D	RMS_Reduction		
		Max (%)	Min (%)	Mean Forward (%)
ERA interim	0.1924	6.04	−3.84	2.40
GLDAS	0.1973	4.56	−2.90	1.97
LSDM	0.1945	5.43	−5.59	2.25
GEOSFPIT	0.1682	7.10	−3.60	2.72
MERRA2	0.1881	7.29	−4.17	2.62

The RMS reduction results are shown in Figure 5 and Table 3. The RMS reduction ranges of vertical GNSS time series corrected by EOST_ERA interim, EOST_GLDAS, GFZ_LSDM, IMLS_GEOSFPIT and IMLS_MERRA2 were calculated as −3.84–6.04%, −2.90–4.56%, −5.59–5.43%, −3.60–7.10%, and −4.17–7.29% , respectively, with mean RMS reduction values of forward correction results of 2.40%, 1.97%, 2.25%, 2.72%, and 2.62%, respectively. The GLDAS/Noah model is mainly composed of soil moisture and excludes the influence of surface water such as lakes and rivers. According to the correlation between five HYDL products and GNSS data and the reduction of RMS value after correction, HYDL (IMLS_GEOSFPIT) products have the best positive effect to correct seasonal changes.

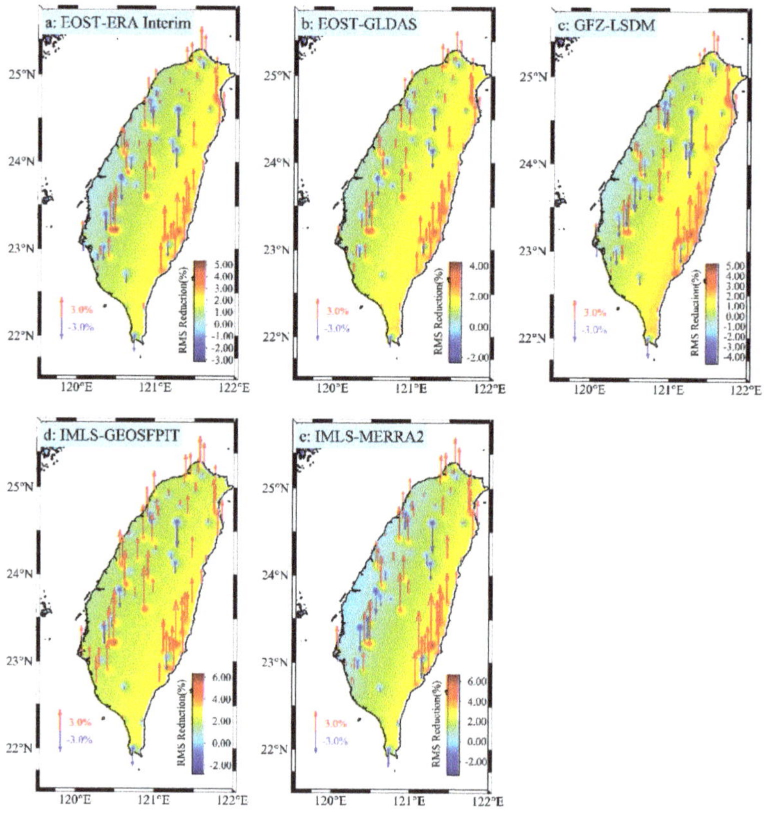

Figure 5. The RMS reduction value after GNSS time series correction of five HYDL products. The red (up) arrow indicates the positive response of ATML to GNSS, and the blue (down) arrow indicates the negative response. (**a**–**e**) respectively represent EOST_ERA interim, EOST_GLDAS, GFZ_LSDM, IMLS_GEOSFPIT, and IMLS_MERRA2.

3.1.3. NTOL Effects

As shown in Figure 3c, using NTOL provided by GFZ_EMPIOM as an example, the annual vertical movements caused by non-tidal ocean loading were calculated as 0.5–1.2 mm, with phases (peak uplift) occurring from June to July. Similar to the atmospheric loading model, according to the phase differences between non-tidal ocean loading and GNSS time vertical series. The correction effect is better for western Taiwan than for eastern Taiwan.

We used Equation (5) to quantify the spatiotemporal correlation between three NTOL products and vertical GNSS time series. The results are shown in Table 4. The RMS reductions of the vertical GNSS time series corrected by EOST_ECCO2, GFZ_EMPIOM, and IMLS_MPIOM06 are shown in Figure 6 and Table 4. The RMS values were decreased by −2.59–2.96%, −2.51–6.41%, and − 2.24–5.14%, respectively, with mean RMS reduction values of forward correction results of 0.95%, 2.00%, and 1.87%, respectively. There was obvious difference between the non-tidal ocean loading provided by GFZ_EMPIOM and EOST_ECCO2 to correct the vertical time series of GNSS. NTOL (GFZ_EMPIOM) products exhibited the best positive effect to correct seasonal changes.

Table 4. Consistency between NTOL products and GNSS vertical time series.

NTOL Model	D	RMS_Reduction		
		Max (%)	Min (%)	Mean Forward (%)
ECCO2	0.2135	2.96	−2.59	0.95
EMPIOM	0.2049	6.41	−2.51	2.00
MPIOM06	0.2055	5.14	−2.24	1.87

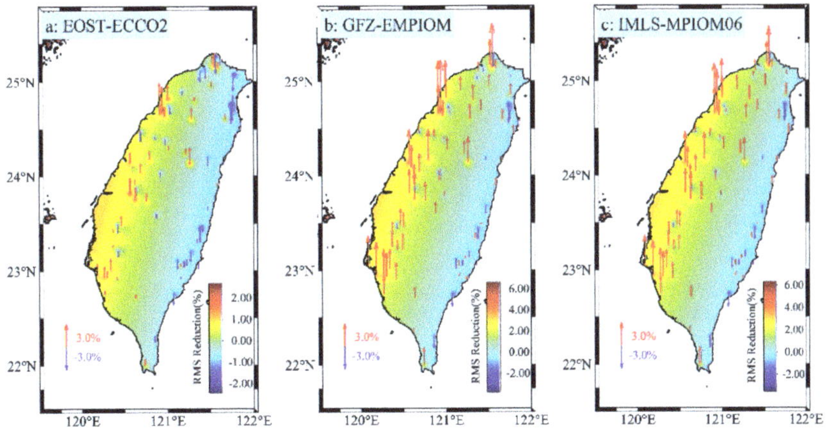

Figure 6. The RMS reduction value after GNSS time series correction of three NTOL products. (**a**–**c**) respectively represent EOST_ECCO2, GFZ_EMPIOM and IMLS_MPIOM06.

3.2. Quantitative Assessment of Thermal Expansion Effects

We used finite element analysis software MIDAS to model the pedestal above the ground surface for DDB and SDB GNSS stations, and then added a temperature field. For a temperature difference of +10 °C, total displacement in the vertical direction was calculated for the two pedestals as 0.252 mm (DDB, Figure 7a) and 0.219 mm (SDB, Figure 7b). Combined with Equation (3), the daily displacements change of each GNSS station was recorded as TEM. According to Equation (4), the vertical displacement change caused by thermal expansion of bedrock was calculated and is denoted as TEB. The phases of TEB lag 45 d behind those of TEM from Equation (4).

As shown in Figure 7c,d, the annual vertical total displacements caused by TEE were calculated as about 0.3–0.6 mm, with peak uplift occurring around August. At this time, the summer temperature is the highest in the northern hemisphere, but Taiwan Province is located at middle and low latitude, so the annual average temperature difference exhibits little change. In addition, because the phases difference between TEE and GNSS coordinate time series is more than 90°, the RMS values are negative after deducting the TEE. This is similar to the analysis of Yunnan Province of the Chinese mainland, similarly located at middle and low latitude.

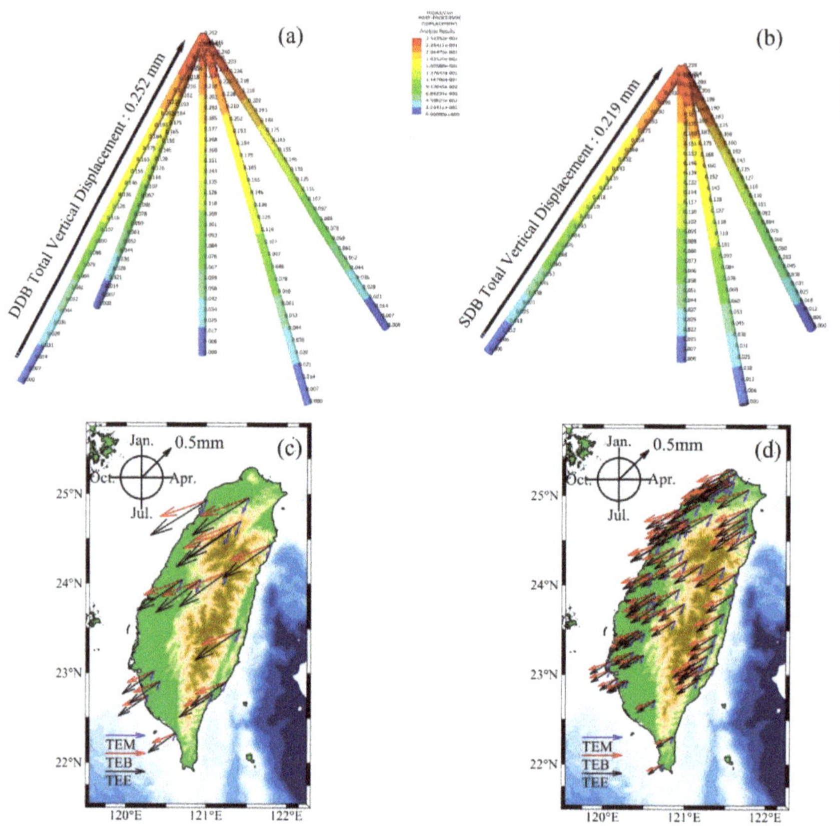

Figure 7. (**a**,**b**) MADIS software models DDB and SDB monuments and calculates the surface vertical displacements of the two monument bases under the condition of temperature difference of +10 °C. (**c**,**d**) The annual average amplitude and annual phase of the total vertical displacements of DDB and SDB monuments caused by thermal expansion effect.

4. Discussions

4.1. Optimal Combination of Environmental Loading Models

We quantitatively evaluated the correction of GNSS coordinate time series by three individual mass-induced displacements. Atmospheric loading and non-tidal ocean loading show similar spatial distribution patterns, with peak uplift phases from June to July and slightly smaller annual amplitudes of non-tidal ocean loading compared to atmospheric loading. This is because Taiwan Province is in a subtropical monsoon climate zone, with many tropical cyclones from July to September. Tropical cyclones cause atmospheric pressure changes and also lead to abnormal migration of seawater quality under the action of wind and atmospheric pressure changes. Non-tidal loading caused by tropical cyclones is the result of the combined action of non-tidal atmospheric loading and non-tidal ocean loading [42]. For the time series correction of GNSS stations, non-tidal atmospheric loading and non-tidal ocean loading are obviously corrected at stations in western Taiwan. This is because this region is mostly plains and GNSS stations are densely distributed, but there are many mountainous areas in the eastern part of Taiwan that block the central mountain range in the middle. Due to the influence of topographic factors, the improvement effect of atmospheric loading and non-tidal ocean loading on eastern stations is not good. Hydrological loading is corrected well at most stations in Taiwan. It is worth noting that, compared with the western Taiwan where the stations are densely distributed, the maximum subsidence and uplift of GNSS stations along the Huadong Valley in eastern Taiwan occurred in August and February, respectively, which is consistent with the precipitation cycle. After correction for hydrological loading, the reductions of RMS values are equivalent to or even greater than for western stations because the Huadong Rift Valley is formed by the deposition of river channels and alluvial fans and the groundwater layer is thicker.

As we can see, there is no obvious overall difference of different models in the three mass loadings for the correlation between amplitudes and phases and GNSS time series or the decrease of RMS values after correction. This is because the GNSS network in Taiwan Province differs from the global or CMONOC networks and belongs to a small-scale area. In addition to the comparison between individual loaded products, we created all possible combinations of different organization loaded products. We analyzed 90 different combinations in total and they are sorted in descending order according to its mean RMS reduction. Table 5 lists the mean RMS reduction for the first five combinations. Finally, we chose ATML (GFZ_ECMWF IB) + HYDL (IMLS_MERRA2) + NTOL (IMLS_MPIOM06) combination as the best combination mass loading model (CML) suitable for GNSS network in Taiwan. We then calculated the amplitudes and phases of CML according to Equation (1). As shown in Figure 8a, the annual vertical movements caused by CML were calculated as 2–3 mm, and uplift phases were concentrated in mid-May. CML modification of the vertical coordinate time series of GNSS stations is shown in Figure 8b, with the maximum forward RMS reduction of 13.87% (SHLU), and the mean RMS reduction of 5.28%.

Table 5. Vertical RMS reduction statistics for selected combinations.

Combinations	Mean RMS Reduction (%)
GFZ-ECMWF IB and IMLS-MERRA2 and IMLS-MPIOM06	5.28
IMLS-MERRA2 and IMLS-MERRA2 and IMLS-MPIOM06	5.21
GFZ-ECMWF IB and IMLS-MERRA2 and GFZ-EMPIOM	5.20
EOST-ECMWF IB and IMLS-MERRA2 and IMLS-MPIOM06	5.17
IMLS-MERRA2 and IMLS-MERRA2 and GFZ-EMPIOM	5.12

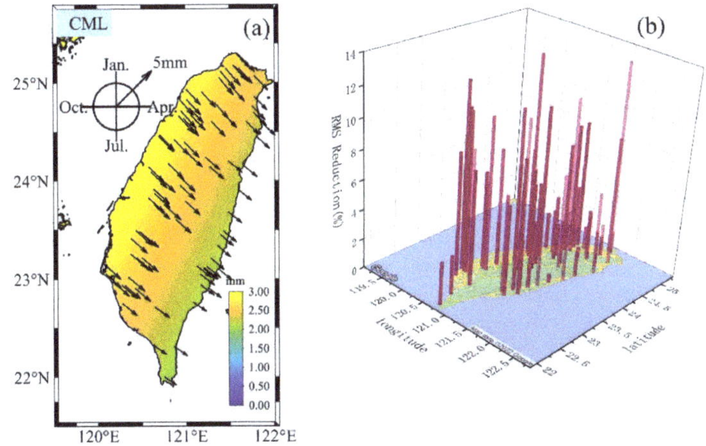

Figure 8. (**a**) Average annual amplitude and annual phase distribution of combined mass loading model. (**b**) The positive RMS reduction value of GNSS time series corrected by combined mass loading model.

4.2. The Effect of the Thermal Expansion

In Section 3.2, we calculated the vertical displacements of SDB and DDB pedestals caused by thermal expansion. The annual average amplitude of vertical displacement is 0.5 mm, accounting for 1.67% of the annual average amplitude of GNSS vertical displacement and about 10% of the annual average amplitude of hydrological loading displacement. Because of the phase difference, the GNSS time series increases after deducting the thermal expansion effect. This may be because the displacements are out of phase, which could be caused by the migration of some surface mass and the thermal expansion effect (Figure 9a), leading to a lack of loading displacements. Lai et al. [23] used GNSS to infer that the terrestrial water storage (TWS) reached the annual maximum and minimum values around August and February, respectively. This is the same timing as the phase mode of vertical displacements caused by thermal expansion effect, but out of phase with the hydrological model. Hsu et al. [24] used geodetic (GNSS and GRACE) and hydrological (precipitation, GLDAS and LSDM assimilation models and groundwater level) datasets to systematically study temporal and spatial water storage changes in Taiwan, and found significant seasonal water load fluctuation in Taiwan, but the total water storage in hydrological assimilation model was underestimated compared with that obtained by GNSS estimation. The difference comes from the complexity of transient water storage due to changes in rainfall patterns, infiltration rates, soil saturation, and runoff.

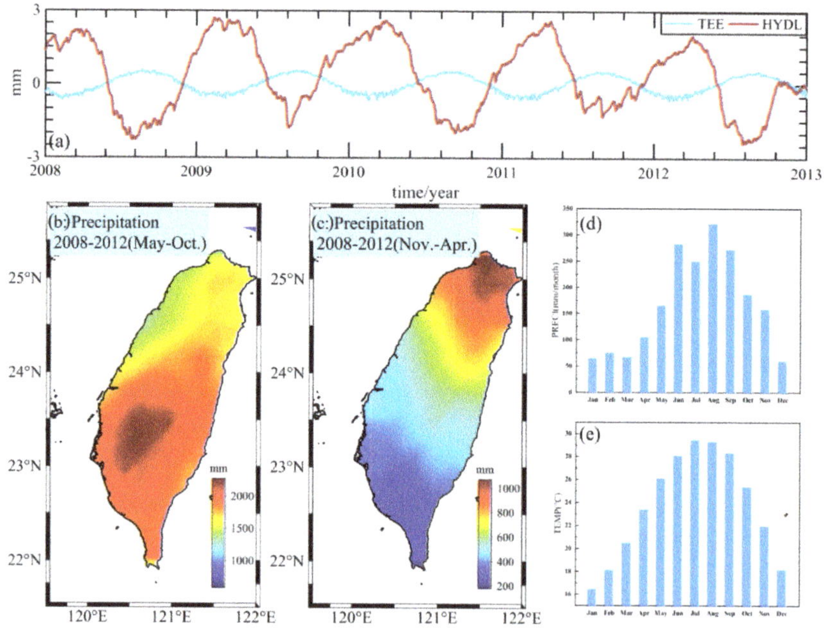

Figure 9. (**a**) Mean hydrological loading displacements and displacements caused by thermal expansion at 84 GNSS stations. (**b**,**c**) Mean precipitation in the wet (May to October) and dry (November to April) seasons from 2008 to 2012. (**d**,**e**) The monthly average precipitation and temperature in Taiwan Province, respectively.

The main source of hydrological loading is the annual rainfall in monsoon (May to June) and typhoon (July to October) seasons. A total of 70% of annual precipitation occurs during the wet season (May to October, Figure 9b), with the remaining 30% occurring during the dry season (November to April, Figure 9c). Although there is abundant annual rainfall, due to the steep terrain of Taiwan, most precipitation either evaporates directly or flows to the sea. The peak rainfall occurs from June to July (Figure 9d). Controlled by infiltration rate and capacity, the peak subsidence of GNSS lags one to two months behind precipitation, which is also consistent with the predicted value of hydrological model. At the same time, this period corresponds to the time when the highest temperatures occur in the northern hemisphere (Figure 9e). Surface water and shallow groundwater produce downward subsidence displacements on the ground, but the thermal expansion effect caused by temperature change will cause slight uplift phenomenon. This would partially counteract the observed hydrological loading displacements, and cannot be neglected in the study of water storage capacity using GNSS.

5. Conclusions

We studied the influence of non-tectonic deformation signals on the vertical coordinate time series of 84 GNSS stations in Taiwan Province. To do this, we compared the calibration effects of ATML, HYDL, and NTOL products provided by EOST, GFZ, and IMLS based on different models for Taiwan GNSS vertical time series. The differences between models were analyzed. Finally, we determined a set of model combinations suitable to correct for vertical displacements for data from GNSS stations in Taiwan Province, namely ATML (GFZ_ECMWF IB) + HYDL (IMLS_MERRA2) + NTOL (IMLS_MPIOM06). We have several interesting findings from our work. Because of the special geographical location of Taiwan Province, there is an obvious monsoon alternation in Taiwan Strait, and there are many tropical cyclones from July to September every year. The tropical storms and typhoons cause abnormal changes in atmospheric and seawater movement. Because of this, non-tidal atmospheric loading and non-tidal ocean loading work together in Taiwan. Non-tidal ocean and atmospheric loading exhibit similar temporal and spatial action patterns, with better correction effect of GNSS stations in the western plain of Taiwan to that for stations in eastern Taiwan. This may be due to a central barrier of mountains as well as additional mountains in the eastern part of Taiwan.

Tropical cyclones bring abundant rainfall, which plays a vital role in groundwater recharge in Taiwan Province. However, due to the relatively small area of Taiwan, no high-resolution hydrological data model has been developed that can accurately evaluate the temporal and spatial changes of water storage capacity in Taiwan, especially due to the variations of rainfall pattern, infiltration rate, and soil saturation in temporal and spatial domains. Although existing hydrological loading data have good corrections for most GNSS stations, there are limits to the use of these data to accurately quantify hydrological changes in Taiwan Province.

We quantified the vertical displacements of different stations due to temperature changes. Because Taiwan includes middle and low latitudes, the temperature difference relative to the annual average temperature changes less than 10 °C, so the average annual amplitude vertical total displacement caused by thermal expansion effect is only 0.5 mm. However, maximum annual subsidence and uplift caused by hydrological loading occur in August and February respectively, with the annual phase change of displacement caused by temperature change is opposite. Thus, the counteracting effect between the thermal expansion displacement and hydrological loading displacement cannot be neglected, and the consideration of thermal expansion effect would help accurately quantify the change of water storage capacity in Taiwan Province using GNSS vertical time series.

Author Contributions: Conceptualization, B.L. and X.X.; methodology, B.L. and X.M.; software, B.L. and X.M.; validation, X.M.; formal analysis, B.L. and X.M.; investigation, X.M.; data curation, X.M.; writing—original draft preparation, B.L. and X.M.; writing—review and editing, J.T., W.P., and L.Z.; visualization, X.M.; supervision, B.L. and X.X.; funding acquisition, B.L. and X.X. All authors have read and agreed to the published version of the manuscript.

Funding: This work is funded by the National Natural Science Foundation of China (Grant No. 41904003), the Natural Science Foundation of Hunan Province, China (Grant No. 2020JJ5571), Open Fund of Engineering Laboratory of Spatial Information Technology of Highway Geological Disaster Early Warning in Hunan Province (Changsha University of Science and Technology) (Grant No. KFJ190602), Graduate research and innovation project (Changsha University of Science and Technology) (Grant No. CXCLY2022018), and Undergraduate research and innovation project (Changsha University of Science and Technology) (Grant No. 20210070).

Data Availability Statement: The GNSS data used in this work can be obtained from the Institute of Earth Sciences, Academia Sinica, Taiwan. The time series of the GNSS stations used in this paper are uploaded to Figshare, and readers can obtain time series data of GNSS sites classified by different monuments, as well as the vertical velocities with their uncertainties from: https://doi.org/10.6084/m9.figshare.20448234.v3 (accessed on 7 August 2022). The environmental loads data provided by IMLS are available at http://massloading.net/#Download (accessed on 7 August 2022). The environmental loads data provided by EOST are available at http://loading.u-strasbg.fr/displ_maps.php (accessed on 7 August 2022). The environmental loads data provided by GFZ are available at http://esmdata.gfz-potsdam.de:8080/repository/entry/show/ (accessed on 7 August 2022). The temperature reanalysis data provided by the European Center for Medium Range Weather Forecasts are available at https://apps.ecmwf.int/datasets/data/interim-full-daily (accessed on 7 August 2022). Rainfall data provided by global precision measurement mission (GPM) comes from: https://disc.gsfc.nasa.gov/datasets?keywords=GPM&page=1 (accessed on 7 August 2022).

Conflicts of Interest: The authors declare no conflict of interest.

References

1. Gross, R.S.; Blewitt, G.; Clarke, P.J.; Lavallée, D. Degree-2 harmonics of the Earth's mass load estimated from GPS and Earth rotation data. *Geophys. Res. Lett.* **2004**, *31*, 1029–1031. [CrossRef]
2. Ray, J.; Altamimi, Z.; Collilieux, X.; van Dam, T. Anomalous harmonics in the spectra of GPS position estimates. *GPS Solut.* **2008**, *12*, 55–64. [CrossRef]
3. Steigenberger, P.; Rothacher, M.; Schmid, R.; Rülke, A.; Fritsche, M.; Dietrich, R.; Tesmer, V. Effects of different antenna phase center models on GPS-derived reference frames. In *Geodetic Reference Frames*; Drewes, H., Ed.; Springer: Berlin/Heidelberg, Germany, 2009; Volume 134, pp. 83–88. [CrossRef]
4. King, M.A.; Watson, C.S. Long GPS coordinate time series: Multipath and geometry effects. *J. Geophys. Res. Solid Earth* **2010**, *115*, B04403. [CrossRef]
5. Petrie, E.J.; King, M.A.; Moore, P.; Lavallée, D.A. Higher–order ionospheric effects on the GPS reference frame and velocities. *J. Geophys. Res. Solid Earth* **2010**, *115*, B03417. [CrossRef]
6. van Dam, T.; Wahr, J.; Milly, P.; Shmakin, A.; Blewitt, G.; Lavallée, D.; Larson, K. Crustal displacements due to continental water loading. *Geophys. Res. Lett.* **2001**, *28*, 651–654. [CrossRef]
7. Dong, D.; Fang, P.; Bock, Y.; Cheng, M.; Miyazaki, S.I. Anatomy of apparent seasonal variations from GPS-derived site position time series. *J. Geophys. Res. Solid Earth* **2002**, *107*, 2075. [CrossRef]
8. Schuh, H.; Estermann, G.; Crétaux, J.-F.; Bergé-Nguyen, M.; van Dam, T. Investigation of hydrological and atmospheric loading by space geodetic techniques. In *Satellite Altimetry for Geodesy, Geophysics and Oceanography*; Hwang, C., Shum, C.K., Li, J., Eds.; Springer: Berlin/Heidelberg, Germany, 2003; Volume 126, pp. 123–132. [CrossRef]
9. Munekane, H.; Tobita, M.; Takashima, K. Groundwater-induced vertical movements observed in Tsukuba, Japan. *Geophys. Res. Lett.* **2004**, *31*, L12608. [CrossRef]
10. Yan, H.; Chen, W.; Zhu, Y.; Zhang, W.; Zhong, M. Contributions of thermal expansion of monuments and nearby bedrock to observed GPS height changes. *Geophys. Res. Lett.* **2009**, *36*, 88–97. [CrossRef]
11. Jiang, W.; Li, Z.; van Dam, T.; Ding, W. Comparative analysis of different environmental loading methods and their impacts on the GPS height time series. *J. Geod.* **2013**, *87*, 687–703. [CrossRef]
12. Hu, S.; Chen, K.; Zhu, H.; Xue, C.; Wang, T.; Yang, Z.; Zhao, Q. A Comprehensive Analysis of Environmental Loading Effects on Vertical GPS Time Series in Yunnan, Southwest China. *Remote Sens.* **2022**, *14*, 2741. [CrossRef]
13. Li, C.; Huang, S.; Chen, Q.; van Dam, T.; Fok, H.S.; Zhao, Q.; Wu, W.; Wang, X. Quantitative Evaluation of Environmental Loading Induced Displacement Products for Correcting GNSS Time Series in CMONOC. *Remote Sens.* **2020**, *12*, 594. [CrossRef]
14. Wu, S.; Nie, G.; Meng, X.; Liu, J.; He, Y.; Xue, C.; Li, H. Comparative Analysis of the Effect of the Loading Series from GFZ and EOST on Long-Term GPS Height Time Series. *Remote Sens.* **2020**, *12*, 2822. [CrossRef]
15. Romagnoli, C.; Zerbini, S.; Lago, L.; Richter, B.; Simon, D.; Domenichini, F.; Elmi, C.; Ghirotti, M. Influence of soil consolidation and thermal expansion effects on height and gravity variations. *J. Geodyn.* **2003**, *35*, 521–539. [CrossRef]
16. Yan, H.; Chen, W.; Zhu, Y.; Zhang, W.; Zhong, M.; Liu, G. Thermal effects on vertical displacement of GPS stations in China. *Chin. J. Geophys.* **2010**, *53*, 825–832. (In Chinese) [CrossRef]
17. Tan, W.; Xu, X.; Dong, D.; Chen, J.; Wu, B. Thermoelastic Seasonal Deformation in Chinese Mainland. *Acta Geod. Cartogr. Sin.* **2017**, *46*, 1080–1097. (In Chinese) [CrossRef]
18. Jia, L.; Wang, Y.; Lian, W.; Xiang, L. Comparison and Analysis of Crustal Vertical Deformation in Mainland China Observed by GPS from CMONOC and GRACE. *Acta Geod. Cartogr. Sin.* **2018**, *47*, 899–906. (In Chinese) [CrossRef]
19. Boccara, G.; Hertzog, A.; Basdevant, C.; Vial, F. Accuracy of NCEP/NCAR reanalyses and ECMWF analyses in the lower stratosphere over Antarctica in 2005. *J. Geophys. Res. Atmos.* **2008**, *113*, D20115. [CrossRef]

20. Rao, R. *Noise in GPS Coordinate Time Series II*; Central Weather Bureau of the Ministry of Communications: Taipei City, Taiwan, 2014; Volume 64, pp. 281–327, (In Chinese). Available online: https://scweb.cwb.gov.tw/Uploads/Reports/MOTC-CWB-103-E-04.pdf (accessed on 7 August 2022).
21. Kumar, U.; Chao, B.F.; Chang, E.T.Y. What Causes the Common-Mode Error in Array GPS Displacement Fields: Case Study for Taiwan in Relation to Atmospheric Mass Loading. *Earth Space Sci.* **2020**, *7*, e2020EA001159. [CrossRef]
22. Ma, X.; Liu, B.; Dai, W.; Kuang, C.; Xing, X. Potential Contributors to Common Mode Error in Array GPS Displacement Fields in Taiwan Island. *Remote Sens.* **2021**, *13*, 4221. [CrossRef]
23. Lai, Y.R.; Wang, L.; Bevis, M.; Fok, H.S.; Alanazi, A. Truncated singular value decomposition regularization for estimating terrestrial water storage changes using GPS: A case study over Taiwan. *Remote Sens.* **2020**, *12*, 3861. [CrossRef]
24. Hsu, Y.-J.; Fu, Y.; Bürgmann, R.; Hsu, S.-Y.; Lin, C.-C.; Tang, C.-H.; Wu, Y.-M. Assessing seasonal and interannual water storage variations in Taiwan using geodetic and hydrological data. *Earth Planet. Sci. Lett.* **2020**, *550*, 116532. [CrossRef]
25. Herring, T.; King, R.; McClusky, S. *Documentation for the GAMIT Analysis Software, in Release 10.0 ed*; Massachusetts Institute of Technology: Cambrideg, MA, USA, 2002.
26. Altamimi, Z.; Collilieux, X.; Métivier, L. ITRF2008: An improved solution of the international terrestrial reference frame. *J. Geod.* **2011**, *85*, 457–473. [CrossRef]
27. Bock, H.; Beutler, G.; Schaer, S.; Springer, T.; Rothacher, M. Processing aspects related to permanent GPS arrays. *Earth Planets Space* **2000**, *52*, 657–662. [CrossRef]
28. Böhm, J.; Heinkelmann, R.; Schuh, H. Short note: A global model of pressure and temperature for geodetic applications. *J. Geod.* **2007**, *81*, 679–683. [CrossRef]
29. Nikolaidis, R. Observation of Geodetic and Seismic Deformation with the Global Positioning System. Ph.D. Thesis, University of California, San Diego, CA, USA, 2002.
30. Bitharis, S.; Fotiou, A.; Pikridas, C.; Rossikopoulos, D. A New Velocity Field of Greece Based on Seven Years (2008–2014) Continuously Operating GPS Station Data. In *International Symposium on Earth and Environmental Sciences for Future Generations. International Association of Geodesy Symposia*; Freymueller, J.T., Sánchez, L., Eds.; Springer: Cham, Germany, 2016; Volume 147. [CrossRef]
31. He, X.; Montillet, J.P.; Fernandes, R.; Bos, M.; Yu, K.; Hua, X.; Jiang, W. Review of current GPS methodologies for producing accurate time series and their error sources. *J. Geodyn.* **2017**, *106*, 12–29. [CrossRef]
32. Dobslaw, H.; Bergmann-Wolf, I.; Dill, R.; Poropat, L.; Thomas, M.; Dahle, C.; Esselborn, S.; Koenig, R.; Flechtner, F. A new high-resolution model of non-tidal atmosphere and ocean mass variability for de-aliasing of satellite gravity observations: AOD1B RL06. *Geophys. J. Int.* **2017**, *211*, 263–269. [CrossRef]
33. Petrov, L. The international mass loading service. In *REFAG 2014*; Springer: Cham, Switzerland, 2015; pp. 79–83.
34. Petrov, L.; Boy, J. Study of the atmospheric pressure loading signal in very long baseline interferometry observations. *J. Geophys. Res.* **2004**, *109*, 409. [CrossRef]
35. Rienecker, M.M.; Suarez, M.; Todling, R.; Bacmeister, J.; Takacs, L.; Liu, H.; Gu, W.; Sienkiewicz, M.; Koster, R.; Gelaro, R. *The GEOS-5 Data Assimilation System: Documentation of Versions 5.0. 1, 5.1. 0, and 5.2. 0*; Technical Report Series on Global Modeling and Data Assimilation Volume 27; NASA: Washington, DC, USA, 2008.
36. Mémin, A.; Boy, J.-P.; Santamaría-Gómez, A. Correcting GPS measurements for non-tidal loading. *GPS Solut.* **2020**, *24*, 1–13. [CrossRef]
37. Dill, R.; Dobslaw, H. Numerical simulations of global-scale high-resolution hydrological crustal deformations. *J. Geophys. Res. Solid Earth* **2013**, *118*, 5008–5017. [CrossRef]
38. Jiang, W.; Wang, K.; Deng, L.; Li, Z. Impact on Nonlinear Vertical Variation of GNSS Reference Stations Caused by Thermal Expansion. *Acta Geod. Cartogr. Sin.* **2015**, *44*, 473–480. (In Chinese) [CrossRef]
39. Yan, J.; Dong, D.; Bürgmann, R.; Materna, K.; Tan, W.; Peng, Y.; Chen, J. Separation of Sources of Seasonal Uplift in China Using Independent Component Analysis of GNSS Time Series. *J. Geophys. Res.-Solid Earth* **2019**, *124*, 11951–11971. [CrossRef]
40. Chen, Q. Analyzing and Modeling Environmental Loading Induced Displacements with GPS and GRACE. Ph.D. Thesis, Universität Stuttgart, Stuttgart, Germany, 2015. [CrossRef]
41. Smith, W.; Wessel, P. Gridding with continuous curvature splines in tension. *Geophysics* **1990**, *55*, 293–305. [CrossRef]
42. Yang, T.; Wang, Q. The non-tidal loading induced by tropical cyclones MATTHEW. *Chin. J. Geophys.* **2020**, *63*, 123–130. (In Chinese) [CrossRef]

remote sensing

MDPI

Article

Comprehensive Remote Sensing Technology for Monitoring Landslide Hazards and Disaster Chain in the Xishan Mining Area of Beijing

Runcheng Jiao [1,*], Shengyu Wang [1], Honglei Yang [2], Xuefei Guo [1], Jianfeng Han [2], Xin Pei [2] and Chi Yan [1]

1 Beijing Institute of Geological Hazard Prevention, Beijing 100120, China
2 School of Land Science and Technology, China University of Geosciences, Beijing 100083, China
* Correspondence: 1012151216@cugb.edu.cn

Abstract: The Xishan coal mine area in Beijing, China has a long history of mining. Many landslide hazards, in addition to collapses and ground fractures, have occurred in this area. This study used multi-temporal satellite images to extract this region's deformation information, identify landslides and analyze the deformation evolution process of these landslides. Taking the Anzigou ditch as an example, we investigate the "Quarry–Landslide–Mudslide" disaster chain model. We found that the landslide evolution process is closely related to the geological conditions, and usually goes through four stages: initial deformation, slope front swelling and collapsing, rear part connecting and rupturing, and landslide creeping. The surface deformation can be identified and tracked by high-resolution optical images and InSAR monitoring. Under the combined effects of rainfall and topographic conditions, medium and large landslides may occur and trigger a "Quarry–Landslide–Mudflow" disaster chain. The identification and analysis of these landslide hazards and the disaster chain help with geological disaster prevention, and provide reference for early identification and research of similar disasters.

Keywords: mining area; hazard identification; landslide evolution pattern; hazard chain

Citation: Jiao, R.; Wang, S.; Yang, H.; Guo, X.; Han, J.; Pei, X.; Yan, C. Comprehensive Remote Sensing Technology for Monitoring Landslide Hazards and Disaster Chain in the Xishan Mining Area of Beijing. *Remote Sens.* **2022**, *14*, 4695. https://doi.org/10.3390/rs14194695

Academic Editors: Linlin Ge, Hsing-Chung Chang, Alex Hay-Man Ng and Zheyuan Du

Received: 2 August 2022
Accepted: 16 September 2022
Published: 20 September 2022

1. Introduction

Underground mining causes ground subsidence, inducing ground cracks, ground collapse and other disasters, which lead to great damages and losses for the local people. The Xishan coal mines in Beijing, China, had been exploited underground for about a thousand years, until 2008, when all coal mines in Beijing were shut down. After such a long period of underground exploration, many ground fractures and ground collapse events occurred. Presently, 22 regions have recorded densely distributed collapse pits and fractures. The total affected area is about 2000 hectares. The Beijing Institute of Geological Disaster Prevention and Control has conducted several rounds of surveys in the Xishan area. They noted the distribution of the collapses, analyzed the deformation and metamorphic characteristics of each coal formation, and summarized the geohazard development characteristics. They also predicted the ground collapse hazards in the mining area [1–5].

In recent years, the government performed sudden geological disaster monitoring and provided early warnings for the collapse areas. The results showed that after the mines were closed, the growth of ground fractures and collapse pits in the collapse area gradually slowed. From 2015 to 2021, only two new small-scale collapses of coal mining lanes occurred, both in 2016. However, underground mining, especially the operation of some small coal companies that featured shallow mining depth, high mining intensity and high disturbance, destroyed the stability of the mountain; therefore, landslides and other geohazards began to appear.

In mountainous areas, landslides and their induced disaster chains are destructive, but hard to spot, due to strong concealment [6,7]. Much research has been done on the failure modes and monitoring methods of landslides [8–15]. Beijing is the political and cultural center of China, and the Xishan mining area is densely populated. Therefore, the identification of geological and other potential hazards is of great importance. However, few studies have examined the landslides in the Xishan mining area.

The Xishan coal mine area has been exploited since the Liao dynasty (907 AD). Due to historical wars, no useful exploration data were recorded. Besides, there were a lot of private mines, which roughly dug out coal without any regulatory oversight. Thus, we cannot get precise information regarding the underground mining in this region. In recent years, remote sensing observations have become increasingly abundant, and related technologies, such as multi-temporal high resolution optical image analysis technology, unmanned aerial vehicle (UAV) photography and oblique photogrammetry, interformetric synthetic aperture radar (InSAR) and 3D laser scanning, have been widely applied for geological hazard detection [16–25]. Multi-temporal high spatial resolution optical images have wide spatial coverage and abundant storage, thus, they are suitable for detecting geological hazards with obvious deformation. UAV photography and tilt photography can obtain high spatial resolution images and high precision 3D models. They are flexible and convenient, and have high recognition efficacy for geological hazards with signs of deformation. InSAR technology performs well at recognizing geological disasters in areas with high surface coherence and moderate deformation rates. It is suitable for large-scale geological disaster detection. 3D laser scanning can reduce the influence of surface vegetation and directly obtain a 3D model of the target area, which has a better recognition effect for geological disasters with relatively obvious features in the vegetated area.

This study collects multi-temporal remote sensing data from the Xishan mine area, and extracts deformation information from optical images and time-series InSAR, by which we identify landslides. On this basis, we classify the landslide evolution stages and analyze the triggering factors and disaster patterns of the disaster chain. This study provides a reference for regional geological hazard research and prevention.

2. Study Area and Data Sources

2.1. Geological Background

The study area is in Fangshan District, Beijing, China, and is geologically located in the south-east flank of the Baihuashan Oblique. The fracture structures are small, with a general north-east orientation and a dip angle of 50° to 80°. Figure 1 shows the geological background of the study area. The coal resources are abundant, mainly distributed in two stratigraphic units: the Lower Permian Shanxi Formation and the Lower Jurassic Yaopo Formation. There are also coal seams in the Longmen Formation.

2.2. Status of the Disaster

The mining activities in the study area destructed the ground surface and forest resources, and induced many geological hazards, such as collapses and debris flows. According to the geological disaster list for Beijing, published in 2021, there were four collapses, nine subsidence areas and seven debris flows in the study area.

2.3. Data Sources

In this paper, 12 optical satellite images and 65 RadarSat-2 images of the study area acquired between 2003 and 2021 were collected for the slope hazard identification experiment (see Table 1).

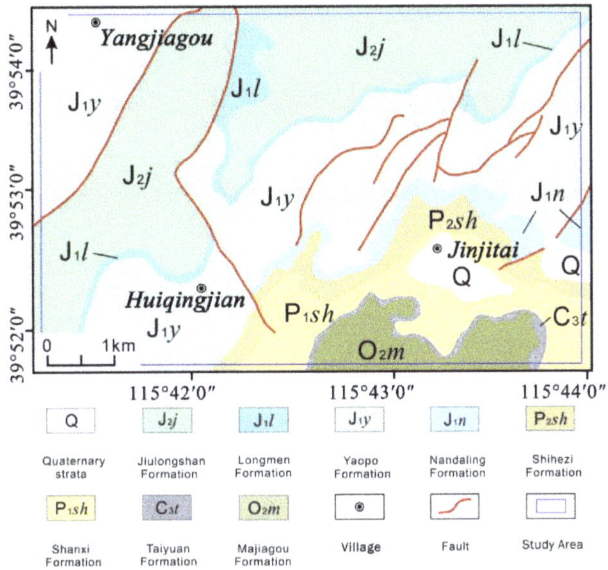

Figure 1. Geological Map of the Study Area.

Table 1. Information of the images used in this study.

Data	Resolution/m	Years	Number
Quickbird	0.61	2003	1
Quickbird	0.61	2006	1
Aerial Photo	0.5	2009	1
GeoEye-1	0.41	2011	1
GeoEye-1	0.41	2012	1
Aerial Photo	0.5	2013	1
Worldview-2	0.5	2014	1
Worldview-2	0.5	2016	1
Pleiades	0.5	2017	1
Pleiades	0.5	2018	1
BJ-2	0.8	2019	1
BJ-2	0.8	2020	1
RadarSat-2 (descending)	5.0	2016-2021	65

3. Research Methodology

3.1. Optical Remote Sensing Landslide Interpretation Approach

Presently, optical images are manually interpreted for identifying landslide hazards. This method has high reliability for areas with complex geological conditions, as it incorporates the expertise and experience of interpreters. Object-oriented classification, DNVI change detection and deep learning have shown high aptitude for landslide identification from remote sensing images [26–30], but their identified objects should have obvious deformation. However, landslides usually have small deformations, which are not obvious on optical images. Thus, optical-based automatic identification methods have limited accuracy for landslide identification.

This paper compares multi-temporal optical remote sensing images to find the differences in tone and texture between landslides and the surrounding areas, extracts the specific location of tension cracks and collapses caused by the swell of the lower slope, and obtains the changes of vegetation, geomorphology and slope (Figure 2).

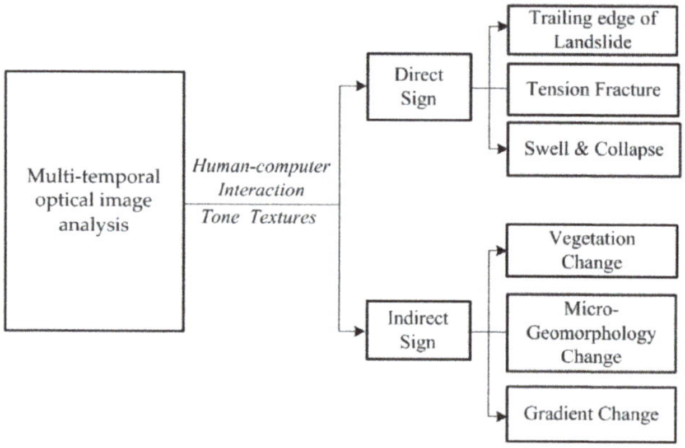

Figure 2. Flowchart of the proposed landslide interpretation method of optical images.

3.2. InSAR Time-Series Analysis

In this paper, InSAR, the time-series analysis process, is as follows: firstly, the master image is selected to register the SAR data. Then, we select the PS points using a combination of amplitude departure index and correlation coefficient. We construct a triangular network for the PS points. After getting the interferometric pairs, differential interference is applied to the PS points. The linear deformation rate of the PS points is obtained by the solution space. Considering the different frequencies of the atmospheric delay phase, nonlinear deformation phase and noise in the time and space domains, different filtering methods are used to obtain the nonlinear deformation rate. Finally, the deformation information of PS points is obtained.

4. Results

The interpretation results of the remote sensing images show 32 collapses and 19 landslides in the study area (see Figure 3 and Table 2).

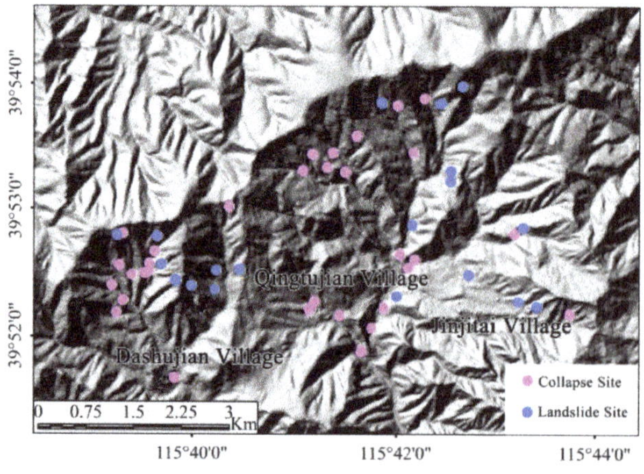

Figure 3. Collapses and landslides distributed in the study area. The base map is shaded relief map.

Table 2. Information of collapses and landslides in the study area.

Disaster	Number of Small Size	Number of Medium Size	Subtotal
Collapse	32	0	32 (11)
Landslide	3	16	19 (3)
Total			51

The numbers in parentheses are the number of disasters that have been included in the geological disaster list for Beijing, published in 2021. Small-sized collapse: volume is less than 10,000 m^3; Small-sized landslides: volume is less than 100,000 m^3; Medium-sized landslide: volume is between 100,000 m^3 and 1,000,000 m^3.

We compare the interpretation results with the geological disaster list for Beijing, published in 2021 (hereafter referred to as the hazard list), and conduct field checks on those not included in the list. We find that the above 51 collapses and landslides are correct. The collapses are small and widely distributed at the slope feet. Their locations coincide with the mining caves or old collapse pits/fissures. A few collapses occurred along new roads or the front edge of a slope. The landslides are medium-sized and concentrated in mountainous areas. The height differences of the landslide bodies are larger than 100 m each. Some landslide rear edges are close to the crests.

In 2003, some collapses and landslide hazards occurred in the Shijiaying mine area, but they did not get public attention at that time, nor were they included in the hazard list. After the mines were closed in 2009, new collapses and landslides occurred in the study area. Presently, 16 landslides of medium size (1×10^5 m^3) were identified in the area (see Figure 4 and Table 3), and the time-series InSAR results show that all of these 16 landslides had continuous deformation since 2016.

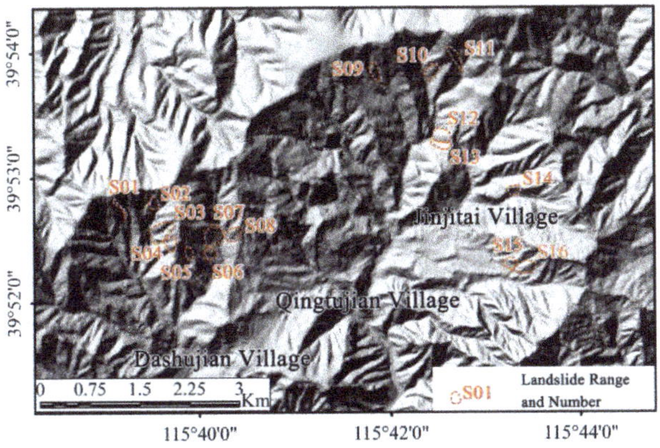

Figure 4. Medium-sized landslides in the Study Area. The base map is shaded relief map.

4.1. Deformation Characteristics of the Landslides Extracted from the Remote Sensing Images

In Table 3, except for S01 and S11, the remaining 14 landslides have the same characteristics: obvious fissures in the rear edge and widespread collapses on the front edge and the body of the landslide.

Figure 5 shows the remote sensing images and deformation evolution process of three points in landslide S02. The images from 2009 show that fissures appeared on the rear edge of the landslide, and small-scale depressions were formed. In 2012, the fissures developed further, extending about 100 m. In 2017, the cracks on the rear edge were further developed, the depression was enlarged, and cracks appeared on the eastern boundary. In 2019, the cracks on the rear edge and those on the eastern boundary connected, forming a long crack with a total length of about 270 m. Additionally, the crack on the western boundary appeared, with a length of 100 m. The remote sensing measurement showed that the landslide had the maximum height difference of 174 m, slope of 48°, and the estimated mass volume of 4.5×10^5 m^3. The 2016–2021 InSAR time series (Figure 5) show that the deformation rates vary significantly over different parts of the landslide body. Point1, point2 and point3 are the positions with large deformations on the landslide. The maximum LOS deformation is about 40 mm, and the deformation rate is close to 10 mm/year.

Table 3. Basic information of landslides.

No.	Developmental Formation	Estimated Volume/10,000 m^3	Features
S01	Jiulongshan Formation	10.00	Type A
S02	Jiulongshan Formation	45.00	Type B
S03	Jiulongshan Formation	30.00	Type B
S04	Jiulongshan Formation	15.00	Type B
S05	Jiulongshan Formation	10.00	Type B
S06	Yaopo Formation	50.00	Type C
S07	Yaopo Formation & Longmen Formation	100.00	Type C
S08	Yaopo Formation	55.00	Type C
S09	Jiulongshan Formation	16.00	Type C
S10	Longmen Formation	34.00	Type C
S11	Jiulongshan Formation	30.00	Type A
S12	Yaopo Formation	50.00	Type C
S13	Yaopo Formation	26.00	Type C
S14	Yaopo Formation	15.00	Type B
S15	Shanxi Formation	25.00	Type C
S16	Shanxi Formation	33.00	Type C

Type A: High-locality landslide, with continuous deformation since 2016. Type B: landslide with initial deformation or slope front swells and collapses since 2016. The deformation is small. Type C: landslide after failure, since 2016.

The above analysis indicates that landslide S02 body is developing, and the sliding surface has not connected yet. The developmental stages and developmental characteristics of the S03, S04, S05, and S14 landslides are basically the same as those of the S02 landslide. The InSAR time series results show that these landslides have experienced some deformation since 2016, with a deformation rate of about 10 mm/year (see Figure 6).

Figure 7 shows the remote sensing images and deformation evolution process of landslide S06. Satellite images in 2003 showed that the slope had not started sliding, and only two collapses were observed near the hill foot. In 2009, the size of the two collapses increased significantly, but there were no signs of failure. In 2017, satellite images showed that the two collapses grew larger, and large tensile fissures and slip cliffs appeared in the upper mountain. The sliding surface of the landslide had merged. In 2019, the collapse at the slope foot and the deformation of the rear edge increased further, resulting in a landslide body with a maximum height difference of 107 m, a slope of 37° and an estimated volume of 5×10^5 m^3.

The above analysis shows that the deformation of landslide S06 developed from 2003 to 2016. In 2016, the rear part of the slope suddenly cracked and slid downward. After the failure, the landslide continued to creep. The InSAR results show that the whole landslide had obvious deformation during 2016–2021. Point1, point2 and point3 are the positions with large deformations on the landslide. The maximum cumulative LOS deformation is about 110 mm, and the velocity exceeds 20 mm/year.

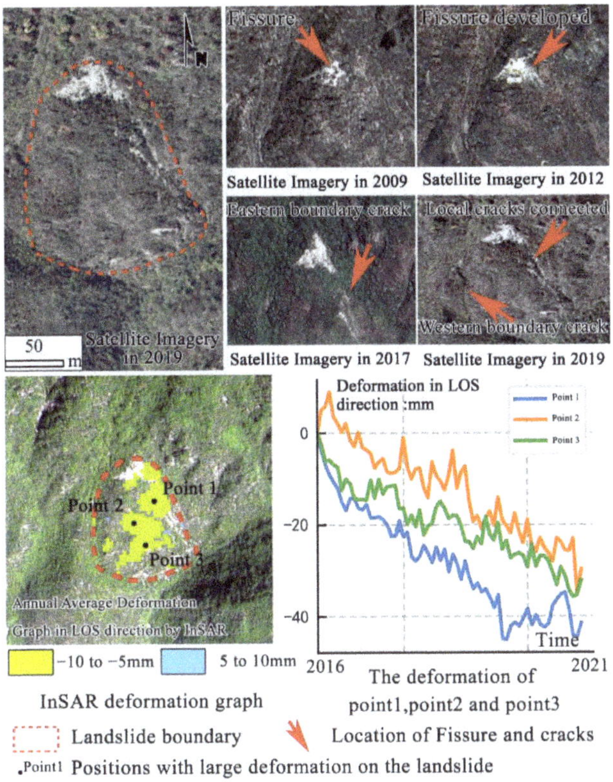

Figure 5. Optical images of landslide S02 acquired in 2009, 2012, 2017 and 2019 (the upper five panels). Deformation velocity of landslide S02 got by InSAR (left lower panel). The deformation in LOS direction of the three points during 2016–2021 are shown in the right lower panel.

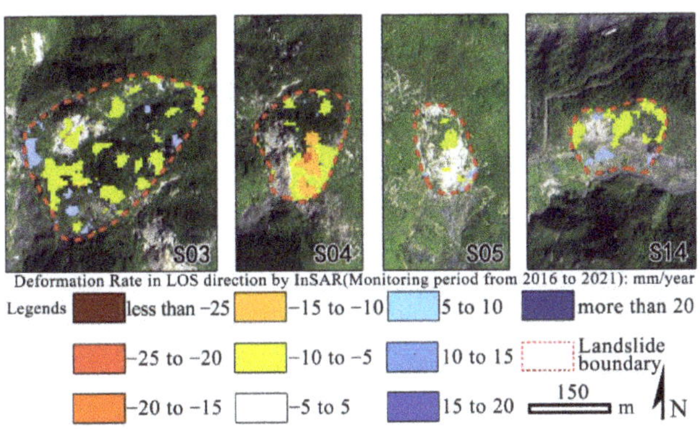

Figure 6. InSAR time series results of landslide S03, S04, S05 and S14.

Similarly, S07, S10, S15, and S16 also experienced the process from initial deformation to failure from 2003 to 2021. The InSAR time series results show that since 2016, these landslides have had continuous deformation, and the deformation rate is significantly higher than that of landslide S02. (See Figure 8).

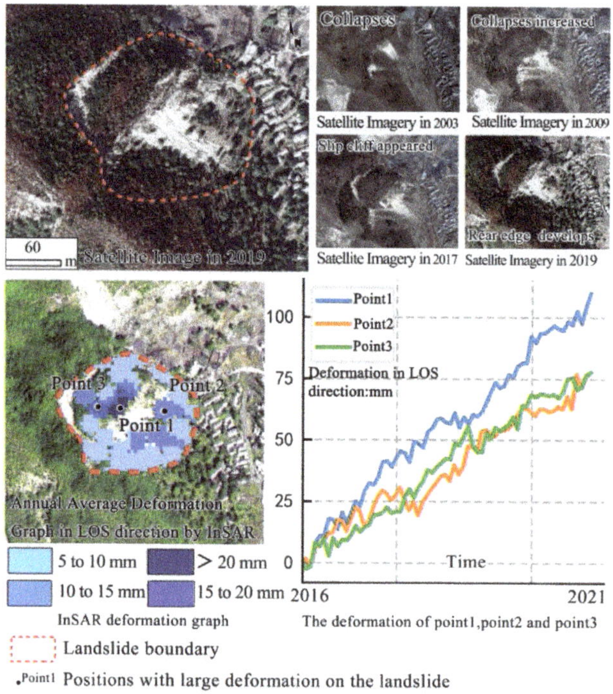

Figure 7. Optical images of landslide S06 acquired in 2003, 2009, 2017 and 2019 (the upper five panels). Deformation velocity of landslide S06 got by InSAR (left lower panel). The deformation in LOS direction of the three points during 2016–2021 are shown in the right lower panel.

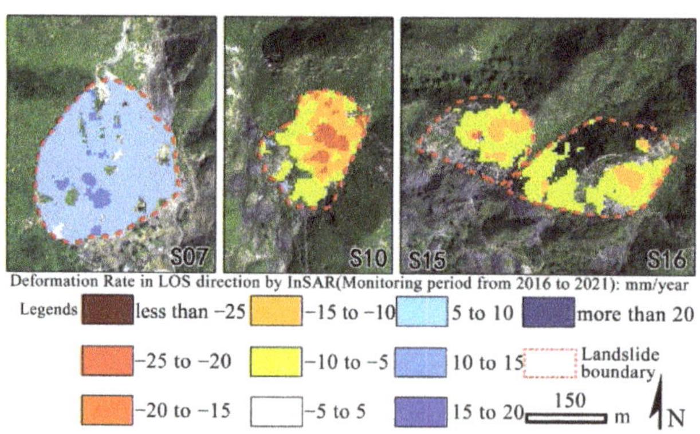

Figure 8. InSAR time series results of landslide S07, S10, S15 and S16.

In Figure 9, satellite images in 2003 showed that landslides S12 and S13 developed fissures at the rear edge. S12 had obvious slip cliff and platform on the rear part. The tension cracks and boundary are clear. Three rockslides occured at S12-1, S13-1 and S13-2 (Figure 9). In 2019, no obvious deformation was observed on the rear edge of the two landslide bodies, but the rockslides at S12-1, S13-1 and S13-2 expanded, and new collapses appeared at S12-2, S13-3 and S13-4. The maximum height difference, slope and estimated volume of landslides S12 and S13 are 175 m, 156 m, 35°, 33°, 5×10^5 m^3 and 2.6×10^4 m^3, respectively.

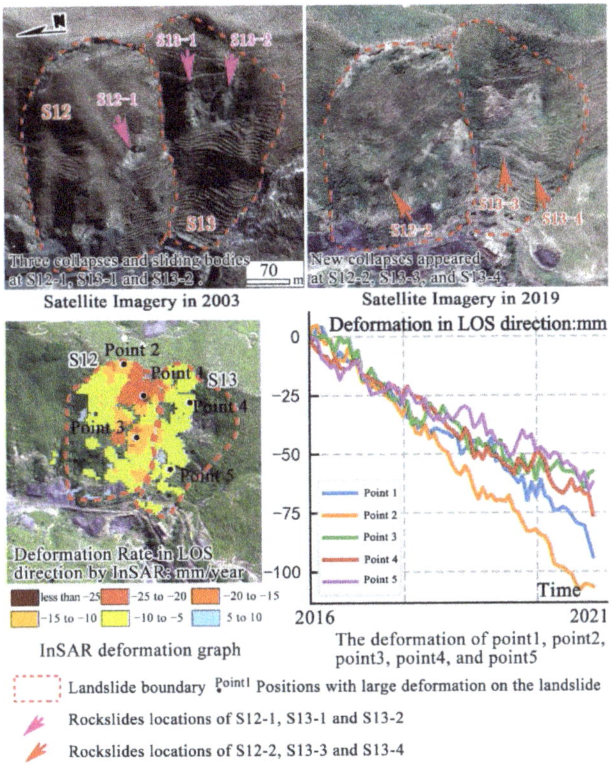

Figure 9. Optical images of landslide S12 and S13 acquired in 2003, and 2019 (the upper two panels). Deformation velocity of landslide S12 and S13 got by InSAR (left lower panel). The deformation in LOS direction of the five points during 2016–2021 are shown in the right lower panel.

The 2016–2021 InSAR time series results show that the landslide body has obvious deformation. Point1, point2, point3, point4 and point5 are the positions with large deformations on the landslide. The maximum LOS deformation rate exceeds 20 mm/year, and the cumulative LOS deformation is about 110 mm. This indicates that the slope is creeping. Landslides S08 and S09 have basically the same deformation patterns as those of landslides S12 and S13 (see Figure 10).

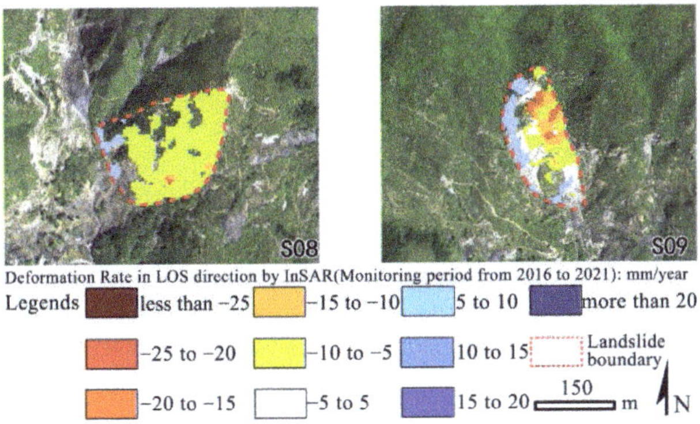

Figure 10. InSAR time series results of landslide S08 and S09.

Among the above 14 landslides, S06, S07 and S12 were developed in coal-bearing strata, with small goaf depth. But S02, S03, S04, S05 and S14 were developed in the strata above the coal-bearing strata, with deep goaf. Cracks develop faster on the rear edge of the former landslides.

4.2. Remote Sensing Characteristics of Landslide Caused by High-Level Collapse

Among the 16 landslides (Table 3), the bodies of S01 and S11 are composed of the rock mass from the upper part of the slope, due to rockfall. In Figure 11, the areas A1 and A2 are the rockfall, B1 and B2 are the shoveled mountain range, and C1 and C2 areas are the accumulation range.

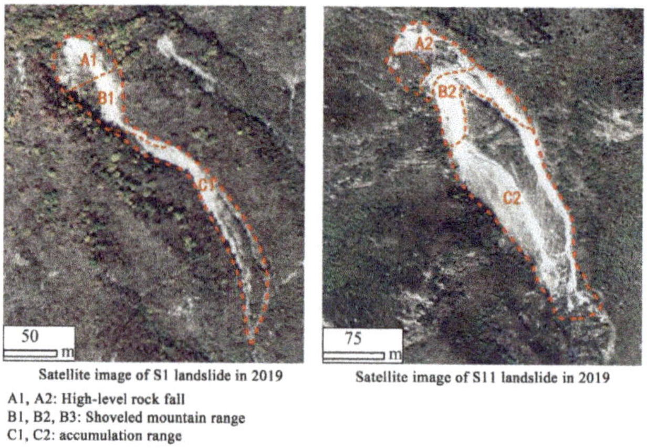

Figure 11. Optical images of landslide S01 (**left**) and S11 (**right**) acquired in 2019. The red dotted line delineates the landslide boundary. A1 and A2 are the rockfall, B1 and B2 are the shoveled mountain range, and C1 and C2 areas are the accumulation range.

In 2006, cracks and collapses appeared in landslide S01. The maximum height difference of the sliding body is 244 m, the slope is 43° and the estimated volume is 5×10^4 m^3. In the same year, cracking and rock sliding also occurred in landslide S11. The high-level collapse areas gradually expanded, and the size of the sliding mass also gradually increased (see Figure 12).

In 2006, the upper part of S11 (A1-1 rock mass) collapsed, scraped off part of the mountain mass and disintegrated. The loose mass composed the early landslide body. In 2009, the A1-1 rock mass collapsed further. On its southwest side, a part of A1-2 rock mass collapsed. The mass loosely piled on the west side of the slide. Images from 2012 to 2019 showed that A1-1 and A1-2 continued collapsing. The landslide body gradually enlarged, with a maximum height difference of 314 m, a slope of 39° and an estimated volume of 3×10^5 m^3. The landslide mass had continuous deformation from 2016 to 2021, and the maximum accumulative LOS deformation is nearly 100 mm.

4.3. Evolution of the Landslides in the Goaf

This study identified 16 landslides in the study area. Except for S01 and S11 that were induced by high-level collapse, the remaining 14 landslides were induced by underground goafs and have similar evolution processes. The satellite images collected in this study completely recorded the evolution process of landslides S06 and S07; therefore we use the time series of the two landslides, combined with the regional coal mining background to discuss the problems caused by coal mining.

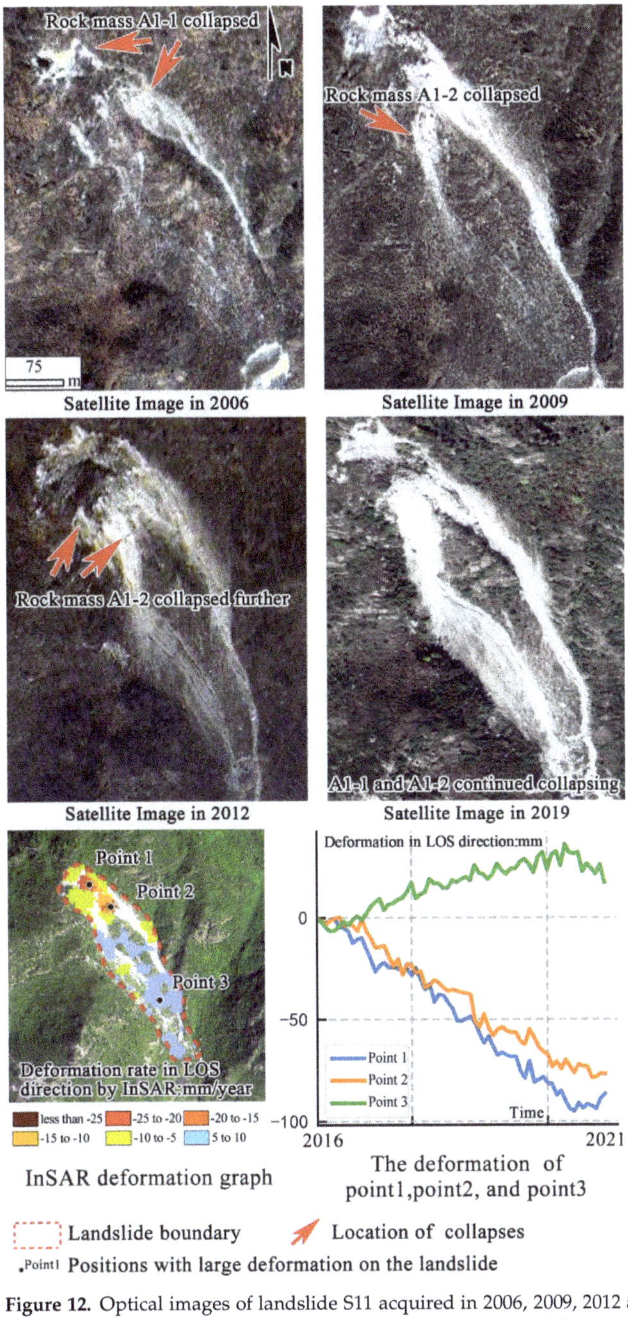

Figure 12. Optical images of landslide S11 acquired in 2006, 2009, 2012 and 2019 (the upper four panels). Deformation velocity of landslide S11 got by InSAR (left lower panel). The deformation in LOS direction of the three points during 2016–2021 are shown in the right lower panel.

Landslides S06 and S07 are mainly developed in the Yaopo Formation and Longmen Formation, which are the two major coal-measure strata in Xishan. The lithology is mainly dark siltstone, with multiple layers of mineable coal and gravel-bearing medium-coarse sandstone. When the coal is mined out, the roof of the goaf will bend and sink due to gravity, then crack and break down. Due to the shallow burial depth and multiple layers of coal, the goaf range is large. The deformation of the overlying rock mass leads to cracks on the surface, which further leads to the deformation of the whole slope, and finally, causes a landslide [31–35]. According to the optical images of the study area from 2003 to 2021, the formation process can be divided into four stages.

Stage 1: initial deformation

In this stage, ground fissures and collapse pits appeared on the mountain surface, due to the goaf. At that time, the stability of the mountain had been destroyed, but the deformation degree is relatively small. Figure 13 shows the satellite image of landslides S06 and S07 in 2003. L1 and L2 are the ground fissures developed on the slope. Many collapse pits can be found at the slope foot (red arrows). The overall deformation of the slope is not obvious.

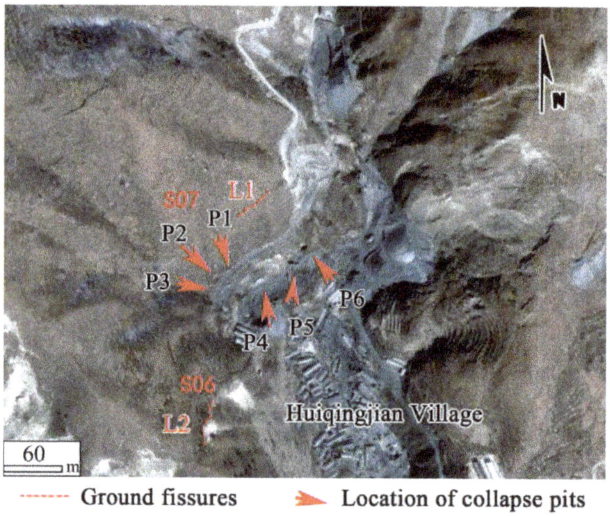

Figure 13. Optical image of landslides S06 and S07 in 2003. L1, L2 denotes ground fissures. P1, P2, P3, P4, P5 and P6 denote collapsed pits.

Stage 2: slope front swelling and collapsing

In this stage, the deformation of the overlying rock mass in the goaf developed. Some rock mass fell down and squeezed the surrounding rock. The lower part of the slope bore greater force, so the deformation was larger. The mountain began to swell. After the deformation had developed for some time, partial collapses occurred. Figure 14 shows the satellite images of S06 and S07 in 2009. The cracks on the rear part of the slopes had not merged, but many collapses occurred at the slope toe (Figure 14), indicating that the two slides had started rapid deformation.

Stage 3: Rear part connecting and rupturing

As the collapses in the lower part of the mountain developed, the stability of the slope was reduced and the deformation increased. The accumulated deformation increased the cracks on the rear part of the slope. These cracks extended, connected and eventually led to the occurrence of failure. As Figures 15 and 16 show, the fissures at the rear parts of the S06 and S07 connected in 2012 and 2016, respectively. According to the field investigation, the failure of landslide S07 occurred on 21 July 2012 after a heavy rain, forming a cliff with a height of 9–12 m. The failure of landslide S06 occurred on 20 July 2016 after a heavy rain as well, forming a cliff with a height of 30–35 m.

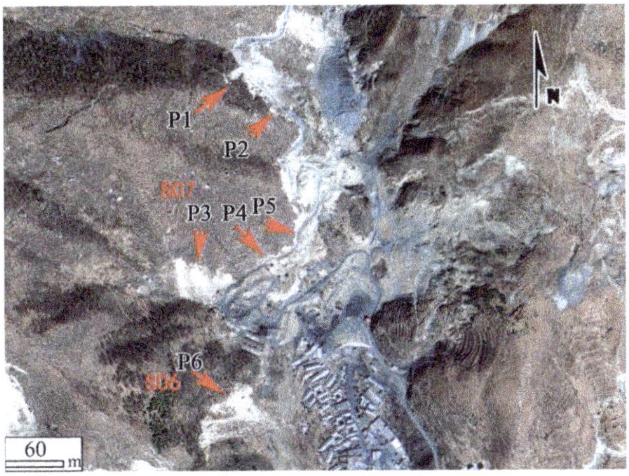

Figure 14. Optical image of Landslide S06 and S07 in 2009. P1, P2, P3, P4, P5 and P6 denote collapses that occurred on the toe.

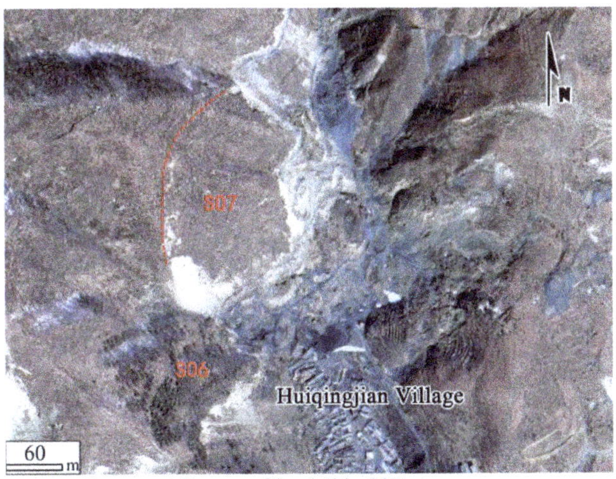

Figure 15. Optical image of landslides S06 and S07 in 2012.

Stage 4: Landslide creeping

After the failure, the internal forces of the landslide body reached a state of equilibrium. The landslide entered a "relatively quiet period." In the absence of external interference, the deformation of the landslide in this stage was only caused by its own gravity. The deformation was slow, and the slope crept. Under the effects of external interference, such as an earthquake and extreme weather, the landslide was reactivated.

The InSAR time series results show that between 2016 and 2021, landslides S06 and S07 both had continuous deformation. The cumulative deformation exceeded 40 mm. Some parts deformed quickly, with the accumulation exceeding 100 mm. But the deformation rate of most parts remained stable, without obvious acceleration. The landslide body showed no signs of disaster (Figure 17).

Figure 16. Optical image of landslides S06 and S07 in 2017. F1 denotes the rear part fissure of landslide S07. F1 denotes the rear part fissure of landslide S06.

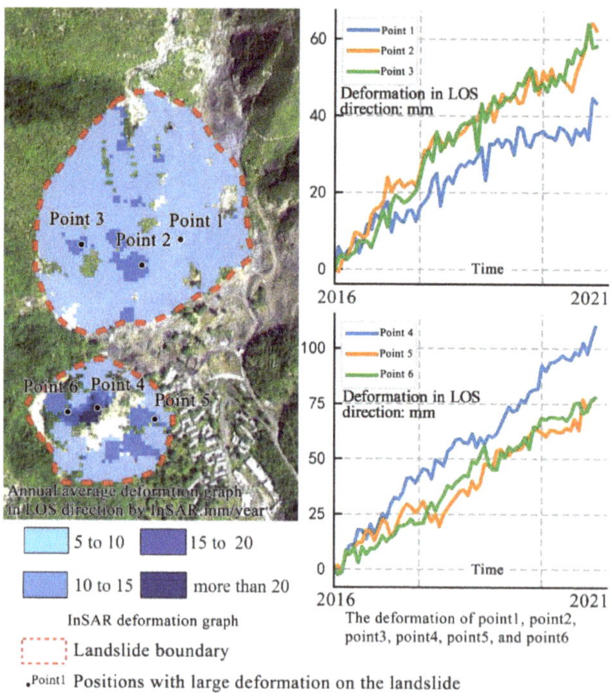

Figure 17. Deformation velocity of landslides S06 and S07 found by InSAR (left panel). The deformation in LOS direction of the six points during 2016–2021 are shown in the right panel.

The InSAR technology has limitations in monitoring the deformation of the areas with large height differences and high vegetation coverage [36–43]. Besides, sudden and large magnitude deformation leads to severe decoherence. However, it can still be used for the identification of landslides in the creep stage, which have slow deformations but are large in size.

The above analysis shows that the goaf-induced landslide has a long development period, which can be divided into four stages. In the first stage, the slope is stable. If there are no drastic changes to the external environment, the hazard risk is small. In the second stage, the lower part of the slope bulges and collapses, but the possibility of failure is still small. The hazard risk mainly comes from rockfall in the lower part of the slope. In the forth stage, if the deformation rate is stable, the slope body is temporarily stable. If the deformation accelerates, the hazard risk increases, and corresponding preventive measures should be taken. According to these features, the deformation stages of the landslide can be determined, to guide the prevention and control measures.

4.4. The Disaster Chain of Goaf-Landslide-Debris Flow

A total of 16 landslides were identified in the study area. All these landslides were developed in the coal mining area. Besides large-scale goafs, a large amount of coal gangue and other materials were loosely piled in the ditches. There were floods and debris flow disasters in this area [44]. Once these landslides become unstable during heavy rainfall, the transported material, together with coal gangue, may block the channel, form a dam, and then form a debris flow.

4.4.1. Mode of the Disaster Chain

We take landslides S06, S07 and S08 as examples to analyze the disaster chain mode. The three landslides are concentrated in the middle reaches of Anzigou ditch. S06 and S07 are located on the west side, and S08 is located on the east side (Figure 18). The three landslides have an estimated total volume of about 2 million m^3. They are located on a Jurassic coal-bearing stratum, which is relatively soft. At present, the landslide bodies are creeping.

Figure 18. 3D map of Anzi Ditch.

According to the monitoring data of the Beijing Institute of Geological Hazard Prevention and Control, heavy rainfall occurs frequently in the flood season in the study area. S06 and S07 experienced heavy rainfall in 2016 (daily rainfall of 176 mm) and 2012 (daily rainfall of 234 mm), respectively, and they slid downward for 30 m and 10 m, respectively. The deformation of the landslide is increasing, and the stability of the slope mass continues decreasing. In the case of heavy rain, a large-scale slide may occur.

A large amount of coal gangue accumulated in Anzigou ditch, with a total volume of about 7×10^5 m^3. No debris flow has occurred yet. The channel is V-shaped. The width of the channel section, where the three landslide bodies are located, is less than 100 m. Once the failure occurs, a mass as large as 2 million m^3 will be washed down, which will block the channel and form a dam. If the dam breaks, debris flow will occur.

4.4.2. Terrain Condition

In addition to precipitation and material, topography also affects the occurrence of debris flows. We combined historic records and field investigations and found that 110 debris flows have occurred in Beijing since 1920. Among them, 53 ditches are distributed in the coal mining area. Using high-precision remote sensing images and DEM, we calculated the channel slope, the average slope of the hillside and the main channel bending coefficient of these ditches (Figure 19).

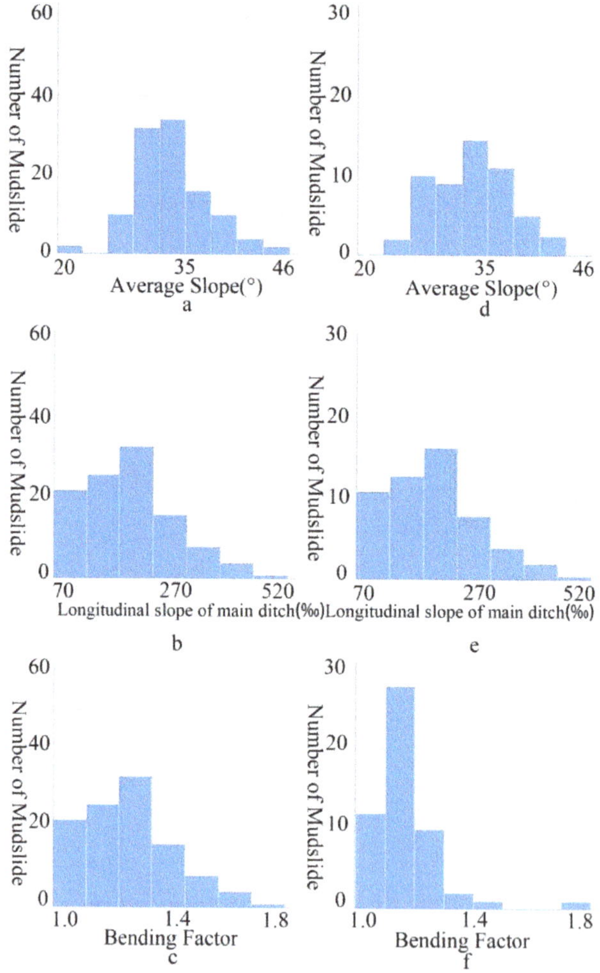

Figure 19. Average slope, longitudial slope of main ditch, and bending factor information of 110 debris flows in Beijing (Panel **a**–**c** in the left side). Average slope, longitudial slope of main ditch, and bending factor information of 53 debris flows in the coal mining area (Panel **d**–**f** in the right side).

The results show that, if the rainfall and mass sources are not considered, in the Beijing area, debris flows are most likely to occur in the ditches with a slope gradient of 130–250‰, an average slope of 29–38°, or a main channel curvature coefficient less than 1.3. The debris flow ditches in the coal mining area of Beijing have the same average slope and longitudinal slope as those of the Beijing area, but the bending coefficient is around 1.1 (d–f in Figure 19).

The Anzigou ditch watershed area is 2.04 km^2, the slope gradient is 221.19‰, the average slope is 31.65° and the curvature coefficient of the main channel is 1.12. Such terrain is favorable for the occurrence of debris flow.

Landslides S06, S07 and S08, together with the Anzigou ditch, form a potential "goaf–landslide–debris flow" disaster chain.

Landslide S10, S11, S12, and S13 developed in the Dongjiang ditch of Jinjitai Village (Figure 20). S12 and S13 are located downstream of the ditch, and their estimated total volume is about 7.6×10^5 m^3. The current deformation stage of the landslides, rainfall, mass source and topographic conditions of the channel all are favorable for the formation of a "goaf–landslide–debris flow" chain disaster, posing a threat to Jinjitai Village (Figure 20).

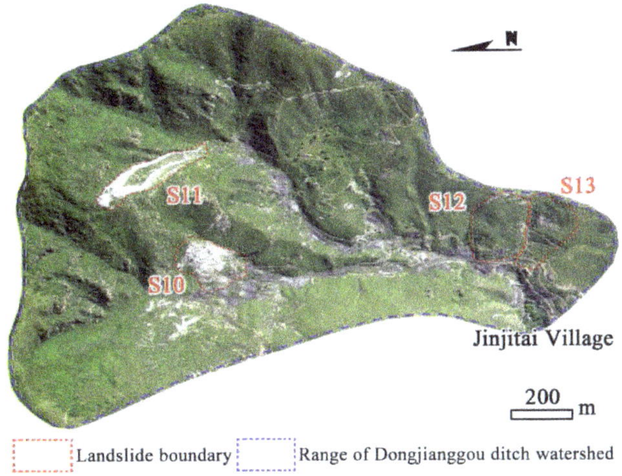

Figure 20. 3D Map of Dongjiang Ditch.

Upstream from the Dacunjian ditch (Figure 21), landslides S03 and S04 are currently in the second deformation stage, when the slope front swells and collapses. Additionally, landslide S01 and S02 have continuous deformation as well. As the deformation develops, the fissures on the rear edges will expand. The landslide body gradually grows, and may cause chain disasters.

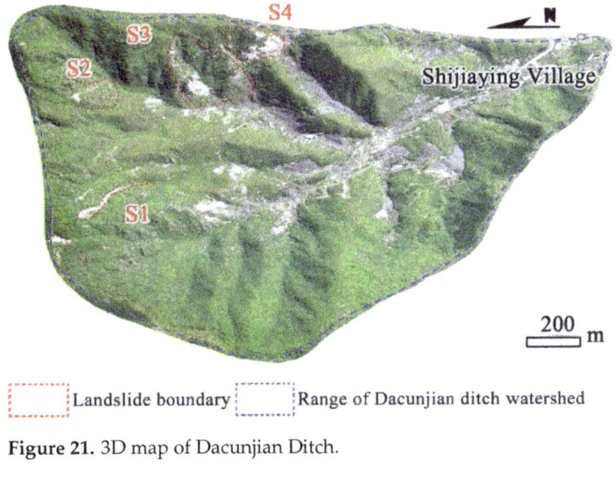

Figure 21. 3D map of Dacunjian Ditch.

Considering the size, location and topography, the remaining landslides cannot currently cause chain disasters.

5. Conclusions

In this paper, the multi-temporal high spatial resolution optical satellite images and RadarSat-2 data are used to identify the geological hazards, such as landslides, in the Xishan mining area, Beijing, China. We also discuss the possibility of a "goaf–landslide–debris flow" chain disaster in the study area. The following conclusions are drawn.

(1) The landslides in the Xishan mining area show obvious surface deformation during their development, which consists of the following four stages: initial deformation, slope front swelling and collapsing, rear part connecting and rupturing, and landslide creeping. The entire evolution process takes longer than ten years. The deformation process can be identified and quantified using optical remote sensing monitoring and time series InSAR technology.

(2) A total of 32 collapses and 19 landslides were identified in the study area. The distribution of collapses is related to topography and mining activity. The landslides are concentrated. There are 16 medium sized landslides, and they are divided into three types according to the development strata, the sliding features and the deformation stage. The first type is caused by high-level rock mass collapse. It develops in the overlying strata of the coal-bearing stratum. After failure, the landslide continues to deform. The other two types are caused by the rupture of goaves. The second type is mostly developed in the overlying strata of coal-bearing strata. The depth of the goaf is large, and the sliding surfaces are not connected. These landslides are mostly in either the initial deformation stage or the slope front swelling and collapsing stage. The third type is developed in coal-bearing strata. The depth of the goaf is small, and the sliding surfaces are all connected. These landslides are in the stage of creeping, and the deformation rate is significantly higher than that of the second type.

(3) The topography and precipitation of the Anzigou ditch and the Dongjianggou ditch are favorable for the occurrence of debris flows. In case of heavy rain, landslides S06, S07, S08, the Anzigou ditch, S10, S11, S12, S13 and the Dongjianggou ditch, as well as S01, S02, S03, S04 and the Dacunjiangou ditch all have the risk of a "goaf–landslide–debris flow" disaster chain.

(4) In a goaf, surface and underground disturbances can be serious, and their underground structures are complex. In those areas, besides ground fissures and ground subsidence, landslides and chain disasters may occur. To prevent and reduce the damage caused by geological disasters in those areas, we can identify areas with geohazard risks using multi-platform and multi-sensory remote sensing technologies, and build a comprehensive monitoring system for geological disasters using ground, airborne and spaceborne observations.

Author Contributions: All authors contributed to the study conception and design. Material preparation, data collection and analysis were done by R.J., S.W., H.Y., X.G., J.H., X.P. and C.Y. The first draft of the manuscript was written by R.J. and all authors commented on previous versions of the manuscript. All authors have read and agreed to the published version of the manuscript.

Funding: This research was funded by Beijing Institute of Geology and Mineral Exploration: Project of Beijing sudden geological disaster monitoring and early warning system (stage I) (11000022T000000439502), Beijing sudden geological disaster monitoring and early warning system (stage 2) (11000022T000000439510), Early identification and early warning of typical geological disasters in Xishan, Beijing Demonstration study (11000022T000001362678).

Data Availability Statement: The datasets presented in this study are available public datasets.

Conflicts of Interest: The authors declare no conflict of interest.

References

1. Shan, Q.; Hu, R.; Deng, Y.; Ji, Y. Study on the surface collapse disasters of Xishan region, Beijing. *Geol. Inst. Beijing Manicipality* **1994**, *5*, 124–129+165.
2. Ji, Y. The hazard and Origin of the Well-bore type collapse in the coal mined-out area of Western Hills, Beijing. *Beijing Geol.* **2002**, *14*, 28–33.
3. Ji, Y. Deformation and metamorphic characteristics of Carboniferous Permian coal bearing formation and related geological hazards in Xishan (Western Hills)Beijing. *Beijing Geol.* **2004**, *16*, 1–17.
4. Li, J. Applications of Geophysical Techniques in Preventing Surface Collapse in the Western Hills of Beijing. *Urban Geol.* **2008**, *3*, 30–33.
5. Li, X.; Qi, G.; Zhang, C. Hazard Prediction of Mining Subsidence. *Jiangxi Coal Sci. Technol.* **2016**, *3*, 57–59.
6. Xu, Q.; Li, W.; Dong, X.; Xiao, X.; Fan, X.; Pei, X. The Xinmocun Landslide on June 24,2017 in Maoxian, Sichuan: Characteristics and Failure Mechanism. *Chin. J. Rock Mech. Eng.* **2017**, *36*, 2612–2628.
7. Li, W.; Xu, Q.; Lu, H.; Dong, X.; Zhu, Y. Tracking the Deformation History of Large-Scale Rocky Landslide and Its Enlightenment. *Geomat. Inf. Sci. Wuhan Univ.* **2019**, *44*, 1043–1053.
8. Conte, E.; Pugliese, L.; Troncone, A. Post-failure stage simulation of a landslide using the material point method. *Eng. Geol.* **2019**, *253*, 149–159. [CrossRef]
9. Szafarczyk, A. Stages of Geological Documentation on the Example of Landslides Located on the Slopes of the Dam Reservoir "Swinna Poreba" (Poland). *IOP Conf. Series Earth Environ. Sci.* **2019**, *221*, 012037. [CrossRef]
10. Szafarczyk, A. Kinematics of mass phenomena on the example of an active landslide monitored using gps and gbinsar technol. *J. Appl. Eng. Sci.* **2019**, *17*, 107–115. [CrossRef]
11. Meng, X.; Guo, Z. Susceptibility assessment of the precursor stage of a landslide threatening Haivan Railway Station, Vietnam. *Landslides* **2018**, *15*, 309–325.
12. Deng, L.; Yuan, H.; Chen, J.; Sun, Z.; Fu, M.; Wang, F.; Yan, S.; Li, K.; Yu, M.; Chen, T. Correlation between Acoustic Emission Behaviour and Dynamics Model during Three-Stage Deformation Process of Soil Landslide. *Sensors* **2021**, *21*, 2373. [CrossRef] [PubMed]
13. Miao, F.; Wu, Y.; Trk, A.; Li, L.; Xue, Y. Centrifugal model test on a riverine landslide in the Three Gorges Reservoir induced by rainfall and water level fluctuation. *Geosci. Front.* **2022**, *13*, 14. [CrossRef]
14. Miao, F.; Zhao, F.; Wu, Y.; Li, L.; Xue, Y.; Meng, J. A novel seepage device and ring-shear test on slip zone soils of landslide in the Three Gorges Reservoir area. *Eng. Geol.* **2022**. [CrossRef]
15. Miao, F.; Wu, Y.; Xie, Y.; Li, Y. Prediction of landslide displacement with step-like behavior based on multialgorithm optimization and a support vector regression model. *Landslides* **2018**, *15*, 475–488. [CrossRef]
16. Xu, Q.; Dong, X.; Li, W. Integrated Space-Air-Ground Early Detection, Monitoring and Warning System for Potential Catastrophic Geohazards. *Geomat. Inf. Sci. Wuhan Univ.* **2019**, *44*, 957–966.
17. Zhao, C.; Liu, X.; Zhang, Q.; Peng, J.; Xu, Q. Research on Loess Landslide Identification, Monitoring and Failure Mode with InSAR Technique in Heifangtai, Gansu. *Geomat. Inf. Sci. Wuhan Univ.* **2019**, *44*, 996–1007.
18. Lu, H.; Li, W.; Xu, Q.; Dong, X.; Dai, C.; Wang, D. Early Detection of Landslides in the Upstream and Downstream Areas of the Baige Landslide, the Jinsha River Based on Optical Remote Sensing and InSAR Technologies. *Geomat. Inf. Sci. Wuhan Univ.* **2019**, *44*, 1342–1354.
19. Zhang, M.; Jia, J.; Wang, Y.; Niu, Q.; Mao, Y.; Dong, Y. Construction of Geological Disaster Prevention and Control System Based on AI. *Northwest. Geol.* **2019**, *52*, 103–116.
20. Li, M.; Zhang, L.; Shi, X.; Liao, M.; Yang, M. Monitoring active motion of the Guobu landslide near the Laxiwa Hydropower Station in China by time-series point-like targets offset tracking. *Remote Sens. Environ.* **2018**, *221*, 80–93. [CrossRef]
21. Tang, Y.; Wang, L.; Ma, G.; Jia, H.; Li, H.; Yin, H. Disaster Monitoring and Application Prospect Analysis of the Jinsha River Landslide Based on "Gaofen+". *Geomat. Inf. Sci. Wuhan Univ.* **2019**, *44*, 1082–1092.
22. Ge, D. Application of Remote Sensing in Early Recognition and Monitoring and Early Warning of Geological Hazards. *City Disaster Reduction* **2018**, *6*, 53–60.
23. Lai, Z.; Chen, J.; Xiang., J.; Xu, B. Landslide extraction based on fractal theory and LiDAR data. *Sci. Surv. Mapp.* **2018**, *43*, 112–117.
24. Ma, X.; Wu, Z.; Li, J. LiDAR Technology and Its Application and Prospect in Geological Environment. *J. Geomech.* **2016**, *22*, 93–103.
25. Dai, K.; Li, Z.; Tomás, R.; Liu, G.; Yu, B.; Wang, X.; Cheng, H.; Chen, J.; Stockamp, J. Monitoring Activity at the Daguangbao Mega-Landslide(China) Using Sentinel-1 TOPS Time Series Interferometry. *Remote Sens. Environ.* **2016**, *186*, 501–513. [CrossRef]
26. Liu, Z.; Song, Y.; Ye, R. An analysis of rainstorm-induced landslides in northeast Chongqing on August 31, 2014 based on interpretation of remote sensing images. *Remote Sens. Nat. Resour.* **2021**, *33*, 192–199.
27. Ding, Y.; Zhang, Q.; Yang, C.; Wang, M.; Ding, H. Landslide identification in Jinsha River basin based on high-resolution remote sensing:taking Wangdalong village of Batang county as an example. *Bull. Surv. Mapp.* **2022**, *25*, 51–55.
28. Guo, Q.; Zhu, L.; Li, A.; Gu, L. Landslide Identification Method based on NDVI Change Detection. *Remote Sens. Technol Appl.* **2022**, *27*, 17–23.
29. Hu, Z.; Xu, H.; Wang, C.; Zhou, Z.; Li, Z. A Landslides Detection Method Based on Time Series Remote Sensing Images. *Spacecr. Recovery Remote Sens.* **2018**, *39*, 104–114.

30. Huang, J.; Li, X.; Chen, F.; Cui, R.; Li, H.; Du, B. A deep learning recognition model for landslide terrain based on multi-source data fusion. *Chin. J. Geol. Hazard Control.* **2022**, *33*, 33–41.
31. Wang, Y.; Ju, N.; Zhao, J.; Xiang, X. Formation Mechanism of Landslide Above the Mined Out Area in Gently Inclined Coal Bens. *J. Eng. Geol.* **2013**, *21*, 61–68.
32. Li, B.; Wang, G.; Feng, Z.; Wang, W. Failure Mechanism of Steeply Inclined Rock Slope Induced by Undergroound Mining. *Chin. J. Rock Mech. Eng.* **2015**, *34*, 1148–1161.
33. Cristina, R.; Jorge, P.; Marcos, M.; Adrián, R.; Patricia, R.; Agustín, M.; Teresa, T.; Roberto, S.; José, V.; Anna, B.; et al. Rapid Characterisation of the Extremely Large Landslide Threatening the Rules Reservoir (Southern Spain). *Landslides* **2021**, *18*, 3781–3798.
34. Paolo, M.; Paolo, C.; Alessandro, B.; Francesco, I.; Francesca, B. Short-term Geomorphological Evolution of the Poggio Baldi Landslide Upper Scarp Via 3D Change Detection. *Landslides* **2021**, *18*, 2367–2381.
35. Deepak, R.; Janusz, V.; Matylda, T. Extended Residual-State Creep Test and Its Application for Landslide Stability Assessment. *Materials* **2021**, *14*, 1968.
36. Ge, D.; Dai, K.; Guo, Z.; Li, Z. Early Identification of Serious Geological Hazards with Integrated Remote Sensing Technologies: Thoughts and Recommendations. *Geomat. Inf. Sci. Wuhan Univ.* **2019**, *44*, 949–956.
37. Zhu, J.; Yang, Z.; Li, Z. Recent Progress in Retrieving and Predicting Mining-induced 3D Displacements Using InSAR. *Acta Geod. Et Cartogr. Sinica.* **2019**, *48*, 135–144. [CrossRef]
38. Yusuke, S.; Taro, U.; Ikushi, H.; Kazunari, T.; Yasumasa, F. Interrelated Impacts of Seismic Ground Motion and Topography on Coseismic Landslide Occurrence Using High-resolution Displacement SAR Data. *Landslides* **2022**, *19*, 2329–2345.
39. Roy, P.; Martha, T.R.; Khanna, K.; Jain, N.; Kumar, K.V. Time and Path Prediction of Landslides Using InSAR and Flow Model. *Remote Sens. Environ.* **2022**, *271*, 112899.
40. Lorenzo, S.; Matteo, D.; Federico, R.; Anna, B.; Silvia, B.; Pierluigi, C.; Nicola, C.; Michele, C. Review of Satellite Interferometry for Landslide Detection in Italy. *Remote Sens.* **2020**, *12*, 1351.
41. Paolo, M.; Paolo, C.; Riccardo, M. Sliding Time Master Digital Image Correlation Analyses of CubeSat Images for landslide Monitoring: The Rattlesnake Hills Landslide (USA). *Remote Sens.* **2020**, *12*, 592.
42. Varvara, T.; Athanassios, G.; Ioannis, K.; Eirini, E.; Ioannis, K.; Efthimios, S. Kinematics of Active Landslides in Achaia (Peloponnese, Greece) through InSAR Time Series Analysis and Relation to Rainfall Patterns. *Remote Sens.* **2022**, *14*, 844.
43. Liu, Y.; Yang, H.; Wang, S.; Xu, L.; Peng, J. Monitoring and Stability Analysis of the Deformation in the Woda Landslide Area in Tibet, China by the DS-InSAR Method. *Remote Sens.* **2022**, *14*, 532. [CrossRef]
44. Liu, L.; Wei, J.; Ye, C.; Zhang, C.; Zhao, Z.; Liu, Y.; Zhang, L.; Jia, S.; Wang, D.; Sun, X.; et al. *Geological Hazards in Beijing*; China Dadi Press: Beijing, China, 2008; pp. 22–27.

remote sensing

Article

A Deep Learning Application for Deformation Prediction from Ground-Based InSAR

Jianfeng Han [1], Honglei Yang [1,*], Youfeng Liu [1], Zhaowei Lu [1], Kai Zeng [1] and Runcheng Jiao [2]

1 School of Land Science and Technology, China University of Geosciences, Beijing 100083, China
2 Beijing Institute of Geological Hazard Prevention, Beijing 100120, China
* Correspondence: hongleiyang@cugb.edu.cn

Abstract: Ground-based synthetic aperture radar interferometry (GB-InSAR) has the characteristics of high precision, high temporal resolution, and high spatial resolution, and is widely used in highwall deformation monitoring. The traditional GB-InSAR real-time processing method is to process the whole data set or group in time sequence. This type of method takes up a lot of computer memory, has low efficiency, cannot meet the timeliness of slope monitoring, and cannot perform deformation prediction and disaster warning forecasting. In response to this problem, this paper proposes a GB-InSAR time series processing method based on the LSTM (long short-term memory) model. First, according to the early monitoring data of GBSAR equipment, the time series InSAR method (PS-InSAR, SBAS, etc.) is used to obtain the initial deformation information. According to the deformation calculated in the previous stage and the atmospheric environmental parameters monitored, the LSTM model is used to predict the deformation and atmospheric delay at the next time. The phase is removed from the interference phase, and finally the residual phase is unwrapped using the spatial domain unwrapping algorithm to solve the residual deformation. The predicted deformation and the residual deformation are added to obtain the deformation amount at the current moment. This method only needs to process the difference map at the current moment, which greatly saves time series processing time and can realize the prediction of deformation variables. The reliability of the proposed method is verified by ground-based SAR monitoring data of the Guangyuan landslide in Sichuan Province.

Keywords: GB-InSAR; LSTM model; deformation prediction; real-time processing

Citation: Han, J.; Yang, H.; Liu, Y.; Lu, Z.; Zeng, K.; Jiao, R. A Deep Learning Application for Deformation Prediction from Ground-Based InSAR. *Remote Sens.* **2022**, *14*, 5067. https://doi.org/10.3390/rs14205067

Academic Editors: Alex Hay-Man Ng, Linlin Ge, Hsing-Chung Chang and Zheyuan Du

Received: 2 August 2022
Accepted: 1 October 2022
Published: 11 October 2022

1. Introduction

After 20 years of continuous development of ground-based SAR, many scholars and experts have conducted research on its application in different fields, so that ground-based SAR can continue to develop and be suitable for monitoring in different scenarios [1–5]. In 2018, S. Kuraoka successfully monitored the eruption process of Hakone volcano in Japan using the IBIS-L system, proving that GB-InSAR technology is a feasible method for monitoring volcanic activity [6]. In 2018, in order to study the relationship between the evolution of volcanic landforms and slope instability, Di Traglia et al. used the joint monitoring method of COSMO satellite and ground-based SAR to monitor volcanoes in a certain area of Italy. This method emphasized ground-based SAR equipment and spaceborne SAR. The feasibility of joint monitoring volcanic activity provides a feasible method for the subsequent study of volcanic evolution [7]. In 2019, T.carla et al. adopted the method of joint GNSS, spaceborne SAR and ground-based SAR for a landslide in the alpine region of Italy; the experiment proved the feasibility of the joint measurement of GNSS technology and InSAR technology, and has more advantages for GNSS technology in snow-covered areas, which provides ideas and methods for subsequent joint data and InSAR monitoring in alpine regions [8]. In 2021, T. Carla et al. used GB-InSAR technology to process and analyze the monitoring data of a large landslide in the Italian Alps for ten years, and analyzed the impact of hydrological changes on the surface deformation rate by combining with precipitation and snowfall data, and constructed a finite element method. The model models the deformation rate of the landslide mass at different heights and depths, and experiments demonstrate the importance of long-term GB-InSAR monitoring [9].

GB-InSAR has a higher sampling frequency, which can describe the dynamic deformation process of the monitoring target more finely [10–15]. The high sampling frequency also brings the problem of a large amount of data. How to quickly calculate the deformation-monitoring results is a hot issue in current research, which is of great significance to the early warning and prediction of geological disasters. The current GB-InSAR time series methods mainly use permanent scatterer technology and small baseline set technology. According to the idea of statistics, modeling and analysis are carried out for points that maintain stable and reliable phase quality in the time dimension, and finally the time series monitoring results of the points are obtained. However, these methods require analysis of the entire data set which is inefficient, and with the accumulation of data, the hardware requirements are also high, and the future cannot be predicted, which is not conducive to the prevention and control of geological disasters. In recent years, China and other countries have also carried out relevant research. In 2019, Wang proposed a real-time ground-based synthetic aperture radar processing method [16], (real-time ground-based SAR, RT-GBSAR), which is based on SBAS-InSAR. The data set is divided into different units for processing, which has the characteristics of small system resource occupation and fast processing. Izumi et al. adopted the method of packet processing [17] to realize real-time data processing. Group processing solves the problem of real-time processing to a certain extent, but it does not have the function of deformation prediction, and there are still certain limitations in the monitoring of ground disasters.

Long short-term memory (LSTM) is an improvement in the recurrent neural network (RNN) model, which solves the phenomenon of gradient disappearance and gradient explosion, and can be used to solve the problem of long-term sequence prediction problems. In recent years, the LSTM model has also been used in InSAR data processing. In 2021, Liu Qinghao et al. built an LSTM model to predict the land subsidence results obtained by InSAR time series processing. However, the application of combining the LSTM model with GB-InSAR to realize deformation prediction and real-time processing is seldom [18]. GB-InSAR has high frequency and regular sampling frequency, so it is suitable to use the LSTM model for time series analysis.

Aiming at the problems of high resource occupation, long consumption time and poor prediction effect in GB-InSAR time series data processing, this paper proposes a real-time GB-InSAR slope deformation processing method based on the LSTM model. The first step is to build a deep learning sample library for the deformation variables obtained in the pre-processing, then to train the LSTM deep learning model, use the LSTM model to predict the deformation variables at the next moment, and remove them from the current differential phase. At the same time, the deep learning sample library is constructed for the atmospheric parameters obtained in the previous stage, the training model is used, and the atmospheric phase at the next moment is predicted, and the same is removed from the current differential phase. Finally, the residual phase is unwrapped using the spatial domain unwrapping algorithm, and the residual deformation is calculated. The predicted deformation and the residual deformation are added to obtain the deformation amount at the current moment. This method only needs to process the difference map at the current moment, which greatly saves time series processing time and can realize the prediction of deformation variables.

2. Materials and Methods

2.1. Desciption of Guangyuan Landslides

On 7 October 2021, a landslide occurred near Yanwo Village, Guangyuan City, Sichuan Province. The landslide was located near Dashiyan, Yanwo Village, Guangyuan City, specifically at 106°0′41.53″ east longitude, 32°22′28.449″ north latitude. The highest point is located on the top of the mountain in the north, with an altitude of 756 m, and the lowest point is located at the bottom of the mountain, with an altitude of 540m, and the height difference between the two is 216 m. The upper part of the landslide is steep, with a gradient of about 15° to 30°, and the lower part is relatively gentle, with a gradient of about 5° to 20°, and the overall main sliding direction is about 223°. The landslide starts from the upper part of the hillside, the trailing edge of the landslide is bounded by bedrock, and the west and east sides are bounded by a little exposed bedrock. The possibility of the landslide continuing to expand cannot be ruled out. The landslide is about 851 m long, 120 m to 160 m wide, about 2 m thick on average, and has a distribution area of about 77,297 square meters. It is a large soil landslide. The landslide extends in the southwest direction as a whole, and the lateral extension is short. At present, the deformation of the landslide is mainly concentrated in the top trailing edge, local slow displacement occurs at the soil layer accumulation from the middle to the bottom, and the overall structure tends to be stable. If the landslide deforms significantly and this causes the landslide to become unstable, then the landslide will pose a threat to the residents in the danger zone and the three high-voltage power line towers. Contact monitoring is dangerous, and the satellite time baseline is long, which is not suitable for emergency monitoring. Therefore, GB-InSAR with short time baseline, high resolution and flexible observation angle is used for continuous monitoring. The location of the study area and the field equipment is shown in Figure 1.

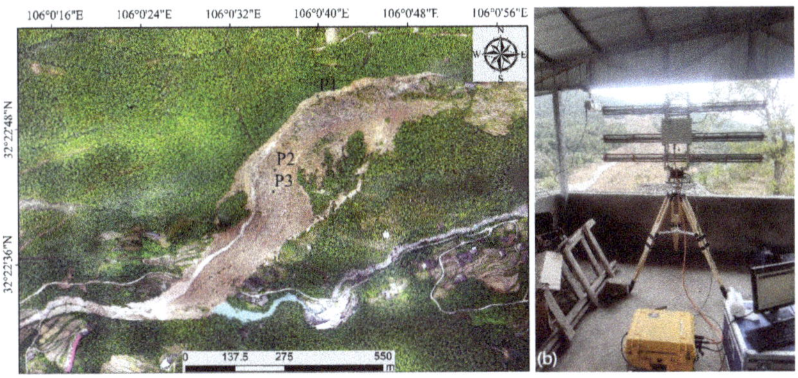

Figure 1. (**a**) is an orthophoto image processed by the DJI Royal 2 drone on 30 October 2021, showing the geographic location of the study area. P1, P2 and P3 are three study points selected from different locations of the landslide. P1 is the rear wall of the landslide, P2 is the landslide platform, and P3 is the secondary step. (**b**) is a photo of the GPRI-II equipment and the landslide mass.

The GAMMA Portable Radar Interferometer (GPRI-II) was set up 500 m south of the landslide. At the same time, the FSR-8 automatic weather station with the same time baseline was used to collect the necessary meteorological parameters for a total of 16 days, of which there was no interruption in the first 14 days. Day 15 was suspended for 8 h due to power issues. The main parameters during measurement are show in Table 1:

Table 1. SAR data acquisition parameters.

Radar elevation angle	0°
Radar gain	54 dB
Measuring distance range	30 m–1400 m
Measuring angle range	−90°–60°
Range resolution	0.7495 m
Azimuth resolution (−3 dB)	6.8 m (1 km)
Single imaging time	15 s
Time interval between two images	5 min

2.2. GB-InSAR Phase Model

After GBSAR was installed, the monitoring target was observed at a certain frequency (minute level). Due to the stable observation attitude, a zero baseline could be achieved. Compared with the spaceborne interference phase model, there is no reference plane phase and elevation phase, and the corresponding interference phase model can be represented by the following formula:

$$\varphi = \varphi_{def} + \varphi_{atm} + \varphi_{noise} + 2k\pi \tag{1}$$

Among them, φ is the interferometric phase of the wrapping, φ_{def} is the deformation corresponding to the radar line of sight, φ_{atm} is the atmospheric delay phase in the radar line of sight, φ_{noise} is the interference phase noise, and k is the number of integer cycles [19]. Among them, the atmospheric delay phase is the most important error source affecting the monitoring accuracy of GB-InSAR. The research shows that at a distance of 1 km from the GBSAR device, when the temperature is 20 °C, a 1% humidity change can lead to a measurement error of 2 mm [20,21].

According to the propagation mode of electromagnetic waves in the air, a mathematical model and a physical model were constructed for the atmospheric delay phase. The mathematical model is based on the propagation principle of electromagnetic waves. The radar emits electromagnetic waves with a wavelength of 1 at all times, and the electromagnetic waves reach the target pixel after the propagation distance return. The echo phase model for this pixel is [22]:

$$\varphi_P(i,j) = \frac{4\pi}{\lambda}\int_0^{r(i,j)} n(r,t)dr \tag{2}$$

n is the refractive index related to distance and time.

The physical model assumes that in the ground-based SAR monitoring, the spatial distribution of the atmospheric medium is assumed to be uniform within the monitoring area of 300–500 m, and the refractive index is independent of distance and only related to time, so the atmospheric phase of the pixel can be expressed as:

$$\varphi(t) = \frac{4\pi r \cdot n(t)}{\lambda} \tag{3}$$

In the formula, the relationship between the refractive index n and the refractive index N can be expressed as:

$$N = (n-1) \times 10^6 \tag{4}$$

In atmospheric refraction studies, the refractive index is usually expressed by temperature, air pressure, and relative humidity [23]:

$$N = \frac{0.2589}{T} + \frac{3.73 \times 10^5 \times e}{T^2} + \frac{71.7 \times e}{T} \tag{5}$$

In the formula, h is the atmospheric pressure and e is the water vapor pressure, which can be expressed as [24] according to the Magnus saturated water vapor pressure formula:

$$e = \frac{h}{100} e_{sat} \tag{6}$$

$$e_{sat} = 6.1016 \cdot 10^{\left(\frac{7.5 \times T}{T+237.3}\right)} \tag{7}$$

Through the above analysis, the range-to-homogeneous atmospheric phase during GBSAR monitoring can be modeled and eliminated from the interference phase, while reducing the difficulty of phase unwrapping in the space domain.

2.3. Long Short-Term Memory for Sequence Modeling

For the phenomenon of gradient disappearance and gradient explosion in RNN, Hochreiter proposed a network structure to improve the gradient disappearance problem, which works in a similar way to RNN. The difference is that a memory storage mechanism is established in some units within RNN to achieve more effective information storage and updates. Figure 2 shows the internal structure of the LSTM. By analyzing Figure 2, it can be seen that LSTM is obviously different from RNN in that it adds a gating unit, and the network transmits the gating unit information and the cell state at the last moment to the next unit in the network.

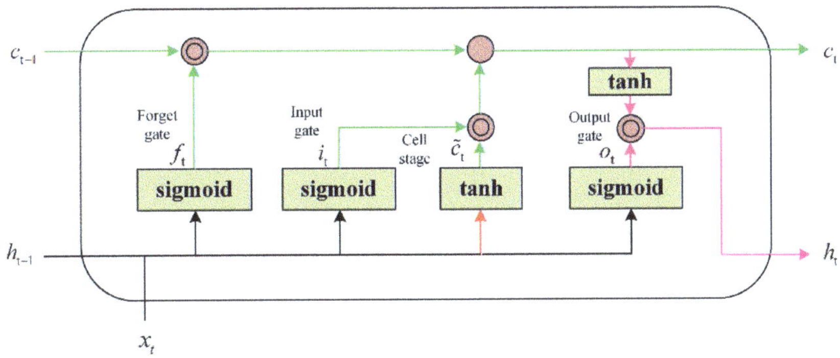

Figure 2. Schematic diagram of the LSTM cell structure. There are four types of gating units in LSTM cells: Forget gate, Input gate, Cell stage and Output gate. c_t and c_{t-1} are the state of the cell at time t and time $t-1$. h_t and h_{t-1} are the state of the hide at time t and time $t-1$. x_t is the input variable at time t.

After the input data are processed by the LSTM model, the predicted value at the next moment is obtained. The LSTM model can be expressed as:

$$f_t = \sigma\left(W_f \cdot [h_{t-1}, x_t] + b_f\right) \quad (8)$$

$$i_t = \sigma(w_i \cdot [h_{t-1}, x_t] + b_i) \quad (9)$$

$$\widetilde{C}_t = \tanh(W_C \cdot [h_{t-1}, x_t] + b_c) \quad (10)$$

$$C_t = f_t \times C_{t-1} + \mathrm{i} \times \widetilde{C}_t \quad (11)$$

$$o_t = \sigma(W_O \cdot [h_{t-1}, x_t] + b_o) \quad (12)$$

$$h_t = o_t \times \tanh(C_t) \quad (13)$$

3. Proposed Methodology

Firstly, this paper uses the time-series InSAR technology to obtain the deformation variable data set, and organizes and models the meteorological parameter data to obtain the atmospheric phase data set. After that, the LSTM prediction model is trained with two kinds of data sets, and the prediction accuracy of the model is evaluated by the prediction effect of the two kinds of data. Finally, the prediction model is used to realize the real-time processing of GB-InSAR monitoring and obtain the real-time processing results. The accuracy of the method is evaluated by the precision of inner coincidence and a computational efficiency evaluation. The technical roadmap for this paper is shown in the Figure 3 below:

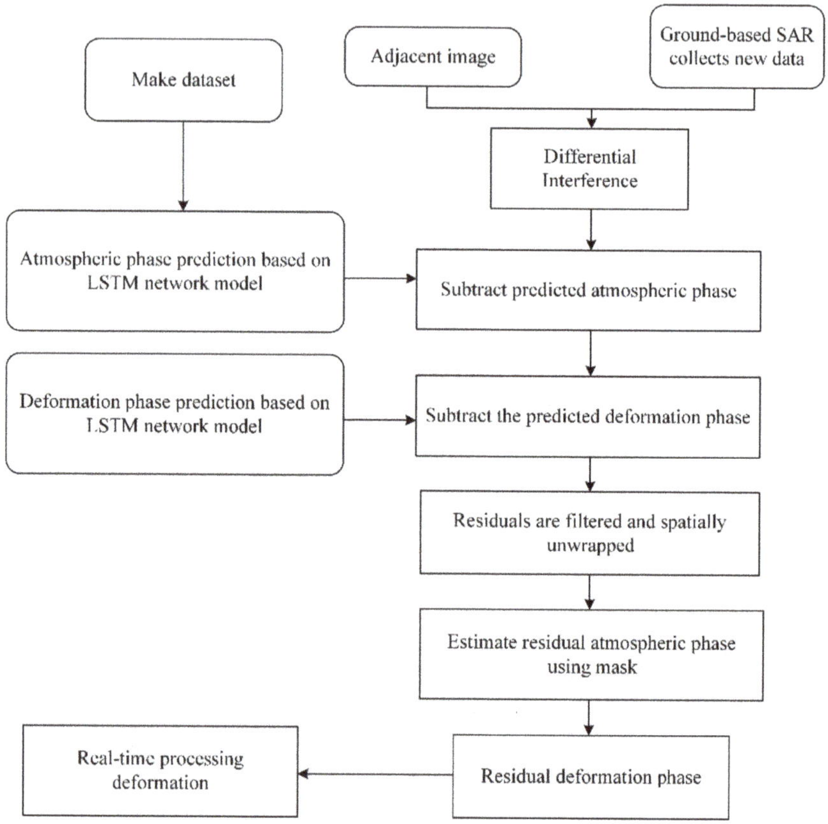

Figure 3. The technical roadmap of GB-InSAR time series processing method based on LSTM model.

3.1. Deformation Prediction

3.1.1. Data Normalization

In order to make the neural network model converge faster and obtain higher prediction accuracy, the variable values entering the neural network model should belong to the same range; that is, the value range of the data set should be consistent, and it is necessary to manually adjust the range of variables according to a certain ratio, the variable is scaled, and the following are two common processing methods:

Normalization refers to the scaling of variable values in the range zero to one or negative one to one [24]. First, make the difference between the maximum value X_{max} and the minimum value of the variable X_{min} to obtain the variation range of the variable X, and then divide each observation X_i value in the variable X with the minimum value X_{min} of the variable X and divide it by the variation range of the variable X to scale the observed value X_i to zero to one; the process is called maximum and minimum normalization, as shown in the following formula:

$$X_{i(\text{nornalized})} = \frac{X_i - X_{min}}{X_{max} - X_{min}} \tag{14}$$

The normalized atmospheric phase value and deformation phase value are selected and shown in Tables 2 and 3.

Table 2. After normalization of deformation data.

Num \ Time	2021.10.17 16:13:05	2021.10.17 16:33:05	2021.10.17 16:53:05	2021.10.17 17:13:05	2021.10.17 17:33:05
1	0.875	0.884	0.885	0.858	0.842
2	0.875	0.882	0.852	0.837	0.839
3	0.875	0.880	0.881	0.858	0.845
4	0.875	0.876	0.878	0.858	0.848

Table 3. After normalization of atmospheric phase data.

Unit \ Time	2021.10.17 16:13:05	2021.10.17 16:33:05	2021.10.17 16:53:05	2021.10.17 17:13:05	2021.10.17 17:33:05
Radian	0.695	0.696	0.686	0.702	0.685

3.1.2. LSTM Network Design

The data set is divided into a training set, validation set and test set according to the ratio of 6:2:2. After the network design is completed, the training set can be input to train the model. In the training process of the model, due to the complex network structure of the neural network, as the number of iterations increases, the model is prone to overfitting. Therefore, in order to have a high generalization ability of the model and avoid overfitting, the model's dropout layer needs to be added in the LSTM layer. The main function of this layer is to discard the neurons in the LSTM layer according to a certain proportion in each training [25].

After experimental testing, the increase in the number of LSTM layers does not change the prediction accuracy of the model, and the selection of timesteps is equally important. Selecting timesteps is three, that is, predicting the deformation variables at the next moment according to the deformation variables of the first three moments, which corresponds to the experiment in this paper. That is, it is recommended to use the one-hour deformation data to predict the deformation data after 20 min as the output. The Units parameter in the LSTM layer represents the number of neurons in this layer, which is the tensor output dimension of the LSTM layer. Activation represents the activation function method. The ReLU activation function is selected in the LSTM layer, and the sigmoid activation function is selected in the Dense layer. This combination method enables the model to obtain higher accuracy. The dropout layer is set to zero point two, that is, twenty percent of neurons are randomly dropped during each training [26–28].

3.1.3. Hyperparameter Selection

For neural network models, there are many hyperparameters in the model, which control the structure, function, and efficiency of the model, and are the key to obtaining high-precision results for the model. The hyperparameter selection of the model in the experiment in this paper is adjusted using the grid search method. By setting the range of hyperparameters, the model is evaluated according to a certain growth rate of hyperparameters, so as to select the combination of hyperparameters with the best model performance.

In the evaluation index of the model, we use the loss rate to measure; the loss rate is mainly the calculation method of the MSE (mean square error), and its calculation method is shown in the following:

$$\mathrm{MSE} = \sum_{i=1}^{n} \frac{(\mathrm{Predicted_i} - \mathrm{Actual_i})^2}{n} \tag{15}$$

In the training process of the model, the fitting degree of the model is judged by measuring the size of the loss function. When the loss rates of the training set and the validation set finally reach the same level, it proves that the model has achieved a good fitting effect. The combination of hyperparameters of the model determined using the grid search method is shown in Table 4:

Table 4. LSTM model hyperparameter combination table.

Hyperparameter Name	Value
LSTM_units	128
Dropout	0.2
Batch_size	128
epochs	400

For deformable data, this paper conducts first-order difference to remove the auto-correlation of the data itself and then trains the model. For atmospheric phase data, since the data itself have periodic components, this paper uses the STL (seasonal trend loss) method to decompose the periodic term, trend term and residual term. After that, the three components are modeled separately. The predicted values of the three components are then added to obtain the overall predicted value. The training set and validation set loss rate curves of the model are shown in Figures 4 and 5:

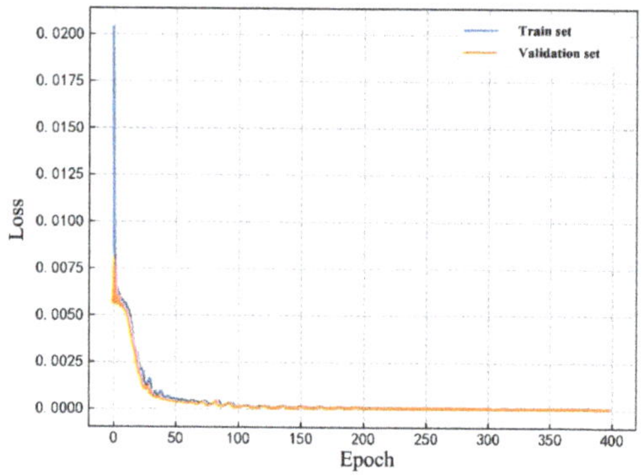

Figure 4. The deformation loss rate curve of the training set and the validation set. The loss value approaches 0 after 400 epochs.

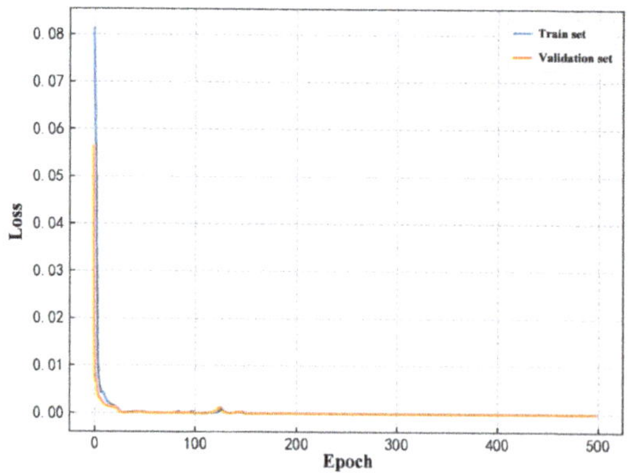

Figure 5. The atmosphere phase loss rate curve of the training set and the validation set. The loss value approaches 0 after 500 epochs.

In the Figure 5, the horizontal axis represents the epoch, and the vertical axis represents the size of the model loss. When the epoch reaches 500 times, the loss rate converges well, and finally reaches a stable value near 0.000002, indicating that the prediction accuracy of the deformation prediction model is very high.

3.2. Real-Time Processing

3.2.1. Obtaining the Initial Deformation Value

PS-InSAR technology is a method of selecting PS (Permanent Scatterer) point targets for time series analysis by statistically analyzing the SAR time series data in the study area according to the amplitude and phase. Different from the D-InSAR method [29], it only models and analyzes point targets with high coherence, strong scattering and stability, which effectively reduces the impact of decoherence on data processing. Sequence analysis can clearly obtain the change in the deformation of the target area over time, and more intuitively understand the change trend of the surface, which is helpful for the investigation and prevention of geological disasters. The technical process of PS-InSAR is shown in Figure 6 below:

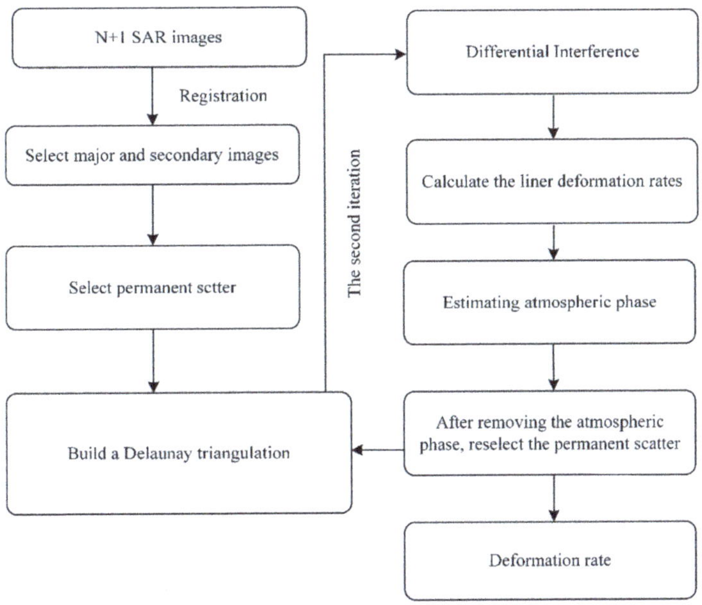

Figure 6. The framework of PS-InSAR analysis. The final deformation rate results of N + 1 SAR data are obtained by the flow in the figure.

In Figure 6, in the monitoring of GBSAR, the radar transmitter moves along a fixed orbit in each measurement. After the main image is selected, PS points are selected by setting thresholds for the phase dispersion index and coherence coefficient, and then the auxiliary image and the main image are subjected to interference processing, respectively, to construct a phase model function. Then, build a Delaunay triangulation for all adjacent PS points, use the solution space search method for the phase difference of PS points on the same baseline, iteratively solve the linear deformation rate, analyze the residual phase, and estimate the atmospheric phase component. After the atmospheric phase is separated from the phase, the PS point is reselected and the above operation is repeated to obtain the deformation time series results of the monitoring area.

3.2.2. Real-Time Processing

For time $t-1$, use the scene data before $t-1$ to analyze the time series of the monitoring area, and obtain the deformation variable data within this time range, and use these data as the data set of the LSTM model to predict the deformation variable at time t. The meteorological parameters [30] collected by the weather station at the previous time $t-1$ are used to fit and predict the atmospheric phase at time t, which can be:

$$\varphi_{t\_pred} = \varphi_{atm\_pred} + \varphi_{def\_pred} \tag{16}$$

The formula represents the predicted phase at time t, including the predicted value of the atmospheric phase and the predicted value of the deformation phase. At time t, if you want to perform time series analysis, you need to remodel and solve the data at the previous time t, which will waste more time and resources. Interference is performed at time t, and the phase includes:

$$\varphi_t = \varphi_{atm} + \varphi_{def} + \varphi_{noise} + 2k\pi \tag{17}$$

The atmospheric phase change and deformation phase change at time t relative to time $t-1$ are subtracted from the difference result between time $t-1$ and time t. At this time, there are only residual atmospheric phase and residual deformation phase in the phase. bit and noise phase. At this time, the spatial domain unwrapping of the interference phase is performed after the spatial domain filtering, which is easier than the original differential phase, and a better unwrapping result can be obtained. After the unwrapped phase is obtained, according to the deformation result at time $t-1$, a mask is generated to cover up the area with a large deformation amount. For the stable area, the residual phase is the atmospheric phase of the area. The residual atmospheric phase of the entire monitoring area is obtained by interpolation, and the residual deformation phase can be obtained by subtracting the phase from the previous step. After being converted into a deformation variable and added to the predicted deformation variable at time t, the predicted deformation variable at time t can be corrected, and the time for time series analysis can be saved in order to achieve the effect of real-time monitoring.

4. Results

In this section, the GBSAR monitoring data of Guangyuan City, Sichuan Province were used for experiments, and 20% of the total data were used to predict the atmospheric phase and shape variables. Finally, 40 views of GBSAR data from 14:39 on October 28 to 03:59 on October 29 were used for the real-time processing experiment.

Figures 7 and 8, respectively, show the interferogram and the selected PS point diagram at 16:00 on the 17th and 16:00 on the 19th. Analysis of Figure 7 shows that the landslide body has undergone large deformation, and the coherence is poor in the vegetation coverage area.

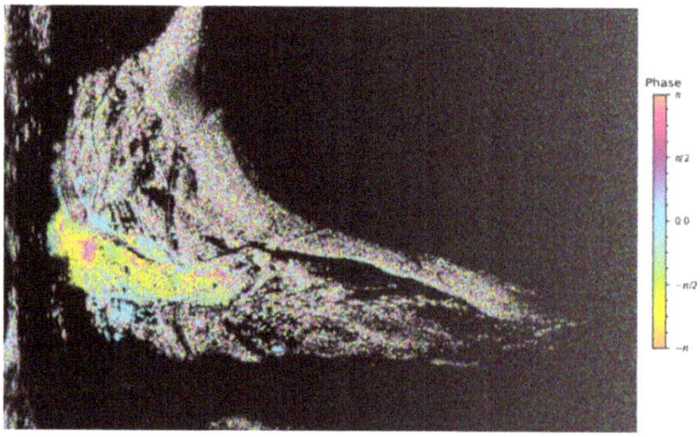

Figure 7. Interferogram result between 16:00 on the 17th and 16:00 on the 19th.

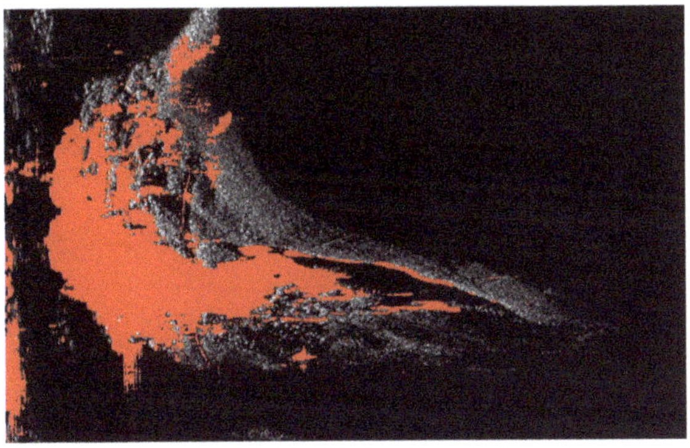

Figure 8. Schematic diagram of PS points selected in the target area.

4.1. Atmosphere Prediction

In this paper, the monitoring area was monitored for 16 days of meteorological parameters. By ensuring that the sampling frequency of GBSAR was consistent with that of GBSAR, the changes in atmospheric parameters during the observation period were obtained. According to the atmospheric phase model of GBSAR, the atmospheric phase based on the distance direction was obtained. The data set was constructed using this data, and the LSTM prediction result of the monitoring area was obtained by applying the LSTM network model, which is plotted in Figure 9.

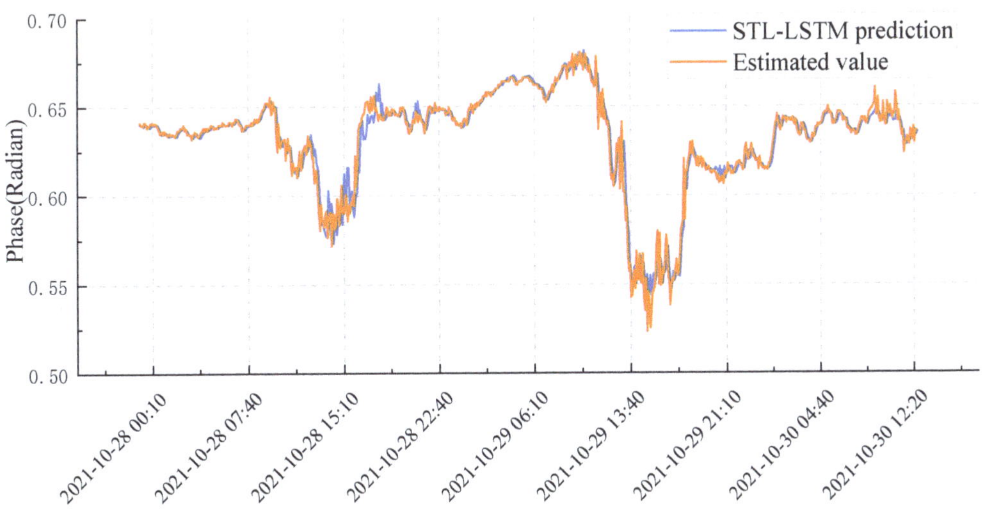

Figure 9. Time series change diagram of atmosphere phase prediction.

4.2. Deformation Prediction

In this paper, a total of 1,398,120 PS points and the deformation amount of 59 h in the entire monitoring area were predicted, and the deformation prediction results are partially shown in Figure 10 below according to a certain time interval. To make a difference, we drew the prediction residuals as shown in Figures 11 and 12, and drew the prediction residuals in Figures 13 and 14. By analyzing the above results, it can be seen that the LSTM network model has high prediction accuracy in the entire monitoring area, and the prediction results of the key deformation area and the stable area are in line with the facts.

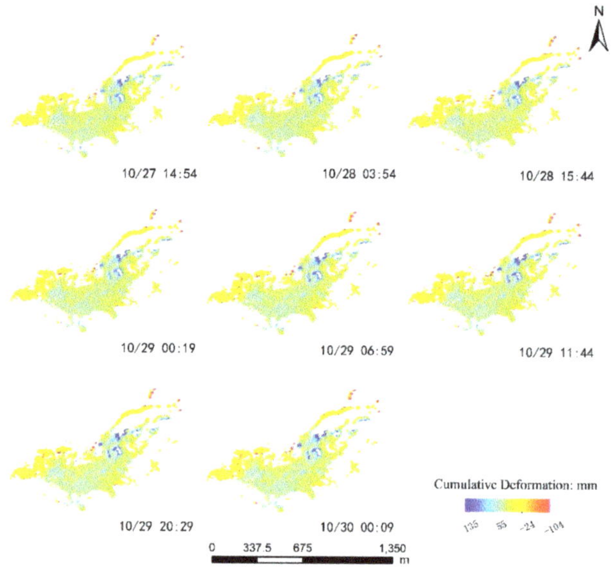

Figure 10. Historical results of deformation prediction along LOS direction. The displacement map was generated between the date that is marked in the low right corner and 17 October 2021 (the first acquisition date of GBSAR image). The deformation of the landslide is obvious. The color bar ranges from −104 mm to 135 mm.

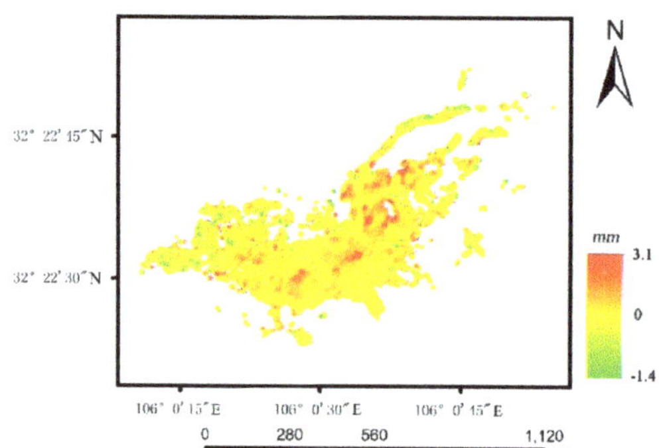

Figure 11. 10/29 20:29 Residual plot of real value and predicted value.

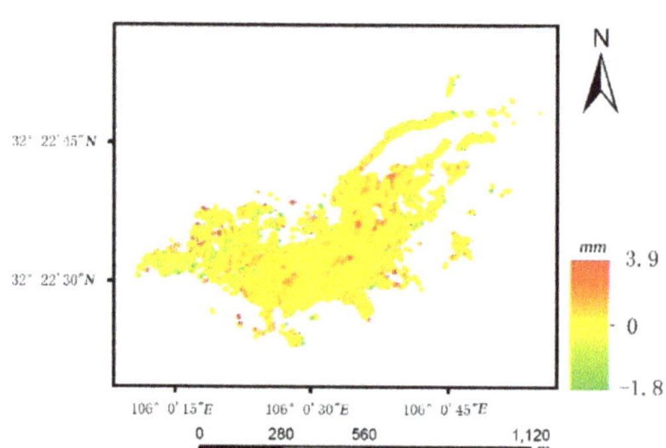

Figure 12. 10/30 00:09 Residual plot of real and predicted values.

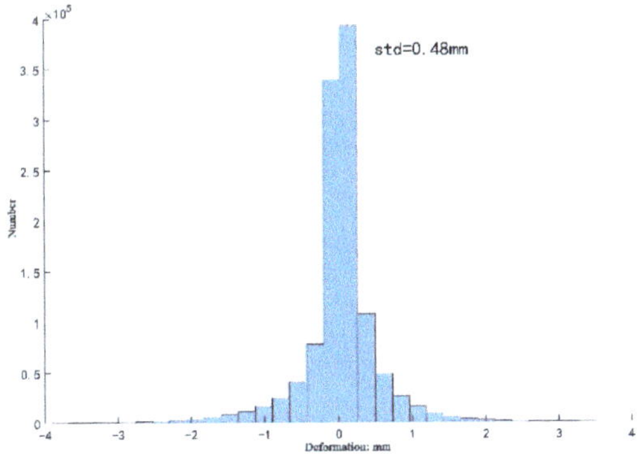

Figure 13. 10/29 20:29 residual histogram.

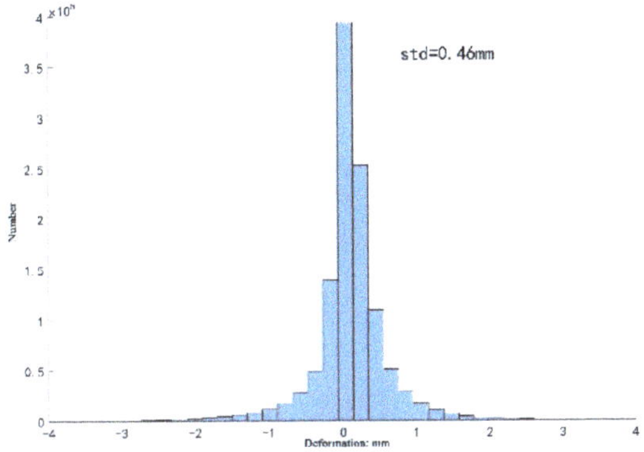

Figure 14. Histogram of residuals at 10/30 00:09.

4.3. Real-Time Processing

This experiment uses the ground-based SAR data of 40 scenes from 14:39 on 28 October to 03:59 on 29 October. First, the difference results between time $t-1$ and time t are obtained as shown in Figure 15. From the difference map, it can be clearly seen that there is an atmospheric phase with a trend term, and there is relatively large atmospheric disturbance in the area farthest from the radar center. The atmospheric phase and deformation phase predicted by the LSTM model are subtracted from the interference phase, and the results are shown in Figure 16. Analysis of the results shows that when fitting the atmospheric phase in the horizontal direction in this area, there is still atmospheric turbulence in the elevation direction and in a small area. For this part of the atmospheric phase, the method of mask interpolation after unwrapping is used to remove it. For the phase residuals of adjacent moments, spatial domain filtering is first performed, and the filtering results are shown in Figure 17. The filtering will remove the noise interference in the residuals. After that, the minimum cost flow method is used to unwrap, and the unwrapping results are shown in Figure 18. After unwrapping, the atmospheric phase of the trend term can be clearly seen, and then the stable region is interpolated after masking, and the fitted atmospheric phase is shown in Figure 19. After subtracting the residual atmospheric phase from the residual, the residual deformation is obtained as shown in Figure 20. Analysis of Figure 18 shows that the residual deformation of most regions is predicted to be zero, which proves that the LSTM model can be used to complete large regions with high accuracy. In the deformation prediction, the real-time processing deformation variable at time t is obtained by adding the deformation residual to the predicted value at the end.

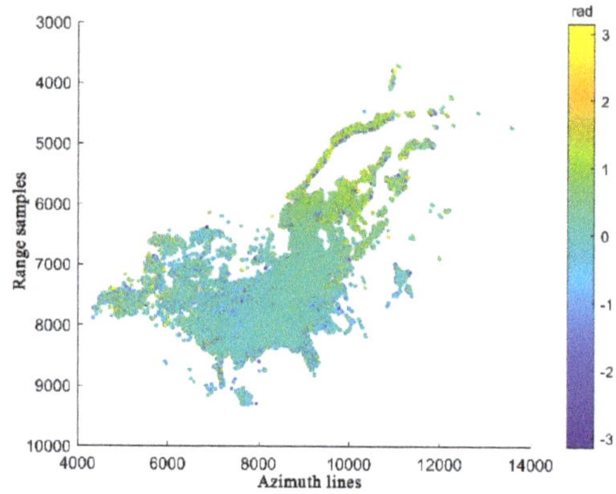

Figure 15. Interferogram of adjacent images. The interference phase includes the deformation phase, atmospheric phase, noise and whole cycle.

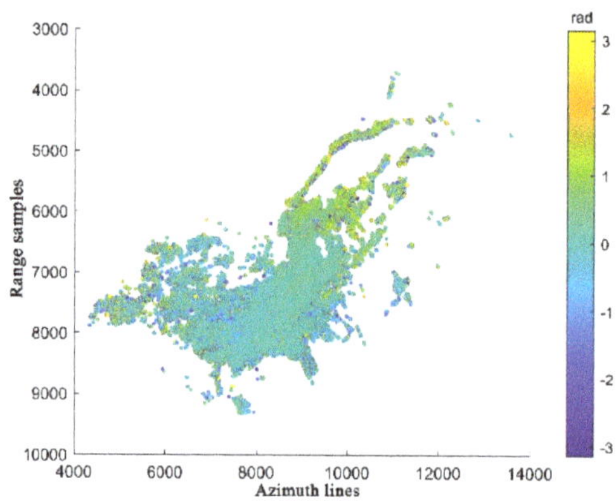

Figure 16. Interferogram residual phase diagram. The phase includes the remaining deformation phase, the remaining atmospheric phase and the whole cycle.

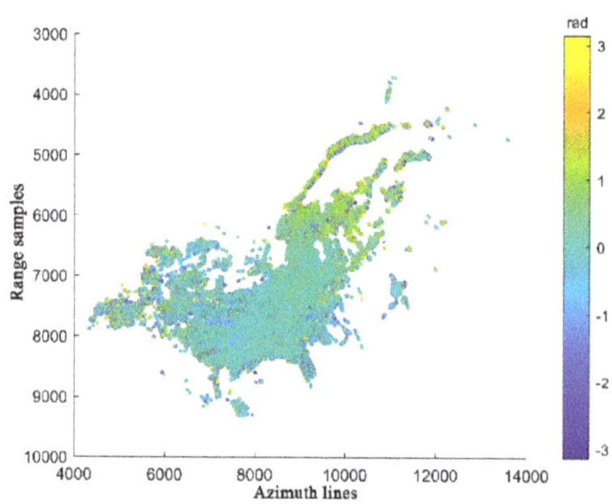

Figure 17. Residual phase map after filtering. The residual phase is filtered by the spatial domain filtering algorithm, and the filtered phase is obtained.

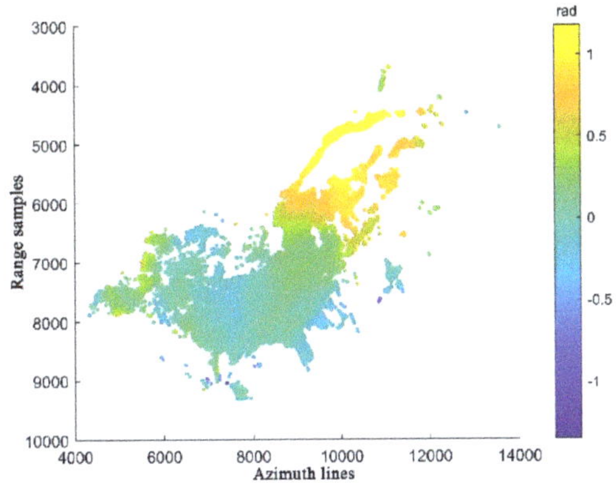

Figure 18. Residual phase diagram after unwrapping. The interference phase contains only a small amount of deformation phase and atmospheric phase, so it is easier to untangle and reduce the unwinding error.

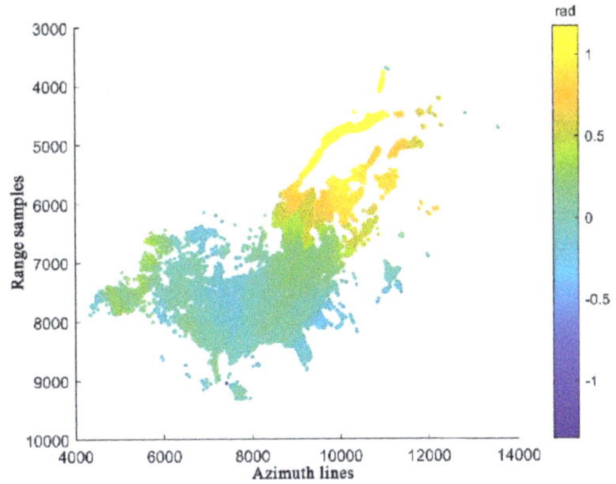

Figure 19. Interpolated atmospheric phase diagram. In the study area, the stable region is selected by taking the shape variable as the threshold, and the phase of the stable region is interpolated to obtain the overall atmospheric delay phase.

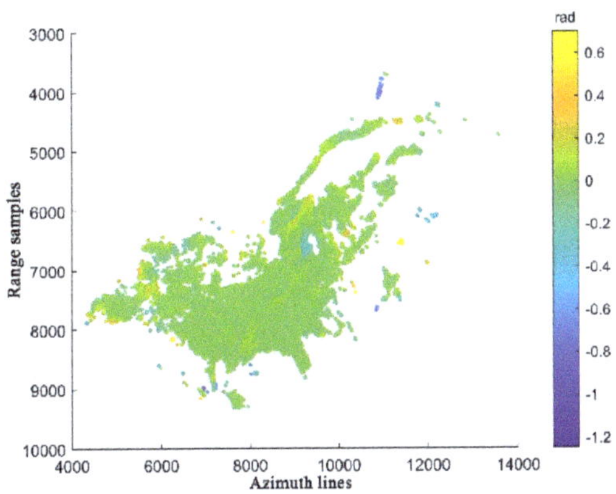

Figure 20. Residual deformation phase diagram. The residual deformation phase can be calculated by subtracting the interpolated atmospheric phase from the residual.

By processing 40 scene data to simulate real-time processing in the monitoring environment, the final deformation rate graph is obtained as shown in Figure 21:

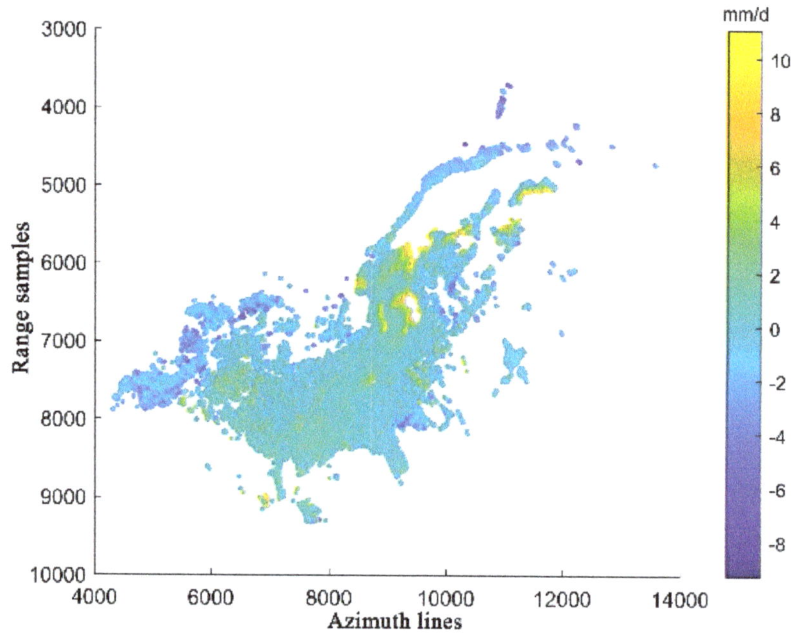

Figure 21. Final deformation rate graph. The slide body can be clearly identified from the figure and the deformation of the slide body can be analyzed.

According to the points P1, P2 and P3 selected in Figure 1a, they are used as the research objects for time series analysis, and their time series change curves within 800 min are drawn as shown in Figures 22 and 23:

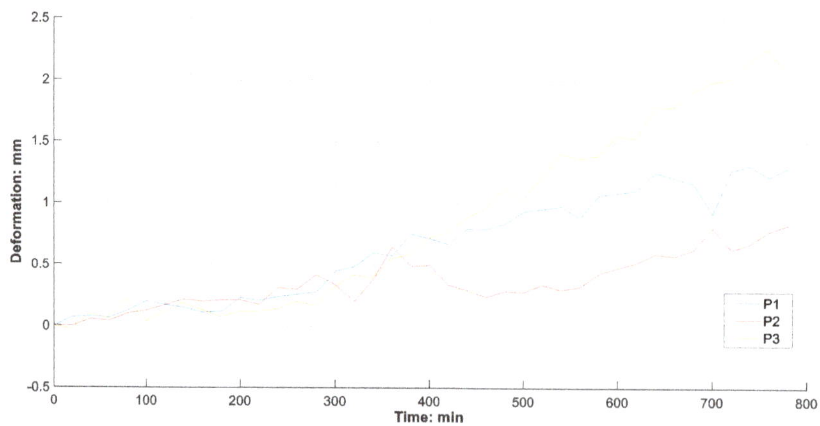

Figure 22. Real deformation curves of three research points. P1 is located at the crown cracks, P2 is located at the head, and P3 is located at the minor scarp. According to the analysis curve, the deformation rate of the three points is different, but all of them keep rising.

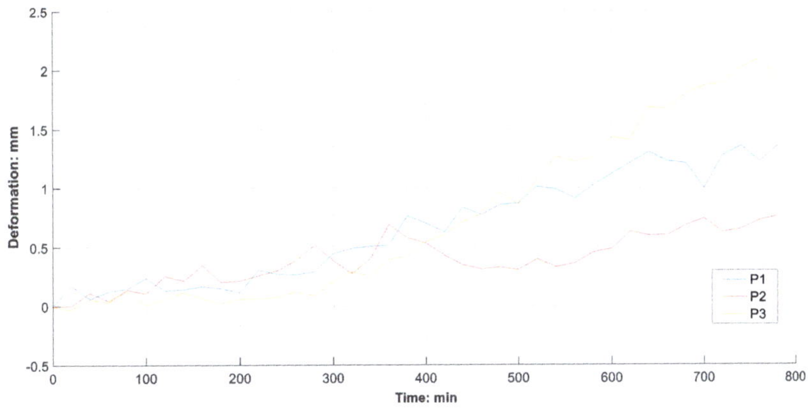

Figure 23. Real time processing method for deformation curves.

By analyzing the change curves of the three points in Figure 23, the time series results obtained by this real-time processing method are consistent with the GB-InSAR time series analysis results, which are in line with the real deformation variable changes, which proves that this method can realize real-time deformation processing. The process is as follows: select the deformation variable in the stable region for statistical analysis, and draw the frequency statistical histogram of the deformation variable in this region as shown in Figure 24:

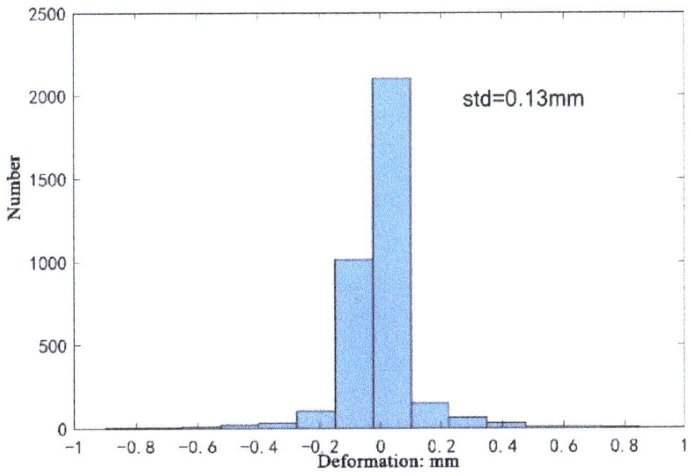

Figure 24. Deformation statistics histogram of the real-time processing method.

Generally speaking, the GB-INSAR deformation results need to be obtained through timing processing. Timing processing requires modeling and analysis of the entire data, using PS-InSAR or SBAS-InSAR methods. Due to the large amount of data in this method, the processing needs a lot of time, and the data set needs to be recalculated every time an additional scene is added. In this paper, the time of sequence analysis of 40 scenes of GBSAR data is counted, and the computer configuration used in the experiment is shown in Table 5. The time consumed by the real-time processing method proposed in this paper is plotted in Table 6.

Table 5. Experimental environment configuration.

Operating System	Ubuntu 20.04
Development Platform	Google Tensorflow
GPU	TITAN RTX 32G
CPU	Inter Xeon(R) Silver 4116 @2.10GHz × 48
Memory	128G

Table 6. Time-consuming statistics table.

Number of Data Scenes	33	34	35	36	37	38	39	40
Consume Time (min)	4.72	4.77	4.75	4.7	4.76	4.71	4.71	4.73
Original Time (min)	35.4	35.4	35.4	35.5	35.5	35.5	35.6	35.6

According to Table 5, GB-InSAR real-time monitoring based on LSTM network model has high processing efficiency, which can realize real-time monitoring and obtain high-precision results in slope monitoring. The average time of obtain deformation results by the real-time processing method is 4.73 min, which is much faster than the ordinary time series processing method. It is proved that the method can obtain results quickly and play a greater role in site landslide monitoring.

5. Discussion

The results show that the application of the LSTM model can effectively predict the deformation variables of the landslide in Guangyuan City, Sichuan Province. By evaluating the accuracy of the prediction results, the results show that the LSTM model can provide accurate prediction results. Compared with the linear prediction model, using the LSTM model for prediction can mine the correlation between historical data and make more accurate predictions for future data; compared with the machine learning model, it has the advantages of faster model convergence and higher prediction accuracy. At the same time, the distance from the study area to the atmosphere was predicted, and good prediction results were also obtained. GB-InSAR technology has the advantages of high precision, high spatial and temporal resolution, and flexible observation. The combination of the LSTM model and GB-InSAR provides more means for ground observation.

Based on the LSTM model, the GB-InSAR real-time processing method proposed in this paper can complete the extraction of deformation information in a short time. In the case of fewer computer resources and shorter time consumption, GB-INSAR time series processing is completed, and the time series results of deformation variables are obtained. The method proposed in this paper can provide case reference and verification for subsequent GB-InSAR monitoring technology and data-processing methods.

6. Conclusions

As a non-contact measurement technology, GB-InSAR technology can measure landslides with high precision and high spatial and temporal resolution. Compared with SAR satellites, ground-based SAR has a short time baseline, no spatial baseline, high resolution, and high-resolution observation. The advantages of flexible angle are widely used in landslide monitoring. Based on this, this paper proposes a GB-InSAR timing processing method based on the LSTM network model, aiming at the importance of deformation prediction in GBSAR monitoring, and the shortcomings of GB-InSAR timing analysis, such as slow processing efficiency, high memory consumption and inability to carry out real-time monitoring. The experimental results show that the method can quickly obtain the deformation prediction results in the monitoring area, and the inner coincidence accuracy evaluation proves that the method has high deformation prediction accuracy. At the same time, the real-time processing module can obtain the shape variables in real time, and the hardware occupancy rate is low, which proves that the method has high-precision real-time processing ability. The following conclusions are drawn through the experiment:

(1) The LSTM model has high deformation prediction accuracy for the deformation results of GB-InSAR, and the standard deviation of the deformation prediction residual is 0.46 mm, which has the advantages of high prediction accuracy and fast model convergence speed.

(2) The proposed method can quickly obtain deformation results without occupying a large amount of memory, and the real-time processing time of a single scene is better than 5 min.

(3) Compared with the traditional GB-InSAR time series processing method, the standard deviation of the proposed method in the slow deformation region is between 0.1–0.3 mm, and the average deviation from the traditional method is within 0.13 mm.

Author Contributions: All authors contributed to the study conception and design. Material preparation, data collection and analysis were done by J.H., H.Y., Y.L., Z.L., K.Z. and R.J. The first draft of the manuscript was written by J.H. and all authors commented on previous versions of the manuscript. All authors have read and agreed to the published version of the manuscript.

Funding: This research was funded by the National Natural Science Foundation of China (42174026), and the National Key Research and Development Program of China (2021YFE011004).

Data Availability Statement: Not applicable.

Acknowledgments: We thank the good advice and comments from anonymous reviewers to help improve the quality of the paper.

Conflicts of Interest: The authors declare no conflict of interest.

References

1. Lingua, A.M.; Piatti, D.; Rinaudo, F. Remote monitoring of a landslide using an integration of GB-INSAR and LIDAR techniques. In Proceedings of the 21st Congress of the International Society for Photogrammetry and Remote Sensing, Beijing, China, 25 June 2008.
2. Broussolle, J.; Kyovtorov, V.; Basso, M.; Ferraro, D.S.; Figueiredo, M.J.; Giuliani, R.; Oliveri, F.; Sammartino, P.; Tarchi, D. MELISSA, a new class of ground based InSAR system. An example of application in support to the Costa Concordia emergency. *ISPRS J. Photogramm. Remote Sens.* **2014**, *91*, 50–58. [CrossRef]
3. Emanuele, I.; Giovanni, G.; Massimiliano, N.; Luca, L.; Francesco, M.; Francesco, F.; Nicola, C. Sinkhole monitoring and early warning: An experimental and successful GB-InSAR application. *Geomorphology* **2015**, *241*, 304–314.
4. Calvari, S.; Di Traglia, F.; Ganci, G.; Giudicepietro, F.; Macedonio, G.; Cappello, A.; Nolesini, T.; Pecora, E.; Bilotta, G.; Centorrino, V.; et al. Overflows and Pyroclastic Density Currents in March-April 2020 at Stromboli Volcano Detected by Remote Sensing and Seismic Monitoring Data. *Remote Sens.* **2020**, *12*, 3010. [CrossRef]
5. Barla, M.; Antolini, F. An integrated methodology for landslides' early warning systems. *Landslides* **2016**, *13*, 215–228. [CrossRef]
6. Kuraoka, S.; Nakashima, Y.; Doke, R.; Mannen, K. Monitoring ground deformation of eruption center by ground-based interferometric synthetic aperture radar (GB-InSAR): A case study during the 2015 phreatic eruption of Hakone Volcano. *Earth Planets Space* **2018**, *70*, 181. [CrossRef]
7. Di, T.F.; Nolesini, T.; Ciampalini, A.; Solari, L.; Frodella, W.; Bellotti, F.; Fumagalli, A.; Rose, G.D.; Casagli, N. Tracking morphological changes and slope instability using spaceborne and ground-based SAR data. *Geomorphology* **2018**, *300*, 95–112.
8. Carla, T.; Tofani, V.; Lombardi, L.; Raspini, F.; Bianchini, S.; Bertolo, B.; Thuegaz, P.; Casagli, N. Combination of GNSS, satellite InSAR, and GB-InSAR remote sensing monitoring to improve the understanding of a large landslide in high alpine environment. *Geomorphology* **2019**, *335*, 62–75. [CrossRef]
9. Carla, T.; Gigli, G.; Lombardi, L.; Casagli, N. Monitoring and analysis of the exceptional displacements affecting debris at the top of a highly disaggregated rockslide. *Eng. Geol.* **2021**, *294*, 106345. [CrossRef]
10. Dario, P.; Luisa, O.; Gianfranco, N.; Michele, C.; Jordi, R.; Pere, B.; Marc, J. Investigating the kinematics of the unstable slope of Barberà de la Conca (Catalonia, Spain) and the effects on the exposed facilities by GBSAR and multi-source conventional monitoring. *Landslides* **2021**, *18*, 457–469.
11. Long, S.; Tong, A.; Yuan, Y.; Li, Z.; Wu, W.; Zhu, C. New Approaches to Processing Ground-Based SAR (GBSAR) Data for Deformation Monitoring. *Remote Sens.* **2018**, *10*, 1936. [CrossRef]
12. Martinez, A.; Aguasca, A.; Lort, M.; Broquetas, A. Micrometric deformation imaging at W-Band with GBSAR. *Eur. J. Remote Sens.* **2016**, *19*, 719–733. [CrossRef]

13. Crosetto, M.; Monserrat, O.; Luzi, G.; Cuevas-Gonzá, M.; Devanthéry, N. Discontinuous GBSAR deformation monitoring. *ISPRS J. Photogramm. Remote Sens.* **2014**, *93*, 136–141. [CrossRef]
14. Wang, Y.; Zhang, Q.; Lin, Y.; Zhao, Z.; Li, Y. Multi-Phase-Center Sidelobe Suppression Method for Circular GBSAR Based on Sparse Spectrum. *IEEE Access* **2020**, *8*, 133802–133816. [CrossRef]
15. Wang, Z.; Li, Z.; Mills, J. Modelling of instrument repositioning errors in discontinuous Multi-Campaign Ground-Based SAR (MC-GBSAR) deformation monitoring. *ISPRS J. Photogramm. Remote Sens.* **2019**, *157*, 26–40. [CrossRef]
16. Wang, Z.; Li, Z.; Liu, Y.; Peng, J.; Mills, J. A New Processing Chain for Real-Time Ground-Based SAR (RT-GBSAR) Deformation Monitoring. *Remote Sens.* **2019**, *11*, 2437. [CrossRef]
17. Izumi, Y.; Zou, L.; Kikuta, K.; SATO, M. Iterative Atmospheric Phase Screen Compensation for Near-Real-Time Ground-Based InSAR Measurements Over a Mountainous Slope. *IEEE Trans. Geosci. Remote Sens.* **2020**, *58*, 5955–5968. [CrossRef]
18. Liu, Q.; Zhang, Y.; Deng, M.; Wu, H.; Kang, Y.; Wei, J. Time series prediction method of large-scale surface subsidence based on deep learning. *Acta Geod. Cartogr. Sin.* **2021**, *50*, 396–404.
19. Yang, H.; Peng, J.; Cui, H. Slope of Large-scale Open-pit Mine Monitoring deformations by Using Ground-Based interferometry. *Prog. Geophys.* **2012**, *27*, 1804–1811.
20. Liu, J. *The Research of Atmospheric Correction Method for GB-InSAR*; China University of Geosciences: Beijing, China, 2020.
21. Rödelsperger, S.; Läufer, G.; Gerstenecker, C.; Becker, M. Monitoring of displacements with ground-based microwave interferometry: IBIS-S and IBIS-L. *J. Appl. Geod.* **2010**, *4*, 41–54. [CrossRef]
22. Zhao, X.; Lan, H.; Li, L.; Zhou, C. A Multiple-Regression Model Considering Deformation Information for Atmospheric Phase Screen Compensation in Ground-Based SAR. *IEEE Trans. Geosci. Remote Sens.* **2020**, *58*, 777–789. [CrossRef]
23. Iannini, L.; Monti, G.A. Atmospheric Phase Screen in Ground-Based Radar: Statistics and Compensation. *IEEE Geosci. Remote Sens. Lett.* **2011**, *8*, 537–541. [CrossRef]
24. Yu, H. A Study on Information Extraction of Water Body with the Modified Normalized Difference Water Index (MNDWI). *Natl. Remote Sens. Bull.* **2005**, *9*, 589–595.
25. Wantong, C.; Hailong, W.; Ren, S. CM-LSTM Based Spectrum Sensing. *Sensors* **2022**, *22*, 2286.
26. Jiao, F.; Huang, L.; Song, R.; Huang, H. An Improved STL-LSTM Model for Daily Bus Passenger Flow Prediction during the COVID-19 Pandemic. *Sensors* **2021**, *21*, 5950. [CrossRef] [PubMed]
27. He, J.; Wu, X.; Cheng, Z.; Yuan, Z.; Jiang, Y. DB-LSTM: Densely-connected Bi-directional LSTM for human action recognition. *Neurocomputing* **2021**, *444*, 319–331. [CrossRef]
28. Chaodong, F.; Li, Y.; Yi, L.; Xiao, L.; Qu, X.; Ai, Z. Multi-objective LSTM ensemble model for household short-term load forecasting. *Memetic Comput.* **2022**, *14*, 115–132.
29. Liu, Y. *Study on Monitoring Method of Surface Subsidence in Filling Mining Area Based on DS-InSAR*; China University of Mining and Technology: Beijing, China, 2021.
30. Du, J. *Application of Kalman Filtering in GB-InSAR Slope Deformation Monitoring*; China University of Geosciences: Beijing, China, 2021.

MDPI
St. Alban-Anlage 66
4052 Basel
Switzerland
Tel. +41 61 683 77 34
Fax +41 61 302 89 18
www.mdpi.com

Remote Sensing Editorial Office
E-mail: remotesensing@mdpi.com
www.mdpi.com/journal/remotesensing

www.ingramcontent.com/pod-product-compliance
Lightning Source LLC
Chambersburg PA
CBHW061758210926
51850CB00006B/587
9783036564425